AT RISK

Natural hazards, people's vulnerability
and disasters

Second edition

Ben Wisner, Piers Blaikie,
Terry Cannon and Ian Davis

Routledge
Taylor & Francis Group

LONDON AND NEW YORK

First published 1994
by Routledge
11 New Fetter Lane, London EC4P 4EE

Simultaneously published in the USA and Canada
by Routledge
29 West 35th Street, New York, NY 10001

Routledge is an imprint of the Taylor & Francis Group

Typeset in Times by Taylor & Francis Books Ltd
Printed and bound in Great Britain by The Cromwell Press, Trowbridge,
Wiltshire

British Library Cataloguing in Publication Data
A catalogue record for this book is available from the British Library

Library of Congress Cataloging in Publication Data
At risk : natural hazards, people's vulnerability and disasters / Ben
Wisner ... [et al].– 2nd ed.p. cm.
Includes bibliographical references and index.
1. Natural disasters. I. Wisner, Benjamin.
GB5014.A82 2003
363.34–dc21
2003009175

ISBN 0-415-25215-6 (hbk)
ISBN 0-415-25216-4 (pbk)

CONTENTS

CONTENTS

CONTENTS

ILLUSTRATIONS

Figures

Tables

FOREWORD

The Great Wave

The cover of our book reproduces a very well-known image. In 1830 Katsushika Hokusai created a woodblock which was to become the most famous Japanese work of art in the West and a major influence on nine-teenth-century impressionist painters. This is commonly called *The Great Wave*, a simplified version of the Japanese title: *In the Hollow of a Wave off the Coast of Kanagawa*. The picture is one of the greatest depictions of a vulnerable situation in the history of art, and it conveniently contains the key elements associated with reducing vulnerability that forms the substance of this book.[1]

Element 1: disaster

The first element is inferred, *not* depicted. This is a *disaster* in which the boats are smashed and the fishermen drowned. However, rather than create this image, the artist is concerned to portray the struggle of people whose livelihoods and property are 'at risk'.

Element 2: hazards

The second element is the portrayal of two *hazards*, both very well known in Japan. Firstly, a storm or a tsunami.[2] Secondly, in the background Hokusai has reproduced the classic volcanic form of Mount Fuji. The image of Fuji has deep symbolic and religious significance in Japan, and volcanic erup-tions are also a hazard.

Element 3: vulnerability

The third element concerns *vulnerability* to storm conditions. This is expressed in three closely linked ways. The first is the 'social vulnerability' of the fishermen; the second is the 'economic vulnerability' of their livelihoods; third, the viewer can imagine the 'physical vulnerability' of their boats and fishing nets. The viewer cannot see the wives, children and extended families back on shore. The loss of a fishing boat or a net will have an impact on their well-being and livelihoods. However, the contributions by those on shore to economic and social life may also help to buffer any such loss. Social networks may assist in repairing a boat or replacing a net. Social arrangements may provide for widows and orphaned children. Farming and other economic activities may supplement income from fishing. In addition, one cannot see in the painting the moneylender or the fish wholesaler who may take advantage of a disaster at sea to profit at the fishermen's expense.

Element 4: capacity and resilience

The final element relates to *capacity and resilience*, sometimes called 'coping mechanisms' or 'capacity'. This dynamic human capacity can be observed in two ways: one appears to be non-structural whilst the other is structural. Put in more precise terms, one approach depicts a pattern of behaviour while the other shows adaptive design.

The first concerns the skills of the fishermen. Clearly the helmsman in each boat is making certain that each boat is facing directly into the great wave, rather than lying in an exposed position parallel with the wave crest. In storm conditions this is standard maritime practice to avoid a giant wave breaking over boats and smashing them to pieces. If the viewer looks carefully at the oars, it is also possible to see a further adaptation to the storm threat. The oarsmen appear to have interwoven their oars into a lattice, perhaps to prevent them being smashed by the giant wave, or as a device to form a 'sea anchor' extending on each side of the boat to maintain them at a right angle to the waves. The fishermen are bent over in a crouched position to protect their faces from the impact of the waves.

The second adaptation to note may be in the design of the boats. The curved bow may well have been an adaptation to the wave conditions '*off the Coast of Kanagawa*'. Traditionally the design of boats has made good use of the 'strength-gain' achieved from using curved planes and forms.

Element 5: culture

Hokusai's great work of art is a reminder of the awareness of such hazards in Japan as well as the way in which all households, groups and societies cope with and adapt to such threats to their everyday lives and livelihoods. This reality has received wide expression in art, poetry and music. And in traditional architecture there have been extensive adaptations to seismic risks and fire threats. However, cultures are never static, and whilst there may have been a cultural awareness of storms and storm protection in Japan in 1830, this knowledge may have been lost and may be in need of renewal. Today fishermen need not go out to sea when advanced satellite warning systems and weather radar can warn that storms are imminent.

Notes

1 Another view of the picture is available at http://www.ibiblio.org/wm/paint/auth/ hokusai/.
2 Tsunami is the Japanese name for a 'harbour wave', a great wave generated by an undersea earthquake or subterranean landslide. It seems that it is unclear whether Hokusai intended to portray a tsunami, or simply a storm, although many interpreters have assumed it is a tsunami.

PREFACE TO THE 2004
EDITION

The world, and the authors, are 15 years older since a book called *At Risk* was first discussed. We are happy that the first edition has been widely used, translated into Spanish[1], and generally been well received. We are less happy about some of the likely reasons for the growing popularity of the notion of vulnerability. One concern is that the term is being used indiscriminately, in a similar manner to 'sustainability'. It is in danger of becoming a catch-all term, with its analytical power and significance diminished. But the term is also becoming more popular as a simple reflection of the growth in people's vulnerability: it is increasing, and more people want to know why. During the past 15 years millions more people have been affected by disasters triggered by natural hazards. These millions are at risk in part because of global economic and political processes that were only becoming recognised in 1988. Looking back, the era in which we first collaborated seems antediluvian. The flood of post-Cold War economic globalisation, the wars and tensions generated by the break up of the Cold War political order, and growing impacts of global environmental change have all contributed to vulnerability. Thinking about all these changes, it was clear we needed to revise our book.

Time has brought changes in our lives as well. Three of us are now retired from full-time teaching. Davis, Visiting Professor at Cranfield University, retired from the Cranfield Disaster Management Centre and continues to make contributions on post-disaster shelter and risk reduction activities. Blaikie has retired from his post at the School of Development Studies (University of East Anglia) and continues to consult and to publish research on resource management in Africa and South Asia. Wisner retired from his position as Director of International Studies and Professor of Geography (California State University, Long Beach) and works as a researcher with the Crisis States Programme of the Development Studies Institute, London School of Economics, as well as consulting for international agencies and NGOs. Cannon is still at the University of Greenwich, where he is Reader in Development Studies in the School of Humanities, and also carries out research and consultancy for the Natural Resource Institute, also at University of Greenwich, London.

The first edition was conceived at the same time that the seeds were sown of what was to become the 1990–1999 UN International Decade for Natural Disaster Reduction (UN-IDNDR). *At Risk* was published in the middle of that decade. The second edition comes well after the end of the IDNDR. Lives lost to disasters, and the economic costs of disasters, increased steadily during that period. How could so much international attention be devoted to a subject with so little to show for it? This question haunts us because we know and respect many of the people who worked tirelessly during the IDNDR. Why is vulnerability increasing despite the best efforts of many scientists, policy makers, administrators and activists? We hope that our second edition will help to answer this question.

As before, yet so much more so, we are indebted to such an enormous number of people we cannot possibly name them all. So we hope they will know that they are included when we express our immense gratitude to many of the same people we thanked the first time and now number, together with many new ones, among the friends, colleagues, publishers, students, conference participants, members of NGOs, and government and UN officials that helped us. We must also thank Kim Allen and Carol Baker for precise and helpful editorial suggestions, and Angela Allwright for the art and science she lavished on many of our illustrations. Also as before, we are particularly grateful to our families for their patience with the ups and downs of our common project over these last three years.

Royalties from this edition are being donated to three groups that are all active in promoting vulnerability reduction approaches to disaster reduction in developing countries: La RED in Latin America, Peri Peri in southern Africa and Duryog Nivaran in South Asia (for more details, see Chapter 1, note 39).

April 2003

Note

1 (P. Blaikie, T. Cannon, I. Davis and B. Wisner, 1996. *Vulnerabilidad, el entorno social, político y económico de los desastres*, La RED/IT Development Group, Peru. Available for download on the internet at:
http://www.desenredando.org/ public/libros/1996/vesped/)

PREFACE TO THE 1994 EDITION

Long before they met each other, the four authors encountered many of the hazards discussed in this book as they worked and visited in Asia, Africa and Latin America. They shared a dissatisfaction with then prevailing views that disasters were 'natural' in a straightforward way. They shared an admiration for the ability of ordinary people to 'cope' with poverty and even calamities, and this perspective has strongly influenced the book.

The authors brought to this project complementary skills and expertise. Blaikie had written on the socio-economic background to land degradation[1] and poverty in Nepal,[2] and more recently on the AIDS epidemic in Africa.[3] Cannon has incorporated teaching on hazards into his work for many years, and is active in the Famine Commission of the International Geographical Union. He has edited a collection of studies of famine[4] and published extensively on development and environmental problems in China. Davis had spent many years studying shelter following disasters[5] and the growth of disaster vulnerability with rapid urbanisation;[6] he also brought to the project years of practical work in training government officials in disaster mitigation. Wisner had been concerned with rural physical and social planning since the mid-1960s. This took the form of land-use studies,[7] research on drought coping,[8] and work on rural energy[9] and health care delivery.[10] As the project began he was about to draw these themes together in a systematic study of 'basic needs' approaches to development in Africa.[11]

This book has taken a long time to complete, with all the complications of multiple authorship, and the added difficulty that the four were for much of the time in three different countries. We met about six times for several days, and progressed from sketched outlines to substantial drafts at each meeting. Much paper and many electrons and floppy disks sailed back and forth among us. A great deal of ideological baggage was stripped away as a consensus view of hazards, vulnerability and disasters emerged. We provided crash courses for each other in areas of our own expertise. The result is a fully co-authored book, although we are aware that some idiosyncrasies of style and variations in point of view may still be visible here and there.

The process had lots to recommend it, even if speed is not one of them. Yet the book has managed to appear midpoint in the International Decade for Natural Disaster Reduction (IDNDR). It arrives in the context of this decade (with its overemphasis on technology and hazard management), in the hope that it will establish the vital importance of understanding vulnerability in the context of its political, social and economic origins. The book reasserts the significance of the human factor in disasters. It tries to move beyond technocratic management to a notion of disaster mitigation that is rooted in the potential that humans have to unite, to persevere, to understand what afflicts them, and to take common action.

While the book was being written, there has been growing awareness of vulnerability to disasters, and of the range of causal factors. We welcome this groundswell of changing awareness, and appreciate the insights of many other contributors to the analysis of disasters. There are so many people to thank for their encouragement and help in the production of this book that it would be impossible to compile a list that did not offend by erroneously omitting some. So if all those people, including those affected by disaster, friends, colleagues, publishers, students, conference participants, members of NGOs, government and UN officials, will forgive us for not including their names, let us express our thanks to them in this way. In particular we must also thank our families for their patience, for enabling our meetings and providing a great deal of help and moral support.

Notes

1 Blaikie (1985b); Blaikie and Brookfield (1987).
2 Blaikie et al. (1977, 1983).
3 Barnett and Blaikie (1992).
4 Bohle et al. (1991).
5 Davis (1978).
6 Davis (1986, 1987).
7 O'Keefe and Wisner (1977).
8 Wisner (1978b, 1980).
9 Wisner et al. (1987); Wisner (1987b).
10 Wisner (1976a, 1988b, 1992a); Packard et al. (1989).
11 Wisner (1988a).

ABBREVIATIONS AND ACRONYMS

$	Indicates US dollars unless otherwise stated
BFAP	Bangladesh Flood Action Plan
CPRs	Common Property Resources
CRED	Centre for Research in the Epidemiology of Disasters (Louvain)
DFID	Department For International Development (UK foreign aid ministry)
ENSO	El Niño Southern Oscillation
FAD	Food Availability Decline
FAO	Food and Agricultural Organisation
FCDI	Flood Control Drainage and Irrigation
FED	Food Entitlement Decline
FEMA	Federal Emergency Management Authority (USA)
FEWS	Famine Early Warning System
G7/G8	Group of 7 (now Group of 8) major economic powers
GM	Genetically Modified
GoB	Government of Bangladesh
HDI	Human Development Index
HEP	Hydro-Electric Power
HIPCs	Highly Indebted Poor Countries
HYV	High Yielding Varieties (of food grain plants)
IDNDR	International Decade for Natural Disaster Reduction
IDP	Internally Displaced Person
IFPRI	International Food Policy Research Institute (Washington DC)
IFRC	International Federation of Red Cross and Red Crescent Societies (Geneva)
IMF	International Monetary Fund
ISDR	International Strategy for Disaster Reduction
LDCs	Less Developed Countries
MDCs	More Developed Countries
MSF	Médecins Sans Frontières

NGO	Non-Government Organisation
OAS	Organisation of American States
OFDA	(US) Office of Foreign Disaster Assistance
PAHO	Pan American Health Organisation
PAR	Pressure and Release (model)
PRC	People's Republic of China
RENAMO	Resistência Nacional Moçambicana (Mozambican National Resistance)
SAP	Structural Adjustment Programme
UNDP	United Nations Development Programme
UNEP	United Nations Environment Programme
UNICEF	United Nations Children's Fund
UN OCHA	United Nations Office for the Coordination of Humanitarian Affairs
UNRISD	United Nations Research Institute for Social Development
USAID	United States Agency for International Development
WFP	World Food Programme
WHO	World Health Organisation

Part I

FRAMEWORK AND THEORY

1

THE CHALLENGE OF
DISASTERS AND
OUR APPROACH

In at the deep end

Disasters, especially those that seem principally to be caused by natural hazards, are not the greatest threat to humanity. Despite the lethal reputation of earthquakes, epidemics and famine, a much greater proportion of the world's population find their lives shortened by events that often go unnoticed: *violent conflict, illnesses, and hunger* – events that pass for normal existence in many parts of the world, especially (but not only) in less developed countries (LDCs).[1] Occasionally earthquakes have killed hundreds of thousands, and very occasionally floods, famines or epidemics have taken millions of lives at a time. But to focus on these (in the understandably humanitarian way that outsiders do in response to such tragedies) is to ignore the millions who are not killed in such events, but who nevertheless face grave risks. Many more lives are lost in violent conflict and to the preventable outcome of disease and hunger (see Tables 1.1 and 1.2).[2] Such is the daily and unexceptional tragedy of those whose deaths are through 'natural' causes, but who, under different economic and political circumstances, should have lived longer and enjoyed a better quality of life.[3]

Table 1.1 Hazard types and their contribution to deaths, 1900–1999

Hazard type in rank order	Percentage of deaths
Slow onset:	
Famines – drought	86.9
Rapid onset:	
Floods	9.2
Earthquakes and tsunami	2.2
Storms	1.5
Volcanic eruptions	0.1
Landslides	<0.1
Avalanches	Negligible
Wildfires	Negligible

Source: CRED at www.cred.be/emdat

3

Table 1.2 Deaths during disasters, listed by cause, 1900–1999

Cause of death [a]	Numbers killed (millions)	Percentage of deaths
Political violence	270.7	62.4
Slow-onset disaster [b]	70.0	16.1
Rapid-onset disaster	10.7	2.3
Epidemics	50.7	11.6
Road, rail, air and industrial accidents	32.0	7.6
TOTAL	434.1	100

Notes:

[a] the source for political violence data is Sivard (2001). For all other causes, data is summarised from that available at www.cred.be/emdat

[b] this figure has been increased by us to an estimate of 70 million, much higher than the official data, which would give a total of around 18 million. This is to compensate for large-scale under-reporting of deaths from drought and famine. There are several reasons why this can occur. For instance, it is often the case that governments conceal or refuse to acknowledge famine for political reasons. The Great Leap Forward famine in China (1958–1961) was officially denied for more than 20 years, and then low estimates put the number of deaths at 13 million and higher ones at up to 30 million or more (see Chapter 4). A further problem is that sometimes recorded deaths in famine are limited to those who die in officially managed feeding or refugee camps. Many more are likely to die unrecorded at home or in other settlements.

However, we feel this book is justified, despite this rather artificial separation between people at risk from natural hazards and the many dangers inherent in 'normal' life. Analysing disasters themselves also allows us to show why they should *not* be segregated from everyday living, and to show how the risks involved in disasters must be connected with the vulnerability created for many people through their normal existence. It seeks the connections between the risks people face and the reasons for their *vulnerability* to hazards. It is therefore trying to show how disasters can be perceived within the broader patterns of society, and indeed how analysing them in this way may provide a much more fruitful way of building policies, that can help to reduce disasters and mitigate hazards, while at the same time improving living standards and opportunities more generally.

The crucial point about understanding why disasters happen is that it is not only natural events that cause them. They are also the product of social, political and economic environments (as distinct from the natural environment), because of the way these structure the lives of different groups of people (see Box 1.1).[4] There is a danger in treating disasters as something peculiar, as events that deserve their own special focus. It is to risk separating 'natural' disasters from the social frameworks that influence how hazards affect people, thereby putting too much emphasis on the natural hazards themselves, and not nearly enough on the surrounding social environment.[5]

4

Many aspects of the social environment are easily recognised: people live in adverse economic situations that oblige them to inhabit regions and places that are affected by natural hazards, be they the flood plains of rivers, the slopes of volcanoes or earthquake zones. However, there are many other less obvious political and economic factors that underlie the impact of hazards. These involve the manner in which assets, income and access to other resources, such as knowledge and information, are distributed between different social groups, and various forms of discrimination that occur in the allocation of welfare and social protection (including relief and resources for recovery). It is these elements that link our analysis of disasters that are supposedly caused mainly by natural hazards to broader patterns in society. These two aspects – the natural and the social – cannot be separated from each other: to do so invites a failure to understand the additional burden of natural hazards, and it is unhelpful in both understanding disasters and doing something to prevent or mitigate them.

Disasters are a complex mix of natural hazards and human action. For example, in many regions wars are inextricably linked with famine and disease, including the spread of HIV-AIDS. Wars (and post-war disruption) have sometimes coincided with drought, and this has made it more difficult for people to cope (e.g. in Afghanistan, Sudan, Ethiopia and El Salvador). For many people, a disaster is not a single, discrete event. All over the world, but especially in LDCs, vulnerable people often suffer repeated, multiple, mutually reinforcing, and sometimes simultaneous shocks to their families, their settlements and their livelihoods. These repeated shocks erode whatever attempts have been made to accumulate resources and savings. Disasters are a brake on economic and human development at the household level (when livestock, crops, homes and tools are repeatedly destroyed) and at the national level when roads, bridges, hospitals, schools and other facilities are damaged. The pattern of such frequent stresses, brought on by a wide variety of 'natural' trigger mechanisms, has often been complicated by human action – both by efforts to palliate the effects of disaster and by the social causation of vulnerability.

During the 1980s and 1990s, war in Africa, the post-war displacement of people and the destruction of infrastructure made the rebuilding of lives already shattered by drought virtually impossible. In the early years of the twenty-first century conflict in central and west Africa (Zaire/Congo, Liberia, Sierra Leone) has displaced millions of people who are at risk from hunger, malaria, cholera and meningitis.[6] The deep indebtedness of many LDCs has made the cost of reconstruction and the transition from rehabilitation to development unattainable. Rapid urbanisation is putting increased numbers of people at risk, as shown by the terrible toll from the earthquake in Gujarat, India (2001) and mudslides in Caracas, Venezuela (1999).

Box 1.1: Naturalness versus the 'social causation' of disasters

When disasters happen, popular and media interpretations tend to focus on their *naturalness*, as in the phrase 'natural disaster'. The natural hazards that trigger a disaster tend to appear overwhelming. Headlines and popular book titles often say things like 'Nature on the Rampage' (de Blij 1994), and visually the physical processes dominate our attention and show human achievements destroyed, apparently by natural forces. There have been numerous television documentaries in Europe, North America and Japan which supposedly examine the causes of disasters, all of which stress the impact of nature. Much of the 'hard' science analysis of disasters is couched in terms that imply that natural processes are the primary target of research. The 1990s was the UN International Decade of *Natural* Disaster Reduction (our italics).

The diagram shown in Figure 1.1 illustrates why this is a very partial and inadequate way of understanding the disasters that are associated with (triggered by) natural hazards. At the top of Figure 1.1, Boxes 1 and 2, the natural environment presents humankind with a range of opportunities (resources for production, places to live and work and carry out livelihoods [Box 3]) as well as a range of potential hazards (Box 4). Human livelihoods are often earned in locations that combine opportunities with hazards. For example, flood plains provide 'cheap' flat land for businesses and housing; the slopes of volcanoes are generally very fertile for agriculture; poor people can only afford to live in slum settlements in unsafe ravines and on low-lying land within and around the cities where they have to work. In other words, the spatial variety of nature provides different types of environmental opportunity and hazard (Box 2) – some places are more at risk of earthquakes, floods, etc. than others.

But crucially, humans are not equally able to access the resources and opportunities; nor are they equally exposed to the hazards. Whether or not people have enough land to farm, or adequate access to water, or a decent home, are determined by social factors (including economic and political processes). And these same social processes also have a very significant role in determining who is most at risk from hazards: where people live and work, and in what kind of build-ings, their level of hazard protection, preparedness, information, wealth and health have nothing to do with nature as such, but are attributes of society (Box 5). So people's exposure to risk differs according to their *class* (which affects their income, how they live and where), whether they are *male or female*, what their *ethnicity* is, what *age group* they belong to, whether they are *disabled* or not, their *immigration status*, and so forth (Box 6).

Box 1.1 continued

Thus it can be seen that disaster risk is a combination of the factors that determine the potential for people to be exposed to particular types of natural hazard. But it also depends fundamentally on how social systems and their associated power relations impact on different social groups (through their class, gender, ethnicity, etc.) (Box 7). In other words, to understand disasters we must not only know about the types of hazards that might affect people, but also the different levels of *vulnerability* of different groups of people. This vulnerability is determined by social systems and power, not by natural forces. It needs to be understood in the context of political and economic systems that operate on national and even international scales (Box 8): it is these which decide how groups of people vary in relation to health, income, building safety, location of work and home, and so on.

In disasters, a geophysical or biological event is implicated in some way as a trigger event or a link in a chain of causes. Yet, even where such natural hazards appear to be directly linked to loss of life and damage to property, there are social factors involved that cause peoples' vulnerability and can be traced back sometimes to quite 'remote' root and general causes. This vulnerability is generated by social, economic and political processes that influence how hazards affect people in varying ways and with differing intensities.

This book is focused mainly on redressing the balance in assessing the 'causes' of such disasters away from the dominant view that natural processes are the most significant. But we are also concerned about what happens even when it is admitted that social and economic factors are the most crucial. There is often a reluctance to deal with such factors because it is politically expedient (i.e. less difficult for those in power) to address the technical factors that deal with natural hazards. Changing social and economic factors usually means altering the way that power operates in a society. Radical policies are often required, many facing powerful political opposition. For example, such policies might include land reform, enforcement of building codes and land-use restrictions, greater investment in public health, provision of a clean water supply and improved transportation to isolated and poor regions of a country.

The relative contribution of geophysical and biological processes on the one hand, and social, economic and political processes on the other, varies from disaster to disaster. Furthermore, human activities can modify physical and biological events, sometimes many miles away (e.g. deforestation contributing to flooding downstream) or many years later (e.g. the introduction of a new seed or animal, or the substitution of one form of architecture for another, less safe, one). The time dimension is extremely important in another way. Social, economic and political processes are themselves often modified by a disaster in ways that make some people more vulnerable to an

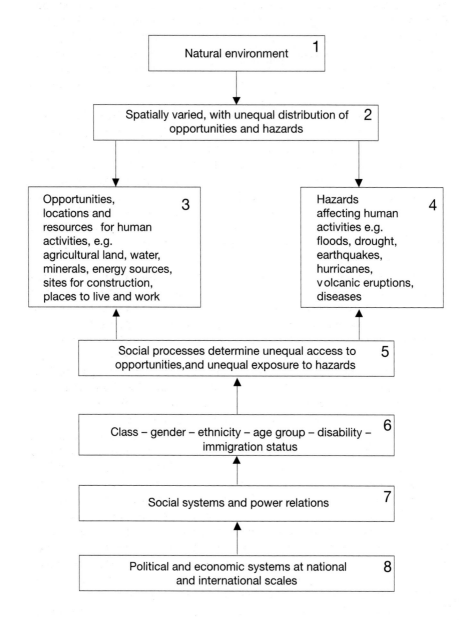

Figure 1.1 The social causation of disasters

extreme event in the future. Placing the genesis of disaster in a longer time frame therefore brings up issues of intergenerational equity, an ethical question raised in the debates around the meaning of 'sustainable' development (Adams 2001). The 'natural' and the 'human' are, therefore, so inextricably bound together in almost all disaster situations, especially when viewed in an enlarged time and space framework, that disasters cannot be understood to be 'natural' in any straightforward way.

This is not to deny that natural events can occur in which the natural component dominates and there is little place for differential social vulnerability to the disaster other than the fact that humans are in the wrong place at the wrong time. But such simple 'accidents' are rare. In 1986 a cloud of carbon dioxide gas bubbled up from Lake Nyos in Cameroon, spread out into the surrounding villages and killed 1,700 people in their sleep. In the balance of human and natural influences, this event was clearly at the 'natural' end of the spectrum of causation. The area was a long-settled, rich agricultural area. There were no apparent social differences in its impacts, and both rich and poor suffered equally.[7]

One example of a natural event with an explicitly inequitable social impact is the major earthquake of 1976 in Guatemala. The physical shaking of the ground was a natural event, as was the Cameroon gas cloud. However, slum dwellers in Guatemala City and many Mayan Indians living in impoverished towns and hamlets suffered the highest mortality. The homes of the middle class were better protected and more safely sited, and recovery was easier for them. The Guatemalan poor were caught up in a vicious circle in which lack of access to means of social and self-protection made them more vulnerable to the next disaster. The social component was so apparent that a journalist called the event a 'class-quake'.

It is no surprise that poor people in Guatemala live in flimsier houses on steeper slopes than the rich and that they are therefore more vulnerable to earthquakes. But what kind of social 'fact' is differential vulnerability in a case such as this? Above all, we think this case involves historical facts. Referring to a long history of political violence and injustice in the country, Plant (1978) believed Guatemala to be a 'permanent disaster'. The years of social, economic and political relations among the different groups in Guatemala and elsewhere have led some to argue that such histories 'prefigure' disaster (Hewitt 1983a). In Guatemala, after the 1976 earthquake, the situation deteriorated, with years of civil war and genocide against the rural Mayan majority that only ended in 1996. During this period, hundreds of thousands of Mayans were herded into new settlements by government soldiers, while others took refuge in remote, forested mountains and still others fled to refugee camps in Mexico. These population movements often saw marginal people forced into marginal, dangerous places.

This book attempts to deal with such histories and to uncover the deeply rooted character of vulnerability rather than taking the physical hazards as

9

the starting point, thereby allowing us to plan for, mitigate and perhaps prevent disaster by tackling all its causes. The book also builds a method for analysing the actual processes which occur when a natural trigger affects vulnerable people adversely.

Conventional views of disaster

Most work on disasters emphasises the 'trigger' role of geo-tectonics, climate or biological factors arising in nature (recent examples include Bryant 1991; Alexander 1993; Tobin and Montz 1997; K. Smith 2001). Others focus on the human response, psychosocial and physical trauma, economic, legal and political consequences (Dynes et al. 1987; Lindell and Perry 1992; Oliver-Smith 1996; Platt et al. 1999). Both these sets of literature assume that disasters are departures from 'normal' social functioning, and that recovery means a return to normal.

This book differs considerably from such treatments of disaster, and arises from an alternative approach that emerged in the last thirty years. This approach does not deny the significance of natural hazards as trigger events, but puts the main emphasis on the various ways in which social systems operate to generate disasters by making people vulnerable. In the 1970s and early 1980s, the vulnerability approach to disasters began with a rejection of the assumption that disasters are 'caused' in any simple way by external natural events, and a revision of the assumption that disasters are 'normal'. Emel and Peet (1989), Oliver-Smith (1986a) and Hewitt (1983a) review these reflections on causality and 'normality'. A competing vulnerability framework arose from the experience of research in situations where 'normal' daily life was itself difficult to distinguish from disaster. This work related to earlier notions of 'marginality' that emerged in studies in Bangladesh, Nepal, Guatemala, Honduras, Peru, Chad, Mali, Upper Volta (now Burkina Faso), Kenya and Tanzania.[8]

Until the emergence of the idea of vulnerability to explain disasters, there was a range of prevailing views, none of which dealt with the issue of how society creates the conditions in which people face hazards differently. One approach was unapologetically naturalist (sometimes termed physicalist), in which all blame is apportioned to 'the violent forces of nature' or 'nature on the rampage' (Frazier 1979; Maybury 1986; Ebert 1993; de Blij 1994). Other views of 'man [sic] and nature' (e.g. Burton et al. 1978; Whittow 1980) involved a more subtle environmental determinism, in which the limits of human rationality and consequent misperception of nature lead to tragic misjudgements in our interactions with it (Pelling 2001). 'Bounded rationality' was seen to lead the human animal again and again to rebuild on the ruins of settlements destroyed by flood, storm, landslide and earthquake.

According to such views, it is the pressure of population growth and lack of 'modernisation' of the economy and other institutions that drive human conquest of an unforgiving nature. This approach usually took a 'stages of economic growth' model for granted (Rostow 1991). Thus, 'industrial' societies had typical patterns of loss from, and protection against, nature's extremes, while 'folk' (usually agrarian) societies had others, and 'mixed' societies showed characteristics in between (Burton et al. 1978, 1993).[9] It was assumed that 'progress' and 'modernisation' were taking place, and that 'folk' and 'mixed' societies would become 'industrial', and that we would all eventually enjoy the relatively secure life of 'post-industrial' society.

The 1970s saw increasing attempts to use 'political economy' to counter modernisation theory and its triumphalist outlook, and 'political ecology' to combat increasingly subtle forms of environmental determinism.[10] These approaches also had serious flaws, though their analyses were moving in directions closer to our own than the conventional views.

Now we try to reintroduce the 'human factor' into disaster studies with greater precision, while avoiding the dangers of an equally deterministic approach rooted in the political economy alone. We avoid notions of vulnerability that do no more than identify it with 'poverty' in general or some specific characteristic such as 'crowded conditions', 'unstable hillside agriculture' or 'traditional rain-fed farming technology'.[11] We also reject those definitions of vulnerability that focus exclusively on the ability of a system to cope with risk or loss.[12] These positions are an advance on environmental determinism but lack an explanation of how one gets from very *widespread conditions* such as 'poverty' to very *particular vulnerabilities* that link the political economy to the actual hazards that people face.

What is vulnerability?

The basic idea and some variations

We have already used the term *vulnerability* a number of times. It has a commonplace meaning: being prone to or susceptible to damage or injury. Our book is an attempt to refine this common-sense meaning in relation to natural hazards. To begin, we offer a simple working definition. By vulnerability we mean *the characteristics of a person or group and their situation that influence their capacity to anticipate, cope with, resist and recover from the impact of a natural hazard* (an extreme natural event or process). It involves a combination of factors that determine the degree to which someone's life, livelihood, property and other assets are put at risk by a discrete and identifiable event (or series or 'cascade' of such events) in nature and in society.

Some groups are more prone to damage, loss and suffering in the context of differing hazards. Key variables explaining variations of impact include class (which includes differences in wealth), occupation, caste, ethnicity, gender, disability and health status, age and immigration status (whether 'legal' or 'illegal'), and the nature and extent of social networks. The concept

11

of vulnerability clearly involves varying magnitudes: some people experience higher levels than others. But we use the term to mean those who are more at risk: when we talk of vulnerable people, it is clear that we mean those who are at the 'worse' end of the spectrum. When used in this sense, the implied opposite of being vulnerable is sometimes indicated by our use of the term 'secure'.[13] Other authors complement the discussions of vulnerability with the notion of 'capacity' – the ability of a group or household to resist a hazard's harmful effects and to recover easily (Anderson and Woodrow 1998; Eade 1998; IFRC 1999b; Wisner 2003a).

It should also be clear that our definition of vulnerability has a time dimension built into it: vulnerability can be measured in terms of the damage to future livelihoods, and not just as what happens to life and property at the time of the hazard event. Vulnerable groups are also those that also find it hardest to reconstruct their livelihoods following disaster, and this in turn makes them more vulnerable to the effects of subsequent hazard events. The word 'livelihood' is important in the definition. We mean by this the command an individual, family or other social group has over an income and/or bundles of resources that can be used or exchanged to satisfy its needs. This may involve information, cultural knowledge, social networks and legal rights as well as tools, land or other physical resources.[14] Later we develop this livelihood aspect of vulnerability in an 'Access model'. The Access model analyses the ability of people to deal with the impact of the hazards they face in terms of what level of access they have (or do not have) to the resources needed for their livelihoods before and after a hazard's impact (see Chapter 3).[15]

Our focus on vulnerable people leads us to give secondary consideration to natural events as determinants of disasters. Normally, vulnerability is closely correlated with socio-economic position (assuming that this incorporates race, gender, age, etc.). Although we make a number of distinctions that show it to be too simplistic to explain all disasters, in general the poor suffer more from hazards than do the rich. Although vulnerability cannot be read directly off from poverty, the two are often very highly correlated. The key point is that even a straightforward analysis on the basis of poverty and wealth as determinants of vulnerability illustrates the significance we want to attach to social forms of disaster explanation. For example, heavy rainfall may wash away the homes in wealthy hillside residential areas of California, such as Topanga Canyon (in greater Los Angeles) or the Oakland–Berkeley hills (near San Francisco), just as it does those of the poor in Rio de Janeiro (Brazil) or Caracas (Venezuela).[16]

There are three important differences, however, between the vulnerability of the rich and the poor in such cases. Firstly, few rich people are affected if we compare the number of victims of landslides in various cities around the world. Money can buy design and engineering that minimises (but of course does not eliminate) the frequency of such events for the rich, even if they are living on an exposed slope.

12

Secondly, living in the hazardous canyon environment is a choice made by some of the rich in California, but not by the poor Brazilian or Philippine job seekers who live in hillside slums or on the edge of waste dumps.[17] Without entering the psychological or philosophical definitions of 'voluntary' versus 'involuntary' risk taking (see Sjöberg 1987; Adams 1995; Caplan 2000), it should be clear that slum dwellers' occupancy of hillsides is less voluntary than that of the corporate executive who lives in Topanga Canyon 'for the view'. The urban poor use their location as the base for organising livelihood activities (e.g. casual labour, street trading, crafts, crime, prostitution). If the structure of urban land ownership and rent means that the closest they can get to economic opportunities is a hillside slum, people will locate there almost regardless of the landslide risk (Hardoy and Satterthwaite 1989; Fernandes and Varley 1998). This, we will argue, is a situation in which neither 'voluntary choice' models nor the notion of 'bounded rationality' (Burton et al. 1993: 61–65) are applicable.

Thirdly, the consequences of a landslide for the rich are far less severe than for the surviving poor. The homes and possessions of the rich are usually insured, and they can more easily find alternative shelter and continue with income-earning activities after the hazard impact. They often also have reserves and credit. The poor, by contrast, frequently have their entire stock of capital (home, clothing, tools for artisan handicraft production, etc.) assembled at the site of the disaster. They have few if any cash reserves and are generally not considered creditworthy (despite the rapid development of 'micro-credit' schemes in a number of countries – see Chapter 9). Moreover, as emphasised above, the location of a residence itself is a livelihood resource for the urban poor. In places where workers have to commute to work over distances similar to those habitually covered by the middle class, transport can absorb a large proportion of the budget for a low-income household. The poor self-employed or casually employed underclass finds such transport expenses onerous. It is therefore not surprising that large numbers of working-class Mexicans affected by the 1985 earthquake refused to be relocated to the outskirts of Mexico City (Robinson et al. 1986; Poniatowska 1998; da Cruz 1993; Olson et al. 1999; Olson 2000; see also Chapter 8).

Multiple meanings of 'vulnerable'

Just before and since the publication of the first edition of *At Risk*, there has been a very welcome increase in the writing about vulnerability (Wilches-Chaux 1992a; Jeggle and Stephenson 1994; Davis 1994; Buckle et al. 1998/99; Buckle et al. 2000; Currey 2002). In this revised edition we happily take on board much of what has been added. There are at least four streams of recent work we should acknowledge.

Firstly, some recent studies give more emphasis to people's 'capacity' to protect themselves rather than just the 'vulnerability' that limits them. Earlier work (including, to some degree, our own) tended to focus most attention on the social, economic and political processes that make people 'vulnerable'. Understandably, it was necessary to use terminology that emphasised the *problem* that is generated by social processes – if people's capabilities were all working properly then there would be few disasters. This kind of analysis is essential, but it tends to emphasise people's weaknesses and limitations, and is in danger of showing people as passive and incapable of bringing about change. There is a need to register the other side of the coin: people do possess significant capabilities as well. Perhaps because of the influence of public health and social work professions, 'socially vulnerable groups' tended to be treated as 'special needs groups'. This approach can reduce people to being passive recipients, even 'victims' (Hewitt 1997: 167), and individuals without relationships. Usually, almost everyone has some capacity for self-protection and group action: the processes that generate 'vulnerability' are countered by people's capacities to resist, avoid, adapt to those processes, and to use their abilities for creating security, either before a disaster occurs or during its aftermath.

Secondly, there is now more interest in trying to quantify vulnerability as a tool of planning and policy making (Gupta et al. 1996; Davidson et al. 1997, 2000; Hill and Cutter 2001; UNDP 2003; Yarnal et al. 2002; Gheorghe 2003). With this has come debates about the correct balance between quantitative and qualitative data, and a deeper question concerning whether it is actually possible to quantify vulnerability. These efforts have been promoted by international agencies such as the Organization of American States (NOAA and OAS 2002), the United Nations Development Programme (UNDP 2003), DFID (Cannon et al. 2003), Emergency Management Australia (Buckle et al. 2001) and a large group of institutions led by FAO (FAO/IWAG 1998; UN-ACC 2000; WFP n.d.).

Thirdly, an increasing number of authors remind us of the cultural, psychosocial and subjective impacts of disasters. Definitions of vulnerability, including our own, usually include the notion of a potential for 'ill-being' (often expressed as an objectively assessed statistical probability) multiplied by the magnitude of the combined impacts of a particular trigger event. Thus, the conversion of risk is turned into a common metric, which enables different hazards to be compared (Rosa 1998), and this is the main analytical route taken by this book. Disaster impact is measured by a range of etic (external) and objectively verifiable indicators, such as mortality, morbidity, damage to property and physical assets, reduction in savings and so on.

While certainly necessary, these indicators are not sufficient, and we are aware that they tend to under-emphasise the cultural, the psychosomatic and subjective aspects of disaster impact (Perry and Mushkatel 1986; Oliver-Smith and Hoffman 1999; Johns 1999; Tuan 1979). Contemporary livelihood analysis must take conventional impact measures further to include notions of resilience and sensitivity, social capital and collective action. This conceptualisation of the drawing down of different 'capitals' and the conversion of one to another offers a more holistic view of well-being and decision making, particularly under conditions of 'normal' life, and this is a contemporary development of disaster theory which we elaborate on at length in Chapter 3. However, even this approach tends to make many untested and simplistic assumptions about preferences, choices and values, particularly under conditions of acute stress and extraordinary circumstances. The disaster event itself alters both capabilities and preferences, in the short term (e.g. grieving, trauma, acute deprivation, sleep, shelter, child care and other intimate relations, with implications for making decisions and carrying them out) and in the longer term (alterations in the access qualifications required to satisfy preferences, the rules of collective action). It provides a shock to expectations that in turn are shaped by people's social constructions of the likelihood of a disaster event (Beck 1992). The individual, household, kinship network and larger collectivities may develop implicit or explicit strategies to manage risk, which themselves constitute an important element in well-being and provide the basis for action when vulnerability is made a reality by the disaster event itself.

Fourthly, overlapping with the previous point, there is a movement away from simple taxonomies or checklists of 'vulnerable groups' to a concern with 'vulnerable situations', which people move into and out of over time. 'Vulnerability', as we use the word, refers only to people, not to buildings (susceptible, unsafe), economies (fragile), nor unstable slopes (hazardous) or regions of the earth's surface (hazard-prone).[18] Typically, social characteristics such as gender, age, health status and disability, ethnicity or race or nationality, caste or religion, and socio-economic status are the focus of attention.[19] Special interest non-governmental organisations (NGOs) have produced detailed checklists to take account of the particular needs and vulnerabilities of such groups as elderly people or unaccompanied children, both in vulnerability/capacity assessments as well as post-disaster needs assessments (see Chapter 9). These post-disaster tools are very useful as *aides mémoires* for busy administrators and case workers in the chaotic situation of a refugee camp or large-scale disaster such as the earthquakes in Gujarat (2001) or north-western Turkey (1999). For example, religion and caste had to be taken into account as they had an impact on the distribution of relief in Gujarat, where there were fears by aid workers that Muslims and Dalits (untouchables) were not receiving an equitable share (Harding 2001).[20]

But the use of post-disaster checklists does not *in itself* help one to understand *why* and *how* those characteristics have come to be associated with a higher probability of injury, death, livelihood disruption and greater difficulty in recovery. The checklists now widely used by international agencies and NGOs are based on some combination of the agency's own empirical observations and the results of a growing number of post-disaster studies and audits, many of them by sociologists. However, the empirical discovery of an association or correlation does not explain the process that gave rise to

15

the association. For example, the finding that domestic violence against women increased after hurricane Andrew has to be understood in process terms. It is not female gender itself that marks vulnerability, but gender *in a specific situation*. These gender relations between women and men were played out in the context of the growth boom of south Florida in the1980s and early 1990s, weak regulation of the building industry, downsizing and restructuring that left many working-class men anxious about future employment. Such male anxieties and frustration were acted out as domestic violence following the hurricane (Peacock et al. 2001).

In contrast, the process of pre-disaster vulnerability/capacity assessment is undertaken in a more reflective state of mind, without the urgency of a typical disaster situation. Thus, within these contexts it is possible to investigate causal factors as well as the symptoms, assuming that political leaders permit such probing analysis.

Many vulnerability situations are temporary, and change as life stages do (marriage, child bearing, old age) or with changes in occupation, immigration status or residence. For example, one study found that there were large numbers of low-income, young, immigrant, non-English-speaking, single mothers living in an area bordering San Pedro harbour (part of greater Los Angeles). This specific geographical location has a higher probability than other parts of San Pedro (or surrounding areas) for cargo explosions, liquefaction and amplified shaking because of soil factors in an earthquake, and exposure to a toxic plume from refinery fires (Wisner et al. 1999). The concatenation of income, age, immigration status, language and single parenthood significantly shifts the meaning of 'gender' as a simple category or box-to-tick in a taxonomy of vulnerability. Only two miles away from San Pedro, other women live in mansions overlooking the Pacific Ocean from the heights of Rancho Palos Verde. They share the socially constructed identity of 'woman' with these young Guatemalan single mothers, but in most other respects, they inhabit a separate universe (Wisner 1999; Wisner et al. 1999).

Risk society?

There is a large and growing literature on risk that we acknowledge but do not directly engage with in this book. The main reason is that it focuses primarily on technological hazards facing the more developed, industrial countries and the condition of late modernity in which they find themselves. In contrast, we direct most of our attention to risk as experienced and interpreted in less developed countries. One influential author writing about risk during the 1980s and 1990s is Ulrich Beck. His books *Risk Society: Toward a New Modernity?* (1992) and *Ecological Politics in the Age of Risk* (1995), amongst a number of others, have been profoundly influential. In these publications he seeks the 'root causes' of environmental crisis just as we in this book look for the 'root causes' of vulnerability to disaster. Beck (like many other researchers) finds those roots in the rampant consumerism of contemporary rich societies. But also (and this is of more interest to disaster studies) in two forms of social control of the consequences of over-consumption. One is 'ecological modernisation', by which the technicians of the 'risk society' attempt to 'fix' environmental problems without ever addressing root causes. The other is a form of amnesia or denial of environmental problems that he terms 'organized irresponsibility' (Beck quoted in Goldblatt 1999: 379).

Beck maintains that the more developed world is in a transitional state between industrial society and 'risk society': with so much wealth also come risks. With an increasingly complex and technologically driven society come new threats: 'hazards and insecurities induced and introduced by modernisation itself' (Beck 1992: 21). Many of these are treated by more affluent societies with a high degree of ambivalence, since a number of risks can no longer be directly experienced in a sensory manner (touched, seen or smelt as in the case of industrial society). Instead, there are risks of nuclear radiation, carcinogens in foodstuffs, toxicity from pesticides and risks associated with lifestyle. In addition, there is a background level of anxiety from a bewildering number of often ill-defined risks, some of them involving lifestyle and others involving incalculable horrors of unknown statistical probability, such as nuclear war or, we might add, since 11 September 2001, terrorist attack. Castel goes further to argue that modernity is involved in 'a grandiose technocratic rationalizing dream of absolute control of the accidental ... an absolute reign of calculative reason' (Castel 1991: 289, quoted in Lupton 1999: 7).

Thus, industrial, affluent society is increasingly protected against the uncertainties faced in LDCs through the application of technology and higher levels of income. Yet it is none the less increasingly preoccupied with incalculable and diffuse risks, which have somehow eluded all the advances of science and medicine. Others have noted a correlation between the emergence of 'environmental' concerns (e.g. with the quality of water and air) and increased affluence of the middle class in the USA and Europe (Hays 1987). In addition, more discrete and dramatic 'surprises' continue to occur in more developed countries, such as the unanticipated scale of the devastation of Kobe by the Great Hanshin earthquake in Japan in 1995 (despite all of Japan's scientific and engineering prowess); the contamination of a large area following the explosion of the Chernobyl nuclear reactor in 1986; the outbreak of BSE (bovine spongiform encephalopathy or 'mad-cow disease') in Britain in 2001; or the loss of the Space Shuttle Columbia and the outbreak of SARS in 2003. This cultural environment of risk, it will be clear to the reader, overlaps with but is different from the concerns we address in this book.

Beck considers the ways in which people in highly developed societies involve themselves in 'reflexive modernity', an institutionalised activity and state of mind involving constant monitoring and reflection upon and (according to Jacobs 1998) confrontation with these risks – whether they objectively exist or not. In particular, reflexive modernisation of risk can involve consideration of risks at the global level, an awareness that is a major incentive for international co-operation and practice, and leads to the globalisation of the meaning of risk. Thus transferred to the global scale, new concepts have been constructed and initiatives undertaken to 'manage' risk: for example, 'conserving biodiversity', 'reversing global warming' and 'disaster reduction' are forms of ecological modernisation conducted by the combined technocracy of rich, consuming nations (Sachs 1999). By extension, international efforts to 'manage' aspects of the impacts of hurricanes, droughts and volcanoes on behalf of poor, former colonial countries could also be considered a form of ecological modernisation. However, the fatal flaw in ecological modernisation is that it never deals with root causes. It is therefore never-ending and self-perpetuating. Later, we will return to several classic cases of this sort, such as the 'management' of the volcanic eruption in Montserrat (see Chapter 8).

Beck's work and the discussions it has stimulated are important and do, in some ways, overlap with our approach (Giddens 1990; Jacobs 1998; Lupton 1999). However it is rather remote from the dynamics of hazard, vulnerability and risk in LDCs that is our principle focus in this book. Nevertheless, there is another use of Beck's notion of reflexive modernisation that we find much closer to our purposes of the analysis of disasters in LDCs. While it can lead to perpetual anxiety and the self-defeating approach of ecological modernisation discussed above, reflexive modernisation can result in more focused political demands on authorities to address what we could call the 'root causes' of vulnerability. This pressure from below on authorities and corporations is that of citizens organised into what Beck calls an 'ecological democracy' (Beck 1995, 1998; Beck et al. 1994). Agreeing in large part with Beck's views, we place considerable emphasis on lay people, citizen groups and the vulnerable themselves as an important target audience of this book. Giddens (1992) has elaborated on the insights of Beck by exploring the relationship between 'risk' and 'trust'. Used in a different context, we also find that trust between, for example, citizen-based organisations and municipal governments, is critical in mobilising human resources for mitigating disaster loss and reducing vulnerability (Wisner 2002a) (see also Part III).[21]

Deconstructive approaches

The writings on risk, as in other subjects in social science, are distributed along a continuum of epistemological positions (Stallings 1997). At one end, there is a realist approach that takes risk as an objective hazard that exists and can be measured independently of social and cultural processes. Theories and methods associated with this epistemology are technoscientific, statistical and actuarial. Moving across the continuum, there are what could be termed 'weak constructionist' approaches, where risk is an objective hazard but is always mediated through social and cultural processes (Oliver-Smith and Hoffman 1999). Finally, there is the strong constructionist approach, where nothing is a risk in itself but is a contingent product of historically, socially and politically created 'ways of seeing' (Lupton 1999: 35). This book broadly takes a realist, and at times a weak constructionist, approach to risk. Many of the concerns and anxieties about which Beck and Giddens write so persuasively are a product of a late modern society in the more developed countries (MDCs), while the risks faced by many in developing countries are different. That is not to say that culturally constructed risks are any less apparent in LDCs. It is rather that they do not have the luxury of indulging in the anxieties found in MDCs, but instead face famine, flood, biological hazards, high winds and earthquakes – without the protection offered (to some) by affluent, industrial countries.

We part company with strong social constructionist approaches because we believe they do not lead, in any direct way, to an improvement in practice – either in disaster prevention or in post-disaster management. Therefore, for example, we acknowledge Bankoff's (2001) approach to famine as interesting but not useful from our perspective. He considers the historical roots of the discursive framework within which hazards are presented, and how that might reflect particular cultural values to do with the way in which certain regions of the world are usually imagined.[22] He characterises modernist approaches to disasters, risk and vulnerability as a historically constructed neo-colonial discourse which denigrates large regions of the world as 'tropical' (the unhealthy and dangerous 'other'), poverty-stricken and disaster-prone (ibid.). Although this view is accurate, we feel it is difficult to use it to contribute to the prevention or mitigation of disasters and improvement of relief and reconstruction. We acknowledge it but leave it to one side.

As noted above, the origins of the vulnerability approach we take in this book can be located in the 1970s when authors began to question the 'naturalness' of 'natural disaster' (O'Keefe et al. 1976). To that extent we have already been where Bankoff would ask us to go, and we now wish to provide more precise advice on linkages that transmit root causes into very specific unsafe conditions. Indeed, deconstructive critique is not new within geography and environmental studies, where for some time authors have pointed out that 'land degradation' and other environmental management categories come loaded with the assumptions and biases of the observer (Adams 2001; Leach and Mearns 1996; Gadgil and Guha 1995). The critique of structuralist, determinist methods is also well established within development studies (Crush 1995; Escobar 1995; Rahnema and Bawtree 1997) and has already had some influence on students of disaster.

There is, however, a heuristic aspect of such a post-structural critique of disaster discourse that we believe provides a valuable caution and corrective (Mustafa 2001). It could be argued that notions such as 'disaster management cycle', and terms such as 'relief', 'rehabilitation' and 'recovery' are technical constructs imposed on different cultural, economic, political and gender realities (Oliver-Smith and Hoffman 1999; Enarson and Morrow 2001). Such constructs fail to comprehend the lived reality of disaster and, to that extent, can fail to engage the co-operation of local people.

Vulnerability and normal/daily life

We argue in this book that feasible and informed practice in reducing disaster risk as well as a better theoretical understanding of disasters are possible only if one places the phenomenon of disaster 'in the mainstream' of policy and practice. Hewitt made this point twenty years ago when he wrote of how disasters had been mentally exiled to an 'archipelago' of exceptionalism (Hewitt 1983b). Agreeing wholeheartedly with Hewitt, we show how 'normal' historical processes contribute to the causation of disasters. We also show how 'normal' pressures in global, regional and national systems of economic, social and political power contribute to creating vulnerability to disaster. The material conditions of daily life, what one might call 'normal life', also underlie or, as Hewitt put it, 'prefigure' disasters (ibid.: 27). These material conditions are, above all, biological in the sense of our access to food, water and the air we breathe. We treat these material underpinnings of existence in some detail in Chapters 3 to 5. The Access model presented in Chapter 3 provides insight into how such material conditions of daily or normal life change with circumstances. It shows how major stress, such as an extreme natural event, can reverberate through a household's livelihood system, playing havoc with its ability to meet its needs, and, moreover, its ability to recover and protect itself against other, perhaps unrelated, stresses and crises at a later time.

Changes since the first edition

Nearly a decade has passed since the first edition of *At Risk* was completed. It has been ten years of very great change and, in some ways, unfortunate continuity. Much theoretical, practical and institutional work has been done on disaster 'vulnerability'. An entire United Nations International Decade for Natural Disaster Reduction (IDNDR) has passed (1990–1999). The language of major development agencies and banks has changed. Yet more and more costly and deadly disasters continue to occur.

The International Decade for Natural Disaster Reduction (IDNDR)

Not long after the publication of *At Risk*, in May 1994, the IDNDR held its mid-decade conference in Yokohama, Japan. This was an important watershed (see Chapter 9). Dissatisfaction emerged with the top-down, technocratic approach to disasters that had characterised the first half of the decade's activities. The resulting 'Yokohama Message' contained much that parallels the arguments we made in the first edition of *At Risk*. In particular, two prerequisites for disaster risk reduction are emphasised:

1 ... [A] clear understanding of the cultural and organizational characteristics of each society as well as of its behavior and interactions with the physical and natural environment.

2 ... [T]he mobilization of non-governmental organizations and participation of local communities.

(Ingleton 1999: 320)

The 'Yokohama Message' warned of the danger of 'meagre results of an extraordinary opportunity given to the United Nations and its Member States' during the first half of the IDNDR.

During the second half of the IDNDR considerable efforts were made to involve NGOs and communities. A popular magazine, *Stop Disasters*, was published. Annual themes for 'World Disaster Day' included social issues, for example a focus on women in disasters. Perhaps the most important development was a turn toward cities during the last three years of the IDNDR. This began with an international electronic conference in 1996 that reached out to many practitioners and NGOs, as well as academics and government officials (IDNDR 1996). An ambitious pilot programme for urban earthquake risk assessment and mitigation was run from 1997 to 2000. This 'Risk Assessment Tools for Diagnosis of Urban Areas Against Seismic Disasters' programme (mercifully known by the short acronym RADIUS) involved a core of nine medium-sized cities in different parts of the world, with a total of 84 cities as observers participating in various ways.[23]

RADIUS displayed the mark of the 'Yokohama Message' very clearly, because work in the nine core cities involved a broad cross-section of sectors, citizens and scientific disciplines. It was focused on mitigation of loss, and it used accessible technologies. RADIUS began in each city with a study of earthquake hazard and vulnerability, and progressed through the development of city-wide action plans that, once again, involved many diverse sectors and institutions.

Urban growth and the growth of urban concerns

The IDNDR's urban turn reflected a judgement that rapid progress in reducing loss of life could be made by focusing on cities. Indeed, another major change since the first publication of *At Risk* is the speed with which the world's population is rapidly becoming urban.[24] The IDNDR's focus on cities was also co-ordinated to provide a contribution to 'Habitat II', a major world conference on urban settlements held in Istanbul, Turkey in 1997 (twenty years after Habitat I). How should we explain the decision to focus IDNDR activity on earthquake risk reduction in cities, as opposed to any one of other possible urban hazards (e.g. flood, storms, volcanic eruptions)? Part of the explanation is found in the origins of the IDNDR. Earthquake engineers were very prominent in its creation and remained influential. Also important was the fact that two costly earthquakes had recently surprised authorities and experts alike in the USA (Northridge, California in 1994, costing $35 billion) and Japan (Kobe in 1995, with losses of over $147 billion).

Changes in earth care

The language of 'sustainable development' had entered development studies and policy documents from the late 1980s, with the publication of *Our Common Future* (WCED 1987). The 'Earth Summit' was held in Rio de Janeiro in 1992, near the start of the IDNDR. Since then, at least on paper, disaster risk reduction has been included as an element of many of the national and local efforts to implement *Agenda 21*, the Rio Summit's plan of action. However, the processes undermining any positive moves to make concrete such diplomatic consensus were soon in evidence after the Summit. In 1998, hurricane Mitch struck several Central American countries and made it obvious that it was underlying processes of land degradation and de-vegetation that made people vulnerable (see Chapter 7). The death toll from this hurricane is estimated to have been 27,000 people, most of these in Honduras and Nicaragua. The majority of these deaths were from floods and landslides that could have been prevented if so much of these countries had not been stripped of their forest cover.

In 2002, the Johannesburg World Summit on Sustainable Development reaffirmed the place of disaster risk reduction within its notion of 'sustainable development'. In the run up to the Johannesburg Summit, ten years after the Rio Summit, the third Global Environmental Outlook report by the UN Environment Programme (UNEP 2002) included a substantial chapter on disasters (see Chapter 9 below). It noted some uneven progress in reducing disaster risk, mostly concentrated in the richer countries. But, on balance, it considered the significance of what it called a 'vulnerability gap', 'which is widening within society, between countries and across regions with the disadvantaged more at risk to environmental change and disasters' (ibid.: 297).

Since the original publication of *At Risk*, the science of global climate change has improved, while the political consensus behind the Kyoto Treaty[25] (on reducing greenhouse gas emissions) has made only slow progress, largely because of US opposition).[26] It appears that the severe impact of hurricane Andrew (which devastated much of Miami in 1992) and the huge floods in the Mississippi basin the following year have not convinced the Bush administration of the possible connection between greenhouse gas emissions and climate change. This is despite strong advocacy for 'sustainable development' by prominent US disaster researchers (Mileti 1999; Burby 1998). Perhaps another dose of rough weather from the next El Niño cycle will wake up the US government to the need for a 'war on wasteful consumption' to parallel its 'war on terrorism'.

In the run up to the Johannesburg summit numerous authors and institutions have revisited the connections between land use and disaster. They recalled the lessons of hurricanes Mitch (1998) and Andrew (1992), the Mississippi floods (1993) and floods throughout many parts of Europe during the 1990s, as well as almost annual huge floods in China. Deforestation and other kinds of land-use problems have been implicated in all of these disasters (Gardner 2002; Burby 1998). They also wrote of the wildfires in Indonesia, the USA, Australia, Mexico and Brazil. They reminded us of the great loss of lives in the flooding and mudslides in Venezuela in 1999, Algeria and Brazil in 2001, and a deadly landslide triggered by an earthquake in El Salvador, also in 2001 (Abramovitz 2001; ISDR 2002a; Wisner 2001f, 2001c). In all these cases, better land-use planning and enforcement could have prevented the extreme natural event becoming a disaster. We are also reminded that a population displaced by a large-scale dam is not likely to understand the hazards of the terrain, climate and ecosystem in the area in which they are resettled. It will be harder for them to protect themselves against natural hazards that are new to them (World Commission on Dams 2000b).

The emergence of the 'precautionary principle'

Natural scientists from many disciplines have begun to discuss the problems of uncertainty in their analysis of various natural phenomena (Handmer et al. 2001). In situations where human actions may be causing catastrophic harm to natural systems on a global scale, a prudent 'precautionary science' is needed. This may apply especially to situations where the probability of a catastrophic outcome may be low but the magnitude of the catastrophe very large (Johnston and Simmonds 1991; O'Brien 2000). A more conventional and optimistic view is that it is possible to 'manage the planet' if there is sufficient knowledge of all the interactions in such large-scale physical systems as the atmosphere, hydrosphere, lithosphere, asthenosphere[27] and biosphere (Clark 1989). Such a technocratic and managerial approach has received increasing criticism over the past ten years. Our book will also challenge this latter line of thinking. Our effort is necessary in part because faith in simple technological fixes is still pervasive. As Zimmerman (1995: 175) notes: 'Too many of us blithely assume that we need not deal with the base causes of our environmental problems because soon-to-be-discovered technological solutions will make those problems obsolete'.

Critiques of economic globalisation

Another major change since this book first appeared is the increase in public and academic opposition to aspects of economic globalisation (including the street protests of Seattle and Genoa) (Hardt and Negri 2000; Sklair 2001; Wisner 2000a, 2001a; Pelling 2003a; Hines 2000; Monbiot 2003). In the first edition of this book, we dealt with the impact of such neo-liberal economic policies as 'structural adjustment' as a dynamic pressure leading to vulnerability. In the 1980s there was evidence that cutbacks in public expenditure on health and social protection were undermining the resilience of poor people to natural hazards. Since then the critique of neo-liberalism has been broadened to include the ideology of free trade and the institutions of economic globalisation such as the World Trade Organisation. In this new edition we recognise fully the role of economic globalisation as a 'dynamic pressure' affecting vulnerability to disasters (see Chapter 2). The scale of globalisation is enormous. As Friedman puts it:

> [G]lobalization is not simply a trend or a fad but is, rather, an international system. It is the system that has now replaced the Cold War system, and, like the Cold War system, globalization has its own rules and logic that today directly or indirectly influence the politics, environment, geopolitics and economics of virtually every country in the world.
>
> (2000: ix)

Starting in 2000 (in Porto Alegre, Brazil), the World Social Forum meets annually to act as a counterpoint to the business and governmental elite who meet at the World Economic Conference. The 2003 World Social Forum attracted 100,000 delegates (Wainwright 2003). Positive proposals are emerging for 'another globalisation' that is not based on dogmatic neo-liberal formulae for 'structural adjustment' of economies and 'free trade'. With widespread support by citizens' groups, churches and NGOs having caused governments to accept the notion of reducing the international debt of the least-developed nations, proposals such as a 'Tobin Tax' on international financial transactions may no longer be seen as utopian or fringe ideas.[28] In the face of rapidly accelerating privatisation of water supplies, others have begun to argue that as a basic need and human right, water should not be considered a commodity among other commodities.[29] Our concern about control of water supplies by multinational corporations is especially about whether 'the market' is sufficient to guarantee resilience of water, drainage and sanitation systems in the face of natural hazards such as earthquakes, floods and storms; and if not, who bears the losses and costs?

Academic support for the critique of blind belief in economic growth as the predominant goal of development has been building up since the UNDP began to publish its *Human Development Report* (HDR) in 1990. Its Human Development Index (HDI) measures equity, health and education, and not just economic activity. In 1995 the HDR added gender-specific measures, and in 1997 two separate measures of human poverty: one for more developed countries and one for the less developed. Other international institutions have responded to the reintroduction of social and other human goals into the development discourse (UNRISD 2000). In 2001 the World Bank devoted two chapters to poverty and disaster vulnerability in its *World Development Report* (the annual publication which had tended to give priority to economic growth and which, to some extent, the *Human Development Report* was designed to counter) (World Bank 2001; however, compare Cammack 2002).

In its *World Disasters Report 2001*, the International Red Cross presented data from the UNDP and Centre for Research in the Epidemiology of Disasters (CRED) that compares the impacts of extreme natural events on countries with high, medium, and low scores on the HDI (IFRC 2001a: 162–165). They looked at data for 2,557 disasters triggered by natural events between 1991 and 2000. Half of these disasters took place in countries with medium HDI, but two-thirds of the deaths occurred in countries with low HDI. Only 2 per cent of the deaths were recorded in the countries with a high HDI. When tabulating deaths and monetary losses per disaster, the relationship with HDI is even clearer (Table 1.3).

UNDP took this analytical work even further in 2002 by commissioning the quantitative study of more than 200 possible indicators of disaster risk vulnerability and producing a vulnerability index for use in its *World*

Table 1.3 Level of human development and disaster impacts

	Deaths per disaster	Loss per disaster ($ millions)
Low HDI	1,052	79
Medium HDI	145	209
High HDI	23	636

Source: based on IFRC (2001a: 162, 164)

Note:

HDI is Human Development Index (see text for explanation).

25

Vulnerability Report. The worldwide results (for the years 1980–1999) are striking (UNDP 2003). The HDI again turns out to be the best predictor of deaths triggered by extreme natural events.

Changes in human development and well-being

In parts of the world (especially in many African countries), the improvements in access to education, health care and the greater longevity achieved in the 1960s and 1970s continued to decline in the 1990s (UNDP 2003b). We noted this trend in the first edition of *At Risk*, and argued that the programmes for managing international debt imposed on many of these countries by the World Bank and IMF had increased people's vulnerability to disaster. Despite reformulating, renaming and giving a 'human face' to these 'structural adjustment programmes' (SAPs) during the 1990s, the effects have continued.

Gardner (2002: 10) observed that health officials in the 1970s believed that the era of infectious disease was about to come to an end worldwide. However, today we find that '20 familiar infectious diseases – including tuberculosis, malaria, and cholera – [have] re-emerged or spread ... and at least 30 previously unknown deadly diseases – from HIV to hepatitis C and Ebola – [have] surfaced' (ibid.: 10–11). HIV-AIDS deaths have grown from 500,000 worldwide in 1990 to nearly 3 million in 2000 (Barnett and Whiteside 2001). Most of the deaths from HIV-AIDS occur in the LDCs (the distribution is similar to that presented above for disaster deaths), and four-fifths of these are in sub-Saharan Africa (ibid.: 12). At the end of 1999, there were 34 million people living with HIV, of whom 25 million (74 per cent) lived in sub-Saharan Africa (1 million of them children). Over 12 million children had been orphaned by HIV-AIDS. The magnitude of this disaster dwarfs anything else we take up in this book, and the numbers are staggering. HIV-AIDS in Africa represents great complexity in its long-term consequences for production, social relations and vulnerability to future crises, including the effects of global climate change (see Chapters 2 and 5 on this series of interlinked problems, and Chapter 5 in particular for more on Africa and African HIV-AIDS). Although in 1998 the UNDP was able to conclude that, on average, health had improved in the previous 30 years (UNDP 1998: 21–23), in many African countries this was certainly not the case.

War and humanitarian relief

Since the first publication of *At Risk*, dozens of violent conflicts have broken out and many civilians have been killed, maimed (especially by land mines), injured, deliberately mutilated, starved, occasionally enslaved and displaced by the belligerent parties. So great has been the need for humanitarian relief in these conflict and post-conflict situations that some 'normal' development assistance has been diverted, and opportunities for self-generated development delayed or destroyed, further worsening the position of marginal and vulnerable populations in the longer term. Furthermore, there has been confusion among development NGOs about how to act in regard to:[30]

- civilian/military relations during 'complex' emergencies;
- relations with war lords, local elites and the army;
- ways to move from relief to recovery, and to development;
- internationally acceptable standards of assistance;
- mobilisation of international support for relief.

Conflicts have continued to exacerbate natural extreme events such as drought in Afghanistan (2002; see Christian Aid 2002; World Food Programme 2002c) and the volcanic eruption in eastern Congo (2002). However, since the mid-1990s, the possible role of 'disaster diplomacy' in peace making has also been noted, and at least a dozen 'windows' for conflict resolution that opened during a natural hazard event have been documented.[31]

Violent conflict interacts with natural hazards in a wide variety of ways:

- It is often one of the main causes of social vulnerability.
- Displacement of large numbers of people in war and other violent conflicts can lead to new risks (exposure to disease, unfamiliar hazards in new rural or urban environments) (US Committee for Refugees 2002).
- Socially vulnerable groups in extreme natural events are often also vulnerable to abuse (injury, death, rape, forced labour) during violent conflict.
- Violent conflict can interfere with the provision of relief and recovery assistance.
- Participatory methods meant to empower and engage socially vulnerable groups may be difficult or impossible during violent conflicts.
- The application of existing knowledge for the mitigation of risk from extreme natural events is often difficult or impossible during violent conflict.
- Violent conflict often diverts national and international financial and human resources that could be used for the mitigation of risk away from extreme natural events (Brandt 1986; Stewart 2000).
- Conflict sometimes destroys infrastructure, which may then intensify natural hazards (e.g. irrigation systems, dams, levees) or compromises warnings and evacuations (e.g. land mines on roads).
- The failure of sustainable development can result in conflict over resources that can lead to violent confrontation.
- Violent confrontations often wreak havoc on vegetation, land and water, and this undermines sustainable development.

- Some economic development strategies and policies can lead to marginalisation and exclusion, and hence the creation of social vulnerability to extreme natural events, and may simultaneously provoke social unrest, e.g. food riots (Walton and Seddon 1994).

Media and policy selectivity

Another change since the first edition of our book is a growing concern about the highly selective treatment of disasters by the Western media, their tendency to overlook significant disasters, and a general decline in interest in the rest of the world. Even when such disasters are noticed, there is little follow up. Typically the most underreported humanitarian crises listed by Médecins Sans Frontières (MSF) for 2001 tend to be slow onset, long-term disasters, most often linked to war or post-war situations. We attempt to redress this balance in this edition of the book. Below is a list of 'missing' crises according to MSF (2001), some of which are dealt with in subsequent chapters:[32]

Malaria epidemic in Burundi: 3 million cases in a population of 6.5 million because of the severe spatial dislocation and displacement of people due to war since 1993.

Precarious situation of Chechnyan refugees in Ingushetia, where mafia-like business groups control the flow of food and other survival goods to the refugees (Agence France-Press 2002d).

North Korean famine refugees in People's Republic of China (PRC): brutality against hundreds of thousands of Koreans fleeing across the remote border with PRC.

Rural violence and urban marginalisation in Colombia: 2 million people have become internally displaced in Colombia since 1985; 300,000 alone in 2000. Rural health services have been destroyed. In urban areas these displaced persons live in very dangerous places. This is a recipe for increasing exposure to flood, landslide, earthquake and epidemic disease.

Breakdown of health care services in the Democratic Republic of Congo: MSF estimates that there are 2.5 million internally displaced persons (IDPs) in Congo. The volcanic eruption in the east added to this number (see Chapter 8). Camp environments are hazardous in many ways, as is isolated survival on the margins of the ongoing conflicts (see Chapter 5).

Continuing violence in Somalia: Despite inter-clan peace talks in Djibouti and other diplomatic initiatives, war lords continue to dominate Somalia. People there are exposed to drought, flood, cyclones and even earthquakes. Without a viable state, their vulnerability to these natural hazards will remain high.

20 years of war in Sri Lanka: 60,000 people have died in 20 years of war, and there are hundreds of thousands of IDPs. During 2001 there was both drought and flood in various parts of the country, and the conflict hampers mitigation of these hazards, response to their impacts, and recovery – as noted in Chapter 2.

Many displaced people in West Africa: Liberia, Sierra Leone, Guinea Bissau, Senegal, Nigeria and Angola have all been affected by severe internal, organised violence. In all these countries the result is to exacerbate vulnerability to 'normal' hazards such as flooding (e.g. Senegal in 2001), drought and outbreaks of human epidemic and animal epizootic disease (see Chapters 5 and 6).

Refugees and displaced people worldwide: MSF estimates that in 2001 there were 22 million refugees in the world (who had taken refuge across a national border) and another 20–25 million IDPs. Even before additional risk factors associated with gender, class, ethnicity, age, disability, etc. are taken into account, the very fact of being a refugee or internally displaced raises a person's vulnerability to some natural hazards.

Neglected diseases: MSF concludes its list of the top ten underreported humanitarian crises with an account of chronic diseases of the poor that had not made the headlines in the same way that HIV-AIDS has done. These include tuberculosis, malaria, human sleeping sickness (of which there are African and Latin American varieties) and Kala Azar (visceral leishmaniasis).[33] All four of these chronic, debilitating and potentially lethal conditions are linked to living conditions and there is considerable disease-agent resistance to available medication. Debilitation and disability mean that people have less time to invest in protecting themselves from other hazards by, for example, constructing or maintaining terraces, fire and wind breaks, farm or community wood lots, or carrying out irrigation works (see Chapter 5 and other chapters in Part II).

Convergence and critique

Convergence

During the 1990s there has certainly been a convergence of thinking – and to a limited degree, practice – concerning natural hazards, people's vulnerability and disasters. The IDNDR put vulnerability squarely on the development agenda. Work by many institutions on urban disasters in particular helped to focus and clarify our view of vulnerability: its causes, effects and remedies. A decade-long attempt to implement Agenda 21 – the programme of action following the Earth Summit – provided many illustrations of the strengths and weaknesses of sustainable development, a very slippery, ambiguous concept. Finally, the notion of human development and its measurement using the HDI has offered new opportunities for planners and scholars to place disaster risk reduction in the mainstream. The evidence indicates that high levels of death and disruption of livelihoods by disasters are closely associated with low scores on the HDI at the national level. Whilst much of the analysis of *At Risk* is focused on the level of the household, neighbourhood or rural community, our understanding of vulnerability is consistent with these new results.

Critique

Commentary on *At Risk* has, on the whole, been positive. Some reviewers have suggested that we need to link more closely the two models presented in Chapters 2 and 3 and to use them more consistently in the chapters that make up Part II. Others have suggested ways to make the book more readable. Some have questioned whether we make enough allowance for human and social factors such as creativity and innovation (Haghebaert 2001, 2002). There have also been questions about whether we have 'thrown the baby out with the bath water' by not concentrating enough on the potential for actually affecting the natural and geophysical 'triggers' of hazards (Lavell 2001; Turner et. al. 2003). Haghebaert (2001) also wonders if our focus on 'root causes' distracts us from the less ambitious, but none the less life-saving, efforts of the state in providing safety. We have read this advice carefully and, where we concurred with it, applied it in the revision process.

A less approving critique involves what some see as the political implications of our approach. Some feel that our focus on root causes and social relations is of no practical use, and amounts to a call for social revolution. Smith (1996: 51) states that work such as ours, belonging to what he calls the 'structuralist school', 'can be criticized for rather stridently expressed views which, at worse, simply call for overall social revolution'.

Others take the opposite tack and believe we have abandoned the political struggle for justice in an unequal world. For example, Middleton and O'Keefe (1998) assert that we neglect political causes of disaster vulnerability on the national and international scale; that we limit ourselves in this way because of our desire to address multiple audiences, especially practitioners; and that we therefore rely exclusively on small-scale, incremental changes and improvement as solutions. Accusing us of sending a message 'of self-defeating counsel of prudence' (ibid.: 145), Middleton and O'Keefe write:

> ... *At Risk* stops short of tackling the larger complex in which the world's poor are so vulnerable. (p. 11)

> ... confining their examinations to unquestionably important detail, the authors add the fateful words that they do so in order not to

oversimplify and not to produce 'a theory that is of little use to managers, planners, and policy-makers'. (p. 11)

[The authors of *At Risk*] feel that sufficient attention to the smaller details will eventually force changes in the macro-economic and social conditions leading to the problems. (p. 162)

We do not propose to occupy a great deal of space giving a detailed defense of the first edition, but to focus on those criticisms which lead, however intentioned by the critics, to potential improvements to this edition. At the outset it must be said that Middleton and O'Keefe set out to write a very different sort of book from *At Risk*. Theirs is more focused on the political aspects, especially the politics of complex emergencies. They lay little claim to build theory; their main claim is to be 'radical'. Their book exposes rather than explains. One of the purposes of such a trenchant criticism of *At Risk* might have been to push aside an established book which occupied the central ground at the time, by differentiating the two different approaches. The issue of our preoccupation with detail at the expense of 'tackling the larger picture' is one way of excusing any author (including themselves) of taking the trouble to analyse in detail different approaches and theories of disasters. The Pressure and Release (PAR) model and the Household Access model, originally presented in the first edition of *At Risk* and re-introduced in an improved format in this edition, are not inconsequential details but tools that allow a carefully crafted explanation of disasters at different levels.

As the reader will soon see, Chapter 2 begins with 'root causes' that are truly global in scope and deeply rooted in history. In our schema we first break down 'root causes' into processes that are driven by ideology and that produce, reproduce and sustain political and economic systems. Secondly, we separate these into factors that distribute access within societies to power, structures and resources. In the schematic presentation of the model outlined in Chapter 2, we explain in the first edition that our intent is to show in detail how 'war, foreign debt and structural adjustment, export promotion, mining, hydropower development, and deforestation work through to localities' (p.24 of 1st edn).

True to our intention, in the first edition we took up, *inter alia*, the role of IMF structural adjustment programmes in undermining health in Nigeria and Zimbabwe (p.114), the role of international aid agencies in promoting a 'tech-fix' solution to flooding in Bangladesh (pp.138–143), the role that absentee land ownership plays in raising the stakes in coastal disaster risk (p.153) and the part played by inflation in Mexico in the lead up to its earthquake disaster of 1985 (pp.174–181). In the face of this evidence, how can our critics claim that we have 'a distaste for the large political issues'? All of these examples fit precisely that class of processes which Middleton and

O'Keefe claim falls outside the scope of *At Risk*: the macro-economic and the political.

These critics claim that the combination of our two models (outlined in Chapters 2 and 3) is capable of producing no more than the following tautology:

> People are vulnerable because they are poor and lack resources, and because they are poor and lack resources, they are vulnerable.
>
> (Ibid.: 12)

They mock this 'triumph of reason' but are kind enough to put it down not to our stupidity, but (returning to their favourite theme) to the fact that we are trapped in a 'fault in the logic of [our] models' (p. 12). This is an important source of misinterpretation. Poverty is not synonymous with vulnerability. The terms both imply relationships, but in the case of poverty it is relations with others in society which reproduces this state, while vulnerability implies causal relations with both society and also the physical environment at particular times. What Middleton and O'Keefe term circular reasoning is nothing of the kind. Our analysis often reveals the kind of vicious circle already mentioned earlier. Each time a disaster takes place, those most vulnerable are likely to be made even more vulnerable to the next extreme occurrence or stress.[34] Middleton and O'Keefe point out such vicious circles themselves in a number of their own case studies. Whether called the 'ratchet effect', 'underdevelopment trap' or 'marginalisation', this phenomenon is well established in the theoretical and empirical literature of development studies (Chambers 1983; Blaikie and Brookfield 1987). A vicious circle is not a tautology.

Audiences

This book will inevitably first come to the attention of academics and students in higher education whose work interests them in disasters, development and LDCs. We hope it will appeal to anthropologists, economists, sociologists, political scientists, geographers and others in social science. We also hope that the book will be read by engineers and natural scientists: physical geographers, geologists, oceanographers, seismologists, volcanologists, geomorphologists, hydrologists and climatologists.

Because we see this book as being useful for action as well as study, we want to identify other groups we hope will use this book. Normally, the discussion about a book's supposed readership is found in the preface, where it seems neutral and less significant. We would rather discuss our potential readers here, in relation to their own role in the social processes involved in making people vulnerable to hazards and in reducing vulnerability. By doing so we may assist in doing something to intervene in those processes to reduce that vulnerability. Such groups may include professionals involved in disaster work as an essential element in their day-to-day activity (e.g. public health workers, architects, engineers, agronomists, urban planners, civil servants, business executives, bankers and investors, community activists and politicians).

The sociologist C. Wright Mills once wrote that there are three audiences for social analysis: those with power who are aware of the consequences of their acts on others; those with power who are unaware of the consequences; and the powerless who suffer those consequences (Mills 1959). In a similar way, we identify three other broad audiences for this book. There are, firstly, those with power who create vulnerability, sometimes without being aware of their actions. Secondly, we address those with power who are attempting to do something about hazards, but may be unable to make their work effective enough because of a failure to incorporate vulnerability analysis. Thirdly, we write for those who are operating at the grassroots level, who suffer the consequences of disasters, or who are working with people to reduce their vulnerability and increase their power.

The first is the group that creates and maintains the vulnerable condition of others. Such groups include major owners of resources at international, national and local levels (whose activities have significant effects on how and where other people live), foreign agribusiness firms, investment bankers, civil engineering contractors and land speculators. In some cases they may be unaware of the consequences their decisions have for the vulnerability of others.

The second audience is extremely broad, and consists of those who attempt to address and to reduce the impact of natural hazards. It includes a variety of levels in government, and people with a range of interests in government activity, whose normal work is not specifically aimed at disasters as such. However, in almost every country, governments and other bodies have assumed some sort of responsibility for dealing with disasters, and this often involves measures to mitigate hazards.

At the apex of political power, leaders will take decisions on disasters, possibly on the advice of their senior civil servants. At this policy formulation level, directives are developed on economic, financial or political grounds, and will involve decisions affecting planning, agriculture, water resources, health, etc. The implementation stage will not necessarily address vulnerable conditions in relation to hazards, and indeed some policies may increase vulnerability. We hope to demonstrate that it is not enough simply to deal with the hazard threat, so that policies will be designed to reduce vulnerability and therefore disasters. There is considerable opportunity to improve policy making and implementation at national, sub-national, and especially at municipal levels in many countries in these early years of the twenty-first century because of the emphasis given by the World Bank and other influential bodies to the question of 'good governance'.

The implementation of policy extends beyond government ministries and agencies. Many voluntary agencies that have provided relief for disasters now see the need to address the pre-disaster conditions which give rise to patterns of repeated disaster and people's failure to cope. The Red Cross system is an example, and for ten years now it has published a *World Disasters Report* which (although not official policy) conveys a great deal of information and analysis on root causes and dynamic pressures.[35] Following an initiative by the Swedish Red Cross (Hagman 1984), many voluntary bodies have attempted to redefine their roles in terms of 'preventing' disasters rather than just alleviating their effects. We hope our book helps to enhance their future contribution.[36]

It is also possible to find representatives of the commercial sector among those involved with vulnerability who might be in a position to introduce mitigation measures. For example, a typical international civil engineering firm may include in its portfolio the design of large-scale engineering projects, such as high dams and flood defences that frequently exacerbate downstream flood hazards and thus increase vulnerability. But the same engineers may also create cyclone-resistant structures. Another example can be found in the logging industry, which can both increase risk (falling into the first category listed above) or it can work to reduce risk through measures such as selective cutting and replanting (Poore 1989; Fire Globe 2003). The same can be said of large-scale commercial agriculture and the mining industry, and parastatal firms such as electrical utilities (or their recently privatised descendants), for example in river basin management, including the construction and maintenance of dams. The construction industry can also, through its practices, either increase or decrease risk. A common perception that may motivate this second wide audience is that it is cheaper in the long run (in economic, social and political senses of the word) to prevent or mitigate disasters than to fund recovery (Anderson 1990). This is certainly the point of view of the World Bank, where its Disaster Management Facility has done the maths and shown without doubt that prevention is less costly than recovery (Gilbert and Kreimer 1999; Freeman et al. 2002). Now a consortium of banks and development agencies exists to promote prevention in the commercial as well as public sector – the ProVention consortium.[37]

The third group of readers are those who are vulnerable, or who at grassroots level are trying to deal with the processes that create vulnerability. We hope this book will assist organisers and activists who are part of grassroots struggles to improve livelihoods, for instance in the face of land deals and projects conceived by outsiders. Such locally organised pressure groups have proliferated rapidly during the 1980s and 1990s. They are now recognised as a major force for social change in general and disaster mitigation in particular (Anderson and Woodrow 1998; Twigg and Bhatt 1998; Fernando and Fernando 1997; Pirotte et al. 1999; Maskrey 1989).[38] This audience includes members of regional NGOs and networks devoted to action research in partnership with vulnerable groups of people. The three groups to which we have donated the royalties from this edition of *At Risk* are part of this audience: La RED in Latin America, Peri Peri in southern Africa and Duryog Nivaran in South Asia.[39]

Scope and plan of the book

Chapters 2 and 3 set out the perspective of our book in detail. They describe how our view of disasters differs from the conventional wisdom, and also where they coincide. It is plainly wrong to ignore the role of hazards themselves in generating disasters, and the framework we are suggesting does not do so. Likewise, we are not suggesting that vulnerability is always the result of exploitation or inequality (just as it is not equivalent to poverty). It is integrally linked with the hazard events to which people are exposed. We also want to acknowledge that there are limits to this type of analysis. It is not always possible to know what the hazards affecting a group of people might be, and public awareness of long-return period hazards may be lacking. For instance, Mount Pinatubo in the Philippines erupted in 1991, but had been dormant for 600 years.

Chapter 2 introduces a simple model of the way in which 'underlying factors' and root causes embedded in everyday life give rise to 'dynamic pressures' affecting particular groups, leading to specifically 'unsafe conditions'. When these underlying factors and root causes coincide in space and time with a hazardous natural event or process, we think of the people whose characteristics have been shaped by such underlying factors and root causes as 'vulnerable' to the hazard and 'at risk to disaster'. This will be referred to as the 'Pressure and Release' (PAR) model, since it is first used to show the pressure from both hazard and unsafe conditions that leads to disaster, and then how changes in vulnerability can release people from being at risk.[40]

We consider that certain characteristics of groups and individuals have a great deal to do with determining their vulnerability to hazards. Some of these, such as socio-economic class, ethnicity and caste membership have featured in analyses since the 1970s. Others, especially gender and age, are more recent research categories, and have developed in part because of the influence of social movements such as feminism.[41] For example, in a classic example of the importance of gender, Vaughan (1987: 119–147) uses the oral evidence provided in women's songs and stories in Malawi to reconstruct a women's history of the 1949 famine that is strikingly different from the men's account:

[Women], along with the very old and very young, were more likely than men to end up relying on government handouts ... [W]omen

stress how frequently they were abandoned by men, how harrowing
it was to be left responsible for their suffering and dying children,
how they became sterile, and how they were humiliated by the
feeding system.

(Ibid.: 123)

During the 1990s a large amount of work on gender and disaster yielded
much more valuable evidence of this kind (Fernando and Fernando 1997;
Enarson and Morrow 2001).[42] Others have emphasised the special needs,
lack of status and access, and hence special vulnerability of the frail elderly,
especially widows (Guillette 1991; Feierman 1985; Wilson and Ramphele
1989: 170–185).

Daily life comprises a set of activities in space and time during which
physical hazards, social relations and individual choice become integrated as
patterns of vulnerability.[43] These patterns are guided by the socio-economic
and personal characteristics of the people involved. Here are found, sometimes (but not always), the effects of gender,[44] age,[45] physical disability,[46]
religion,[47] caste[48] or ethnicity,[49] as well as class. All of these may play a role,
in addition to poverty, class or socio-economic status. Although we include
class in our analysis, we fully recognise the role of this wide range of social
relations and do not dwell exclusively on class relations.

Chapter 3 adds to our alternative framework by focusing on patterns of
access to livelihood resources. We expand the discussion there of 'underlying
factors and root causes', identified in Chapter 2. In doing so we seek to shift
the focus of our analytical method further in the social direction, without
oversimplifying or producing a theory that is of little use to managers, planners and policy makers.

Part I concludes with a discussion of coping. We believe that too little
attention has been given to the strategies and actions of vulnerable people
themselves. In large part their 'normal' life is evidently (at least to outsiders)
a continual struggle in which their conditions may resemble a disaster.
People become braced to cope with extreme natural events through the
stress of making ends meet, in avoiding the daily hazards of work and home,
and of evading the predations of the more powerful. They form support
networks, develop multiple sources of livelihood access and 'resist' official
encroachments on livelihood systems in a variety of ways (Scott 1985, 1990,
1998). People learn rather cynically, yet realistically, not to rely on services
provided by authorities (Robinson et al. 1986; O'Riordon 1986; Maskrey
1989; Oliver-Smith and Hoffman 1999). Our discussion of 'coping' will
neither romanticise the self-protective behaviour of ordinary people, nor
dismiss it.[50]

Having set out our alternative framework in Part I (Chapters 1–3), Part II
presents case material organised by hazard type – those linked with drought,
biological hazards, flood and landslide, cyclone, earthquake and volcano

(Chapters 4–8). In each chapter we follow a similar method in tracing the causes of vulnerability, making use of both PAR and Access models. It may appear to contradict our approach to deal with disasters through different natural hazard types. However, we have deliberately chosen to do this because users of this book may themselves be concerned with particular hazards, or may find it difficult to accept our approach without seeing it interpreted more concretely in the context of nature.

Part III (Chapter 9) draws out lessons for recovery and for preventive action. We provide a holistic view of recovery and review the mixed history of narrow relief and reconstruction efforts, paying special attention to whether and how 'dynamic pressures' and 'root causes' of disaster vulnerability can be addressed during what has been called the 'window of opportunity' for policy change created by disasters. We end the book with a series of objectives that link human development and vulnerability reduction, emphasising issues of governance and livelihood resilience and local capacity that have begun to be accepted as desiderata in mainstream development circles.

Limits and assumptions

Limitations of scale

There are logical grounds for limiting our book to certain sorts of disaster. Disasters cannot, of course, be neatly categorised either by type or scale. At one extreme, it seems that there have been five mass extinctions over the last 400 million years in which up to half of the life forms on the planet disappeared (Wilson 1989: 111). The best known of these is the disappearance of the dinosaurs. The scale of such disasters (and even the use of the term is perhaps inappropriate) is clearly so many orders of magnitude greater than those with which we are concerned that we exclude them. Such events are beyond the present scale of human systems.

More recently, there have been two or three occasions when a large proportion of the human inhabitants of this planet died with apparently little distinction in regard to the relative risk of different social groups. Many millions died during the pandemics of bubonic and pneumonic plague known as the Plague of Justinian (AD541–93) and the Black Death (1348–1353). More recently the influenza virus that swept the world during and after the First World War killed 22 million in less than two years (1918–1919). This was approximately four times the total of military casualties during that war. The demographic and socio-economic consequences of the first two events had epochal significance. The current HIV-AIDS pandemic could equal them in its widespread socio-economic consequences unless a vaccine is found or sexual practices change. Despite the great significance of biological disasters, we shall address such events only tangentially (see Chapter 5), in part to illustrate the limits of the vulnerability approach. Catastrophic epidemics may be limiting cases that shed light on 'normal' disease disasters, such as outbreaks of cholera and malaria in Latin America and Africa, meningitis and Ebola in Africa, or plague in India.

Nuclear war is another type of disaster that we do not consider because it is produced directly by humans, although some research on the 'nuclear winter' has been inspired by threats from natural events such as massive volcanic explosions or asteroid impacts. There is also considerable climatological, astrophysical and palaeontological work on mass extinctions which links some of these to severe interference with received solar radiation. Atmospheric phenomena of a similar scale of magnitude, such as global warming, will be treated as part of the more remote 'dynamic pressures' of the PAR model, shaping patterns of vulnerability. We also consider war itself (in its non-nuclear form) to be a significant 'root cause' of disaster and will address it several times throughout the text.

We devote only a little attention to what might be called 'social hazards', especially to terrorism. The events of 11 September 2001 in New York City have caused disaster researchers to reflect upon the lessons that twenty-first century terrorism might have for their own work on other kinds of hazards (and vice versa). If the official US position is correct – that the attack on the World Trade Center constituted the beginning of a war (the 'war on terrorism') – then, in fact, such a disaster is not new.[51] Millions of civilian lives have been lost in wars during the twentieth century (Hewitt 1994, 1997). An alternative position is that the attack was not an act of war but a crime (albeit with a large number of victims). If this alternative view is correct, then there are also precedents, such as the gas attack on the Tokyo subway in 1995 and the bombing of the Murrah Federal Building in Oklahoma City. In either case, our book cannot expand to include such disasters, and we might simply offer the observation that those seeking to understand such 'acts of war' or 'crimes' should, as we do, look for root causes and not for quick (including massive military) fixes.

Technological hazards

Vulnerability assessment is also relevant to analysing disasters resulting from technological hazards. However, we restrict the scope of this book and exclude technological hazards, for the simple reason that they are not natural in origin. One of our purposes in this book is to deal with natural hazards, because of the inadequacy of explanations of disasters that blame nature. Our aim is to demonstrate the social processes that, through people's vulnerability, generate human causation of disasters from natural hazards. So there is little point in looking at specifically human-created hazards.

Failure of technology, such as that which occurred at the Chernobyl nuclear facility in Ukraine in 1986 and the chemical factory at Bhopal, India in 1984, massive oil and toxic spills and the dumping of nuclear waste in polar regions (UNEP 2002: 297), fall outside the scope of our book because they are chiefly failures of techno-social systems.[52] Later, there will be some tangential discussion of the Bhopal disaster, which involved explosions and the release of toxins from a fertiliser and chemical factory. The same locational factors responsible for generating hillside slums already mentioned in other countries led to dense squatter settlement around the plant. Such a case is at the limits of our type of analysis, and overlaps with a related literature concerning technology and society (Perrow 1984; Weir 1987; Piller 1991) and environmental justice (see below).

What happens to poor and other vulnerable people who find themselves in the path of rapid industrialisation, de-industrialisation, industrial deregulation or the importation of toxic waste is clearly of concern to us. But it is not a central issue in this book. Some overlap with a critical appraisal of technological risk and what Beck calls 'ecological modernisation' will nevertheless occur in the chapters that follow. Flooding caused by the failure of a dam is a good example (Chapter 6). The web of cause and effect in the connections between society, nature and technology is often impossible to disentangle (Abramovitz 2001).

Another point of similarity between our approach to natural hazards and studies of technological and more pervasive environmental risks is a concern with bottom-up, grassroots activism. The environmental justice movement has grown rapidly since its origins in the study of racial disparities in the location of US hazardous waste facilities during the late 1980s (Bullard 1990; Hofrichter 1993; Shiva 1994; Heiman 1996; Johnston 1997; Faber 1998).[53] One question, to which we will return in Part III, is whether a similar worldwide movement is possible through which citizens assert their human right to protection from avoidable harm in extreme natural events.[54]

We will be concerned with the impact of technology on vulnerability, particularly technology in its apparently simplest and benign forms.[55] For example, a new road may link a previously isolated rural community with sources of food that may reduce vulnerability in times of drought. That same road may also lead away able-bodied youth in search of urban income, reducing the labour available to maintain traditional earth and stone works constructed to prevent erosion, or to build or repair houses adequately to withstand earthquake. The result may be a reduction of crop yield during drought years because of additional soil loss or deaths from an earthquake which otherwise would be preventable.

The same road may introduce mobile clinics that immunise children against life-threatening diseases, or it may provide the channel through which 'urban' diseases such as tuberculosis and sexually transmitted diseases arrive via the men who have gone to work in city, mine or plantation. It may also provoke landslides that kill people or reduce the available arable land. All these contradictory effects of technological change are possible. The same may be said of the introduction of new water or energy sources, new seed varieties, construction of a dam or a new reinforced concrete building.

There are several ways in which such questions of technological change arise in relation to disaster vulnerability. One of the most frequent responses to disaster by outsiders is the provision of various technologies to the affected site during relief and rehabilitation activities. These include temporary housing, food supplies, alternative water supplies and sanitation facilities, seeds and tools to re-establish economic activities. In all such cases, the new or temporary technology may play a role in increasing or decreasing the vulnerability of a particular social group to a future hazard event. The controversy over the use of genetically modified maize when offering famine relief in southern Africa in 2002 is a dramatic example.[56]

Development planners sometimes introduce technology at the so-called 'leading edge' of whatever version of rapid, systemic change they define as 'development'. This may be irrigation technology in the form of a large dam that displaces thousands of families in what economists call 'the short run'. It might take the form of low-income housing or the development of an industrial complex. Such development initiatives can have a series of unintended, unforeseen consequences.

The people displaced following the flooding behind a large dam may not benefit from resettlement in the areas that are fed by the irrigation water. If they are included among settlers, they may end up at the bottom of the water distribution system, where water is scarce.[57] Women on such new schemes may lose conventional rights to land on which they used to grow food for their families (Rogers 1980) or their knowledge and skills may be rendered 'obsolete' (Shiva 1989). Nutritional levels among children may fall, paradoxically, as cash income from the marketed product of irrigation increases (Bryceson 1989).

The introduction of technology can modify and shift patterns of vulnerability to hazards. For example, the Green Revolution varieties of grain have shifted the risk of drought and flood from an emergent class of 'modern' farmers to the increasing number of landless and land-poor peasants. These latter have become more vulnerable because they are denied access to 'commons' that formerly provided livelihood resources and because they are highly dependent on wages earned in farm labour to purchase food and other necessities (Jodha 1991; Chambers et al. 1990; Shiva 1991). They are also vulnerable because they now depend for food and other basic necessities on wages from farm employment that can be interrupted by flood, hail, drought or outbreaks of pests and disease (Drèze and Sen 1989; see Chapter 4).

The change in technology brought about by the Green Revolution has affected the resource-poor in rural areas because the pre-existing social and economic structure has not been able to distribute benefits properly, and this has led to a realignment of assets and income. The losers may consequently be subject to new hazards. For example, in order to find somewhere to farm, they may migrate into low-lying coastal land that is exposed to storms (see Chapter 7). They may have little choice but to live in poorly constructed housing as urban squatters. In Bhuj, Gujarat (India) many thousands of such people died in the earthquake of February 2001.[58] The literature on development is full of studies of such unintended consequences.[59] This book will focus on such technological developments and their consequences where they can be seen to impinge on people's vulnerability to extremes of nature, or where they affect the ability of groups to sustain their livelihoods in the aftermath of environmental extremes.

Notes

1 We use the term LDC for 'less developed country' (including such extremes as 'least developed' and 'highly indebted, least developed') in keeping with UN practice. LDCs are contrasted to 'more developed' countries (MDCs). In the first edition we used the term 'Third World' to refer to LDCs, but that term has a history. It connotes the historical process (usually one form of colonialism or another) by which a country was impoverished or 'underdeveloped' (as a transitive verb). We still find merit in this view, and our 'Pressure and Release' model often has processes set in motion during the colonial past as 'root causes' of vulnerability. However, the term 'Third World' also carries overtones of the logic of the Cold War, during which period there existed two opposing 'worlds' and a third, non-aligned world. But with the collapse of the Soviet Union, many of its constituent republics (which are now independent), and even some central and eastern European countries that were part of the Soviet bloc, are now clearly seen to be 'less developed' and have many people who share vulnerabilities in common with inhabitants of countries previously designated Third World. Since the first publication of this book, the changes that began in 1989 have so reshaped the geopolitical map that use of the term Third World may be confusing.

2 We used diverse sources in estimating these numbers, which, especially for the earlier part of the century and for specific kinds of conflicts, must be considered only the roughest approximations. *For estimates of deaths due to war and political violence* we are most grateful to Professor Kenneth Hewitt, Wilfred Laurier University, Canada, for time spent in personal communication with Ian Davis during July 2002. Hewitt's book, *Regions of Risk* (1997), and an earlier 1994 article, were also helpful sources as well as Sivard (2001) and White (1999). *Drought/famine death statistics* are based on the authors' approximate calculations that expand on the official reports that are regarded as gross underestimates, since entire famines, such as the 'Great Leap Forward Famine' in China (1958–1961), which may have killed 30 million people (Yang 1996; Becker 1996; Heilig 1999), are omitted from official databases. Discussions were held between Ian Davis and researchers at the CRED, Université Catholique de Louvain, Brussels and the US Office of Foreign Disaster Assistance (OFDA) in July 2002, who confirmed that they are only able to document statistics that governments provide to them. Famine is treated at length in Chapter 4. *For other disaster mortality statistics* we relied on the database maintained by CRED and OFDA called EM-DAT (available at www.cred.be/emdat, which we accessed for this purpose on 11 July 2002). For a critical note on the reliability of disaster statistics, including those for drought and famine, see Chapter 2, Box 2.3. *Traffic*

accident statistics came from the *World Disasters Report 1998* (IFRC 1998: 20–31). *Estimates of deaths due to HIV-AIDS* came from Barnett and Whiteside (2001). For more on HIV-AIDS, see Chapter 5.

3 For example, the World Health Organisation (WHO) estimates that 12 million children under five die *each year* (mostly in LDCs) from easily preventable illnesses such as diarrhoea, measles and malaria (Mihill 1996; Boseley 1999). This is ten times as many as the average deaths from natural hazards in an entire *decade* (see Chapter 5).

4 In our usage, 'social' refers to human-created systems, and so includes economic and political processes. For brevity, from here on when we refer to 'social framework' or 'social environment', we normally mean to include political and economic factors as well.

5 Hewitt (1983b) referred to the segregation of disasters from the normal functioning of society and policy making as creating a 'disaster archipelago'. He maintained and elaborated on this position in subsequent work (Hewitt 1997).

6 In April 2003, the International Rescue Committee reported that as many as 4.7 million people in the Republic of Congo had perished as the result of the combination of injuries sustained in the conflict, starvation and disease. Although there is a margin of error of 1.6 million lives in this estimate, the conflict in the Congo has, according to the report 'claimed far more lives than any other conflict since the second world war' (Astill and Chevallot 2003: 7).

7 Baxter and Kapila (1989); in recent years there have been attempts to prevent this happening again, with projects that have placed pipes in the lake which attempt to trap the carbon dioxide gas and vent it safely to the atmosphere (Jones 2001, 2003). For further background on the lake Nyos disaster, see Chapter 8, note 7.

8 A major watershed for relief agencies was the year 1970, when enormous disasters in Peru, East Pakistan (now Bangladesh) and Biafra (Nigeria) coincided. A new theory of disasters that focused on the vulnerability of 'marginal' groups was suggested by subsequent reflections on these events, plus the Sahel famine (1967–1973) and drought elsewhere in Africa, erosion in Nepal, an earthquake in Guatemala (1976) and a hurricane affecting Honduras (1976) (Meillassoux 1973, 1974; Baird et al. 1975; Blaikie et al. 1977; Davis 1978; Jacobs 1987).

9 In the second edition of the 1978 book *The Environment as Hazard*, the authors have made no fundamental change to their 'stages of development' model (Burton et al. 1993).

10 On the response of 'political economy' and 'political ecology' to both 'modernisation theory' and 'environmental determinism' see Meillassoux (1974); Baird et al. (1975); Wisner et al. (1977); Jeffrey (1980, 1982); Susman et al. (1983); Watts (1983b); Bush (1985); Spitz (1976). Work during this period was heavily influenced by Latin American dependency theory. For a summary of more recent rebuttals, see Adams (2001: chs 7 and 9).

11 For examples of the use of a too-general notion of vulnerability, see Anderson and Woodrow (1998); Parry and Carter (1987); Cuny (1983); Davis (1978). In such cases it is essential to specify the mechanisms by which one gets from generally widespread conditions (e.g. 'poverty' or 'crowded conditions') to particular vulnerabilities (e.g. loss due to mudslide, cyclone, earthquake, famine).

12 Such functionalist views of social system coping include work by sociologists and others influenced by Parsons and Durkheim – Mileti et al. (1975); Timmerman (1981); Pellanda (1981); Drabek (1986); Lewis (1987) – and also the work of self-defined 'sustainability scientists' who have emerged particularly as work on 'adaptation' to global climate change has been funded (Kasperson and Kasperson 2000). While there is some valuable work from these points of view,

on the whole we believe that one has to be more specific. *People* cope, not disembodied systems (see Chapter 3).

13 Since publication of the first edition of our book, development policy has become more concerned with wider notions of 'human security' that encompass reduced vulnerability to disaster as well as social protection from economic crisis and respect for people's human rights in war and violent conflict (see UNDP 1994a).

14 Readers who are familiar with the Sustainable Livelihoods approach of the Department for International Development (the UK foreign aid ministry) will see a parallel here with the five types of capital commonly used in that framework – natural (mainly land, forests, water sources); physical (infrastructure and production resources); financial; human (e.g. education level); and social (e.g. networks and family connections). See Chambers (1995b); Carney (1998); Moser (1998); Rakodi (1999); Sanderson (2000).

15 The World Commission on Environment and Development (the Brundtland Commission) linked the concept of livelihood to the ability of people to protect the environment, and stated that the goal of development should be 'sustainable livelihood security' (WCED 1987). In our view, vulnerability to hazards is likely to increase when livelihoods are pursued at the expense of environmental stability (Abramovitz 2001). So it is not a solution to vulnerability if people seek to increase their access to livelihood resources for short-term gains, even if it is necessary to cope with the immediate impact of hazards. We develop a more accurate view of livelihoods in relation to disasters in Chapters 3 and 4.

16 In 1991 and 1992 there were torrential rains and mudslides in southern California affecting two counties (Ventura and Los Angeles) where 10 million people live. Also in 1991 there was a fire storm that killed twenty-five people and left thousands homeless in the middle income, suburban hills above Oakland and Berkeley in northern California. This fire left the denuded, steep hills subject to landslides. During this same period there were a number of mudslides in Rio de Janeiro and Belo Horizonte in the industrial south of Brazil. More recently, in 1999, flash floods and landslides killed 30,000 poor urban residents on the extreme periphery of greater Caracas who lived in the coastal hills (IFRC 2001b: 82; Dartmouth College 1999; see also Chapter 6).

17 During a rainy night in 2000, a 100 m high pile of solid waste collapsed on hundreds of poor people in Payatas, to the north-east of Manila, the capital of the Philippines. They were permanent residents, some of perhaps 2,000 that make their living by sifting the rubbish and selling scrap metal and other recyclable items. Seven hundred people were confirmed killed or reported missing (Luna 2001; Westfall 2001).

18 As we write this second edition we acknowledge the fact that the term 'vulnerable' and 'vulnerability' are widely used in many disciplines and professions involved with disaster risk reduction. Somewhat quixotically, we believed in the early 1990s that we could reverse this linguistic trend. By now it is so well entrenched that we have put down our lance and sit under a tree with Sancho Panza enjoying the wine and landscape. However, for the sake of clarity, in our book at least we will maintain the convention of reserving the adjective 'vulnerable' for people.

19 Morrow (1999: 10) writes of the urban context of Miami, Florida, in the USA and provides a checklist which identifies the following categories: (1) residents of group living facilities, (2) elderly, particularly frail elderly, (3) physically or mentally disabled, (4) renters, (5) poor households, (6) women-headed households, (7) ethnic minorities (by language), (8) recent residents/immigrants/

migrants, (9) large households, (10) large concentrations of children/youth, (11) the homeless, (12) tourists and transients (homeless people).

20 It is additionally tragic that a year after the earthquake in Gujarat hatred between the two groups led to attacks by Hindus and Muslims on each others' communities (especially in the capital Ahmedabad), with the loss of perhaps 2,000 (mostly Muslim) lives (Harding 2002).

21 There is further discussion of the concept of the 'risk society' in Chapter 5.

22 We have no doubt that stereotypes and images, especially those arising in colonial relations, have profoundly influenced the way that LDCs are viewed today and the kinds of policies that are produced (Blaut 1993; Said 1988; Arnold 1999). We question only whether this kind of analysis is sufficient to provide a purchase on the nexus of economic and political relationships that constitute the root causes of disaster vulnerability.

23 See http://www.geohaz.org/radius.html.

24 In 2000, 47 per cent of the world's population was defined as urban, up from 38 per cent in 1990. In 1950 the world's urban population was only 30 per cent of the total (United Nations 1999: 2; Worldwatch Institute 1998: 33–34); see also Chapter 2, where urbanisation is discussed as a 'dynamic pressure'.

25 At the Johannesburg Summit in September 2002, Russia and Canada announced that they would sign the Kyoto Accord, thus bringing the number of signatories up to the required number for it to come into force. The USA, however, still refused to sign.

26 On the science behind the study of global climate change, see Chapters 2, 4, 5 and 7. Even the controversial author of *The Skeptical Environmentalist*, Bjorn Lomborg (2001), admits that warming of the atmosphere has taken place, but argues that the rate of change is toward the lower rather than higher range suggested by studies by the Intergovernmental Panel on Climate Change. For critiques of Lomborg and his answers, see http://www.lomborg.org.

27 This is the layer of the earth's mantle upon which the lithospheric plates sit. Convection currents in the asthenosphere allow heated material to rise, while cool material sinks, leading to movement of the plates. Understanding of biogeochemical cycling and plate tectonics (including earthquakes and volcanoes) would require study of the asthenosphere as well as the more accessible lithosphere.

28 Tobin has proposed a tax on international financial transfers in order to reduce the flows which are simply used to exploit price differentials (e.g. of currencies) for private benefit. For information see ATTAC, a worldwide network of citizens' organisations lobbying for this tax: http://attac.org/indexen/ and search on 'Tobin'.

29 See Petrella (2001); Barlow and Clarke (2002). The World Bank estimates that private water industry revenue approached $800 billion in 2000; 15 per cent of the water supplies in the USA have been privatised, 88 per cent of UK supplies and 73 per cent of water systems in France (Rothfeder 2001: 102; Petrella 2001: 72). African, Asian and Latin American municipal water systems are also being privatised rapidly, often at the insistence of the International Monetary Fund (IMF) as a condition of its loans, either as direct sales of municipal assets or, more commonly, long-term concessions, leases or management contracts. Large multinational corporations are the major bidders, including Vivendi, Suez Lyonnaise, Bectel-United Utilities, ENRON-Azurix, Bouygues-SAUR and RWE-Thames Water. Under new management, water prices have increased, putting more pressure on the livelihood systems of the poor (see Chapter 3). This has sometimes caused violent protests, as in Cochabamba, Bolivia in 2000 (Rothfeder 2001: 107–114). Although the terms of contracts are becoming more

precise and incorporating details as regards minimum standards and protection for the poor, municipalities are often working with limited information, technical and legal capacity against some of the largest corporations in the world (Lee 1999: 140–183).

30 See Middleton and O'Keefe (1998); Anderson (1999); Pirotte et al. (1999); Cuny and Hill (1999); Sphere Project (2000); Vaux (2001).

31 See Disaster Diplomacy, the website at Cambridge University maintained by Ilan Kelman since 2001: http://www.arct.cam.ac.uk/disasterdiplomacy/

32 The list for 2001 is sadly similar to those compiled by MSF for previous years (as, alas, is the list for 2003). In 2000 their list included displaced persons due to war in Angola, Chechnya, Indonesia, Burma (minority Rohingya Muslims who had fled across the border to Bangladesh), Democratic Republic of Congo, Afghanistan (not much of a story until 11 September 2001), Sierra Leone and Colombia (see MSF-USA 2001).

In 1999 the list included conflict, displacement, and acute vulnerability to environmentally linked disease on the part of hundreds of thousands of people running from conflict in Democratic Republic of Congo, Afghanistan, Angola, Colombia, Sri Lanka, Burundi and Somalia. In addition, MSF list a little-known severe outbreak of cholera in Mozambique (December 1998 to mid-May 1999) that infected 62,263 people and killed 2,063 (see MSF-USA 1999)

33 Kala Azar is caused by infestation by a protozoan transmitted by the bite of the sand fly. It causes fever, weight loss, swelling of the spleen and liver and anaemia. Untreated, it is almost always fatal. See World Health Organisation fact sheet: www.who.int/inf-fs/en/fact116.html.

34 We take up this critique again in more detail in Chapter 3.

35 The International Federation of Red Cross and Red Crescent Societies (IFRC) has its world headquarters in Geneva and member societies in many countries that are involved in hospitals, primary health care, training for public health, safety and emergency response. It is a federation of 178 national societies.

36 Early self-critical evaluations by voluntary agencies included one by a broad coalition that supported 'Operation Lifeline Sudan' (Minear 1991) and the group 'USA for Africa' (Scott and Mpanya 1991). More recent appraisals have been collected by Action Against Hunger (1999, 2001), Anderson (1999), Pirotte et al. (1999) and Vaux (2001).

37 For details go to http://www.proventionconsortium.org/.

38 On NGOs (private voluntary organisations, popular development organisations, development support organisations, etc.) see Conroy and Litvinoff (1988); Holloway (1989); During (1989); Wellard and Copestake (1993); Bebbington and Thiele (1993); Farrington and Lewis (1993); Riddell et al. (1995); Christoplos (2001).

39 The Network for Social Science Research for Disaster Reduction Latin America, headquartered in Panama City, Panama (La RED): www.desenredando.org/;Peri Peri, whose base is in Cape Town, South Africa: www.egs.uct.ac.za/dimp/; Duryog Nivaran, centred in Colombo, Sri Lanka: www.adpc.ait.ac.th/duryog/duryog.html.

40 This view has much in common with other attempts to reconcile an analysis of structural constraints on people's lives with an appreciation of the individual's agency and freedom (Mitchell 1990; Palm 1990; Kirby 1990a; Hewitt 1997; Alexander 2000; Wisner 2003a; Pelling 2003b).

41 The women's movement makes an enormous contribution to our understanding of vulnerability, environmental degradation and the possibilities for restoration, peace making and 'healing'. This often requires redefining what is meant by such terms as 'development' and 'progress'. See Sen and Grown (1987); Momsen and

Townsend (1987); Dankelman and Davidson (1988); Shiva (1989); Tinker (1990); Cliff (1991); Keller-Herzog (1996); WEDO (2002); Kerr (2002); on women and the politics of 'development' and vulnerability, as well as eco-feminist philosophers, see Merchant (1989) and Biehl (1991).

42 See also the Gender and Disaster Network website:
    http://online.northumbria. ac.uk/geography_research/gdn/.

43 Accounts of disaster that try to balance macro- and micro-perspectives include Hewitt (1983a); Oliver-Smith (1986b); R. Kent (1987); Maskrey (1989); Kirby (1990a, 1990b); Palm (1990); Hewitt (1997); Tobin and Montz (1997); Alexander (2000).

44 Studies emphasising the role of gender in structuring vulnerability include Jiggins (1986); Schroeder (1987); M. Ali (1987); Rivers (1982); Vaughan (1987); Drèze and Sen (1989: 55–59); Sen (1988, 1990); Agarwal (1990); Phillips (1990); Kerner and Cook (1991); O'Brien and Gruenbaum (1991); Walker (1994); Wiest et al. (1994); Cutter (1995); Fothergill (1996, 1999); Fernando and Fernando (1997); Fordham (1998, 1999, 2003); Morrow and Phillips (1999); Stehlik et al. (2000); Enarson and Morrow (2001); UN Economic and Social Department and ISDR (2001) and Cannon (2002).

45 The very young are highly vulnerable to nutritional and other health stresses during and after disasters and are vulnerable to emotional disturbance in the post-disaster period (Chen 1973; UNICEF 1989, 1999: 25–46; Goodfield 1991; Cutter 1995; La RED 1998; Harris 1998; Jabry 2003). Jabry (2003) states that 'an estimated 77 million children under 15, on average, had their lives disrupted by a natural disaster or an armed conflict, each year, between 1991 and 2000'. The old are often more vulnerable to extremes of heat and cold, are less mobile, and are therefore less capable of evacuation, and may have medical conditions that are complicated by injury or stress (Bell et al. 1978; Melnick and Logue 1985; O'Riordon 1986: 281; Tanida 1996; Klinenberg 2002; HelpAge International 2000), and are particularly vulnerable to recurrent disasters (Guillette 1991). The elderly can also suffer serious psychological harm following disasters (Bolin and Klenow 1983; Ticehurst et al. 1996). Widows in many parts of the world are especially vulnerable, as in southern Africa (Wilson and Ramphele 1989: 177–178; Murray 1981), and east Africa (Feierman 1985) or in the USA (Childers 1999).

46 Disabilities such as blindness, mental retardation, somatic hereditary defects and post-traumatic injury (such as spinal cord injuries) affect hundreds of millions of people worldwide (Noble 1981). People with disabilities have specific increased vulnerabilities in the face of hazards due to their impaired mobility or interruption of the special attention to their hygiene and continuous health care needs in disasters (UNDRO 1982b; Parr 1987, 1997; Tierney et al. 1988; Kailes 1996; Wallrich 1998; Wisner 2003c); they may also have particular needs when it comes to warnings and evacuation (Van Wilkligen 2001; Norman 2002, 2003).

47 The role of religion has not been as well studied, but consider recent events. The Burmese fleeing into Bangladesh during 1992 were a Muslim minority in their home country. The 400,000 people forced to leave squatter settlements around the city of Khartoum for an uncertain future in 'resettlement camps' in the desert were mostly a Christian or animist minority, refugees from war in the south, in the predominantly Muslim north of Sudan.

48 The role of caste has been most fully explored in studies of famine in India (see Chapter 4); however there is also a suggestion that caste-based locational segregation homes in rural and urban India may have a bearing on vulnerability to riverine flood and cyclone (see Chapters 6 and 7). The Burakumin 'caste' in

Japan is also subject to discrimination and may have suffered disproportionately in the Kobe earthquake (see Chapter 8).

49 Ethnicity and race emerge as an important factors in explaining vulnerability in studies by Regan (1983); Franke (1984); Perry and Mushkatel (1986); Bolin and Bolton (1986); Winchester (1986, 1992); Rubin and Palm (1987); Laird (1992); Miller and Simile (1992); Johnston and Schulte (1992); Bolton et al. (1993); Bolin and Stanford (1998b); Fothergill et al. (1999); Steinberg (2000).

50 Perception, experience and discourse about risk are never straightforward. For example, perceptions of risk are sometimes deeply rooted in cultural understandings of ritual purity and danger (Douglas and Wildavsky 1982) and claims of suffering (or their absence) can sometimes be gambits in games over local political power (Richards 1983; Laird 1992; Steinberg 2000: ch. 1).

51 We do not disregard or underestimate the intellectual challenge of dealing with the complexities and uncertainties vividly brought to mind by the attack on the World Trade Center. There are some who think that an enormously complex system such as a mega-city cannot possibly be fully understood, and hence cannot be protected properly (Mitchell 1999b; cf. Homer-Dixon 2001; Rubin 2000). Perrow (1984) put forward that argument some years ago regarding even 'simpler' systems such as single large jet aircraft or a nuclear power station – a view that was possibly reconfirmed by the 'surprising' destruction of the US space shuttle Columbia in early 2003. It also may be that when one adds the additional level of complexity and uncertainty of a global economy and the relations and histories that constitute 'international relations' among 191 nations, it is impossible to predict the consequences of actions. For example, in a case that falls more within the scope of our book, there was a deadly mudslide in Algiers in 2001 (Wisner 2001b). A key factor was heavy rain, to be sure. However, in addition, in their own 'war on terrorism' the Algerian authorities had cut and burned the forest on the mountain above Algiers and blocked up the storm water drainage system. Both actions were taken to deny 'terrorists' a hiding place. Both official acts exacerbated the flood.

52 Such technological hazards are discussed by other authors, including Ziegler et al. (1983); Perrow (1984); Weir (1987); Kirby (1990c); Shrivastava (1992); Button (1992); Jasanoff (1994); Dinham and Sarangi (2002).

53 A gateway to web sites dealing with environmental justice is: www.ejrc.cau.edu/.

54 See discussions and debates about the relationship between disaster and human rights: http://online.northumbria.ac.uk/geography_research/radix/.

55 For example, it is hard to disentangle risks associated with construction technologies (Chapter 8) or agricultural innovations (Chapter 4) with such hazards as earthquake and famine.

56 The USA offered Zimbabwe, Zambia and Malawi genetically modified maize as part of the international response to a famine in the region that affected 15 million people (see Chapter 4). These countries refused the maize because it was unmilled, and their scientific advisers were concerned that if planted (and not eaten), there might be contamination of local varieties of maize (a staple in the region) with unforeseen, but potentially grave, consequences for the future.

57 The social and ecological consequences of building high dams worldwide have been systematically reviewed by the World Commission on Dams (2000c).

58 There seems to be uncertainty in the figures for the number who died. The UK Disasters Emergency Committee report (DEC 2001a) accepts an official figure of 20,000 deaths as being accurate.

59 The unintended consequences of 'development' are documented by Trainer (1989); Shiva (1989); Wisner (1988a); Lipton and Longhurst (1989); Johnston

(1994, 1997); Adams (2001: ch. 8). Special note should be taken of a 'classic' early paper on disease and development by Hughes and Hunter (1970) and the contrast with the role of other kinds of 'development' in restoring the health of communities (Wisner 1976a).

2

THE DISASTER PRESSURE AND RELEASE MODEL

The nature of vulnerability

Two models

In evaluating disaster risk, the social production of vulnerability needs to be considered with at least the same degree of importance that is devoted to understanding and addressing natural hazards. Expressed schematically, our view is that the risk faced by people must be seen as a cross-cutting combination of vulnerability and hazard. Disasters are a result of the inter-action of both; there cannot be a disaster if there are hazards but vulnerability is (theoretically) nil, or if there is a vulnerable population but no hazard event.[1]

'Hazard' refers to the natural events that may affect different places singly or in combination (coastlines, hillsides, earthquake faults, savan-nahs, rainforests, etc.) at different times (season of the year, time of day, over return periods of different duration). The hazard has varying degrees of intensity and severity.[2] Although our knowledge of physical causal mechanisms is incomplete, some long accumulations of records (for example of hurricanes, earthquakes, snow avalanches or droughts) allows us to specify the statistical likelihood of many hazards in time and space. But such knowledge, while necessary, is far from sufficient for calculating the actual level of risk.

What we are arguing is that the risk of disaster is a compound function of the natural hazard and the number of people, characterised by their varying degrees of vulnerability to that specific hazard, who occupy the space and time of exposure to the hazard event. There are three elements here: risk (disaster), vulnerability, and hazard, whose relations we find it convenient to schematise in a pseudo-equation:

$$R = H \times V.$$

Alexander (2000: 13) distinguished between risk and vulnerability, noting that 'vulnerability refers to the potential for casualty, destruction, damage, disruption or other form of loss in a particular element: risk combines this with the probable level of loss to be expected from a predictable magnitude of hazard (which can be considered as the manifestation of the agent that produces the loss).'

A disaster occurs when a significant number of vulnerable people experience a hazard and suffer severe damage and/or disruption of their livelihood system in such a way that recovery is unlikely without external aid.[3] By 'recovery' we mean the psychological and physical recovery of the victims, and the replacement of physical resources and the social relations required to use them (see Chapter 9).

In order to understand risk in terms of our vulnerability analysis in specific hazard situations, this book uses two related models of disaster. The Pressure and Release model (PAR model) is introduced in this chapter as a simple tool for showing how disasters occur when natural hazards affect vulnerable people. Their vulnerability is rooted in social processes and underlying causes which may ultimately be quite remote from the disaster event itself.

The basis for the PAR idea is that a disaster is the intersection of two opposing forces: those processes generating vulnerability on one side, and the natural hazard event (or sometimes a slowly unfolding natural process) on the other. The image resembles a nutcracker, with increasing pressure on people arising from either side – from their vulnerability and from the impact (and severity) of the hazard for those people. The 'release' idea is incorporated to conceptualise the reduction of disaster: to relieve the pressure, vulnerability has to be reduced. This chapter focuses on the pressure aspect of the PAR model, and the discussion of conditions for creating release are left mainly for Part III.

A second model, referred to as the 'Access model', is discussed in Chapter 3. In effect it is an expanded analysis of the principal factors in the PAR model that relate to human vulnerability and exposure to physical hazard, and focuses on the *process* by which the natural event impacts upon people and their responses. It is a more magnified analysis of how vulnerability is initially generated by economic, social and political processes, and what then happens as a disaster unfolds. The point of application of this second model is indicated on Figure 2.1 by means of a magnifying glass. Later in the book, the Access model indicates more specifically and in more detail how conditions need to change to reduce vulnerability and thereby improve protection and the capacity for recovery. It complements the PAR model, and unites the two sides of the PAR diagram in a detailed process model.

The PAR model might suggest (in its image of two separate sides in the diagram) that the hazard event is isolated and distinct from the conditions which create vulnerability. As will be seen in the Access model described in

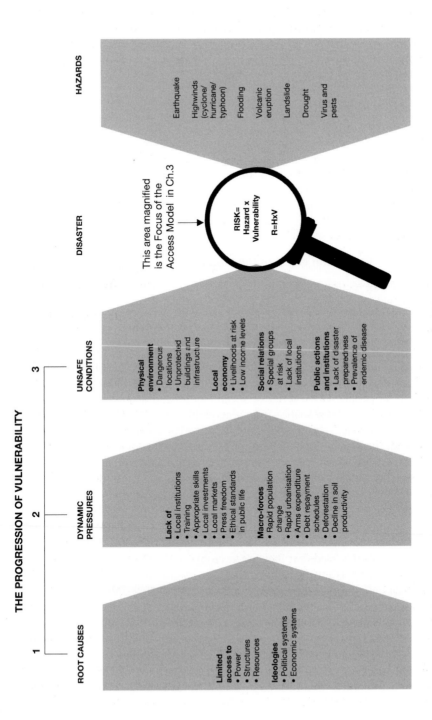

Figure 2.1 Pressure and Release (PAR) model: the progression of vulnerability

Chapter 3, hazard events themselves also change the set of resources available to households (e.g. through the destruction of crops or land by floods), and alter the patterns of recoverability of different groups of people. Hazards sometimes intensify some people's vulnerability, and the incorporation of this insight improves upon those interpretations that see disasters simply as the result of natural events detached from social systems. Conversely, Part II will also show that economic and political circumstances, and the specific situations affecting particular livelihood opportunities, often force or encourage people to engage in practices that worsen the impact of hazards. Such desperate measures, taken in order to survive in the short term, include rapid deforestation, farming inappropriately and speculatively on steep slopes which had hitherto been avoided, overgrazing, living on flood plains (Abramovitz 2001) or subdividing an already crowded apartment.

Cause and effect in the Disaster Pressure model

The following section anticipates Part II, where the chain of explanation of disasters will be related to a series of different types of hazard.

The chain of explanation

Figure 2.1 illustrates the PAR model,[4] and is based on the idea that an explanation of disasters requires us to trace the connections that link the impact of a hazard on people with a series of social factors and processes that generate vulnerability.[5] The explanation of vulnerability has three sets of links that connect the disaster to processes that are located at decreasing levels of specificity from the people impacted upon by a disaster. The most distant of these are *root causes* which are an interrelated set of widespread and general processes within a society and the world economy. They are 'distant' in one, two or all of the following senses: spatially distant (arising in a distant centre of economic or political power), temporally distant (in past history), and finally, distant in the sense of being so profoundly bound up with cultural assumptions, ideology, beliefs and social relations in the actual lived existence of the people concerned that they are 'invisible' and 'taken for granted'.

The most important root causes that give rise to vulnerability (and which reproduce vulnerability over time) are economic, demographic and political processes. These affect the allocation and distribution of resources, among different groups of people. They are a function of economic, social, and political structures, and also legal definitions and enforcement of rights, gender relations and other elements of the ideological order.

Root causes are also connected with the function (or dysfunction) of the state, and ultimately the nature of the control exercised by the police and military, and with good governance, the rule of law and the capabilities of the administration. Military force sometimes has its own impact as an underlying cause of disasters such as famine, especially in prolonged, so-called low-intensity warfare (Clay and Holcomb 1985; Hansen 1987; Leaning 2000) or where denial of food to a civilian population is actually used as a weapon (Article 19 1990; de Waal 1997; Action Against Hunger 1999). The effects of past wars sometimes linger for a very long time, so we feel it appropriate to include them in the category of root causes of vulnerability, while in the next section current wars will appear as a 'dynamic pressure'. Examples are all too common and include Afghanistan, Somalia, Sudan, Ethiopia, Chad, Liberia, Angola, Mozambique, Sierra Leone and Congo, where long drawn-out war and famine have coincided, often exacerbated by an extreme natural event such as drought. Long civil wars may also undermine the ability of central or local governments to prevent or mitigate hazard events. They can also erode the trust between government and citizen that is required for prevention and mitigation to be effective. Here, examples may include Burma, Cambodia, El Salvador and Guatemala. We will return to war as a factor in disaster vulnerability later in this chapter. Wars, of course, are thankfully finite, but militarism and the use of armed force to control a domestic population is a long-standing practice.

Root causes reflect the exercise and distribution of power in a society. People who are economically marginal (such as urban squatters) or who live in environmentally 'marginal' environments (isolated, arid or semi-arid, flood-prone coastal or forest ecosystems; steep, flood-prone urban locations) tend also to be of marginal importance to those who hold economic and political power (Blaikie and Brookfield 1987: 21–23; Wisner 1976b, 1978b, 1980). This creates three often mutually reinforcing sources of vulnerability. Firstly, if people only have access to livelihoods and resources that are insecure and unrewarding, their activities are likely to generate higher levels of vulnerability. Secondly, they are likely to be a low priority for government interventions intended to deal with hazard mitigation. Thirdly, people who are economically and politically marginal are more likely to stop trusting their own methods for self-protection, and to lose confidence in their own local knowledge. Even if they still have confidence in their own abilities, the 'raw materials' needed or the labour time required may have disappeared as a result of their economic and political marginality and low or uncertain access to resources.

Dynamic pressures are processes and activities that 'translate' the effects of root causes both temporally and spatially into unsafe conditions.[6] These are more contemporary or immediate, conjunctural manifestations of general underlying economic, social and political patterns. For example, capitalism is an economic and ideological system that is at least 500 years old, while neo-liberalism is the particular form that capitalist relations have taken since the late 1970s and early 1980s. In the 1980s neo-liberal structural adjustment policies were imposed on many less developed countries (LDCs). Some may have benefited, particularly in south-east Asia where they were able to apply these policies in ways that suited their national circumstances (Stiglitz 2002). But in many others, structural adjustment policies are widely regarded as being responsible for the decline of health and education services which in our parlance suggests they are a root cause of vulnerability.

Dynamic pressures channel the root causes into particular forms of unsafe conditions that then have to be considered in relation to the different types of hazards facing people. These dynamic pressures include epidemic disease, rapid urbanisation, current (as opposed to past) wars and other violent conflicts, foreign debt and certain structural adjustment programmes. Also on the list of dynamic pressures is export promotion, which in some circumstances can undermine food security. It can, for example, encourage mining that destroys local habitats and pollutes water and soil, hydro-electric power development that floods valuable agricultural lands without compensating those affected, and deforestation that can destroy the habitats of forest dwellers, damage farming systems that use the forest for nutrient transfers to agricultural land, and downstream can cause problems such as flooding or the silting of rivers and irrigation canals. It is important to note that these pressures are not labelled 'bad' and vulnera-bility-inducing *per se*. There is a tendency in neo-populist and 'radical' development writing to damn these pressures indiscriminately, without examining their particular historical and spatial specificities. In short, PAR needs thorough research that is locally- and historically based.

The ways in which these dynamic pressures operate to channel root causes into unsafe conditions lead us to specify how the pressures play them-selves out 'on the ground', in a strong spatial and temporal sense. This will allow micro-mapping of unsafe conditions affecting households differen-tially (e.g. wealthy ones, or in distinction, those lacking crucial access to material and human resources) and subsequently groups across households (women, children, the aged, disabled, marginalised ethnic groups, etc.).

This process can be illustrated clearly by examples of endemic disease and malnutrition. People's basic health and nutritional status relates strongly to their ability to survive disruptions to their livelihood system and is an important measure of their 'resilience' in the face of external shock.[7] People who are undernourished and sick succumb sooner in times of famine than those who were previously well-nourished and healthy. There is an important relationship between nutrition and disease, which is often evident after a hazard impact (especially when people are forced to seek refuge and come into close contact with one another). Chronically malnourished people have weaker immune systems and contract illnesses such as measles or dysentery more easily (see Chapter 5). Age is also a significant factor in people's resilience, with children and the frail elderly likely to suffer much more from hunger and hazards such as extreme heat and cold.[8]

Rural–urban migration is another dynamic pressure that arises in many LDCs in response to the economic and social inequalities inherent in root causes. Such migration may follow the loss of land used by poor farmers and pastoralists, discriminatory pricing of crops produced in small quantities by poor farmers and by proletarianisation of the peasantry. Out-migration may lead to the erosion of local knowledge that might serve to prevent disasters and a loss of the skills required for coping in the aftermath of a disaster. An example is given in Box 2.1 below.

Unsafe conditions are the specific forms in which the vulnerability of a population is expressed in time and space in conjunction with a hazard. Examples include people having to live in hazardous locations, being unable to afford safe buildings, lacking effective protection by the state (for instance in terms of effective building codes), having to engage in dangerous livelihoods (such as ocean fishing in small boats, wildlife poaching, prostitution with its attendant health risks, small-scale gold mining in the Amazon and eastern Africa, or small-scale forestry), or having minimal food entitlements, or entitlements that are prone to rapid and severe disruption.[9] Also, unsafe conditions are dependent upon the initial level of well-being of the people, and how this level varies between regions, micro-regions, households and individuals. It is important to consider the pattern of access to tangible resources (e.g. cash, shelter, food stocks, agricultural equipment) and intangible resources (networks of support, knowledge regarding survival and sources of assistance, morale and the ability to function in a crisis) (Cannon 2000a). These aspects of unsafe conditions serve as a bridge between PAR and the Access model discussed in the next chapter.

We propose the following terminology when dealing with unsafe conditions. People, as should be apparent already, are *vulnerable* and live in or work under unsafe conditions ('unsafe' can refer to locations of work or habitation, wherever people spend their daily lives). As we said in Chapter 1, and it bears repeating, we avoid using the word vulnerable in regard to livelihoods, buildings, settlement locations or infrastructure, and instead use terms such as 'fragile', 'unsafe', 'hazardous' or their synonyms.

While all of these are components of people's vulnerability, a building should be regarded as unsafe, rather than vulnerable; a settlement's location is hazardous, not vulnerable. In this way, we retain the term vulnerability for people only. The reason for this is straightforward: already the term vulnerability (and its associate, *vulnerability analysis*) has been appropriated for use in such a wide range of situations that (like 'sustainability') it is in danger of losing its significance in relation to people and hazards. If 'vulnerability' becomes a catch-all term for any aspect of conditions related to disasters, then it will lose its analytical capacity. Moreover, it will lose the focus about which we are very explicit – that it is the *vulnerability of people* that is crucial to understanding disasters and disaster preparedness. It is, of course, absolutely right to be concerned about the condition of buildings, the places where people have to live, crop yields and variability and so on. But if policy is directed at these alone, it is in danger of being compartmentalised (e.g. into issues of building codes, or land-use planning, or production-oriented agricultural programmes).

No single element, particularly the technical (and seemingly a-political) determinants of people's vulnerability, should be taken in isolation from the entire range of factors and processes that constitute this situation. The other danger is that a focus on the 'hardware' aspects of vulnerability will distract from the attention that needs to be given to the political and economic determinants of vulnerability: most people are vulnerable because they have inadequate livelihoods, which are not resilient in the face of shocks, and they are often poor. They are poor because they suffer specific relations of exploitation, unequal bargaining and discrimination within the political economy, and there may also be historical reasons why their homes and sources of livelihood are located in resource-poor areas.

In other words, in many cases reducing vulnerability is about dealing with the awkward issue of poverty in society. That is why there needs to be a clear link between disaster preparedness, vulnerability reduction and the process of development itself (the improvement of peoples' livelihoods, welfare and opportunities). This is illustrated in Figure 2.1, where the vulnerability that arises from unsafe conditions intersects with a physical hazard (trigger event) to create a disaster, but is itself only explained by an analysis of the dynamic processes and root causes which generate the unsafe conditions.

It is important to note that, throughout the causal chain of explanation from root causes to unsafe conditions, we do not imply by the phrase 'cause and effect' that single causes give rise to single effects. In their study of land degradation, Blaikie and Brookfield (1987) refer to such causal sequences as 'cascades'. There are many ways in which dynamic processes (some unique to particular societies, some nearly universal because of the pervasive influence of global forces) channel root causes into unsafe conditions and to specific time–space convergence with a natural hazard. This can be illustrated in the outcome of floods in Bangladesh (see Box 2.1) and landslide and earthquake impacts in part of north Pakistan (Box 2.2).

Box 2.1: Landless squatters in Dhaka

Dhaka, the capital of Bangladesh, is situated in the flood plain of a major river, the Buriganga, a tributary of an even larger river, the Meghna (see Figure 6.2). To the north-west is a large zone of low-lying, flood-prone land in the vicinity of Nagor Konda. Here, squatter settlements grew rapidly in the 1980s as they did in many areas around the capital (Shaker 1987). This area had been densely settled, particularly since 1970, mostly by poor landless families from the south and east of the country (Rashid 1977).

Box 2.1 continued

The former landless people who inhabit this depression are there because of its proximity to Dhaka's vegetable market. Already the chain of explanation of their vulnerability can be seen at work: rural people who are landless have few alternatives, and many seek the economic opportunity provided by the urban vegetable market. But this means living in an unsafe location. As newcomers, and extremely poor, the squatters in these low-lying areas had no access to the structures of power that control marketing. They also had insecure title to land in the depression, and therefore no access to credit to allow them to increase their productivity and compete with better-established market gardeners (A. Ali 1987). This situation meant that they had to grow rice rather than vegetables on their land, and thus the poor were forced into low-income pursuits.

On the eve of the massive floods of August 1988 (see Chapter 6), this relatively powerless group with few assets was living in an economically marginal situation close to the city, on low-lying land prone to flooding. Their children were frequently malnourished and chronically ill. This is precisely how the dynamic pressures arising out of landlessness and economic marginalisation are channelled into a particular form of vulnerability: a lack of resistance to diarrhoeal disease and hunger following the flooding in 1988. Factors involving power, access, location, livelihood and biology come together to create a particular situation of unsafe conditions and enhanced vulnerability. These social, economic and political causes constitute one side of the pressure model. The other – the floods themselves during August 1988 – constitutes the trigger event whose impact on vulnerable people created the disaster.

Box 2.2: Karakoram and house collapse

This case comes from an interdisciplinary study of housing safety in the Karakoram area of northern Pakistan (Davis 1984b; D'Souza 1984). We follow the chain of explanation that links vulnerability to the specific physical trigger that creates a disaster in reverse, starting with 'unsafe conditions'. The PAR model may be constructed equally well in either direction of causality, starting with unsafe conditions and working from the specific to the general or *vice versa* (see, for example, Blaikie's (1989) causal chain of land degradation from the specific site characteristics to more distant causes arising from the global political economy).

The research team carefully examined local dwellings and settlement patterns within the context of a rural economy. They found that the communities were at risk from a wide range of hazards. In this region

Box 2.2 continued

traditional dwellings were built with stone masonry walls. A series of timber bands were set at regular intervals in the height of each wall in order to hold the stones together, and the complex timber roofs were constructed with a very heavy covering of earth to provide much-needed insulation.

These traditional dwellings were built until around the 1960s or early 1970s, and provided some protection against earthquakes. But, subsequently, local building patterns changed in favour of concrete construction. The new houses were intended to be reinforced, but in reality they were built without any real understanding of how to connect steel to concrete or roofs to walls. The siting of most buildings was equally dangerous, since to avoid reducing their meagre land-holdings (all available flat land was used for agriculture), many houses were built on exceedingly steep slopes, putting them at risk from landslides.

The result was an extremely hazardous situation, with a number of factors together producing these unsafe conditions, including reduced concern about building safety and the diversion of money intended for dwellings to fulfil everyday needs. There was also a lack of knowledge of both concrete construction and aseismic (shock-proof) construction techniques, a shortage of skills and a change in the availability of building materials.

In turn, some of these factors (especially the lack of both skills and materials) could be directly attributed to 'dynamic pressures'. Firstly, the shortage of timber for building and other purposes in the region had arisen because of deforestation, mostly due to illegal felling and corrupt practice (Blaikie and Sadeque 2000). In addition, population growth over a long period undoubtedly increased the demand for fuelwood in such a cold climate and for building materials. This led to a rapid increase in tree-cutting and forest clearing to create additional fields for cultivation.

Secondly, there was a serious shortage of skilled carpenters and masons, so buildings were constructed and maintained by farmers and labourers who freely admitted that they knew very little about the task. In trying to piece together the reasons for the absence of knowledge-able builders another dynamic pressure emerged. During the 1970s the Chinese government had built the Karakoram Highway, a major access road into the area. This linked China with the Pakistani capital, Islamabad. The road was built for political and strategic reasons, but it was also intended to bring 'development' to the remote Northern Areas. Risk was 'imported' via the highway to the extent that heavy (unsafe) concrete buildings were developed and considered 'modern'

Box 2.2 continued

and their use increased (Coburn et al. 1984). The road also allowed a migration of carpenters out of the area to Karachi, Islamabad and even to the Gulf region (where earnings were twenty times higher). As so often happens, while the road was being used to bring in medical and educational resources, it also enabled loggers to enter the region for the first time and they had removed vast quantities of timber, a process that continues today in spite of logging bans imposed by government. It is likely that the resulting deforestation contributes to soil erosion and slope instability, which increases on-site hazards when earthquakes occur.

Furthermore, the Pakistani government encouraged their workforce to emigrate so as to attract the foreign currency remittances sent for family support by the workers abroad. This was a policy designed to boost the country's balance of payments deficit.[10] In this way we are led from proximate and specific cause to more remote 'root causes'. The net result was that the families were left to live in dangerous homes, often with a depleted and de-skilled labour force due to out-migration.

Time and the chain of explanation

Our two models function in a variety of time scales (see the section on time in Chapter 3). Root causes, dynamic pressures and unsafe conditions are all subject to change, and in many cases the processes involved are probably changing faster than they have done previously. The changes in building techniques and materials in Pakistan were rapid (see Box 2.2), as were the processes of out-migration and deforestation. This affected communities that had changed little for many years. Even large-scale processes, such as population growth, are rapid by comparison with changes in, say, values and beliefs or legal structures. For example, during the 1970s Kenya had an annual population increase of 4.2 per cent, giving a doubling time of 16 years. This rapid population growth was, at that time, one of the factors channelling the root causes of vulnerability into unsafe conditions during years that saw great suffering by vulnerable groups during droughts (Wisner 1988a). However, by 1995 the growth rate had fallen to 2.9 per cent (UNDP 1998: 177) In other words, even something such as population dynamics can change rapidly.

We should add, however, that we are not invoking a neo-Malthusian explanation for the impact of rapid population growth as a dynamic pressure leading to increased vulnerability. It is only when rapid population growth is combined with other dynamic conditions (such as rapid urbanisation, the need to adapt to rapid agricultural intensification in areas of low resource potential, or incompetent economic management) that very rapid population growth can exacerbate vulnerability (see below).

The location of settlements and livelihoods can change even more rapidly. For example, between 1973 and 1976 about half of the then 12 million rural inhabitants of Tanzania were variously encouraged or coerced into nucleated villages (Coulson 1982). This completely altered settlement patterns and the resource basis of the affected people's livelihoods over a period of only three years.[11] Other instances of such disruptions are common as a result of war. Four million people, one-third of the Mozambican population, were forced by the civil war there in the 1980s and 1990s to flee to refugee camps in Zimbabwe and Malawi, while many lived as internally displaced persons (IDPs) near a few of Mozambique's major towns. The impact of such disruptions of access on the vulnerability of these peoples to drought and other hazards has not been studied; nor has the effect of such population movements on the environment, and thus on the creation of future hazards via land degradation (Black 1998).

Root causes often shift because of disputed power and claims to resources (financial, physical and informational) as well as identities (Platt et al. 1999; Oliver-Smith and Hoffman 1999; Caplan 2000), and vulnerability may therefore change as a result. The converse is also true. Mass suffering due to disaster may contribute to the overthrow of elites and lead to dramatic realignments of power. It can be argued that the cyclone and storm surge in East Pakistan in 1970 contributed to the development of the Bangladesh independence movement, and that governments in Niger and Ethiopia were overthrown as a result of their incompetent and malign behaviour in the 1970s Sahel famine. The revolutionary movement in Nicaragua from 1974 to 1979 derived some of its impetus from the effects of the Managua earthquake of 1972. Hurricane Mitch (1998) did not cause the overthrow of the national governments in the affected countries, but it did contribute to a widespread reassertion of local, municipal political power in Nicaragua and El Salvador, and a significant increase in the political assertiveness of citizen-based groups. We take up the counter-intuitive notion of disaster as opportunity in Chapter 9.

Limits to our knowledge

Vulnerability can be assessed reasonably precisely for a specific group of people living and working at a specific time and place, and the 'unsafe conditions' that contribute to it have been the subject of a great deal of research reviewed in this book. In much of the world, detailed knowledge has been obtained about which sites might be affected in a landslide, which buildings will survive or collapse in an earthquake and why, or about the outcomes of drought in terms of food production and possible shortfalls.

Similarly, dynamic pressures and root causes are reasonably well understood in many situations, although treatments may be highly polemical – indeed they are always political. However, as we move up the chain of explanation from unsafe conditions to root causes, the linkages (and therefore the level of precision in disaster explanation) become less definite. In analysing the linkages between root causes, dynamic pressures and unsafe conditions, it becomes increasingly difficult to have reliable evidence for causal connections, especially as we go further back in the chain of explanation.[12]

The uncertainties and gaps in knowledge concerning how vulnerability is demonstrably and causally linked to underlying causes or pressures have some quite serious implications.[13] The first is that the links can be dismissed as polemic and ideology, particularly by those who treat disasters as a technical issue alone. However, these uncertainties explain in part why policy makers and other important actors at the international and national levels have caused or allowed unsafe conditions to arise and allowed them to persist. At best, lack of understanding and uncertainties are likely to result in policy makers and decision takers, restricted by the scarce resources at their disposal, addressing immediate pressures and unsafe conditions while neglecting both the social causes of vulnerability as well as the more distant root causes.

Yet these gaps exist mainly because of a failure to ask the right sort of questions. It is imperative to accept that reducing vulnerability involves something very different from simply dealing with hazards by attempts to control nature (engineering measures and 'public works') or emergency preparedness, prediction or relief, important though these are. However, most government agencies charged with such responsibilities as 'environment', 'health and welfare' and 'public safety' generally still deal with disasters as though they are *equivalent to* the natural hazards that trigger them; the principal object is the hazard, and the range of underlying reasons for the dangerous situation may be regarded as peripheral, or even irrelevant and immaterial. The factors involved in linking root causes and dynamic processes to vulnerability are seen as too diffuse or deep-rooted to address. Those who suggest they are crucial may be labelled as unrealistic or over-political.[14] As Cannon (2000a: 48) puts it, '[V]ulnerability analysis is avoided as being "irrelevant to science" or "too difficult to get involved in".'

Our view is that there is little long-term value in confining attention mainly or exclusively to hazards, in isolation from vulnerability and its causes. Problems will recur again and again in different and increasingly costly forms unless the underlying causes are tackled. This perspective does not deny the importance of technical or planning measures to reduce physical risks. It simply insists on a concern for a deeper level of analysis which places moves to mitigate hazards within a comprehensive understanding of the vulnerabilities they are supposed to reduce. In this way, efforts to mitigate hazards will be appropriate and will emerge within the supportive environment for implementation provided by the affected people themselves. Disaster research and policy must therefore account for the connections in society that cause vulnerability, as well as for the hazards themselves.

Global trends and dynamic pressures

Although there is still a serious lack of analysis of the linkages between vulnerability and major global processes, it is encouraging that during the last ten years many more authors and institutions have begun asking such questions. For example, it is now possible to identify more precisely how urbanisation increases hazard impact (Mitchell 1999a; Fernandez 1999; Velasquez et al. 1999) (see below).

There is a general consensus in research on disasters that the number of natural hazard events (earthquakes, eruptions, floods or cyclones) has not increased in recent decades.[15] If this is true, then we need to look at the social factors that increase vulnerability (including, but not only, rising population) to explain the apparent increases in the number of disasters (as opposed to hazard events) in terms of the value of losses and the numbers of victims.

Figure 2 2 shows the number of great disasters during the second half of the twentieth century. Some of the increase may be a result of better reporting and improved communications, or the incentive for governments to declare a disaster in an attempt to win foreign aid. But the rising trend seems to be too rapid for these explanations alone (see Box 2.3 below).

Disasters are also becoming more expensive. Economic losses, and especially the share composed of insured losses, are increasing (Figure 2.3).

At this stage, it is important to review in very broad terms how certain of these various dynamic pressures contribute to the increase in disasters. We have chosen seven global processes for further attention: population change, urbanisation, war, global economic pressures (especially foreign debt), natural resource degradation, global environmental change and adverse agrarian trends. These processes are not independent of each other. They are intricately connected in a series of mutually influencing relationships that obscure causes and consequences. Also, it should be remembered that some of these processes appear both as root causes and dynamic pressures: for example, past urbanisation and past war may set up patterns that influence vulnerability hundreds of years later (the decision by the Spanish in 1521 to locate what became Mexico City on the bed of a lake they had drained once their Aztec opponents were conquered; the Second World War that resulted in a new map of Europe). In these cases urbanisation and war can be considered root causes. However, recent or current urban growth and violent conflict should be seen as dynamic pressures.

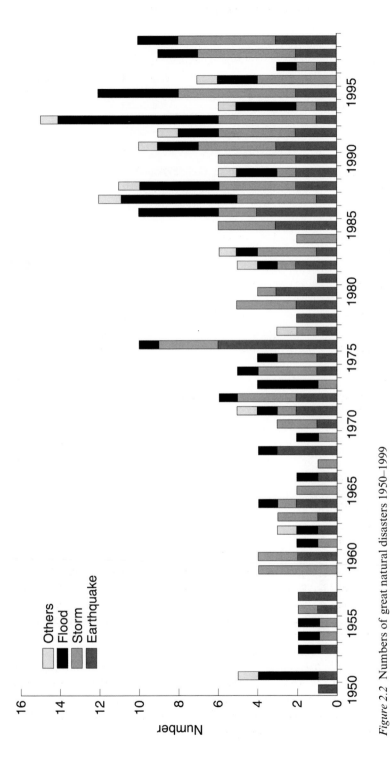

Figure 2.2 Numbers of great natural disasters 1950–1999

Note: The chart shows for each year the number of events defined as great natural catastrophes, divided up by type of event

Source: Munich Re. 2000. Great natural catastrophes – long-term statistics. Available online at http://www.munichre.com./pdf/pm_2000_02_29_anhang3_e.pdf
Adapted by kind permission of Munich Re

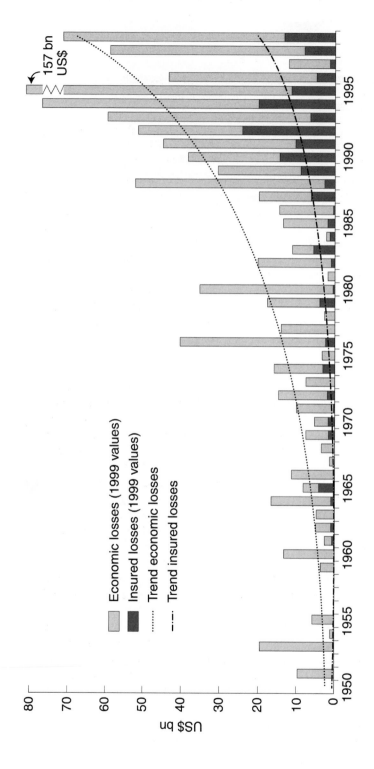

Figure 2.3 Economic and insured losses (with trends) for 1950–1999

Note: The chart presents the economic losses and insured losses – adjusted to 1999 values. The trend curves illustrate the alarming increase in catastrophic losses at the turn of the century

Source: Munich Re. 2000. Great natural catastrophes – long-term statistics. Available online at http://www.munichre.com./pdf/pm_2000_02_29_anhang3_e.pdf Adapted by kind permission of Munich Re

Box 2.3: Problems with disaster statistics

Where do disaster statistics come from?

As with world-wide health and population statistics, disaster statistics are reported by governments to United Nations agencies. These 'official' numbers are supplemented and cross-checked by some groups using the reports of non-governmental organisations (NGOs) and journalists. The pre-eminent of such institutions is the Centre for the Epidemiology of Disaster (CRED) in Belgium (http://www.cred.be). Large reinsurance companies such as Munich Re and Swiss Re also compile international statistics on disasters. The World Bank and some of the UN regional economic commissions, such as the Economic Commission for Latin America (ECLA), have conducted studies of disaster loss and costs. Regional banks such as the Inter American Development Bank (IADB) and Asian Development Bank also study disaster statistics, but from the point of view of economic loss. The World Health Organisation (WHO) and Pan American Health Organisation (PAHO) do not maintain permanent registers of death, injury and post-disaster health consequences, but they do, on occasion, analyse and interpret such numbers.

How good are disaster statistics?

Like all numbers, disaster statistics are as good or bad as the methods used to collect them. Also disaster statistics have other specific weaknesses. Firstly, despite a large academic literature on the subject, there are no universally agreed definitions of the word 'disaster' (Quarantelli 1998) or other critical terms. One of the imprecise statistics often used by governments and aid organisations is the number of people 'affected' by a disaster. Since definitions of what it is to be 'affected' can vary so much, we do not use the number 'affected' at all in our book. 'Injury' is also a term that can have many meanings (Shoaf 2002; Benson 2002). The term 'death', too, can be problematic. For example, in the USA the death toll of the Northridge earthquake varies from 33 to 150+ depending on who defines what an earthquake-related death is: 33 died of direct or indirect earthquake injuries, 57 were defined by the LA County Coroner as dying of causes either directly or indirectly related to the earthquake; FEMA paid death benefits to survivors of more than 150 (Shoaf 2002).

Also, many extreme events that take only a few lives and affect only a local economy go completely unreported. This is an issue that a regional network of disaster researchers in Latin America have recognised by producing free, bilingual (English and Spanish) accounting software to be used to keep track of these 'small' disasters that could well have a highly erosive effect on development (http://www.desinventar.org/desinventar.html). We recommend it highly.

Box 2.3 continued

Secondly, there may be deficiencies in the reporting system itself. Many injuries may go unreported or simply are not recorded by health workers who are too busy because of the volume of care demanded in an emergency. In some countries, or regions of a country, even in 'normal' times there may be poor coverage of vital statistics, with many births and deaths going unrecorded. This could happen in isolated rural areas as well as densely populated squatter settlements in cities. So, some people may die in an extreme natural event whose lives were not even officially recognised as existing. Others are never found, and are 'missing', but are never recorded as 'dead', even after a considerable period of time. There is also wide historical variability in disaster data. Davidson observes (2002), 'This is because of changes in the methods of reporting, the number of people in an affected place, systems and facilities for storing records. This all makes efforts to track historical trends in disasters even more problematic than trying to account for impacts in a single event today. Plus, of course, most records are short compared to the return period of events.'

Thirdly, there can be political pressures either to overstate or to understate casualties. If a government wishes to 'talk up' the level of relief assistance, it might exaggerate the lives lost, homes destroyed, people injured. On the other hand, if a government believes it will be criticised by its citizens for not protecting them, there may be a tendency to understate the impacts of a disaster, or to remain silent about it altogether (some examples of politically expedient silences about famine are given in Chapter 4). However, in fairness, it is very difficult to collect data on losses and damage in a timely way when undergoing the stress of the disaster itself, especially if a country has limited transport and communications. The sheer difficulty of drawing up reliable estimates should therefore be considered a fourth reason why disaster statistics should be handled with care.

Finally, when it comes to economic loss and long-term effects on development, the problem is even murkier (Benson 2003). The longer term 'knock on' effects of a disaster are conceptually difficult to model, and in most cases governments are not set up to study them (Benson and Clay 1998). Davidson (2002) puts the problem this way: '[W]ith economic effects it's difficult to assess which changes are caused by the disaster and which would have happened anyway. That is, there's always the problem that it's easier to compare before and after the disaster, but what we really should be comparing is with and without the disaster'.

Population change

During 2000 the world population passed the six billion mark, yet only 100 years ago it was under two billion. Despite this impressive growth, the predictions of even more rapid population growth forecast in the 1970s by the Club of Rome have not materialised. In fact, there is now evidence that birth rates and total fertility rates (the number of children a woman gives birth to) are declining in India, Indonesia, Iran, Brazil, Mexico and elsewhere (Naik et al. 2003). The UN prediction of two billion more people in the next 25 years, making a total of 8.2 billion by the year 2025, may be too high, and it is even possible that the world's population will stabilise at around nine billion by mid-century (ibid.).

Nevertheless, one thing is certain: populations in many LDCs will continue to grow – 90 per cent of population growth over the next few decades is predicted to occur within developing countries, many of which are subject to frequent extreme natural events (United Nations 2002c; Population Reference Bureau 2002). It is difficult to object to the idea that population growth is a significant global pressure contributing to increasing vulnerability, and yet the linkages remain uncharted except in rather simplistic terms (e.g. there are more people, therefore some have to live in dangerous places). There are difficulties in trying to explain demographic change more carefully. For instance, there is considerable debate about whether population growth is a cause or a consequence of poverty in LDCs (where, for instance, children are needed to provide labour and security). It is more likely a complex interaction of both.

So we still need an analysis of the consequences of growth in numbers. This requires a better understanding of the linkage of population growth to disasters, and of any causality involved (Clarke 1989; Dyson 1996). Demographic processes themselves are largely a reflection of people's (women's and men's) responses to the opportunities and uncertainties presented to them by broader economic processes. Some of the implications of population expansion relative to disaster risks can more easily be related to different age groups (see Box 2.4). Therefore we would not want to accept an overly-simplistic linking of population growth with vulnerability that suggests more people suffer more disasters simply because there are more of them in dangerous places. It is also necessary to explain why people put themselves at risk. This is a process not explained by the increase in numbers alone, but by the differential access to incomes and resources in society.

The apparently illogical behaviour of people who seem to have too many children in hazardous places can be seen to be more logical (if no less risky) in the context of the Access model used in Chapter 3. Is it significant that rapid population growth occurs in some countries with a long record of disasters? It is difficult to be certain about how to equate a rise in the number of vulnerable people with population growth: if population increases *because* people are poor, then in effect rising vulnerability is still a product of poverty and not of population growth.

As a dynamic pressure, population *growth* does not seem to us as important as *change* and *age structure* (see Box 2.4). In southern Africa as a result of the HIV-AIDS epidemic life expectancy and growth rates are falling, not increasing. This instability in population change is causing severe dislocation in the rural economy and complicating recovery from the drought of 2001–2002.

Box 2.4: Age structure and vulnerabilities

In Chapter 1 we reviewed studies which show that in some situations the young and the elderly are more vulnerable to the impacts of natural hazards. At the macro-scale, then, one indication of vulnerability may be provided by statistics on the age structure of national populations. How many and what proportion in a given population are children or elderly? On average, how many children and elderly people does each productive adult have to support (referred to as the 'dependency ratio')?

In many developing countries as much as 50 per cent of the total population is under 15 years of age (compared with 20 per cent in industrialised countries). Although a high proportion of these children and teenagers engage in productive economic activity, it will be increasingly difficult to cater for their basic needs since a relatively small percentage of the adult population has to carry the responsibility for feeding, clothing, housing and educating them. Under the political and economic conditions that create poverty, some households simply cannot support their children. These young people have no option other than to become 'street children', forced to fend for themselves in hostile urban environments where they are even more vulnerable (Ennew and Milne 1989; Hardoy and Satterthwaite 1989). Another implication of such an age structure is the need to focus on the critical importance of making all school buildings resistant to hazards (Wisner 2003d; Spence 2003; OAS/USDE 2000).

At the other end of the age spectrum there is also a growing challenge posed by the ageing populations of Japan, North America and Europe. Studies of disaster casualties have indicated that the young and the old are often most at risk. They are, for example, less mobile (capable of evacuation), more dependent, have less resistance to disease, and often command fewer resources. Increasing casualties in disasters can be anticipated in this age group. The implication is that specific risk-reduction policies will be needed to focus on the protection of the elderly (ICIHI 1988: 16; World Bank 1994). Also, as a population ages there is a smaller proportion of younger adults working and providing payments into social security systems. Social protection in various forms (pensions, health care, etc.) may deteriorate, thus increasing the vulnerability of the elderly.

Box 2.4 continued

In southern Africa, the impact of HIV-AIDS has meant that some rural areas have lost many of their younger adults. The productivity of agriculture has suffered, as has the ability of households to engage in the variety of activities traditionally associated with coping with hazards such as drought (de Waal 2001, 2002).

Bangladesh had a population of 118 million in 1995 and a land area of only 144,836 sq. km. (UNDP 1998). Land shortage is often assumed to be a result of this ratio. But it is really a problem for the poor and powerless, created by inequality in access and ownership, a factor in many forms of vulnerability described in Part II. Of the population, 85 per cent depend on agriculture, and between 40 and 60 per cent own no land (Hartmann 1995; Boyce 1987). The landless and those with little land depend on wage labour and various non-farm activities to make a living. Being labour-intensive, this kind of livelihood strategy encourages large families whose members can work from an early age.

In Chapter 3 we show that livelihood strategies are the key to under-standing the way people 'cope' with hazards. Unequal access to land and the resulting poverty and vulnerability of families is one of the factors that drives population growth (Hartmann and Standing 1989). Vulnerability can be the result. Brammer has noted:

Growing population pressure has increased the number of landless families, ... increased the rate of rural–urban migration and forced increased numbers of people to seek living space and subsistence on disaster-prone land within and alongside major rivers and in the Meghna estuary.

(1990a: 13)

We would add that highly skewed land and income distribution has created many landless and land-poor households who migrate and seek a living space in hazardous locations. 'Population pressure' is, in our view, an effect, and not a cause, in this situation. The consequence of highly unequal access to land is that more and more hazardous land is being settled. This is particularly true of the low-lying islands (known locally as *char*) that emerge as a result of silt deposition in the river estuaries of the delta regions. This poses severe risks to the occupants from both cyclones and river flooding (see Chapters 6 and 7).

Whereas this situation is often considered hopeless from a 'technical' point of view, there are a range of *social* solutions that would both reduce the desire for large families and reduce disaster vulnerability. These solutions could include radical land reform, the empowerment of women, and the

provision of adequate public services (e.g. health, communications, education). China, Sri Lanka and Kerala State in India have all reduced
population growth in this way (Hartmann 1995: 289–304; Franke and
Chasin 1989).

Whilst Bangladesh might be considered an extreme case, it is hard to
define the precise relationship between rural population density and well-
being, ill-being or vulnerability (Cassen 1994). In other cases there is clear
evidence that *higher* population density has triggered rural development that
includes soil and water conservation measures, as in the Close Settled Zone
of Kano, northern Nigeria, and Machakos District of Kenya (Adams 2001:
193–197).

Urbanisation

Urbanisation is a major factor in the growth of vulnerability, particularly of
low-income families living within squatter settlements.[16] The urbanisation
process results in land pressure as migrants from outside move into already
overcrowded cities, so that the new arrivals have little alternative other than
to occupy unsafe land, construct unsafe habitations or work in unsafe environments (Havlick 1986). But the risks from natural hazards are only a part
of the dangers these people face in squatter settlements. There are often the
far greater and more pressing 'normal' risks of malnutrition and poor health
(Richards and Thomson 1984; Pryer and Crook 1988; Cairncross et al.
1990a; Wisner 1997).

Hewitt examined the literature on earthquake impacts and found that
urbanisation was closely related to damage to once-new multi-storey
constructions and in the concentrated poor housing of squatter settlements.

> [W]here older sections of cities are run-down, often they have
> become slums that modernisation passes by. Here, even once solid
> buildings are weakened by neglect and decay to become death traps
> in relatively moderate earthquakes.
>
> (1981/1982: 21–22)

This situation was typified in the earthquake that severely affected decaying
inner-city tenements in Mexico City in 1985 (Cuny 1987) and is discussed in
more detail in Chapter 8.

Maskrey has argued that the inhabitants of such critical areas:

> would not choose to live there if they had any alternative, nor do
> they deliberately neglect the maintenance of their overcrowded and
> deteriorated tenements. For them it is the best-of-the-worst of a

number of disaster-prone scenarios such as having nowhere to live, having no way of earning a living and having nothing to eat.

(1989: 12)

Slum residents often incur greater risks from natural hazards (flood, landslide and mudslide) as a result of having to live in very closely-built structures which can disturb natural land drainage patterns and water-courses (see Chapters 6 and 8). One example was the loss of life in flash flooding and mudslides in the outlying coastal, hillside suburbs of Caracas, Venezuela in 1999. Thirty thousand people died and 100,000 were displaced by this disaster in the densely populated coastal hills where 40 per cent of Venezuela's population of about 24 million is concentrated into less than 2 per cent of the national land area (IFRC 2001b: 82–85; Gunson 2000). Another tragic case, in 2001, was a mass movement of compacted garbage at an open-air solid waste dump that was triggered by heavy rainfall. This dump on the north-eastern edge of Manila is called Payatas, where 2,000 people lived in shacks, working as informal material recyclers – 700 were killed (Westfall 2001).

The rate of informal or unplanned urban growth can rapidly put large numbers of people at risk, as the example of Quito, Ecuador shows. Since the last destructive earthquake affecting Quito in 1949, the city has grown from 50,000 inhabitants to 1.3 million (1997). Many people have settled on steep slopes, where the Swedish Rescue Services Agency (1997) describes how 53 ravines have been filled in so that sewage or water pipes can cross them, or in order to build roads or houses. The vulnerability of the people living under such conditions is very high.

Currently nearly half of all humanity lives in cities – a proportion projected to be 60 per cent by 2030 (United Nations 1999: 2). Since a signifi-cant proportion of this urban population is poor and lives in informal urban settlements, the challenges of urbanisation are likely to grow, and with them the opportunities for disaster reduction. The most recent revision of the *World Urbanisation Prospects* makes several extremely important points related to this. In the period 2000–2030:

- virtually all the population growth in the world will be concentrated in urban areas (UN 1999: 2);
- most of the increase will be absorbed by urban areas of the less devel-oped regions (UN 1999: 2);
- the proportion currently living in small cities is considerably greater than in large cities, although it is growing at a slower pace: in 2000, 29 per cent of the world population lived in cities of less than one million, while by 2015 this percentage is likely to grow only to 31 per cent (UN 1999: 6).

Thus, cities of all sizes are growing, but the very largest urban regions, those over 10 million, are of particular concern. There have been primate cities and metropoli for centuries; however, the new urban regions with more than 10 million inhabitants, the 'mega-cities', are relatively recent. The average size of the world's largest 100 cities increased from 2.1 million in 1950 to 5.1 million in 1990. In developing countries, the number of cities with over one million people jumped six-fold between 1950 and 1995. In the year 2000, worldwide, the number of cities larger than five million was 41, and the UN believes this number will rise to 59 by 2015. This will add another 14 million people to the streets and homes of large cities (accounting for 21 per cent of the world's urban growth) (UN 1999).

In the year 2000 there were 19 cities with more than 10 million residents, a number believed likely to increase to 23 by 2015. Of these, fifteen are in LDCs, and they are all prone to natural hazards of one kind or another (see Table 2.1 for similar rankings as of 1996). Eight of these large urban regions are within moderate-to-high seismic risk zones. These cities contain large numbers of buildings of variable quality, many of them poorly constructed or badly maintained. Since the vast majority of deaths and injuries from earthquakes result from building collapse, the vulnerability of people living or working in such structures is bound to be high. Among the list of 23 cities projected by the UN to be of 'mega-city' size by 2015, 19 will be in LDCs. The four additions to the list are Hyderabad (India), with a history of destructive floods and Tianjin (China) together with Istanbul and Bangkok.

In 2015, if these projections hold up, nine of the largest cities among the LDCs will have a combined population of 148 million people (about the total population of Russia, more than twice the population of Great Britain). Of these, Mexico City and Istanbul are probably at greatest risk (see Chapter 8).

Not only are mega-cities at risk, but smaller cities as well, such as the small coffee marketing centre Armenia in Colombia or Bhuj in Gujarat, India – both severely damaged by earthquakes. Another example is Goma, a city of 500,000 people in eastern Congo. It was cut in half and 40 per cent destroyed by an eruption of the nearby Nyiragongo volcano in January 2002. The vulnerability of the inhabitants was very high because of a history of conflict in this region that had drained their financial reserves and destroyed the local economy and because there was hardly any municipal governance provided by the rebel force that controlled the city in defiance of the national government in Kinshasa. Without a functioning city government, there was no warning of the eruption, no organised evacuation and no shelter plan (Wisner 2002b; see also Chapter 8).

The urbanisation process not only magnifies the dangers of hazard events; it is in itself partly a consequence of a desperate migrant response to rural disasters. There is evidence from Delhi, Khartoum and Dhaka

Table 2.1 Largest cities in hazard areas (ranked by population in 1996)

City/conurbation	Population 1996 (millions)	Projected population 2015 (millions)	Hazard(s) to which exposed
Tokyo-Yokohama	27.2	28.9	Earthquake, cyclone
Mexico City	16.9	19.2	Earthquake, flood, landslide
São Paolo	16.8	20.3	Landslide, flood
New York	16.4	17.6	Winter storm, cyclone
Mumbai/Bombay	15.7	26.2	Earthquake, flood
Shanghai	13.7	18.0	Flood; typhoon
Los Angeles	12.6	14.2	Earthquake; landslide, wildfire, flood
Calcutta	12.1	17.3	Cyclone, flood
Buenos Aires	11.9	13.9	Flood
Beijing	11.4	15.6	Earthquake
Lagos	10.9	24.6	Flood
Osaka	10.6	10.6	Earthquake, cyclone flood
Rio de Janeiro	10.3	11.9	Landslide, flood
Delhi	10.3	16.9	Flood, heat and cold waves
Karachi	10.1	19.4	Earthquake, flood
Cairo-Giza	9.9	14.4	Flood, earthquake
Manila	9.6	14.7	Flood, cyclone
Dhaka	9.0	19.5	Flood, cyclone
Jakarta	8.8	13.9	Earthquake, volcano
Tehran	6.9	10.3	Earthquake

Source: UN Department of Economic and Social Affairs, Population Division.
http://www.un.org/esa/population/pubsarchive/urb/furb.htm

(Bangladesh) that rural families who have become destitute as a result of droughts or floods have moved to these cities in search of food and work. Shakur studied the urbanisation process in Dhaka. His household surveys revealed that:

> the overwhelming majority of Dhaka squatters are rural destitutes who migrated to the city mainly in response to poor economic conditions (37 per cent) (particularly landlessness) or were driven by the natural disasters (25.7 per cent) (floods, cyclones and famines).
>
> (1987: 1)

In a related way, cities have provided safe refuges in Africa and Central America from civil wars and rural warlordism, thus accelerating urban growth. For example, during the last few years of the civil war in Angola, tens of thousands of IDPs risked malaria and other diseases, living in swampy conditions on the edges of Luanda, capital of Angola.

War as a dynamic pressure

There will unfortunately have to be frequent mention of war in the case study chapters in Part II. In 1985 van der Wusten (1985) counted more than 120 wars since the end of the Second World War. Another source that took a narrower definition than van der Wusten gives the total for the twentieth century to be 165 wars that have claimed 180 million lives (White 1999). Whether the definition used is narrow or broad, violent conflicts have had disastrous consequences in their own right for the people caught up in them, and they have also influenced vulnerability to extreme climatic and geological processes (Wisner 2002c). On a regional and local scale, war has disrupted and degraded the environment, for instance, in Vietnam and the Gulf (SIPRI 1976; Kemp 1991; Seager 1992; Austin and Bruch 2000). Bomb craters, burning of forest or wetlands or poisoning with herbicide (SIPRI 1980; Westing 1984a, 1984b, 1985) can either trigger extreme events (such as landslides) or remove people's protection from extremes (such as coastal mangroves as a screen against high winds). Unexploded mines deny people access to arable land, thus reducing food security. Rural people in Afghanistan, Angola, Mozambique, Eritrea and Cambodia have lost limbs trying to farm in heavily mined areas.

The economic impact of war, especially so-called 'low intensity' or 'counter-insurgency' warfare, is very high for isolated rural households, who may often be highly vulnerable to begin with (Stewart and Fitzgerald 2000). Contending forces ebb and flow over such peasant lands, extracting rations or tribute, making life insecure. The influx of refugees following war in a neighbouring territory can have an immediate and dramatic influence on vulnerability by suddenly raising the population density (Hansen and Oliver-Smith 1982; Jacobson 1988). Demands on local services and infrastructure increases, fuelwood and water needs must be met, sometimes with damaging consequences for the local environment (Black 1998). This local population pressure can also increase disaster vulnerability.

As we pointed out in Chapter 1, the interlinkages between conflict and disaster are numerous and complex. The case of the 50 years of civil war in Colombia illustrates such complexities; 2.1 million Colombians were internally displaced at the end of 2000. Many have sought refuge on the edges of cities, where they live in conditions that make them highly vulnerable to shack fires, earthquakes, landslides, floods and epidemic disease, not to mention violence, sexual abuse, hunger, unemployment and despair. Ninety per cent of these people have been displaced by violent conflict. The majority are children, Afro-Colombians and poor women. Only 20 per cent of these IDPs have received any aid from the Colombian state, and even that has been 'minimal and short term' (Lopez 2001: 7). This example shows how violent conflict can affect some of the particular social and demographic groups we have already identified as being highly vulnerable to extreme natural events. It shows, too, how violent conflict increases the pressure of unplanned urbanisation.

Finally, in Colombia the connection between violent conflict, environmental degradation, loss of livelihood and vulnerability are vividly clear. Part of the enormously complex civil war in Colombia includes aerial spraying of pesticide onto coca crops (part of the US 'War on Drugs'). From August 2000 to May 2001 alone there were 1,158 reports of damage to human health, food crops, livestock and the environment (Lopez 2001: 8). Where food crop areas are located near to small plots of coca plants, a household's food has also been destroyed, as have grazing animals including horses, cattle, sheep, goats, rabbits, tortoises and fowl (Lopez 2001).

The manufacturer (Monsanto) of the main fumigation ingredient, Roundup, advises against aerial spraying and recommends that grazing animals do not enter areas where it has been applied for two weeks. Yet the US and Colombian military spray it directly onto small farmers, crops and livestock. In addition, the manufacturer advises against using Roundup near bodies of water because of potential harm to aquatic life. During 2000–2001 spraying was intense in the Putumayo region of Colombia, on the edge of the Amazonian basin. It is a region with intricate waterways and strong currents. Thus unintentional contamination, far from the coca-growing targets, is almost certain (Lopez 2001). In response to the aerial spraying some small farmers have abandoned the land and have joined the urban displaced persons discussed earlier. Others have gone even deeper into the rainforest where they hope they can farm in peace. In this way, the agricultural frontier is being extended, with accompanying deforestation and long-term reduction of biodiversity and forest cover. This, too, contributes in the long run to increased vulnerability to extreme natural events.

Global economic pressures

A further global pressure on vulnerability to disasters involves the workings of the world economy (Castells 1996; Stiglitz 2002; Cavanagh et al. 2002). Since the Second World War the global economic order has changed rapidly. In particular, the pattern of financial relationships between the industrialised MDCs and LDCs has altered following decolonisation. Globally, prices are falling for the agricultural and mineral exports on which LDCs have traditionally had to depend.

Meanwhile, the prices LDCs have to pay for their imported energy and technology have increased. This has created circumstances in which many LDCs face great difficulty in maintaining their balance of payments. In addition, the oil price rises of 1973 and 1979 led many countries to incur foreign debts. These were transformed into repayment crises, especially in light of rapid increases in interest rates in the late 1970s and early 1980s. In many African countries, debt servicing alone (i.e. payments of interest and charges) amounts to 40–50 per cent of export earnings (George 1988; Onimode 1989; ROAPE 1990; Africa World Press 1997). The flow of financial aid into Africa (net of debt payment and repatriated profits) has declined steadily (Cheru 1989; Adedeji 1991; Mengisteab and Logan 1995; Mkandawire and Soludo 1999), and in some cases is exceeded by debt and interest repayments. In 1998, 31 of 48 African countries paid debt service in excess of 50 per cent of their GNP (TransAfrica Forum 2003). Foreign debt amounted to a very high percentage of annual GDP (gross domestic product) in many Latin American countries in 1985: 107 per cent in Bolivia, 99 per cent in Chile, 80 per cent in Uruguay, 77 per cent in Venezuela and 73 per cent in Peru. The Latin American average was 60 per cent of GDP (Branford and Kucinski 1988: 9). This percentage fell to 40 in the period 1996–2000, but this is still a significant burden (IMF 2002: 63).

The outcome of this pressure was to intensify the need to export at any cost. At the national level, this world economic situation added pressure to exploit natural resources to the fullest extent possible to maximise exports. As discussed below, such a 'growth mentality' has resulted in degraded forests and soil that increase vulnerability to disasters (Tierney 1992; Mander and Goldsmith 1996; Burbach et al. 1997; UNRISD 2000).

Since the 1980s many indebted countries have agreed to structural adjustment policies (including IMF 'stabilisation' and World Bank 'restructuring' policies), or initiated their own programmes that involve cutting public spending and a number of other measures discussed below. As a result, services such as education, health and sanitation are often reduced and state-owned enterprises privatised (both these measures leading to unemployment), while food subsidies are reduced. The early effects of these policies on welfare were analysed in studies by Cornia et al. (1987) and Onimode (1989), among others, but there was little discussion of the effect of such programmes on disaster vulnerability.[17] As a result of such studies, various 'safety nets' and other modifications of the structural adjustment policy design were built in, as described by Stewart (1987) and Haq and Kirdar (1987).

There is still controversy over whether these modifications were sufficient to protect vulnerable people and fragile environments. Because there have been a number of different phases of these programmes, with different impacts across time and between different countries, it is more difficult to make an evidence-based case for the impact of structural adjustment policies on vulnerability in a particular country. However, the fundamental characteristics of these programmes have remained, though the targets and means to reach them have changed.

The structural adjustment policies objective was to reduce the debt burden of the poorest countries by inducing them to export their way to economic growth and freedom from crippling debts (Panos Institute 2002: 6). Such policies first insisted on privatisation of the economy and a rolling back of the state's control of sectors such as health, water and electricity supply, marketing and transportation. Secondly, capital market liberalisation was imposed, and any outflows of foreign capital which had invested in the newly privatised sectors had to be persuaded back by raising interest rates (sometimes to 60–80 per cent), with ensuing financial volatility. Thirdly, market-based pricing was insisted upon, which, for consumers, meant greatly increased prices for public utilities and some basic food staples which had hitherto been subsidised by the state. The result was social unrest in many countries during the 1980s. Lastly, conditions for free trade and the dismantling of barriers to foreign investment took place.

Despite the controversy surrounding the short- and long-term effectiveness and side effects of these IMF and World Bank policies, it is now widely recognised that these measures did not produce the desired effects and were particularly onerous for the poor (Rich 1994). Oxfam International estimated that the IMF-imposed cuts (in countries where structural adjustment policies had been implemented) had resulted in 29,000 deaths from malaria and had increased the number of untreated cases of tuberculosis by 90,000 (Brecher 1999). Health care, nutrition of the poorest, investment in human capital through education, all declined. Public infrastructure was neglected and public works programmes (upon which the poorest relied most heavily for safety and for employment) were cut back. Safety regulations at work and pollution standards were reduced to attract foreign investors, further increasing vulnerability. Wage levels were pared down in order to win export orders and attract investments by multinational corporations. This price war between the poorest countries has been described as 'a race to the bottom' (Madeley 1999). Child labour increased in sweatshops.

A new initiative for assisting Highly Indebted Poor Countries (HIPCs) was introduced in 1996 by the IMF. In 1999 an 'enhanced initiative' was introduced to decrease their debt to manageable and sustainable levels, and increased the number of eligible countries from 29 to 36 (22 of which were in Africa). A new condition was introduced for countries to comply with, which required them to prepare a Poverty Reduction Strategy Policy (PRSP). Although this exercise was supposed to be directed by national governments and shaped by widespread civic participation, it is claimed that this has not been the case. None the less, the enforced focus on poverty by the IMF and the World Bank cannot be detrimental to vulnerability, although whether it makes any long-term difference is open to question. The World Bank's *World Development Report 2000/2001* has a whole chapter devoted to 'Managing Economic Crises and Natural Disasters' (World Bank 2001), and there is an analysis of the impacts of natural disasters and their impact upon the poor. Prevention and mitigation measures are suggested which look quite similar in some respects to those we suggest in Chapter 9.

However, our main scepticism concerns whether these PRSPs will ever be implemented. There have been long delays in their acceptance because of the difficulties in complying with complex and stringent conditions (especially over privatisation). To the extent that what is known as the 'decision point' or acceptance of the PRSP is hard to reach, debt relief and the implementation of the programme has been delayed. Furthermore, many of the PSRPs do not have clear policies to reduce poverty, and many of the more radical measures which would help (for example, improvement of labour rights, minimum wages, safety at work and land reform) seem to have been omitted (Marshall and Woodroffe 2001). Instead other, rather vaguer directions have developed such as good governance, careful monitoring, addressing corruption, improving access to education and health services (Panos Institute 2002). However, optimists point to encouraging rhetoric and real efforts to identify positive poverty-reducing policies, while pessimists see the latest reincarnation of structural adjustment policies as *plus ça change...*

In this debate, much of the discussion has been of global pressures on poverty rather than vulnerability, and as our book emphasises, poverty is not synonymous with vulnerability, although the two conditions are often highly correlated. It is difficult to provide hard evidence of such a relationship between structural adjustment policies and vulnerability, although it is much clearer regarding poverty. However, the majority of the deteriorating conditions in the living standards of the poor, as outlined here, can reasonably be assumed also to affect vulnerability adversely, and as such constitute a global pressure.

In some cases, however, it is possible to demonstrate clear links between vulnerability and the operation of the global economy, as exemplified by a case from Jamaica during the 1980s and 1990s (Ford 1989). In the 1980s, the government of Jamaica intervened in the financial sector to try and reduce inflation and stimulate production because of its large foreign debt. This policy was intended to attract foreign capital seeking high interest rates. Interest rates went up to over 20 per cent, and home mortgage rates ran between 14 and 25 per cent. These financial changes took place in a situation where the government enforced rent control and levied an import duty on construction material. As a result of this combination of policies, new residential construction declined rapidly.

Thus, the global economy – acting as a dynamic pressure – worked its way through to specific unsafe conditions in Jamaica. According to Ford there was an immediate increase in vulnerability of a significant proportion of the urban population to hurricanes and earthquakes. This results from the fact that property owners faced with such high mortgage interest rates and little hope of recouping this by increasing their rents (due to the rent restrictions) simply ignored maintenance (Ford 1989). Hurricane Gilbert damaged more than 100,000 low-income homes in 1988, producing costs of $558 million. More than 28,000 homes of the poor were either completely destroyed or had severe damage to roofs and structure (Government of Jamaica 2003).

In the early 1990s, external debt was more than 80 per cent of GDP. By 1997 that ratio had fallen to below 50 per cent (Government of Jamaica

2002). For instance, in 1989, service on external debt amounted to nearly one-third of export income; while by 1999 that had been reduced to 17 per cent (Jamaica at a Glance 1999). Nevertheless, interest rates continued to climb. The average lending rate reached a peak of 54 per cent in 1997. Although the lending rate declined gradually to 26 per cent in 2002 (PSOJ 2002), the pressure on new residential construction and maintenance of old structures described by Ford continued. Low-income housing would therefore still seem to be susceptible to the next big direct hit by a hurricane. The experience of Jamaica during the past 20 years illustrates the linkages that exist between the global economy, national economic policies and vulnerability. The impact of 'structural adjustment' on vulnerability went far beyond the issue of building maintenance. Because of the high cost of finance, builders tried to keep the cost of construction as low as possible so some small profit could be made. Again, safety suffered.

Health and education budgets suffered cuts under Jamaica's structural adjustment policies. Even more crucial is the fact that the government's own programmes to introduce preparedness or mitigation measures were also cut as a result of the economic constraints. It would be difficult to determine whether the severe damage to Jamaica from hurricanes Gilbert in 1988 and Hugo in 1989 were made worse by the economic policies described above, but such potential connections are clearly possible. An additional irony in the Jamaican situation is that part of the foreign debt burden that caused the government to launch its structural adjustment policy was due to loans used to pay for previous hurricane damage (see Chapter 7).

In the years preceeding hurricanes Gilbert and Hugo, an estimated 50,000 children under four years old suffered from malnutrition in Jamaica (Oxfam 1988). More than one-third of the labour force earned less than £5 ($7.50) per week, while four times this sum was needed to feed an average family. In Chapter 5 we argue that such a weak nutritional (and therefore health) status of a population contributes to other forms of vulnerability in the long run. If the Jamaican debt burden has had a negative impact on the poor, it is affecting an already impoverished people, a considerable proportion of whom are vulnerable to local hazards.

Adverse agrarian trends and livelihood diversification

It is becoming increasingly clear that sustainable livelihoods cannot be supported by natural resource-based activities (primarily agriculture) in many parts of the world, particularly in sub-Saharan Africa. Thus, there has been a reduction in the farming component of livelihoods, such that agriculture may provide only 50 per cent of family income, even in very rural areas (Reardon 1997; Bryceson 1999). This process has also acted as a 'dynamic pressure' affecting vulnerability both positively and negatively. Some of the pressures which have driven this adaptation are the following:

- Population growth without a concomitant increase in agricultural production, leading to the sub-division of land holdings and falling food security through local food production.
- Adverse environmental change, including global climate change (long-term desiccation, 'unseasonable' drought, or exceptional rainfall – see below).
- A decline in agricultural markets relative to non-farm wage levels.
- Rises in agricultural input costs due to the removal of subsidies following structural adjustment policies (see above).
- A general decline in access to rural public services due to economic mismanagement, protracted civil war and cost recovery programmes, and (again) under structural adjustment policies (Ellis 2000).

Some critics of structural adjustment policies argue that they have seriously undermined rural livelihoods and increased the risks of destitution and famine, but this is very difficult to verify without reliable longitudinal studies. According to Ellis (2001), a minority of authors such as Booth et al. (1993) have come to the opposite conclusion, namely that structural adjustment policies have enabled a positive diversification of livelihoods, with beneficial reductions in household risks.

There is also a growing literature on the decision-making aspects of diversification and the socio-economic characteristics of different households undertaking diversification. In poorer households, the decision to diversify may be driven by acute food insecurity and risk aversion, where agricultural yields are declining and their variability increasing. Increasing food security through the generation of cash from non-agricultural activities is one clear possibility. In better-off households, investments in education and other human capital has an effective, albeit longer term, benefit (Dercon and Krishnon 1996).

There are, however, marked gender, age and class differences (younger males migrating with a concomitant increase in the burden on the elderly and all women who have to stay at home). Resulting labour shortages on the farm are also a factor and intensify the feminisation of agricultural labour. For example, in Nepal the migration of significant numbers of people from the Middle Hills to the lower *terai* and to India resulted in remittances of cash which improved the food security of most households and brought the opportunity to change agricultural technology (Blaikie and Coppard 1998). The impacts of these pressures and adaptations on vulnerability are clearly diverse and often complex. Remaining a 'local' member of common pool management institutions becomes important, and temporary absence (particularly of males) may lead to a dispossession of rights and increasing insecurity.

Natural resource degradation

Another significant global dynamic pressure is destruction of forest, soil, wetlands and water sources. This is often closely linked with the debt question, since land degradation may result from national policies favouring export production, although this is difficult to prove since *ceteris paribus* conditions seldom exist and the policy effect is difficult to identify (Mearns 1991). In order to service debt, new lands have been cleared (e.g. in Brazil, the Philippines, Indonesia and many African countries) for ranching or commercial cropping, although there are usually other domestic factors at play here too. Coastal areas have been drained and mangrove forests cut in order to accommodate the expansion of tourist hotels and other foreign installations that offer the hope of hard currency earnings. Likewise, much forest has been destroyed by the timber industry in Asia and Africa, where uncontrolled cutting of high-value exportable hardwoods is another way debtor governments can pay.[18]

The connection between land degradation and unsafe conditions can be quite significant (Pryor 1982; Cuny 1983; Abramovitz 2001; ISDR 2002a; UNEP 2002). Deforestation, soil erosion and the mismanagement of water resources can increase hazard intensity or frequency in the long run. The connection between deforestation and slope stability, erosion and the risk of drought, and other issues, will be discussed at various points in Part II of this book. Loss of biodiversity can also affect patterns of vulnerability, and in Chapter 5 we will inquire into the link between the extinction of wild genes (sometimes called 'genetic erosion') and vulnerability to plant pests and diseases.

Deforestation, wetland destruction, over-fishing and destruction of coral reefs all contribute to genetic erosion, leading to the loss of many species, known and unknown (UNEP 2002). The physical growth of cities has caused the destruction of much coastal wetland. Swamps are drained for living space, for urban-fringe gardening, for fish ponds or salt works. Mangroves are cut for building material. Chapter 7 will emphasise the importance of these wetlands as buffers against coastal storms (Maltby 1986; H. John Heinz Center 2000). The growing demand for wood and charcoal in some south Asian and African cities means that fuels are being produced at ever-increasing distances, causing loss of vegetation (Leach and Mearns 1989). In other cities the demand for electricity is satisfied by more and more dams (often large-scale). These dams flood vast areas of forest and other lands, forcibly displacing the inhabitants (Little and Horowitz 1987; World Commission on Dams 2000b). Persons displaced by mega-projects of this sort (and others such as mining and oil extraction) often become more vulnerable to natural hazards because of their unfamiliarity with the environments to which they have been moved. Social and economic dislocation can also play a part in the vulnerability of people displaced in this manner (Watts and Peluso 2001; IIED 2002).

Another important aspect of loss of species and genetic variation is the changes in cropping systems and especially the increasing tendency for farmers to use fewer varieties of crops. Modernisation is accompanied by dietary change, with imported and processed food items replacing traditional varieties of grain, legumes, fruits and vegetables. Farmers grow a more limited number of commercial crop varieties and the traditional ones die out (Juma 1989). When biological hazards strike, there may be no resistant varieties (genetic ancestors of the affected crops) on which to fall back. The Irish 'Potato Famine' of 1845–1848 is a classic example. Irish peasants simply did not have access to (or knowledge of) the South American tubers that might have been imported to improve the disease resistance of the existing land race, which was derived from a very narrow genetic base (see Chapter 5). The destruction of habitats is wiping out the wild ancestors of many crops altogether. In the 1970s farmers in the USA were able to get hold of other seed sources when maize (corn) blight halved the yield of monocropped hybrids on which they had become reliant. In the future the insurance of older varieties of maize may not be available if they have become extinct (Fowler and Mooney 1990; Cooper et al. 1992).

The increasing use of genetically modified (GM) crops is highly controversial, and part of the debate bears directly on disaster vulnerability. The proponents of GM crops believe that in the future new varieties of disease and pest-resistant food crops, and those with other properties such as high levels of the precursors of vitamin A (so-called 'golden rice'), could significantly increase food security worldwide and help to wipe out nutritional deficiencies (Royal Society of London et al. 2000). Another example is the development of crops with a gene for salt tolerance added from mangroves that would allow food production on degraded lands. Opponents advise extreme caution in disseminating seed of this kind because of the fear of contaminating existing varieties – especially those that have been bred in conventional ways by generations of small farmers in Africa, Latin America and parts of Asia and the Pacific.

Recalling unanticipated and negative social and ecological side effects of the widespread use of so-called Green Revolution seeds from the 1960s onward, critics are concerned that premature release could destroy the existing stock of well-adapted staple grains and other food crops before a similar array of 'down-side' effects become evident. By then, critics believe, it might be too late to turn back. These anxieties were in the minds of the scientific advisers to the governments of Zimbabwe, Zambia and Malawi when, during a severe food shortage in southern Africa in 2002, they counselled their governments to reject the donation of GM maize from the USA. An additional line of criticism comes from those who believe that hunger, and in particular famine, is not caused by a shortage of food but by the *distribution* of food (see Chapter 4).

Global environmental change

There is by now little doubt that changes in the interacting systems of the atmosphere, hydrosphere, and biosphere have resulted in the build-up of 'greenhouse gases' (Liverman 1989; Watson et al. 1998; 1999). The dangers are that the changes will increase the intensity and frequency of climatic hazards and enlarge those areas affected by them (McGuire et al. 2002). It is not possible to blame the 'greenhouse effect' in a definitive way for the powerful hurricanes Gilbert, Joan and Hugo (1988 and 1989), Andrew (1992), Mitch (1998) or Georges (2000), or the record storms in Europe in the winter of 1989–1990, or the Australian floods of 1990 onwards. But global climatic change provoked by warming is predicted to increase the number and intensity of storms and cyclones and to amplify the variations in precipitation over much of the earth's surface.

The impact on livelihoods could be immense (especially for farming and fishing peoples), in addition to the dangers from any intensification of the hazards (Downing et al. 2001). Considerable work over the past ten years on the El Niño Southern Oscillation (ENSO) suggests that these cycles of exceptionally wet and exceptionally dry weather, associated with periods of warming of surface water in the Pacific, may be increasing in frequency. El Niño and La Niña episodes were associated in the 1980s and 1990s with failure of the monsoon in South Asia, spiking incidence of malaria and dengue, floods and landslides in the Andes, and wildfires in Australia (Glantz 2001; 2002).

Rising sea-level due to global warming is another dynamic pressure that will increase vulnerability. One result could be the destruction of livelihoods of six million farm workers living in the fertile delta regions of India (Watson et al. 1998). Low-lying areas of many islands, as well as the flood-prone delta regions of Bangladesh, India and Guyana are particularly at risk of a sea-level rise (Brammer 1989). In the Pacific, Tuvalu and Tonga may become uninhabitable (Lewis 1989; Wells and Edwards 1989; Pelling and Uitto 2002), and coral atolls which are home to many people in the Pacific and Indian oceans would experience submergence or destruction by storms. Low-lying parts of coastal cities are also at risk (O'Neill 1990 and see Chapter 7).

Uses of the PAR model

Since the first edition of *At Risk,* in addition to academic and policy applications (e.g. Watanabe 2002; Turner et al. 2003; Haque 1997) several NGOs have made use of the PAR (or 'crunch') model as the basis for community-based self-study of vulnerability and capability. In most of these pilot projects, communities and groups adopt the concept of vulnerability to inquire into their own exposure to damage and loss (Wisner 2003a).

The concept of vulnerability thus becomes a tool in the struggle for resources that are allocated politically. In some parts of Latin America and southern Africa such community-based vulnerability assessment has become quite elaborate, utilising a range of techniques to map and make inventories, seasonal calendars and disaster chronologies.[19] Pilot projects have shown that lay people in citizen-based groups are capable of participating in environmental assessments that involve technology not previously accessible to them, such as geographical information systems (Pickles 1995; Levin and Weiner 1997; Liverman et al. 1998; Maskrey 1998).

In these applications of our PAR model and others like it, the community defines its own vulnerabilities and capabilities, not outsiders. They also decide what risks are acceptable to them and which are not. As Morrow remarks:

> The proposed identification and targeting of at-risk groups does not imply helplessness or lack of agency on their part. ... Just because neighbourhoods have been disenfranchised in the past does not mean they are unwilling or unable to be an important part of the process. There are many notable examples of grassroots action on the part of poor, elderly and/or minority communities..., and of women making a difference in post-disaster decisions and outcomes.... Planners and managers who make full use of citizen expertise and energy will more effectively improve safety and survival chances of their communities.
>
> (1999: 11)

The employment of the concept of vulnerability as a tool in and by the community also involves a thorough analysis with and by the residents of their own resources and capacities. This is the 'other side' of the vulnerability coin. It is in the hands of local people that the logic of their situation, the phenomenology of their living with risks, forces them to be aware of and to discuss their strengths and capacities, as well as their weaknesses and needs (Wisner 1988a; Anderson and Woodrow 1998).[20]

Notes

1 'Risk' may still exist, however, since in the absence of an actual extreme natural event, natural process capable of generating such events may continue. The only point we are making here is that *given* a specific extreme natural event (e.g. an actual hurricane or landslide), if there is little or no vulnerability, there will be no disaster.

2 Good detailed discussion of the physical processes and extreme events themselves (the 'hazards') are available in Alexander (1993), Tobin and Montz (1997) and Smith (2000) as well as US Federal Emergency Management Agency (FEMA 1997) and Organisation of American States (OAS 1991).

3 There are many definitions of 'disaster', 'emergency', 'catastrophe', etc. We adopt our own, which shares much with the most common definitions in use. For more on questions of definition, see Oliver-Smith (1999b) and Quarantelli (1995, 1998).

4 For the sake of convenience, we will sometimes present the PAR model as a specific variation on Figure 2.1 in diagrammatic form, and sometimes in non-diagrammatic, list form. Such a list reads from top to bottom, beginning with 'root causes', proceeding through 'dynamic pressures', etc. For example, see the PAR lists for the Montserrat volcanic eruption or Kobe earthquake in Chapter 8.

5 This way of organising proximate and ultimate causes has been used elsewhere (e.g. in explaining land degradation by Blaikie and Brookfield 1987; Blaikie 1985a, 1985b, 1989).

6 Readers of the first edition of *At Risk* noticed that there is some ambiguity or overlap between processes we call 'root causes' and those we term 'dynamic pressures'. That is true, because a factor that can be considered a 'root cause' in one set of circumstances may be more of a 'dynamic pressure' under different conditions.

7 Smith (1992: 25) defines resilience as 'The measure of the rate of recovery from a stressful experience, reflecting the social capacity to absorb and recover from the occurrence of a hazardous event'.

8 For example, in May 2002, unusually hot, dry winds from the north pushed temperatures up in eastern and central India. More than a thousand people died from heat stress and dehydration, among whom were many children and the frail elderly (Kriner 2002). By January of the same year, temperatures had dropped to just above freezing in northern India, and at least another thousand people died of exposure and hypothermia (Rajalakshmi 2003). Klinenberg (2002) describes a heat wave in Chicago in 1995 that killed more than 700 persons, many of whom were low-income, elderly women.

9 'Entitlement' is a term for the access that people have to food from the sale of their labour, their own food producing activity, or via social networks, or some political claim on state resources (including moral claims on international food aid). Economist and Nobel laureate Amartya Sen introduced the term in a rigorous way in his 1981 book, *Poverty and Famines* (see Chapter 4).

10 In 1988 Pakistan had the second highest balance of payments deficit in the world at $3.5 billion.

11 This 'villagisation' policy in Tanzania had been intended to have a positive impact on livelihoods, by increasing the size of settlements so they could form the basis for providing health and educational services. It was also intended to create better economies of scale in agriculture and co-operatives for producers and marketing (Coulson 1982). Subsequent research has suggested that such radical, often forced, resettlement seriously disrupted patterns of coping with natural hazards such as flooding in the Rufiji river delta (Hoag 2002).

12 We would argue that the difficulty in providing direct evidence of the linkages between unsafe conditions, dynamic processes and root causes does not undermine this method of 'explaining' vulnerability. Quite clearly, if it is accepted that vulnerability is a function of socio-economic processes and not just the characteristics of the hazard itself, then there must be a chain of explanation for that vulnerability which can be traced back to root causes. This is no different from explaining any other social phenomenon: a 'housing shortage' in London cannot simply be explained by saying that there are not enough houses being built or too few people with enough cash: there are 'dynamic processes' and 'root causes' to be taken into account, even if there are political differences as to what these are in any given problem. Dovers et al. (2001) discuss the varieties of ignorance and uncertainty in environmental policy making in very much the same terms we do.

13 Here we focus on the political consequences of uncertainty. However we also recognise and certainly do not underestimate the philosophical consequences of opting for a meso-level theoretical position that spans such widely different

scales of time and space (historical and global affecting present and local). In brief, we do not believe that historical events and well-established social structures 'cause' unsafe conditions in any rigid, deterministic way. There is an interplay of structure and human agency. In much social science, as Abbott (2001: 98) puts it so nicely, 'action and contingency disappear into the magician's hat of variable-based causality, where they hide during the analysis, only to be reproduced with a flourish in the article's closing paragraphs'. We attempt to do justice to human action and agency – even, or especially, that of people who at first (viewed from the outside) appear as very weak and powerless – as well as contingency. That is one of the reasons why we have two models and complement the PAR model in this chapter with a model of household action (coping in adversity) in Chapter 3.

14 See, for example, Bryant (1991: 7–8), who labels those who might want to consider social processes as 'Marxist' as a way of dismissing them.

15 During the decade of the 1990s growing precision in the international study of global climate change now suggests that an increase in the frequency of intense storms could be one result of the warming of the atmosphere. This work emphasises that what one is dealing with is increasing variability in the ocean-atmosphere system, with complex results. One of these seems to be more frequent El Niño events. If correct, this research points to more floods and storms as extreme natural events. See McGuire et al. (2002).

16 See Maskrey and Romero (1983); Davis (1987); Hardoy and Satterthwaite (1989: 146–221); Fernandes and Varley (1998); Mitchell (1999a); Fernandez (1999); Wisner (2002a) and Pelling (2003b).

17 Related problems of environmental degradation have been raised. In a paper for a meeting of CIDIE (Committee of International Development Institutions on the Environment), hosted by the World Bank, Hansen noted that for a number of reasons structural adjustment policies

> often lead to a deterioration of the situation for those with the least resources to adapt to the changed economic circumstances. To the extent that poverty in many regions of the world is the primary cause for environmental degradation, increased poverty caused by structural adjustment policies can lead to further environmental damage.
>
> (1988: 7)

18 It is clear that many of these damaging activities pre-date the debt crisis. The argument is that the response of governments and entrepreneurs to the priority for exports has intensified them. However, it is possible that the intensification of deforestation, for example, does not earn foreign exchange for the government to repay debt. In some circumstances individuals and enterprises control foreign earnings and siphon them off out of the country ('capital flight') without any benefit to the economy. There is also a serious corollary of this: that reduction of the debt burden may not alleviate the destruction of forests or other resources, since the motivation for damage is not always to service the economic problems of the nation (see Little and Horowitz 1987; Faber 1993; Gadgil and Guha 1995).

19 Wisner et al. (1979); Cuny (1983); Maskrey (1989); Wisner et al. (1991); Geilfus (1997); Soto (1998); von Kotze and Holloway (1996); Anderson and Woodrow (1998); Carrasco and Garibay (2000); Plummer (2000); Turcios et al. (2000); Chiappe and Fernandez (2001); Wilches-Chaux and Wilches-Chaux (2001).

20 See an interesting discussion of 'poverty as capability deprivation' in Sen (2000: 87–110) and more on capability deprivation in Chapter 3.

3

ACCESS TO RESOURCES AND COPING IN ADVERSITY

Access to resources – an introduction

In the last chapter, we argued that disasters occur as the result of the impact of hazards on vulnerable people. We suggest two frameworks for explaining this relationship between natural events and the social processes that generate unsafe conditions. The first is the Pressure and Release (PAR) model, which shows in diagrammatic terms how the causes of vulnerability can be traced back from unsafe conditions, through economic and social ('dynamic') pressures, to underlying root causes. PAR is an organising framework outlining a hierarchy of causal factors that together constitute the pre-conditions for a disaster. We can also describe this as a pathway, 'progression of vulnerability' or 'chain of causation'. It is a sequence of factors and processes that leads us from the disaster event and its proximate causes back to ever more distant factors and processes that initially may seem to have little to do with causing the disaster. The 'Release' aspect arises from the realisation that to release the pressure that causes disasters, the entire chain of causation needs to be addressed right back to the root causes, and not just the proximate causes or triggers of the hazard itself or the unsafe conditions of vulnerability.

But the PAR model does not provide a detailed and theoretically informed analysis of the precise interactions of environment and society at the 'pressure point', at the point where and when the disaster starts to unfold. Firstly, any analysis of a disaster must explain differential vulnerability to, and the impacts of, a disaster – why wealthier people often suffer less, and why women and children may face different (and sometimes more damaging) outcomes than men and adults. Particular groups, defined by ethnicity, class, occupation, location of work or domicile may suffer differentially from others. In these senses, the Access model focuses on the precise detail of what happens at the pressure point between the natural event and longer-term social processes, and, to signify this in visual terms, a magnifying glass is drawn on the PAR model (Figure 2.1).

Secondly, the PAR framework is essentially static, and without a series of iterations through the trajectory of a disaster, it cannot suggest nor account for change, either before the onset of a disaster, and more importantly, during and after it. In addition to the PAR model, what is required is a detailed account of 'normal life' before the disaster. We need a complementary model that details the progression of vulnerability to (and through) the pressure point and through the unfolding of the disaster. It must also show how normal life becomes abnormal, and how and when disjunctures between the normal and the exceptional take place. To achieve this, in this chapter we present the Access model, which deals with the amount of 'access' that people have to the capabilities, assets and livelihood opportunities that will enable them (or not) to reduce their vulnerability and avoid disaster.

The purpose of the Access model

The Access model is designed to understand complex and varied sets of social and environmental events and longer-term processes that may be associated with a specific event that is called a disaster. A disaster may be described and labelled according to the natural hazard that triggered it (for example, drought impacting upon vulnerable people leading to famine, or an earthquake impacting on unsafe buildings leading to destruction of life and property). Indeed, the chapter headings of Part II of this book adopt this categorisation, as we have responded to much received wisdom in order to undermine this type of approach. Much of the literature on disasters relates to each type of natural event trigger, and is specific to famines, biological hazards, floods, severe coastal storms and so on. On the other hand, there are generally shared characteristics in the way that vulnerability is generated, how the trigger event and the unfolding of the disaster has its impact, and various responses by different actors, both local, national and international. It is these that the PAR model – and now the Access model – seek to address.

In Figure 2.1 the reader will have noticed the magnifying glass in the diagrammatic representation of the PAR model. This is intended as a visual metaphor for the location of the Access model in the wider explanatory framework: the place within the PAR approach where it magnifies and clarifies the explanation of a disaster. The magnifying glass metaphor is appropriate since the Access model sets out to explain at a micro-level the establishment and trajectory of vulnerability and its variation between individuals and households. It deals with the impact of a disaster as it unfolds, the role and agency of people involved, what the impacts are on them, how they cope, develop recovery strategies and interact with other actors (e.g. humanitarian aid agencies, the police, the landlord and so on).

We first introduce the outline structure of the model, in which boxes are labelled in summary fashion but remain, if not 'black boxes', still somewhat opaque. Then later in the chapter each box will be opened up and explained in detail. There is a risk in this that some repetition will irritate the reader. However, it is hoped that clarity as well as detail can be captured by dealing first with the overall structure and then the more theoretically informed and detailed aspects sequentially. The Access model picks up the state of 'normal life' and explains how people earn a livelihood with differential access to material, social and political resources.[1] This model is shown in Figure 3.1and can be identified in summary form in Box 1

The outline has eight boxes, each representing a set of closely related ideas, an event or distinct process. They are linked by arrows which denote cause and effect linkages. Although the linkages are shown by arrows (from cause to effect), they also iterate, and effects can also shape causes at a later period of time. In the briefest terms, hazards (Box 3) have specific time and space characteristics (Box 4), which can – and in this depiction do – result in a 'trigger event' (Box 5, for example an earthquake, a tropical storm or a

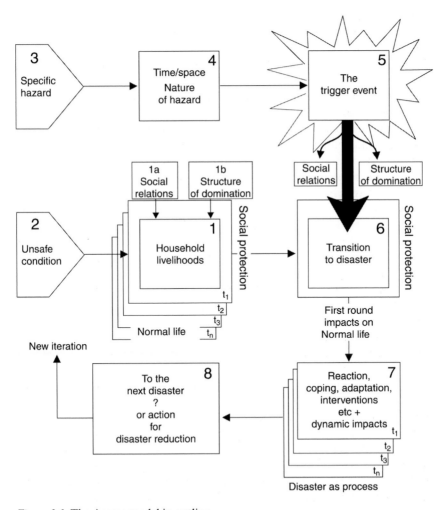

Figure 3.1 The Access model in outline

89

serious drought). Households earn their livelihoods in normal times (Box 1), and are subject to unsafe conditions (Box 2) and the political economy in which they all live is also shaped by social relations and structures of domination (Boxes 1a and 1b). The trigger event occurs and impacts upon social relations and structures of domination and upon households themselves (Box 6). The heavy black arrow is depicted as bursting through an outer layer, called 'social protection' (which, as will be explained below, is both individual, collective and public), and as impacting on different households in a process termed 'transition to disaster' (Box 6). Subsequent iterations of unfolding impacts and human responses occur through time (Box 7). Box 8 asks the question 'To the next disaster?', and indicates altered conditions of vulnerability, social protection and actions for preventing future disasters.

The model stylises the process of earning a living as a set of decisions made at the household level (as will be described below when this process in Box 1 is opened up and explained in Figures 3.2 and Figure 3.3), individual decisions are always made in a political-economic environment, and this is indicated by two boxes (1a and 1b) labelled 'social relations' and 'structures of domination'. Thus, life in normal times is characterised by repeated decisions about how to obtain a livelihood, decisions which are made every season in an agricultural setting (for example, a cropping strategy, investment in new inputs or agricultural equipment) and sometimes irregularly and more frequently in an urban setting (for example, changing the nature of employment, starting up a small shop or handicraft enterprise).

The iterative character of a livelihood is suggested by repeated cycles of livelihood decisions, each on one sheet, arranged in the diagram behind each other and labelled 't1', 't2', indicating subsequent iterations of decision making year by year. There is also an outer border to the household livelihood box called 'social protection'. This symbolises the presence (or absence) of hazard precautions and preparedness that is provided by the state and local collective action. It is the local expression of the more generalised 'unsafe (or safe) conditions' (shown here as Box 2, derived from the PAR model). This links the broader scale of disaster causes to the microcosm of normal life. The resources which define the quality of social protection at household level are varied in scope and may include flood protection embankments, concrete storm shelters, enforcement of building regulations as well as community coping mechanisms, self-help and communal charity. These are discussed below in more detail. It is also worth specifying how to choose the 'households' in the box. This matter (essentially one of scope and sample) must be defined by the local people and the researcher, planner or development professional according to their focus. This can be defined spatially (for example, an area threatened by a specific and severe natural hazard, a particular village or a quarter in a city), or defined by ethnic group, class or other characteristic which may render the chosen group of households more vulnerable.

Let us now turn out attention to the potential hazard in Box 3 in the top left-hand corner of the diagram. The arrow-shaped box should be recognisable from the PAR framework, and introduces the specific hazard(s). The time and place characteristics of the hazard (where, how often, when) are examined in more detail and illustrated in Box 4, labelled 'nature of the hazard'. The scene is now set for the disaster process to start, following the hazard strike (stylised as 'the trigger event', in Box 5). It must be noted here that some disasters occur without a clear-cut, single, natural hazard trigger, nor perhaps with an identifiable event in the political economy. Instead, there are multiple contributing events which together constitute a 'complex emergency', which then unfolds in all its intractable complexity over a long period of time. An example is the situation in 2001–2003 in Zimbabwe, which has gone from a country producing a surplus of food into one facing bankruptcy and impending famine, in a complex mix of drought and political conflict.[2]

The example to be used here involves a disaster that is triggered by a definite hazard event. It is a composite event, but not an unusual one, involving high winds, coastal storms, intense rainfall, landslides, flooding of urban and rural areas, contaminated water supplies and so on, and will be described in more detail later on. This event now attacks 'normal life', and the first round of impacts are shown in Box 6. Some of the immediate consequences are mediated or deflected by the safety measures in place, while other impacts penetrate these safety measures (depicted by the 'impact arrow' striking through the outer protective barrier) and fall upon different households with varying degrees of severity. The hazard event also alters existing social relations as well as structures of domination, as the more detailed explanation of these processes will show.

Within the microcosm of households, adaptations, coping strategies and access to safety become urgent as a potential disaster starts to overtake what is no longer 'normal life'. This is the transition from normal life in Box 1 to transitions from the first round of impacts in Box 6, labelled 'transition to disaster'. Thereafter, there may be interventions into the 'microcosm' of households from the outside, such as disaster relief, or in contra-distinction, a continuation of, for example, military activity and the disruption of relief supplies. Then, in subsequent iterations, the disaster unfolds in a series of 'time sheets' labelled 'disaster as process' in Box 7. The process of recovery, and return to normal life (or, in some cases, to a more vulnerable life, waiting, as it were, for the next disaster) is suggested in Box 8 and is examined in Chapter 9.

Access in more detail

The PAR model, which forms a 'chain of explanation', is an analytical tool, subject to a number of inadequacies which we have tried to illustrate. One of its weaknesses is that the generation of vulnerability is not adequately integrated with the way in which hazards themselves affect people; it is a static model. It exaggerates the separation of the hazard from social processes in order to emphasise the social causation of disasters. In reality, nature forms a part of the social framework of society, as is most evident in the use of natural resources for economic activity. Hazards are also intertwined with human systems in affecting the pattern of assets and livelihoods among people (for instance, affecting land distribution and ownership after floods).

To avoid false separation of hazards from social system, we have proposed a second dynamic framework called the 'Access' model. This focuses on the way unsafe conditions arise in relation to the economic and political processes that allocate assets, income and other resources in a society. But it also allows us to integrate nature in the explanation of hazard impacts, because we can include nature itself, including its 'extremes' (as they are experienced by people with different characteristics, in the workings of social processes and social change). In short, we can show how social systems create the conditions in which hazards have a differential impact on various societies and different groups within society. Nature itself constitutes a part of the resources that are allocated by social processes, and under these conditions people become more or less vulnerable to hazard impacts. In this chapter, the concept of 'access' to resources is explored in a more formal way, and the model within which it can be understood is developed fully.

The notion of access can be illustrated using a narrative taken from the work of Winchester (1986, 1992), which analysed the impact of tropical cyclones in coastal Andhra Pradesh (south-east India) (see also Chapter 7).[3] Cyclones in the Bay of Bengal periodically move across the coast and strike low-lying ground in Andhra Pradesh. They sometimes cause serious loss of life and property, and disrupt agriculture for months or even years afterwards. The damage is done by very high winds, often causing a storm-surge, followed by prolonged torrential rain. Let us compare how the cyclone affects a wealthy and a poor family living only a 100 m apart.

The wealthy household has six members, with a brick house, six draught cattle and over a hectare of prime paddy land. The (male) head of house-hold owns a small grain business for which he runs a truck. The poor family has a thatch and pole house, one draught ox and a calf, a quarter of a hectare of poor, non-irrigated land and sharecropping rights for another quarter hectare. The family consists of husband and wife, both of whom have to work as agricultural labourers for part of the year, and children aged five and two. The cyclone strikes, but the wealthy farmer has received warning on his radio and leaves the area with his valuables and family in the truck. The storm surge partly destroys his house, and the roof is taken off by the wind. Three cattle are drowned and his fields are flooded, their crops destroyed. The youngest child of the poor family is drowned, and they lose their house completely. Both their animals also drown, and their fields are also flooded and the crop ruined.

The wealthy family returns and uses their savings from agriculture and trade to rebuild the house within a week. They replace the cattle and are able to plough and replant their fields after the flood has receded. The poor family, although having lost less in monetary and resource terms, have no savings with which to replace their house (although it would cost less than 5 per cent of the cost of the house of the rich family). The poor family have to borrow money for essential shelter from a private moneylender at exorbitant rates of interest. They cannot afford to replace their ox (essential for ploughing) but eventually manage to buy a calf. In the meantime they have to hire bullocks for ploughing their field, which they do too late, since many others are in the same position and draught animals are in short supply. As a result, the family suffers a hungry period eight months after the cyclone.

Although this story suggests a generalised negative association between wealth and damage, such a result is not automatic. In this area, in some locations that are less protected from tidal surge and are more hazardous to all social classes, higher mortality can occur across the local population, irrespective of the wealth of household. But this example serves to illustrate how access to resources varies between households and the significance this has for potential loss and rate of recovery. Those with better access to information, cash, rights to the means of production, tools and equipment, and the social networks to mobilise resources from outside the household, are less vulnerable to hazards, and may be in a position to avoid disaster. Their losses are frequently greater in absolute terms, since they may have more to lose in terms of monetary value, but they are generally able to recover more quickly. After a famine poor and disadvantaged households can recover but may compromise their resilience to the next famine (Rahmato 1988). In our illustration above, the seeds of further hardship, maybe starvation, have been sown for the household with poor access to resources, but this is not so for the other family.

This example helps to demonstrate the arguments of the first two chapters that variations in level of vulnerability to hazards are central in differentiating the severity of impact of a disaster on different groups of people. In general, rich people (and urban people of all wealth categories) almost never starve. Some avoid hazards completely and many recover more quickly from events that are disastrous for others. However, a major explanatory factor in the creation (and distribution of impacts) of disasters is the pattern of wealth and power, because these act as major determinants of the level of vulnerability across a range of people. We therefore need to understand how this distribution is structured in normal life before a disaster, explaining in detail the differential progression of vulnerability through the triggers of natural and other events into disasters. The idea of 'access' (to resources of all kinds, material, social and political) is central to this task.

Access involves the ability of an individual, family, group, class or community to use resources which are directly required to secure a livelihood in normal, pre-disaster times, and their ability to adapt to new and threatening situations. Access to such resources is always based on social and economic relations, including the social relations of production, gender, ethnicity, status and age, meaning that rights and obligations are not distributed equally among all people. Therefore, it is essential that assets and the patterns of access to them remain central to this project and do not become detached from the underlying political economy which shapes them. For example, private property rights confer upon the owners of buildings and land their ability to control the uses to which they are put. This provides the conditions for the generation of surpluses of cash and food, and collateral for loans – all of which may become crucial in times of disaster. A careful analysis of political economy tends to blur the distinction between *access* and *resources*, because *access* can be understood to be the most critical resource of all (Bebbington and Perrault 1999).

In this Access model, the political economy is modelled in two related systems. The first is called *social relations* (Figure 3.1, Box 1a) and encompasses the flows of goods, money and surplus between different actors (for example, merchants, urban rentiers, capitalist producers of food, rural and urban households involved in various relations of production and endowed with a particular range and quality of access to resources, called an *access profile* [see below]). The second system is termed *structures of domination* (Figure 3.1, Box 1b), and refers to the politics of relations between people at different levels. These include relations within the household, between men and women, children and adults, seniors and juniors. These relations shape, and are shaped by, existing rights, obligations and expectations that exist within the household and which affect the allocation of work and rewards (particularly crucial in terms of shock and stress). The structures of domination also include the wider family and kinship ties of reciprocity and obligation at a more extended (and usually less intensive) level, and those between classes that are defined economically (such as employer and worker, patron and client) and between members of different ethnic groups.

Finally, the structures of domination involve, at the most extended and highest level, relations between individual citizens and the state. These are multifarious and become crucial in times of shocks and stress. They involve issues of law and order and how these are exercised – with partiality and personal discretion, with particular degrees of intensity and efficiency, with differing degrees of coercion, or sometimes with violence.

Relations at this level usually involve standards of governance and the capabilities of the civil service and the police. For example, as Chapter 8 will illustrate, the building codes and bylaws which are applied to the physical planning of a city may be called into question by an earthquake. The degree of administrative competence in disaster preparedness, as well as in disaster relief and recovery, is of great importance. Finally, the state (with its army, police force, and semi-official and sometimes encouraged vigilante groups) may be involved in civil war, systematic persecution or ethnic cleansing, in which case the resolution of what is known as a 'complex emergency' can become virtually intractable, and the state may block or divert international humanitarian assistance altogether (see Chapter 4 on famines).

Structures of domination may draw on dominant and shared ideologies, world views and beliefs for their legitimacy. Such ideologies and world views are often the 'root causes' of vulnerability and are present at the extreme left-hand side of the PAR model. This is one of many points of connection between our two models. The influence of ideology can be seen in the ways that different groups of people perceive risk. Earlier narrow, positivist comparative studies of 'the psychology of risk perception' were baffled by what it labelled 'fatalism' in the face of hazards such as drought in Nigeria (Dupree and Roder 1974). It is critically important for international and national initiatives to understand risk cultures in different contexts so as to be able to improve 'risk awareness', and we shall return to this in Chapter 9.

New thinking since 1994

Since the introduction of the Access model in the first edition of *At Risk*, there have been a number of other developments in this field and also some (mostly) constructive criticism of the model. The most important parallel innovation is the advent of the 'sustainable livelihoods (SL) approach' to development. This appeared in preliminary form in Chambers and Conway (1992) and was promoted by the UK aid ministry (Department for International Development) in Carney (1998), and by others including Drinkwater and McEwan (1994), Leach et al. (1997), Moser (1998), Scoones (1998) and Bellington (1999). This SL approach is very similar to the Access model in the first edition of *At Risk*, and earlier versions of the household Access model on which it was based (Blaikie et al. 1977; Blaikie 1985b).

A livelihood:

> comprises the capabilities, assets and activities required for a means of living: a livelihood is sustainable [when it can] cope and recover from stress and shocks, maintain or enhance its capabilities and assets, and provide sustainable livelihoods for the next generation;

and which contributes net benefits to other livelihoods at local and global levels in the long and short term.

(Chambers and Conway 1992: 7–8)

The SL approach was not developed specifically for the analysis of disasters, but more generally for a wide range of (usually agrarian) policies. None the less, it is implied that the occurrence of a disaster (or in livelihood terminology by 'shock' or 'stress') implies non-sustainability of the affected livelihoods. While the occurrence of a disaster is certainly evidence of non-sustainability, it cannot be treated as conclusive evidence. After all, disasters can be prevented or palliated, and recovery achieved, without necessarily reducing the reproduction of sustainable livelihoods.

Livelihood analysis seeks to explain how a person obtains a livelihood by drawing upon and combining five types of 'capital', which are similar to the assets that are involved in our Access model:

1   human capital (skills, knowledge, health and energy);
2   social capital (networks, groups, institutions);
3   physical capital (infrastructure, technology and equipment);
4   financial capital (savings, credit);
5   natural capital (natural resources, land, water, fauna and flora).

In some ways, the addition of the idea of 'capitals' being drawn down, built up or substituted is illuminating, and is handled in our original Access model in a different way (which the reader will be able to identify below). However, because the original Access model is so similar to the livelihood approach, as it was later developed, the authors decided not to adopt the livelihoods terminology, but to acknowledge it and its contribution separately, and to build upon it wherever it offers new insights.

It is worth emphasising that the Access model is essentially dynamic, and iterates through time to provide a precise understanding of how people are impacted by a hazard event and their trajectories through that event. It is a micro-level, disaggregated model which is shaped by (and shapes) overarching political processes at different levels (from the form of international intervention, the nature of the state, downwards). The Access model remains in this edition an economistic model, which can be as precise, deterministic and quantitative as the user wishes. The political-economic realm is acknowledged to be profoundly important but is not modelled directly, though its structural 'scaffolding' within which households take decisions has to be identified. The root causes, dynamic pressures and unsafe conditions which the PAR framework deals with are treated as qualitative inputs into the Access model and have to be specified in more detail through the dynamic operation of structures of domination and social relations.

Haghebaert (2001, cf. 2002) has made some constructive criticisms of the Access model. These are as follows: (1) the version of the model used in the first edition 'appears to be more designed to analyse general livelihood processes than to investigate specific disaster related processes' and issues of safety are not well defined; (2) non-tangible assets, such as creativity, experience and inventiveness (in short, human agency) are under-emphasised; and (3) the framework does not link up with political and socio-economic processes. Access to safety is important, and this has been strengthened in this version. Safety is partly a matter of what Cannon (2000a) has called 'social protection' and 'self-protection'. Social protection against hazards is (or should be) provided by entities that operate on levels above the household, especially the state, or community, or through collective action, while self-protection is provided by and for the household (to the extent that its assets make this possible, or its attitude makes it willing to do so; there may well also be an intra-household variation in self-protection between men and women, children and adults, and older and younger adults).

Households can to some extent 'buy' safety (e.g. strengthening their house, locating on a plot safe from rising floods, using drought-resistant seeds). Later in this chapter we refer to this self-protection also as 'individually generated safety'. Other aspects of safety – social protection – are possible only at a level above that of the household. They are a function of both non-monetary social relations (for example, mutual aid in a community, neighbourhood, or among extended kin: the equivalent of 'social capital' in the sustainable livelihood approach), and the provision of preventive measures by government and other institutions. The social protection component of safety is determined by the structures of domination: they are a function of the relationship between the members of the household as 'citizens' with 'rights', 'claims' and 'entitlements' in relation to the state (and civil society, which allows social networks to operate). The last mentioned extends to the citizens' 'right to know' (e.g. awareness of risks, warning systems) as well as enforcement of codes and standards, provision and maintenance of lifeline infrastructure, strategic, staple food reserves, etc.

The other two points made by Haghebaert (2001) are more difficult to accommodate. Firstly, our Access model is economistic, implicitly quantitative and structuralist, and, we maintain, there are considerable advantages because of this. It isolates important economic and political economic processes of normal life. It is very difficult to model, predict or find regularities in agency or inventiveness. Coping mechanisms in the face of disasters are discussed later on in this chapter, but these can usually be described in a qualitative manner only. Indeed, Part III of this book picks up this aspect and suggests ways of strengthening, rather than hampering and undermining, local ingenuity. In this sense, the Access model was never designed, nor made claims, to explain all things about all disasters to all people. In similar vein, the political is also unpredictable, although the PAR framework, plus the analysis of structures of domination and social relations all specify the importance of the political, which qualify at different points the iterations of daily life and their outcomes. The Access model in isolation does not directly incorporate political factors, but used with the PAR model, which is much less precise but more holistic, both together provide a satisfactory analytical link with political and socio-economic processes.

'Normal life' – the formal Access model

Let us now turn to the central model, called the Access model. The following explanation will 'unpack' Box 1 of Figure 3.1. We will assume that the people we are concerned with in analysing vulnerability are members of economic decision-making units. These units can sometimes be called 'households', or 'hearth-holds' (Ekejuiba 1984; Guyer and Peters 1984), that is, those who share common eating arrangements which coincide with production units. Admittedly, there are cases where it is difficult to distinguish households at all.[4] There are squatter camps where remnants and fragments of households live together under one roof. There are 'hot bedders', or very poor immigrants who sequentially share a single bed, one sleeping there at night and working the day shift, and the next slipping into the warm bed before working the following night shift. There are hostels for short-stay workers and street populations which may not have much to do with the conventional household at all – indeed, some (abused children, battered wives, orphans living with distant relatives) may be refugees *from* the household.

All these people may also be as vulnerable to hazards as those living in – and, in terms of expectations, obligations and rights – part of a household. However, such examples apart, it is usually possible to identify units that share labour and other inputs and consume meals together under one roof (or compound). We shall label these units 'households', each having a range or profile of resources and assets that represents their particular access level (the boxes numbered 2a in Figure 3.2). Also, each individual has an initial 'state of well-being', primarily defined by physical abilities to withstand shocks, prolonged periods of stress and deprivation specific to the particular disaster being addressed. At later stages in the disaster process, well-being will be affected and is likely to be shifted negatively from the initial conditions, as subsequent discussion will show. Each individual in a household has a collective claim which may be termed as access to resources.

Access to resources may include land of various qualities, livestock, tools and equipment, capital and stock, reserves of food, jewellery, as well as labour-power and specialist knowledge and skills (Figure 3.2, Box 2b). Non-material 'resources' are also essential, such as knowledge and skills, the

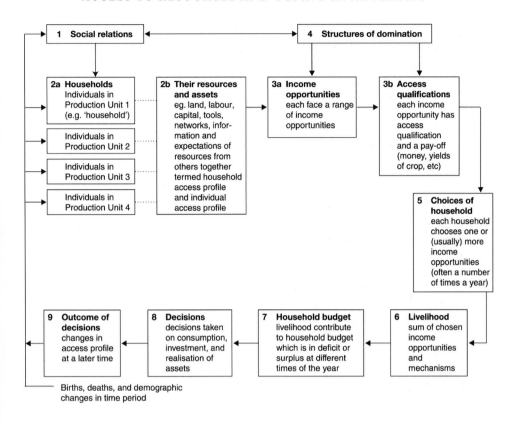

Figure 3.2 Access to resources: 'normal life'

structural position occupied in a society such as gender, or membership of a particular tribe or caste (which can either enable or exclude a person from networks of support, facilitate or prevent access to resources and their utilisation). These are personal attributes (social and human capital in terms of the SL approach) and not material resources. Access to material resources is secured through rights (e.g. property rights, rights accruing to women in marriage, as well as others sanctioned by law or custom) or sometimes criminality. Rights may change, of course, particularly after the shock of disaster, so that the physical resource may still exist, but some individuals may no longer have access to it, or others have greater access in post-disaster periods.

Each household makes choices, within constraints, to take up one or more livelihoods or *income opportunities* (Box 3a). In rural areas, most of these will be the growing of different crops, or pasturing animals, while in urban areas there will probably exist a wide range of opportunities including petty-trading, working in a factory, casual labour and domestic work.[5] Each income opportunity has a set of *access qualifications* (Box 3b). This is defined as a set of resources and social attributes (skills, membership of a particular tribe or caste, gender, age) which are required in order to take up an income opportunity (Box 3a). The notion bears a close resemblance to Sen's idea of capability and to the livelihood lexicon of the five capitals. Here, access qualification describes a more precise and specific facility for a particular income-earning opportunity.

Some income opportunities have high access qualifications such as considerable capital, rare skills or costly physical infrastructure, and therefore bar most people from taking them up. As a result, they typically provide the highest returns. Others are much less demanding (e.g. casual labouring, which requires only an able-bodied person available at the point of employment), and these are usually over-subscribed and poorly paid. Each income opportunity has a payoff in terms of physical product, money or other services, and eventually in health and well-being (as well as affecting the initial or baseline state of well-being before the disaster).

In rural areas, payoffs are often determined by crop or animal yields multiplied by market price. Both yield and price may be particularly prone to large and adverse fluctuations. The mechanisms which set payoffs (the behaviour of markets in particular) for income opportunities of different households or groups or households (e.g. for agricultural food producers, labourers, artisans, fishermen, unskilled industrial workers, and so on) are of crucial importance. The labour market for casual, part-time and unskilled workers in urban areas also shows fluctuations, as do the conditions that determine the profitability of 'informal' activities such as street vending (where harassment and bribes by the police can be as unpredictable as the weather).

Access to all the resources that each individual or household possesses can collectively be called its *access profile* (see Box 2b). This is the level of access to resources and therefore to income opportunities, with some having a much wider choice of options than others (Box 5). Those who possess access qualifications for a large number of income opportunities have a wide choice and can choose those with high payoffs or low risks. Their flexibility also allows them choice in securing a livelihood under generally adverse conditions, to command considerable resources and have reserves of food. Such a household can be said to have a good-resource profile. On the other hand, those whose access profiles are limited usually have little choice in income opportunities, and have to seek the most over-subscribed and lowest paying options, and subsequently have the least flexibility during adverse conditions. Those with a limited access profile often have to combine a number of income opportunities at different times of year as some may provide a livelihood for only part of the year, be only seasonally available or be unreliable because other people are competing for a limited number of employment opportunities.

Each individual or household therefore makes choices, typically during key decision-making times in the agricultural calendar, or more irregularly under urban situations. The resulting bundle of income opportunities (both in kind and in cash), together with the satisfaction of such needs as water and shelter, can be said to constitute a *'livelihood'* (Box 6), which is the sum of the payoffs of the household's constituent income opportunities.

Some households structure their income opportunities in such a way as to avert the risk of threatening events such as drought, flood, loss of employment, failure of food crops or serious illness. They also employ survival strategies and coping mechanisms once a threatening event has occurred, although this usually involves an element of physical or institutional preparation. Grain must be stored and cattle numbers increased in good years to protect the reproductive capacity of the herd in bad years. A network of obligations and rights are also built up in the form of institutions (called social capital) that deal with these events and aim to prevent them from becoming disasters. These become crucial in the transition to disaster and are described in greater detail below.

The flows of income then enter the household as a range of goods and cash: wages, grain, remittances from absent household members, profits from commerce or business, and so on. A *household budget* can be constructed in which expenditures and income are listed, and the account accumulates, is in equilibrium or runs into deficit (Box 7). On this basis, decisions are made about how to cope with deficits, save or invest any surpluses, and what forms of consumption should occur, including 'one-off' arrangements for marrying adolescent offspring, festivals, investing in social capital, having babies, migrating (see Box 8, 'decisions'). If in surplus, the household may decide to invest and improve its access to resources in the future. If the account is in deficit, consumption will have to be reduced, assets disposed of, or the household will have to postpone equity and possibly increase the deficit in the long term by arranging a consumption loan (which may be inadequate in the short and/or longer term). A more detailed representation of the household budget is given in the case of famine in Chapter 4. The outcome of these decisions will result in a change in the access profile of each household in the next period (Box 9). These will in aggregate alter the flows of surplus between groups and households and may alter the social relations between groups (Box 1), so that in the next round the households are in a different set of relations to each other and larger scale structures, and enter Box 2b with different access profiles.

Households and access in a political economy

The outline of the 'household model' above may seem to some an overly mechanistic and economistic treatment of access to resources. We need to include further discussion of 'the rules of the game of the political economy' or social transactions, and specifically of rights and social expectations which may give people access to resources. The dominant relations of production and flows of surpluses provide the main explanation of access to resources. Changes in the political economy at the level of 'root causes' in the PAR model are slow moving but can, as a result of revolution or major realignment in the balance of class forces, lead to fundamental shifts in access to resources and the character of disasters.

What is of more immediate and specific importance are the structures of domination and social relations at the local level (Boxes 1 and 4 in Figure 3.2), or the 'rules of the political economic game' for the microcosm of households selected in this model. Between individuals within a household, these involve the allocation of food, who eats first, who will have to absorb consumption cut-backs in times of dearth or who receives medical treatment. Gender politics within the household are of great importance here, and show how inadequate it is to treat the household as a homogeneous unit. As Rivers (1982) and Cutler (1984, 1985) amongst others have pointed out, women and children sometimes bear the brunt of disasters because of the power of male members of the household to allocate food while in refugee camps.[6]

Among family and kin, an important aspect of resource allocation is embodied in a range of expectations and obligations involving shelter, gifts, loans and employment. Often these linkages reflect and reproduce the structures of domination of households and society in general. Between classes and groups, transactions include patron–client relations, taboos, untouchability, gender division of labour outside the home, sharecropper–landlord relations, and rules about property and theft, amongst many others.[7]

Many of these transactions, we shall see, form an important basis for mutual help or individual survival in times of crisis, and therefore can be looked upon as additional elements in an individual's or household's access profile. However, the rules governing these transactions change (often very quickly) in the face of social upheaval such as war, famine or pandemic. Usually, this means a reduction in obligations and therefore in 'income opportunities' for the receivers of goods and services, and an increase in disposable income for those who forgo these obligations. In a few circumstances, new opportunities can open up. For example, upon occasions of extreme famishment, theft may be sanctioned, grain stores may be broken into and obligatory redistribution mechanisms from rich to poor set in motion.

Markets are another set of social transactions that allocate resources on the basis of price. Their behaviour is crucial in the relative worth of people's resources, and in governing their household budgets. There is a good deal of research into the behaviour of markets, particularly preceding and during famines. The prices of essential goods and services often rise after sudden disasters when the immediately available food, shelter, clothing and medical supplies are destroyed and transport to bring in replacement supplies is disrupted. The behaviour of traders in essential commodities is crucial, as Chapter 4 will show. These rapid changes in the rules of resource allocation brings us to the issue of the time dimension in disasters.[8]

Access to common property resources (CPRs) is also of great importance to household livelihood and vulnerability. At various times, in various broader social and economic settings, a wide range of physical resources may have been excluded from private or state ownership and now exist as common pool resources. These resources might include trees, pasture, ground or surface water, wildlife, marine resources, famine foods and arable land, depending on the region and its history. In some places a proportion of these may be set aside for common management and use by a group larger than the household. Rules governing access to CPRs are highly localised and complex (Jodha 1991) and will be observed in many situations described in Part II of this book. Less is known about the rules governing the use of the urban equivalent of rural CPRs; however, it is clear that scavengers, recyclers and the poor who actually live at or near urban solid waste facilities in many countries follow complex rules of access.

Research on famine has led to the development of other concepts related to the idea of access. Most notable is Sen (1981), whose concept of 'entitlements' in relation to food and hunger has an affinity with the notion of access. This involves the set of resources or livelihood opportunities that may be used to produce food or procure it through various forms of exchange. His formulation is similar to the concept of access in many ways, though it is more specific; his ideas are discussed more fully in Chapter 4.

Transitions from 'normal life' to disaster

The outline structure of the Access model, and the details of 'normal life', are summarised in Figure 3.3 using a sample of households pursuing their livelihoods in conditions of unequal safety and livelihood opportunities. It remains to fill out the detail of the model in order to trace the transition of 'normal life' to disaster. Figure 3.3 repeats the structure of the model in Figure 3.1, but specifies the processes and events *in more detail* which appear only in summary form in Figures 3.1 and 3.2.

Time–space positioning of hazards

Turning attention to the hazard itself, Figure 3.3 depicts some key characteristics of the hazard, or multiple hazards, that threaten a particular area. A hazard has a time–space geography, involving the probabilities of events of significant magnitude to cause potential damage of differing magnitudes over geographical space. In Box 4a, a tide table shows the dates, timings and height above data for high tides and Spring tides, the probabilities of

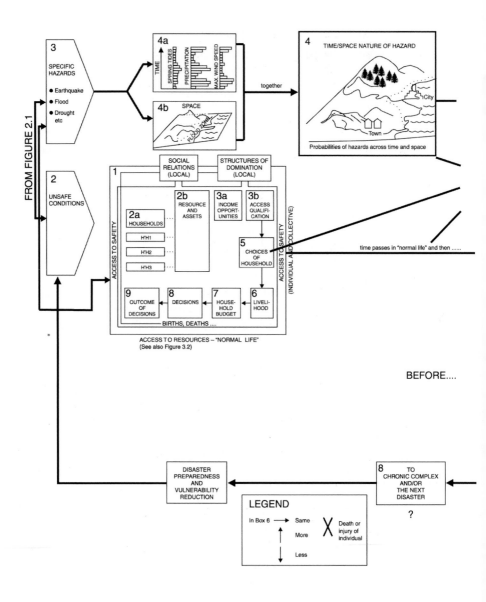

Figure 3.3 The Access model in the transition to disaster

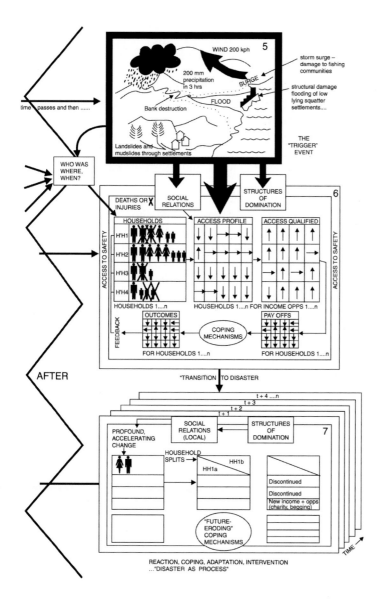

Figure 3.3 continued

tropical cyclones and associated rainfall intensities. Each of these hazards also incorporates a spatial dimension. In the hypothetical case of a tropical cyclone (see real cases discussed in Chapter 7), a number of linked threats are implied:

- fresh-water flooding from inundation of low-lying areas;
- fresh-water flooding following the breaking of river banks;
- sea surges and damage to coastal settlements, fishing boats and equipment;
- mudslides and landslides following exceptionally heavy rainfall;
- high winds causing structural damage to buildings and property, and injury to humans and livestock;
- salinisation of fresh-water supplies for humans and livestock.

These are shown in simplified pictorial form, and constitute part of the basis of conventional hazard assessment.

Below the details of the hazard threat in Figure 3.3, 'normal life' and in-built vulnerabilities continue (already described and shown in Box 1). As Figure 3.3 suggests ('time passes and then...' between Boxes 4 and 5), a hazard event occurs (dramatised in pictorial form in Box 5) and the various potential threats outlined in the list above materialise. The immediate impact can be explained largely by asking the question 'Who was where, when?' At the moment of impact of the event, people were to be found in co-ordinates of time–space as might be expected from their pursuit of 'normal life'. If the event occurred suddenly at night, most people would have been in bed and would have been slower to take avoidance action than during daylight hours. If the storm occurred during a season when fish stocks are to be found offshore, rather than inshore, then fishers will have been exposed to even more serious threat, as it would have been harder to find shelter inshore before the worst of the storm. If wood cutters and charcoal burners choose an income opportunity that takes them to mangrove swamps during a season when there is a high probability of coastal storms, then they occupy, as part of their livelihood strategy, a particular time–space co-ordinate that carries a higher risk.

A similar urban situation was seen on a night in 2000 when heavy rain brought a mountain of solid waste sliding down on the houses of rag pickers and scavengers who lived at the Payatas landfill on the north-east edge of Manila. Their livelihoods had a specific space–time structure that put these people in harm's way (see Chapter 6).

Time and disasters

Leaving aside the time–space element and focusing on the time dimension alone, time is 'of the essence' in an understanding of disasters. So far, time has only been treated in the sense that the framework of 'normal life' permits the succession of events in a process to be analysed, allowing for people's decision-making actions (such as the timing of activities like planting crops, selling assets, migrating, and so on). The importance of time in understanding disasters lies in the frequency of the event, when the disaster occurs (time of day, season) and in the stages of the impact of the disaster after the hazard has occurred.

It may be said that disasters do not happen, they unfold. Our characterisation here chooses a hazard event that has a rapid onset (as do others, for example, *tsunami*, bush or forest fire, earthquake or some floods, and that accurately could be labelled a 'trigger'). However, in the case of 'slow-maturing' or slow-onset disasters such as famine, the even slower HIV-AIDS pandemic or climate change, processes which can unfold over a period as much as 30–80 years or more, this dramatic, time-dependent characterisation is less inappropriate. However, even in sudden-onset cases, the pre-conditions for disasters ('root causes' and 'dynamic pressures' in terms of our PAR model) may have been forming over a long period. Indeed, Oliver-Smith (1994) treats the Peruvian earthquake of 1970 as having 'root causes' that reach back 500 years to the Spanish conquest of the Inca Empire and the ensuing decay of Inca methods of coping with environmental risk. Our treatment of the 1985 earthquake in Mexico City in Chapter 8 follows a similar historical, time-based narrative.

It is therefore important to give a 'temporal frame' to our Access model and transition to disaster, so that the impact of its timing can be understood. The timing (and spatial expression) of hazards has therefore to be superimposed upon the 'temporal frame' of people earning their livelihoods and living out their daily lives. Thus, as we have said, for the shortest time frame, the time of day or night of the onset of a sudden hazard can be important. Ninety per cent of all people killed in earthquakes while occupying buildings die at night. The day of the week (particularly market days, rest, festival or holy days) is also relevant in terms of possible concentrations of people in space and time.

Seasonality is one of the most important rural time factors. Chambers (1983; Chambers et al. 1981) has highlighted the impact of seasonality on health, nutrition and people's capacity for hard work in the 'normal' annual cycle. The coincidence of a sudden hazard with the 'hungry' season (usually the wet season before crops have matured and are ready for consumption) when labour demands are highest, food reserves lowest and some major diseases most prevalent can produce a much more severe disaster impact. The build-up of famines may have a seasonal element, in that crop failures (or a number of successive failures) are sometimes involved. Food prices as well as wage rates for agricultural work have important seasonal dimensions that other factors can exacerbate and combine to precipitate famine (see Chapter 4).

The stages of the impact of a disaster after the hazard strikes are fundamental. The various elements in the vulnerability framework (class relations; household access profiles; income opportunities; household budget; and structures of domination and resource allocation) each iterate at a different speed. Table 3.1 summarises typical time periods of change and gives some examples.

There is a fundamental difference in time between sudden disasters and slow-developing disasters such as famine or pandemics (in which the most acute distress may extend over a period of months and years). In terms of their mortality and damage to homes and livelihoods, the onset of some

Table 3.1 Time periods for components of the Access model

Component of the access framework	Typical time period of change after disaster	Examples
Class relations	Months or years	Nicaragua (1972)
		Portugal (1755)
Change in political regime		Ethiopia (1974)
Household access profile	Sudden, immediate impact	Loss of life and house
	Weeks	Sale of livestock, jewellery
	Weeks or months	Other assets sold
Income opportunities	Sudden if urban employment is disrupted	Rural employment collapses due to drought, flood
	Usually over months	Taboo foods accepted
Household budget	Immediate impact in sudden-onset hazards	Cuts in consumption; reallocation by age, gender
	Months	Food price rises and famine
Structures of domination	Immediate impact in sudden disasters	Sharecroppers refuse to give up landlords' share
	Months or years, with episodic food shortages and high prices	Famine

sudden disasters can be measured in seconds or minutes in terms of earthquakes, in hours or a few days for floods. Affected populations may rearrange their accepted pattern of responsibilities and rights and combine into completely unfamiliar groupings, when strangers, refugees, those temporarily taking shelter, the traumatised and longer-term displaced turn up in a new social environment after a trigger hazard event. On the other hand, slow-onset disasters require careful analysis of social adaptation, the emergence of new rules of inclusion and exclusion regarding networks of support and changing access qualifications for new and existing income opportunities.

Post-event transition to disaster

Returning to the time- and space-bound narrative of a disaster outlined in Figure 3.3, the hazard event has other quick acting and profound impacts on the practice of 'normal life'. Box 6 is labelled 'transition to disaster' and identifies the trajectory to 'abnormal' life. The households in this diagram are represented by individual men, women and children, and some are shown to have suffered injury, become disabled or died (crossed out, in pictorial terms). Household access profiles are suddenly and profoundly altered. Paid employment may cease, and with it access to cash with which to purchase food, medical care, repair shelters or productive equipment such as ploughs, acquire livestock for ploughing and fishing equipment. Land may be inundated; brick factories remain under water; regular customers for haircuts, sweets and petty services in cities and small towns have their minds on other matters and have few funds available to spend on non-essential services. The access qualifications for many income opportunities are raised, sometimes to infinity – they simply are not viable choices any more. Thus the income opportunities chosen by different households in the period immediately after the event are usually drastically reduced for most (although for a few, such as merchants and the wholesalers of essential supplies, they are increased greatly).

The outcomes of these straitened circumstances, for the majority, are not (yet) disastrous. However, they feed back in an iterative manner to subsequent decisions and the asset profiles and access qualifications on which they are based (shown by the circular arrows, labelled 'feedback' in Box 6). There are still a wide range of adaptive strategies, both individual and community-based, coping mechanisms and outside interventions that can avert or palliate the transition to disaster for some or most people.

There are other important aspects of post-event crisis which are not reached by the Access model and which must be mentioned here. Crisis ill-being is increasingly being recognised as a common experience in the wake of disasters, and a concept of vulnerability must incorporate notions of ill-being. Definitions of vulnerability usually include a potential for ill-being (often expressed as an objectively assessed statistical probability, as in this model) multiplied by the magnitude of the combined impacts of a particular trigger event. Thus, the conversion of risk into ill-being is turned into a common metric, which enables different hazards to be compared (Rosa 1998), but brushes out the cultural, psychosomatic and subjective constructions of disaster impact, and little work has yet been done on how these experiences map into post-crisis situations of chronic ill-being.

The disaster event itself alters capabilities and preferences both in the short term (e.g. grieving, trauma, acute deprivation) and also in the longer term, since the aftermath of a disaster sees a reappraisal of previous individual and collective commitments, the strength and nature of trust, and the intensity and diversity of social networks including rules of membership. Thus extreme events are frequently written into the history of social relations and well-being. It is important to complement the economistic and quantitative aspects of our Access model with an understanding of the ways in which the disaster event was experienced by different people, and how it altered their sense of well-being and their strategies to reconstitute that well-being in a new, post-disaster world.

Access, transition and safety

As the discussion in Chapters 1 and 2 demonstrated, vulnerability is a measure of a person's or group's exposure to the effects of a natural hazard, including the degree to which they can recover from the impact of that event. Thus, it is only possible to develop a quantitative measure of vulnerability in terms of a probability that a hazard of particular intensity, frequency and duration will occur. These variable characteristics of the hazard will affect the degree of loss within a household or group, in relation to their level of vulnerability to various specific hazards of differing intensities.

Thus, vulnerability is a hypothetical and predictive term, which can only be 'proved' by observing the impact of the event when, and if, it occurs. By constructing the household Access model for the people affected we can understand the causes and symptoms of vulnerability. This requires analysing the political-economic structures that produce the households' access profile, income opportunities and payoffs (these structures are labelled 'social relations' and 'structures of domination' in the framework). This implies that the question 'Vulnerable to what?' is answerable only in the context of an actual and specific hazard. This raises an important point. Different people will be vulnerable in differing degrees to different hazards, although there may well be households which, if they are vulnerable to one type of hazard, are likely to be vulnerable to others too. Typically, such people will have a poor access profile with little choice and flexibility in times of post-disaster stress.

None the less, in the following chapters, it is necessary to specify *to what hazards* people are vulnerable. Figure 3.3 identified a tropical storm, but the content of a disaster narrative as it unfolds will differ in many important ways even in the same location and with the same political economy with each different hazard. In other words, the conventional aspects of natural hazards (and the ways in which they are treated and monitored) are still essential, particularly the time–space and other technical issues of the hazard in question. In the case of earthquakes, for example, the indicators of vulnerability ('unsafe conditions') will concern housing materials, building standards, skills regarding aspects of building safety, income level, available spare time and the ability to keep habitations in good repair, type of tenure (owner-occupier or rented accommodation in urban areas), location of dwelling relative to zones of seismic activity, ground stability and degree of support networks which could be mobilised after the event (see Chapter 8). Alternatively, in the case of drought, the set of indicators of vulnerability will be quite different, and will concern food entitlement profiles, physical access to markets and market behaviour, and the prospects of earning enough money to buy food or the possibility of exchanging other goods for food. The time–space patterns of households in their 'normal life' will be as important as in the earthquake case, but will be related to the spatial structure of markets and to crop or pastoral production.

Much of the discussion so far has focused on the vulnerability of households earning a living under normal and transitional conditions, and some passing mention has been made of access to safety (or 'social protection' in Figure 3.1) depicted in Figure 3.3 as the protective barrier surrounding individual households' daily lives (Boxes 1 and 6). This barrier – when it is properly in place – may be considered as access to safe conditions that apply to all households collectively.

Many of the conditions of vulnerability are shaped by individual households. Their decisions about income opportunities (or ignorance of risk) may place them in dangerous time–space co-ordinates, facilitating further choice, building reserves, etc., as we have described. These may be considered as individually generated access to safety (self-protection), or as its reciprocal, individually generated vulnerability. However, there are other elements of collectively generated access to safety, and these link the individual household to its surrounding social relations, structures of domination and so on to the generation of unsafe conditions from the PAR framework, which refer to the provision of resources from the community or the state. Some of the large-scale, more generalised state provisions have been discussed in the progression of vulnerability of the PAR model (usually, a lack of appropriate training, relevant skills, press freedom, good governance), leading to unsafe conditions (unprotected buildings and infrastructure, suppression of information which could lead to relief measures). However, the focus on the specific and the micro-level in the Access model requires that access to safety is treated in a more detailed and dynamic way in the 'transition to disaster'.

A large part of this access to safety can be understood in terms of self-generated access, known as coping mechanisms, which are discussed below. Others, which are usually provided by outside institutions, are discussed in a subsequent section.

Coping and access to safety

The Access model provides a dynamic framework of socio-economic change, in which people of different identities (gender, age, seniority, class, caste, ethnic group) avail themselves of the means of securing their livelihoods and maintaining their expectations in life. The model implicitly, rather than explicitly, allows for people to develop strategies to try to achieve these ends. In this sense, the economic and social means to secure their livelihoods are not 'handed down' to them in an economistic and deterministic manner. People must not be assumed to be passive recipients of a profile of opportunities, hedged about by constraints of the political economy of which they are a part.

On the contrary, the pattern of access in any society is subject to (and the result of) agency, decision making under externally created constraints, struggles over resources and also co-operation. The pattern of access is the outcome of those decisions, co-operation and struggles by people of different gender, age, class, and so on. They are a part of daily life and are pursued with ingenuity and resourcefulness. In adverse or disastrous times people are stimulated by circumstances of threat, desperation and loss. As Rahmato (1988) put it, the measures which rural Ethiopian people have taken to enable them to live through the privations of the past two decades indicate ingenuity, strength of character, an effective use of natural resources and communalism.

It has been said that official perceptions of 'disaster victims' usually underestimate their resources and resourcefulness (Chapter 9). Perhaps one of the reasons for this is that indicators of vulnerability based on the measurement of resources are the more easily recognised by outside institutions. They are also more enduring and part of the observable socio-economic structure, while people's struggles and strategies to cope with adverse circumstances, particularly acute ones, are more ephemeral and change quickly (Corbett 1988). Therefore, they remain unnoticed and understudied. It is the purpose of this section to focus on these strategies. Without a proper understanding of them, policy makers are more likely to make stereotyped responses in both preventive measures of vulnerability reduction and relief work. Further, misdirected relief efforts may undermine rather than assist affected people in their attempts to help themselves towards recovery.

Coping defined

Coping is the manner in which people act within the limits of existing resources and range of expectations to achieve various ends. In general this involves no more than 'managing resources', but usually it means how it is done in unusual, abnormal and adverse situations. Thus coping can include defence mechanisms, active ways of solving problems and methods for handling stress (Murphy and Moriarty 1976). 'Resources' in this book have been defined as the physical and social means of gaining a livelihood and access to safety. Resources include labour power, or as Chambers (1989: 4) aptly puts it, able-bodiedness, or the ability to use labour power effectively. The more that poor people rely on physical work, the higher the potential costs of physical disability and ill health (see Chapter 5).

Resources also include land, tools, seed for crops, livestock, draught animals, cash, jewellery, other items of value which can be sold, storable food stocks as well as skills. In order for tangible resources to be mobilised, people must be entitled to command them, and this may be achieved in many ways. These include using the market, the exercise of rights, calling upon obligations (from other household members, kin, patrons, friends, from the general public by appeals to moral duty, as in alms-giving), through theft or even violence.

In many cases specialised knowledge is required for certain resources, for instance in finding wild foods or using timber for rebuilding, knowing the moisture capacity of certain soils, the likelihood of finding wage labour in a distant city or plantation, or finding water sources. This knowledge is similar to that which supports 'normal' rural or urban life, and is passed from generation to generation. However the 'ethnoscience' essential for some coping behaviour can disappear with disuse or be rendered useless by rapid change (O'Keefe and Wisner 1975).[9] We return to this point below.

Often it is assumed that the objective of coping strategies is survival in the face of adverse events. While this is indeed common, it masks other important purposes. These may be examined using Maslow's (1970) hierarchy of human needs. Such a hierarchy involves identifying distinct levels of needs, with each level incorporating and depending on the satisfaction of needs below them in the hierarchy. The need for self-realisation, involving the giving and receiving of love, affection and respect might be said to be the highest in the hierarchy. A lower one, on which the former is founded, may be an acceptable standard of living. Lower ones still may include adequate shelter and food for healthy survival, whilst other needs near the bottom of the hierarchy will include minimum security from violence and starvation. Reviewing 20 years of work since Maslow, Doyal and Gough (1991) conclude that a 'core' of basic human needs can be identified, and that failure to satisfy these means that other needs cannot be met (see also Wisner 1988a).

In adverse circumstances, a retreat to the defence of needs that are lower in the hierarchy implies the temporary denial of those needs higher up. For example, the experience of extreme poverty can cause a loss of self-respect and self-regard (de Waal 1989a). However, it is important not to oversimplify and over-generalise the expectations and priorities in the lives of vulnerable people or those affected by a disaster. Oliver-Smith (1986b) has described very complex motives and ideals among survivors of a dire earthquake tragedy. Scott (1990: 7) reminds us of that 'slights to human dignity' can fester and emerge in surprising demonstrations of 'resistance' against authority. This is certainly relevant for disaster relief and recovery (Chapter 9). Jodha (1991) surveyed people's own criteria for well-being (in this case a list of no fewer than 38) in Gujarat, which attest to a complex set of priorities. Raphael (1986) analysed the psychological trauma of disasters and the adjustments made to loss, grief and the impacts of dislocation (see above for our brief discussion of ill-being and non-measurable aspects of the impact of disasters). Coping in the face of adverse circumstances may, therefore, be seen as a series of adaptive strategies to preserve needs as high up the hierarchy as possible in the face of threat.

However it is common that what may be broadly termed 'disasters' force a retreat down the hierarchy. For example, it may become necessary to engage in demeaning activities (and therefore to lose respect) in order to secure a minimum food supply. Certain activities may be proscribed or discouraged by membership of a social group, caste or by gender. During the drought of 1971–1973, members of the Reddy caste in Medak District, India were reduced to selling vegetables to earn a living, an occupation that was considered below their dignity (Rao 1974); while women not of the shoemaker caste were found making shoes during the 1966–1967 Bihar drought (Singh 1975, quoted in Agarwal 1990). Despite the mutual economic and emotional support that it provides, families may break up to allow its individual members to survive. The survival of the individual in the short term may be the only attainable need and objective of coping.

Famine may be unique or at least extreme among disasters in often provoking social tension and breakdown of this kind. For many years, Quarantelli and his sociology colleagues have studied community responses to disasters such as earthquakes and floods. They find that emergent organisation is much more common than social chaos, and that altruism and stoicism are more common than selfishness and panic (Quarantelli and Dynes 1972, 1977; Quarantelli 1978, 1984; Dynes et al. 1987).

Types of coping strategies

Crisis events occur from time to time in people's lives, as well as in the lives of whole communities and societies, in which case they are often called disasters. Such events call for the mobilisation of resources at various levels to cope with their impact. When people know an event may occur in the future because it has happened in the past, they often set up ways of coping with it (Douglas 1985). We return to these coping strategies as a reference point for building the policies we recommend in Chapter 9, especially in our discussion of the Fifth and Sixth Risk Reduction Objectives.

Such coping strategies depend on the assumption that the event itself will follow a familiar pattern, and that people's earlier actions will be a reasonable guide for similar events. Most disasters have such precedents, particularly in hazardous physical and social environments. However, some hazards have such a long return period that the precedents are imperfectly registered. There are also others which are unprecedented, such as the HIV-AIDS pandemic, and which therefore have no familiar pattern. If this is the case then coping strategies may not apply, and the decision framework (consisting of the social, economic and natural environments) will not be relevant.

The assumptions on which people make their decisions therefore rest on the knowledge that, sooner or later, a particular risk will occur of which people have some experience of how to cope. On the other hand, people do not like conditions of uncertainty where there are no known and familiar ways (such as explicit systems of rights and obligations, providing safety nets and support groups) of coping with a particular event. Thus the unprecedented or unknown event creates a situation of uncertainty. The HIV-AIDS pandemic in certain areas of Africa, or calamities of exceptional severity (for example, what is known in Bengali as *mananthor*, or 'epoch-ending' famines), are cases in point.

Almost all coping strategies for adverse events which *are* perceived to have precedents consist of actions before, during and after the event. Each type of coping strategy is discussed and illustrated below.

Preventive strategies

These are attempts to avoid the disaster happening at all, which are called preventive action elsewhere in this book. Many require successful political mobilisation at the level of the state. This is often easier in the immediate aftermath of a disaster, when public awareness is high and the political payoff for government action is significant. But preventive action at the individual and small group level is also important. It may involve avoiding dangerous time–spaces (such as fishing offshore in small open craft during the storm season), avoiding concentrations of disease vectors (e.g. malaria mosquito, tsetse fly) that have variability by season and/or altitude and choosing residence locations that are less exposed to wind, flood or mass movement of the earth.

Impact-minimising strategies

These are referred to elsewhere as 'mitigation', especially where they are the object of government policy. These strategies seek to minimise loss and facilitate recovery. The range of these strategies is enormous and varies significantly between people with different patterns of access. However, two generalisations may be made. Firstly, the objective of many strategies is to secure needs quite low down the hierarchy, particularly if the risk is perceived to be damaging and probable. It may be preferable to improve access to a minimum level of food, shelter and physical security rather than increase income. This further underscores the important distinction between poverty and vulnerability made earlier.

Secondly, maintaining command of these needs in a socially and/or environmentally risky situation usually implies diversification of access to resources. Under the terms of the Access model, this involves broadening the access profile and seeking new income opportunities. This can be attempted in agricultural and pastoral production by setting up non-agricultural income sources and by strengthening or multiplying social support networks.

Building up stores of food and saleable assets

For those rural people who have access to land, a store of grain or other staple food is a most important buffer against expected seasonal shortages, as well as more prolonged periods of hardship. An accumulation of small stock and chickens is another defensive strategy (Watts 1983a). Pastoralists may follow a strategy of increasing their herd size in years of good rains and grass availability (when calf birth rates rise and mortality falls), in order to maintain the herd size in the inevitable bad years with high mortality (Dahl and Hjort 1976; Thébaud 1988; Odegi-Awuondo 1990).

Diversifying production

Farming people are often regarded as being risk-averse (in the sense of avoiding chances in cultivation that may bring higher rewards but with greater exposure to dangers).[10] Usually their production involves mixed cropping, intercropping, the cultivation of non-staple root crops and use of kitchen gardens. This strategy often results in a 'normal surplus' in good years since it is planned on the basis of meeting subsistence needs even in bad (but not the worst conceivable) years (Allan 1965; Wisner 1978b; Porter 1979).

Planting a greater variety of crops has many advantages apart from providing the best chance of an optimum yield under all variations of weather, plant diseases and pest attack. It represents one of the most important precautionary strategies for coping with food shortages (Klee 1980; Altieri 1987; Wilken 1988). Diversification strategies often make use of environmental variations, including farming at different altitudes, on different soils or utilising diverse ecosystems on the slopes of mountains.

Diversifying income sources

The entirely self-provisioning rural household is an ideal type, and is very rare in the world today. Even the most isolated people in the Amazon rainforest, the Andes or the Himalayas engage in production for sale. In addition, the remittance of income from wage earners who have moved to distant cities, mining camps or plantations is very important to rural livelihoods in many parts of the world. This is sometimes graphically demonstrated by the economic disruption and hardship caused when crises interrupt such systems, as with the hundreds of thousands of guest workers from Egypt, Bangladesh, Nepal and the Philippines who were obliged to leave Iraq in 1991 as a result of war.

Non-farm income becomes even more important following disasters that temporarily disrupt farm and livestock production. Crafts, extractive enterprises such as charcoal making, honey and gum arabic collection have often been noted in studies of drought coping in Africa. Brewing beer is also an important source of income, especially for women, and drought reduction of the grain ingredients can affect their income and nutrition (Kerner and Cook 1991; Murray 1981; Mbithi and Wisner 1973). For urban dwellers a series of 'sidelines', sometimes illegal or quasi-legal (such as hawking on the streets without a licence, waste recycling, pilfering, looting ruined and abandoned shops and buildings), may become a temporary mainstay of post-disaster life. Both production and income diversification strategies can be effective as coping mechanisms in the short run, while they undermine the basis of livelihood in the long run. Cannon (1991) discusses de-vegetation of the landscape in order to provide fodder for livestock in a drought. Charcoal production as an income source is another example. Both can lead to long-term erosion and desertification (A. Grainger 1990; O'Brien and Gruenbaum 1991).

Development of social support networks

These include a wide variety of rights and obligations between members of the same household (e.g. wives and husbands, children and parents), with the extended family and with other wider groups with a shared identity such as clan, tribe and caste. Parents may try to make a strategic choice of marriage partner for their daughter or son into a comparatively wealthy family. This may increase their ability to call on resources in difficult times (Caldwell et al. 1986).

117

Within the household and family, successfully securing resources in potentially disastrous times depends upon the implicit bargaining strength of its members and of their 'fall-back' position (Agarwal 1990: 343), or 'break-down' position as Sen (1988, 1990) terms it, if co-operation in this bargaining process should fail. Women tend to lose these conflicts for scarce resources and are affected by who eats first, the share of available food and the lack of access to cash earned by other family members (e.g. cash from casual male labour). The range of resources controlled by women, and the employment opportunities open to them, tends to be more limited.[11] The disintegration of the family and abandonment of women, children and old people is the expression of the breakdown of such obligations.[12]

There are other forms of support based mainly on non-economic relations. Some writers term these the 'moral economy' (Scott 1976); examples are between patrons and clients, or between rich and poor in times of hardship. These offer a minimum subsistence and a margin of security and constitute what Scott calls 'a subsistence ethic', based on the norm of reciprocity – but at a price of the reproduction and even the deepening of inequality. Examples are legion, but it is widely reported that such obligations are being eroded.[13]

On the other hand, Caldwell et al. (1986: 667) state that, at least for the elderly in a period of extreme food scarcity in south India, 'the support system still worked well'. But it can also be argued that the continued existence of such support systems is responsible for the retention of people in the countryside and for discouraging them from abandoning such local systems. In other cases these obligations of the wealthy are still upheld. For example, a case study in Nepal found that the wealthy were prevailed on not to reduce daily wages for agricultural work, nor to sell grain outside the village at a profit, and to secure a loan from shopkeepers in the nearby bazaar for re-lending to the most needy villagers (Prindle 1979). In another village that is tribal (the village in the previous example being multi-caste), there was an expectation of gifts in times of hardship combined with a powerful ethic of equality, with surpluses being shared. Although reported to be in a state of subtle change, this system was still largely operational.

There are also wider obligations for the whole community to assist and provide for those facing acute adversity. These include alms, for example the giving of a grain tithe in some Muslim societies (Longhurst 1986: 30). *Meskel* is a form of community redistribution in parts of Ethiopia, where credit is given to the needy to celebrate the festival of this name, thereby enabling them to acquire food. Neighbourly assistance, such as rescuing trapped individuals from collapsed buildings and rendering medical assistance, are other examples. These are 'claims' as Swift (1989) calls them, alongside the other two broad categories of assets ('investments', both human and productive, and 'stores' of food, money and 'stores of real value' such as jewellery).

It is probable that throughout the LDCs such networks and moral obligations are in decline. In some areas more exploitative systems are superseding the 'moral economy' of the peasant, where a low-level safety net against starvation and complete destitution operated in some form, albeit a highly exploitative one. Instead, even these safety networks tend to break down. This may involve the provision of food on credit at usurious rates of interest, which exacerbate the 'ratchet effect' and increase the vulnerability of deprived groups in the longer term (Chambers 1983). Unfortunately, given the demise of traditional systems, there is rarely any growth of state-run social security alternatives.

Post-event coping strategies

When there is a potential food shortage and possible famine, the period during which stress develops can be long, allowing for a succession of strategies. A number of studies have found similar sequences.[14] It is clear that a sequence of adaptations in consumption patterns is made very early when shortfalls in food are anticipated. These include the substitution of lower quality and wild foods (or 'famine foods') for more expensive staples. Here the significance of common property resources for allowing access to these is important.[15] Wild foods also feature as famine foods in almost all parts of Africa (de Waal 1989b; McGlothlen et al.1986).

The next step involves calling on resources from others (usually family and kin) that can be obtained without threatening future security. This usually involves reciprocal social interactions, and avoids usurious rates of interest, therefore preserving the longer-term access position of the individual or household. At the same stage, sources of household income other than the dominant one may be tapped, such as wage labour, petty commodity production or artisanal work. Sale of easily disposable items that do not undermine future productive capacity (e.g. small stock) may also take place. As the food crisis deepens, loans from money lenders and sale of important assets such as oxen for ploughing, agricultural implements and livestock may have to be arranged. Finally, when all the preceding strategies have failed to maintain minimum food levels, migration of the whole household to roadsides, towns and possible sources of food often ensues.

Coping and vulnerability analysis

Coping strategies are often complex and involve a number of sequenced mechanisms for obtaining resources in times of adversity and disaster. They grow out of a recognition of the risk of an event occurring and of established patterns of response. They seek not just survival, but also the maintenance of other human needs such as the receiving of respect, dignity and the maintenance of family, household and community cohesion.

119

Outsiders are often surprised by strategies that do not seem to try to maintain an adequate food intake for a household (or perhaps different amounts for various members), but which instead are aimed at preserving the means for continuing the household's livelihood after the difficult period has passed.[16] Many of these strategies have been highly resilient to social and economic changes and are reported to be functioning still throughout the world.

Throughout this book we try to signal the ways in which the 'people's science' or indigenous technical knowledge that provides the basis for much coping behaviour, and patterns of coping themselves, interact with 'official' attempts at disaster prevention and mitigation. Sometimes a sensitive administration or a non-governmental organisation has been able to build on such foundations. Many examples are provided by Maskrey (1989) and Anderson and Woodrow (1998) and others, and we will return to these in Chapter 9.[17] More often than not, however, 'official' relief and recovery practice pays little heed to what the ordinary people do. The result is wasted resources, squandered opportunities and a further erosion of vernacular coping skills.

Coping and transition to disaster

Coping strategies have been discussed at length in this chapter, and will receive more detailed attention in later chapters. Considerable importance is attached to coping since it points clearly to people's agency, ingenuity and abilities to help themselves individually and collectively. It suggests that outside agencies must understand these strategies, otherwise external humanitarian interventions will undermine them, creating aid dependency and all manner of unintended and detrimental outcomes. However, the term 'coping' is not without problems, and can serve to hide a situation in which people are destitute or even dying. As Seaman et al. (1993: 27) wrote, 'in current development jargon, Africans do not starve, they "cope"'. Coping in disaster transition and abnormal times also implies a graduated rack of dearth, difficulty, destitution and maybe, ultimately, death. To return to the final stages of our Access model as illustrated in Figure 3.3, coping is managing under stress, but coping is in essence a strategy reactive to events beyond the immediate control of the individual, household or 'community'. As circumstances deteriorate, these may prove insufficient. As will be seen in the case study of famine in the next chapter, informal support systems among the Dinka communities of south-western Sudan were well developed but they fell apart under extreme pressure. They could not resist the cumulative onslaught over a long period of war, drought, enslavement and displacement. Thus communal coping strategies broke down – and the event became known locally as the 'famine of breaking relationships' (*cok dakrua*) (Deng 1999).

Thereafter, unless outside assistance and disaster preparedness have not already averted the transition, the transition continues in an iterative and descending manner (Box 7, 'reaction, coping, adaptation, intervention ... disaster as process'). The Access model continues to operate but under such dislocating conditions that many of the recognisable patterns of decision making, with associated rules of choice, allocation of work and rewards, break down completely. The mechanistic and predictable process of earning a livelihood, and also the structural conditions of normal life as shaping the initial conditions of vulnerability, become less operational, and less valuable as a way of understanding 'disaster as process'. As Niehof (2001) has noted, there is an important paradigm gap between the analysis of rural development in stable situations and those of disaster situations. The transition from 'normal life' to disaster here in the Access model demonstrates both this discontinuity as a disaster unfolds, but also a bridge between them. The Access model demonstrates well that conditions of vulnerability start with and are explained by the political economy of 'normal life' – that coping and access to safety also develop out of normal life. Niehof explicitly suggests that disasters linger and shape future 'normal life' for a more or less vulnerable future. The iteration of disaster as process to Box 8, the final one shown in Figure 3.3, asks the question 'To the next disaster?', a topic which is examined in Chapter 9.

The Access model as a research framework

The formal access framework has so far been presented as an explanatory and organisational device. It is not a theory, although theories of disaster can and should be inserted into, and therefore be allowed to shape, the general framework, as happens in subsequent chapters. For example, in Chapter 4 competing theories of famine are seen to deal with different parts of the framework.[18] It draws attention to the socio-economic relations that cause disasters or maps their outcomes. While it focuses on those at risk of disasters, it also includes the relations they have with others that keep them in that unfortunate state, independently from any disaster. It also allows for people's response to what is often a rapidly changing situation, either by coping or by more active and permanent efforts to change those relations.

The access framework therefore does not explicitly include national policies or world systems in the way that the PAR model does, although the impact of national and international actions can and should be incorporated in the model. Land reform, food policy, famine relief, food-for-work programmes, rural reconstruction programmes and laws governing urban property are all initiated exogenously to the Access model but their impacts in shaping access profiles, access qualifications, payoffs and a range of income opportunities should be incorporated into the modelling of the disaster process.

As a research framework, therefore, access is useful in charting impacts of policy, for identifying vulnerable populations and for predicting the probable outcomes of extreme natural events. However, the data requirements for using the framework as a research design are very large. After all, it provides a general outline of the material conditions of life for a population, and most aspects of society can potentially be included. Yet we believe that in use the number of factors to be incorporated would be restricted, because, in use, the framework would be informed by theory and *a priori* assumptions. This would lead to the ability to choose the most significant factors and permit selectivity in the use of the framework. The framework provides a dynamic and moving 'map' of a disaster. Readers will choose which aspect they need to visit, and will bring to it the theories they need.

Some of the main criteria for making the choices for readers of different structural and functional positions can be suggested. Firstly, the researcher's emphasis on certain theories and priorities will determine what has to be modelled in detail. For example, if gender relations are empirically an important element in disaster impact and policy, then the individual rather than the household would be the unit of study and the main focus. If a researcher believes that a supply-side theory of famine requires attention, then those income opportunities available in crop production would be emphasised with reference to drought or pest attack, along with other determinants of supply (e.g. the transport network).

Secondly, the scale of the investigation will also be determined in part by choice of theory. Individual, household, class or village, region and nation are not so much alternative objects of analysis, but rather a series of conceptual limits that nest inside each other (like Russian dolls or Chinese boxes), the smaller-scale enclosed by the next highest level and given context by it. None the less, the study will have to choose the major spatial frame appropriate to its purposes. If a seismic zone, a farming system or an administrative area is chosen as the principal scale of the study, other scales can be sketched in through secondary data, rapid rural appraisal and key informants.

Thirdly, the framework is principally an 'externalist' (or etic) approach, in that it imposes the researcher's own interpretation and perception of vulnerability, hazard and risk – or at least the researcher's interpretation of local people's interpretations. Those experiencing a disaster and other actors, such as aid professionals, members of the civil service in a country facing a disaster, have their own interpretations. As Chapter 4 will show, for example, 'famine' is perceived in a variety of ways which differ significantly between the world media, aid or relief agencies (de Waal 1987). Likewise in the urban context, residents in Alexandra Township, in metropolitan Johannesburg, put flash flooding far down their list of concerns, frustrating attempts by a professor of hydrology to 'solve' the flood problem (Wisner 1997). There are many hazards which outsiders could never even imagine without intensive discussion and understanding of the society involved. The collection and study of indigenous interpretations of extreme events and processes can enrich and perhaps alter the framework.

Fourthly, most studies do not examine vulnerability for its own sake, but assist in the prevention or mitigation of disasters. Therefore, many variables mentioned here in the general framework will simply not apply to particular hazards or in particular situations. The Access model is used as a predictive and organising device for this book. Only parts of it, at the discretion of the researcher or policy maker, will be relevant in each case.

Notes

1 Hewitt (1983a) sees the potential for disaster 'prefigured' in 'normal life'; while Wisner (1993) refers to 'daily life'. We use the two phrases interchangeably.

2 See RADIX web site page on 2002 southern African food emergencies (RADIX 2002).

3 This type of hazard, and the Andhra Pradesh case, is dealt with at greater length in Chapter 7. Winchester's work (1986, 1992) is a valuable and rare example of a study of the actual operation of ideas of vulnerability in the analysis of a sudden-onset disaster.

4 For example, Richards (1986) likens the farming venture undertaken by the Mende of Sierra Leone with a ship, with a crew hired and paid off at each point during the voyage. Each voyage takes place over a catenary soil profile and through an agricultural calendar (i.e. through both space and time), involving the labour of women in some areas, senior women in others, men for certain agricultural activities, and it is only at one particular point in the agricultural calendar that anything approaching a 'farm household' appears at all. In cases where larger units are significant, such as production brigades in China from the 1950s–1980s, the household may not be an appropriate unit for all aspects of analysing access. The household may control some of its consumption, and small plots of land for production, but most resources and the accumulated surplus is outside their control.

5 For simplicity we refer to income opportunities, although a better term is probably livelihood, which implies the content of supporting life without the assumption that this is done through access to a cash 'income'. Livelihoods may include activities of self-provisioning (subsistence farming, fishing or pastoralism) in which cash may play an insignificant part.

6 It is the female children who tend to be withdrawn from school when illness with HIV-AIDS removes principal wage or food earners from Ugandan families (Barnett and Blaikie 1992).

7 The role of structures of domination in defining untouchability are highlighted in the bias in relief and recovery against outcaste victims of the 2001 earthquake in Gujarat (India). Also, in the Kobe (Japan) earthquake (1995) there were high losses suffered by the Burakumin ('untouchables'), whose traditional occupation of shoe making (which involves the use of flammable materials) and their densely populated and dilapidated housing resulted in severe fire damage and high mortality (see Chapter 8).

8 The 'rules of the game' can shift very rapidly, as in the new regimes that accompany the establishment of colonial rule or the sudden establishment of private ownership of land (or, conversely, by sudden collectivisation of land as in the Ukraine in the 1920s and 1930s). O'Keefe and Wisner (1975) show how changes to the 'rules of the game' rendered ineffective a number of indigenous African mechanisms for coping with drought, resulting in increased potential for famine during the colonial period.

9 'Ethnoscience' is the term often used for vernacular, local knowledge of the physical environment. Some have used the terms 'people's science' (Wisner et al. 1977), 'folk science', 'folk ecology' (Richards 1975, 1985), 'écologie populaire', 'people's knowledge' (Rau 1991) and 'indigenous knowledge' (Brokensha et al. 1980). Within environmental design and architecture the term 'community design' is common (Wisner et al. 1991). We will use the term 'local knowledge', connoting a broader knowledge base that includes social relations and not just taxonomy, mechanics, chemistry, etc. For a critical review of the use and misuse of local knowledge by outside development agents, see Wisner (1988a: 256–262).

10 Models of the risk-averse farmer abound: see Ellis (1988) for a review.

11 On women's access to resources, see Rogers (1980); Dey (1981); Agarwal (1986); Vaughan (1987); Sen and Grown (1987); Carney (1988); Wisner (1988a: 179–186); Shiva (1989); Downs et al. (1991); Schoepf (1992).

12 Examples are given by Cutler (1984); Greenough (1982), writing on the 1943–1944 Bengal famine; and Vaughan (1987) on Nyasaland in 1949.

13 For south Asia see Agarwal (1990: 367); Fernandes and Menon (1987). On Kenya 1971–1976, see Wisner (1980); Downing et al. (1989).

14 Corbett (1988) has reviewed four major studies of coping mechanisms in the face of famine: these are of northern Nigeria, 1973–1974 (Watts 1983a); Red Sea Province, Sudan, 1984 (Cutler 1986); Wollo Province, Ethiopia, 1984–1985 (Rahmato 1988); and Darfur, Sudan, 1984 (de Waal 1987). Brown (1991) presents another detailed account of the coping sequence in Chad, as do O'Brien and Gruenbaum (1991) from two contrasting sites in Sudan. Agarwal (1990) has also reviewed accounts of coping strategies in south Asia.

15 This is true even in more densely populated regions such as south Asia. On common property resources in Asia, see Blaikie et al. (1985); Agarwal (1990); Chambers et al. (1990).

16 This is especially true in drought onset, when it is impossible to know how long reduced or interrupted rainfall will persist and the initial coping strategy is to preserve the basis for continued existence at normal levels afterwards. See Cannon (1991) for a review of these approaches.

17 For other examples of disaster relief, prevention and mitigation in which vernacular coping and innovations from the outside are combined, see Wijkman and Timberlake (1984: 104–143); Timberlake (1988); Harrison (1987); Maskrey (1989); Anderson and Woodrow (1998); A. Grainger (1990: 276–321); Harley (1990); Pradervand (1989); Rau (1991: 145–205); Eade and Williams (1995).

18 These competing theories of famine causation discussed in Chapter 4 include Sen (1981) and Ravallion (1987), who emphasise the behaviour of markets and their impact on the population, Rangasami (1985, 1986) and Firth (1959), who deal with the structures of domination and the time–space aspects of disasters, and Hellden (1984) who studies the impact of drought upon famine in Ethiopia.

Part II

VULNERABILITY AND HAZARD TYPES

4

FAMINE AND NATURAL HAZARDS

Introduction

Of all disasters, famine is perhaps the most damaging. There have been more references to its occurrence historically than any other type of disaster, and throughout history the state has been involved with famine far more closely than with earthquake, flood, tsunami, storm surges and other types of disaster.[1] Generally the number of people affected has been far greater in famines, and its social and political impact on the affairs of state and rulers has been more profound. Also, behind each case of excess mortality from famine there lies a much wider net of destitution, displacement and impoverishment that may endure for many years after the acute symptoms of famine have subsided.

The worst recorded earthquake disaster caused the deaths of about 240,000 people in 1976 at Tangshan (China), but in the twentieth century alone it is dwarfed by famines that have frequently caused the deaths of more than a million people. For example, 1.5 million perished in the famine of 1974 in Bangladesh, and between 900,000 and 2.4 million in North Korea between 1995 and 1999 (Noland et al. 1999). In the Chinese 'Great Leap Forward' famine of 1958–1961, the death toll is estimated at between 14 and 26 million (Kane 1988), or between 30 and 33 million (Becker 1996), and possibly as high as 40 million (Article 19 1990: 18). In some cases, mortality statistics are unreliable, and the cause of death is disputed (e.g. between bubonic plague and starvation in Europe in the 1340s, and between deaths due to fighting or the slaughter of civilians, and the famine associated with war, e.g. Biafra 1968–1970). However, the scale of mortality dwarfs that of all other disasters. No other type of disaster has caused as many deaths as famine (in the order of 70 million deaths in the twentieth century: Devereux 2000), although it is not surprising that estimates vary greatly.

Today, famines still occur, although the affected regions are substantially different to those of the past. Famine is now unlikely in south, east and south-east Asia (the exceptions being North Korea from the mid-1990s up to the time of writing in early 2003, and Cambodia in the 1970s), but are more common and widespread in Africa, for reasons discussed below. Acute food emergencies, if not formally defined as 'famines', have also begun to affect parts of the world where they have not been seen before, such as Central America in 2001.

The causes of famines have changed too, although people die in the same appalling ways. Starvation is a pre-occupation of many people in Africa, and is alive in the consciousness of many millions of others who have witnessed famine within living memory (e.g. in southern Nigeria, Ethiopia, the Sudan, Bangladesh, Cambodia and China). It is also a vivid reminder of oppression and imperial callousness (Davis 2000). And famines can remain as emblematic political events for nationalist movements many years after their occurrence, as with the Ukraine (which suffered appalling famine under Soviet control in the 1930s: Dando 1980) and Ireland (which endured the loss of over a million people under British rule in the 1840s: Regan 1983).

Famines and their causes

Part I of this book questioned the simplistic perception that disasters are 'natural'. Some preliminary arguments were developed to show the need for 'social' (in its widest sense, including political, economic and cultural) reasons being given much greater prominence in understanding how disasters are caused. Most popular ideas of famine as expressed in the media still tend to blame drought and other 'exceptional' natural events (e.g. floods or epidemics in human or bovine populations). There is more than a grain of truth in this belief. Throughout this book, we emphasise the role of climate change, which has accentuated the magnitude and frequency of hazardous events. Yet even here, anthropogenic causes of climate change, and the social environment which shapes vulnerability, show how disasters are linked to human action rather than capricious nature.[2]

Famine, although often linked with drought, flood or epidemics, can also occur without a well-defined 'trigger' event in nature at all, but instead as a result of war or conscious attempts to use food as a weapon (which may become a device for 'ethnic cleansing').[3] While this second edition still attempts to reduce disasters by examining the links between 'nature' and 'society' in the explanation of disasters, it also reflects new debates and (unfortunately) new experiences.

The year 1983 saw a significant coherent critique of the notion that disasters are to be explained primarily through natural factors. The main target of this critique (*Interpretations of Calamity*, edited by Hewitt 1983a) was the idea that if disasters (particularly famines) were attributed to natural causes, then they could be explained in terms of exceptional events and not as 'normal' or day-to-day social processes.[4] Famines could be attributed instead to *un*precedented, *un*natural and *un*expected events, and therefore appear to be quite separate from normal life. In this framework, sudden natural phenomena, or even slow-onset drought, could provide the explanation because they constitute the new, decisive and therefore principal cause. In Hewitt's view (ibid.: 9–24), the explanation of disasters should rest more fully on a social analysis of the processes which create the conditions under which 'exceptional' natural events can act as the 'trigger' for a disaster.

This analysis focused on the social processes of impoverishment and exploitation which expose people to hazards (make them vulnerable) as a normal and continuing part of life. An explanation of disasters in terms of natural events invites technological solutions (rather than social ones) to the containment of floods, the design of earthquake-resistant buildings and the introduction of more productive technology in agriculture. When this approach demonstrably fails (as it sometimes does), it is perceived to be beyond the powers of technology or, even more culpably, beyond the powers of human agency to prevent famine, allowing the cause to be once again thrown back into the lap of 'exceptional' and 'unprecedented' events – with the result that no blame can therefore attach to anyone.

Hewitt (ibid.: 12–14) argues that a previous generation of academics and practitioners virtually ostracised those who sought explanations of disaster that went deeper than the impact of the natural hazard. Given the dominance of science and technology in the modern era, the authors of any analysis of causes that did not suggest that hazards could be modified, palliated or avoided altogether by the application of technology (e.g. satellite early warning systems and reinforced concrete) were exiled from mainstream social explanations.

Today, twenty years after Hewitt's book, the primacy of natural hazards in explaining the causes of famine has to be questioned with even more vigour, in the light of subsequent events and theoretical developments. There are two issues: firstly, natural hazards (especially drought, but also sometimes flood and other sudden-onset 'natural causes') are much less capable of acting as triggers for famine: the connection has been undermined by improved national and international humanitarian responses, better preparedness and the political will which can enable this type of progress to be made. There is now a growing and encouraging record of averted famines, as in Bangladesh (1979 and 1984), Botswana and Kenya (1984) and southern Africa (1991–1993) (Devereux 2000).

Unfortunately this is not the case everywhere, especially in Africa, where even as this chapter is being written there is famine in Malawi, Zimbabwe and neighbouring countries, and a likely famine in Ethiopia. Many of the famines in Africa in the past 30 years have been attributed to a combination of natural triggers (mostly drought) and civil strife and war. The current crises in southern Africa and the Horn are being partially blamed on drought, while there are also crucial political and economic factors involved in generating vulnerability as well.

The second reason is not so much a shift in the circumstances and location of famines (and potential famine) as changing attitudes about their causes. Devereux (2000, 2001) argues that famines have become more complex, and there is certainly some evidence for this. There is also much more information available about them, through historical research on past events, and reports and analysis from early warning and monitoring systems. With more information, and the means to broadcast it, conspiracies of silence about a famine are more difficult to maintain. There will always be politically engineered silences about famine, sometimes enforced through censorship, and at other times through carefully managed discourses where silences are unnoticed (Article 19 1990, 1993; de Waal 1999). For example, the drought-triggered famine of 1984–1985 in the Sudan had little cause other than the failure of Nimeiri's government to acknowledge the existence of a food crisis. It was an embarrassment and distraction, the existence of which Nimeiri consciously denied, causing the death of about a quarter of a million people (de Waal 1997: 91). Also, closed and totalitarian societies tend to collect information through the eyes of party officials who have their own categories and statistical collection techniques. This has led to incorrect information about food security and food stocks, the silencing of the voices of the hungry, and a dearth of information about the true state of the disaster reaching the outside world (e.g. the famine of North Korea since 1995, and the Great Leap Forward famine in China, 1958–1962 – see below).

Despite these important silences (to which we will return later), the amount of information available about famine has increased enormously, and more detailed information can lead to seemingly greater complexity. Lastly, 'the belated recognition by observers that famines are more complex and open-ended than they had previously thought' (Devereux 2000: 4) is explained by the increasing occurrence of protracted complex emergencies associated with hunger, in which many factors are interwoven. These include a breakdown in social order, prolonged conflict, large numbers of displaced persons, disruption of agriculture and food production, of marketing and the supply of food, and political deadlock between armies or administrations with de facto control of food movements, and of organisations that are in a position to supply and deliver relief measures. It is this nexus of factors, involving a greatly increased number of actors (particularly at the international level), which gives the impression of increased complexity. It may be worth pausing to consider whether the famines of 30 or more years ago were any less complex, or any less political than they are today, or whether it is simply that the politics and structures which shaped the famine situation were often unreported and that the voices of the starving (and already dead) were never heard. When the facts of these events are disinterred by painstaking research (e.g. the classic work by Watts [1983a] in northern Nigeria, and the reports of past famines in M. Davis's Late Victorian Holocausts, 2000), the explanation of famine matches in complexity any contemporary 'complex emergency'.

Box 4.1: Famine in Malawi, 2002

Ninety per cent of Malawi's 11 million population live by subsistence agriculture in the countryside. This country, like all of southern Africa, suffered its third year of drought in 2002 in some parts of the country, while others were still recovering from floods that happened in 2001. Malawi is landlocked, highly indebted and among the least developed countries by UN standards. There were early warnings by aid organisations of a famine that might match the scale of Ethiopia's tragedy in 1984, or an earlier famine that elderly Malawians remember from 1948. Piecing together reports from agencies working in rural areas, as many as 10,000 people, mostly the weak (children, the aged, people living with AIDS), may have died by May 2002. The World Food Programme (WFP) estimated that up to three million were at risk, that 70 per cent of the population was affected[5] and that Malawi would require 600,000 tonnes of food by the end of 2002.

HIV-AIDS, malaria, tuberculosis, diarrhoea and an epidemic of cholera further weakened the population up to and during 2002, intensifying the health consequences of hunger. The 2002 cholera epidemic is the worst in the history of Malawi, with 32,968 cases reported and 980 deaths by 7 April 2002.[6] In June 2002, bubonic plague appeared among the population already weakened by hunger.[7] Many AIDS orphans and those living with HIV already had poor nutrition even before this current crisis.[8] Where farmers have been successful in growing some maize and protecting it from animal or human thieves, hunger has forced them to harvest it before it is mature. It is also clear that vulnerability varies considerably according to social factors. The UN reported that '[s]tudies have shown that smallholders with less than one hectare of land, estate workers and tenants, low income urban dwellers, female-headed households and children suffer from chronic food insecurity. They also found that the majority of the poorest people in Malawi are women and that female-headed households – which make up 34 per cent of households – shoulder the greater burden of poverty' (OCHA 2002c).

Chronic food insecurity for such groups has a long history in the country. Expropriation of small holdings by large estates, and the forbidding (under the 1968 Special Crops Act) of smallholders from growing valuable export crops such as tea and tobacco, were followed by worsening access to fertiliser, agricultural extension services and credit during the period 1987–1994 (Cross 2002). A severe drought in 1992, a doubling of inflation, a 40 per cent decline in the maize crop, and a drawing down of strategic reserves further worsened the conditions for rural producers. The Customary Land Use Survey undertaken in 1994/1995 revealed a dire situation of high vulnerability to food shortages and the danger of famine (Green 1996). Given these

Box 4.1 continued

very adverse circumstances, it is surprising that the World Bank continued with neo-liberal policies, aimed at market liberalisation.

In colonial times, Malawi exported labour to the South African mines, men who were highly prized as 'tropicals' (who could tolerate the heat in the deepest mine shafts) and who remitted income back home. Now Malawi's main exports are agricultural products. Its government has attempted to implement a World Bank structural adjustment package aimed at reducing government expenditure. But much of the rural population remains isolated, without basic services and at the mercy of an uncontrolled market through which the price of maize (the staple food crop) has increased by 400 per cent.[9] This puts the purchase of food beyond the reach of most of the hungry, who have to sell off assets (including bicycles, livestock and even the roof timbers of their houses in 'distress sales') in order to survive.

The government stands accused of failure to intervene to control the staple food price, to import food in a timely manner and to maintain a strategic reserve against such contingencies. This arguably transformed a food shortage into famine. The significance of strategic reserves in this has been especially controversial, particularly as the government was required by the International Monetary Fund (IMF) 'to privatise its agricultural development and marketing corporation – which maintained a central stock of grain and regulated prices' (Burgo and Stewart 2002). A recent report from the UK NGO World Development Movement provides a severe indictment of mismanagement and the pursuit of self-interest by various interest groups:

> The decision to sell Malawi's grain reserves followed advice from the IMF to reduce operational costs and the level of buffer stocks held from 167,000 Metric Tonnes (MT) to 60,000 MT, in order to repay a South African bank for a commercial loan of US$300 million, incurred by the National Food Reserve Agency (NFRA), when it was established as a quasi-independent agency [ABSA Bank]. The IMF further advised that the maize be exported to neighboring countries, in disregard of the impending food crisis in Malawi.[4][10]
>
> In the event, and evidently in clear defiance of the IMF, the Malawi Government sold almost all of the 167,000 MT reserve, much of it on local markets, to private traders, the new agents of the partially liberalised grain market. Traders stockpiled it and later profiteered from hunger. Since then, all parties to the gross mismanagement of the country's food security policy have traded accusations and refused to take responsibility for their failures.
>
> (Owusu and Ng'ambi 2002: 14–15).

Box 4.1 continued

A report from the news agency Reuters gives some indication of the resulting infighting and accusations of blame (Kotch 2002).

When drought on this scale last affected southern Africa in 1991–1993, there was a rare moment in which co-operation among all the affected countries (including South Africa) enabled food aid to be used in a remarkably efficient way.[11] Although at that time some 13 million people needed emergency food, this daunting logistical mission was accomplished.[12] During the recovery period in the early 1990s numerous non-governmental organisations (NGOs) developed projects to build drought coping capacity. The transport infrastructure, communications and political co-operation of the neighbouring countries at that time seemed a good example of the kind of regional resilience that development can provide in the face of disaster.

This time around, countries in central and southern Africa are in the grip of an even more serious epidemic of HIV-AIDS, electoral democracy is under severe pressure in Zimbabwe, Zambia and Malawi, and local state institutions are very weak.[13] In Zimbabwe there is civil unrest associated with the expropriation of white-owned commercial farms and a disputed presidential election in 2002. Angola is just emerging from decades of civil war, and Zambia still hosts many hundreds of thousands of Angolan war refugees, who need to be fed.

Perhaps because of concerns with the Malawi government's capacity and accountability, donor response to the appeal for food was weaker than it was in 1991–1993. The WFP was already feeding 2.6 million people in southern Africa in March 2002 when it appealed for $69 million for this emergency. By the end of April it had received pledges of only $3 million,[14] and by June 2002 it estimated that it had to feed millions of people in the region: 3 million in Malawi, 6 million in Zimbabwe, 2.5 million in Zambia, 1 million in Angola, nearly half a million in Mozambique and 144,000 in Lesotho.[15]

Explanations of famine

Theories of famine can be considered in terms of four main strands. The *first* may be called neo-Malthusian, and is focused on the potential famine-inducing consequences of rapid population growth outstripping the limits of global and regional food production. The *second* relates to environmental limitations on food output, principally through drought. More recently it has involved considering the disruption of production and its uncertainty and variability as a result of global climate change. If the Malthusianism is 'demand'-oriented and centred on population supposedly outstripping production, the environmental approach is on the 'supply' side and looks primarily at supposed natural 'causes' which reduce the capacity of natural resources to provide adequate or reliable food supplies. The *third* strand is dominated by economics, where famine is seen as being caused by a fall in the aggregate supply of food, imperfect markets which fail to supply food, or by the failure of people to generate effective demand to purchase the food that they need. The *fourth* strand centres on the political and human rights aspects of famine and on the emerging complexities of contemporary famines. Let us take each of these in turn, and review their complementarities and contradictions.

Neo-Malthusian explanations

Malthusian and neo-Malthusian explanations of famine have a long history (since Thomas Malthus's original essay on the subject in 1798), an extended period of dominance and a more recent (but long) history of vigorous refutation.[16] Devereux (1993: 46) quotes a view which illustrates the crudest form of neo-Malthusianism expressed by the US Secretary of Agriculture in 1946: 'Some people are going to have to starve … We're in the position of a family that owns a litter of puppies; we've got to decide which ones to drown'. Leaving aside the violence perpetrated by this remark, it succinctly illustrates the kind of politics which can flow from neo-Malthusianism. Other crude forms of neo-Malthusianism, such as Hardin's controversial essays 'The Tragedy of the Commons' (1968) and 'Living on a Lifeboat' (1974), need not be given a full treatment here. Refutations have been based on both empirical and ideological grounds. Part of the empirical ground for refuting the link between famine and 'over-population' is based on the fact that the population of Bangladesh, China and India have greatly increased since earlier famines (supposedly caused themselves by rapid population growth). Since its last famine in 1974, the population of Bangladesh has almost doubled (from 70 to 125 million) and yet, arguably, it is more secure from the threat of famine now than at that time.

There are well-established demographic arguments that predict a stabilisation of the world's population growth through the fertility transition. There is also no clear association (let alone a proven causal link) between rapid population growth and slow economic growth (see Cassen 1994; Kiessling and Landberg 1994; Sen et al. 1994 for a theoretical discussion). Also, as Osmani (1996) has pointed out, famines do little to check population growth in the long run (birth rates soon recover and compensate for losses), and so fail to provide the 'natural' checks on population growth that Malthus predicted. Even the demographic impact of China's famine of the late 1950s and early 1960s (which involved a period during which population declined at a rate of about 1.4 per cent per annum) was recovered from within 15 years (Field 1993).

There are also ideological objections to the neo-Malthusian case, which demonstrate that its political implications are unhelpful at the very least, and may also damage the prospects of preventing famine in the short and longer term. Simply put, objections to the theory are that it: (1) blames the most vulnerable for breeding uncontrollably and bringing famine on their own heads; (2) diverts attention from the global and structural causes of vulnerability to famine, which have much more to do with the restructuring of global capitalism than with overpopulation; (3) makes it appear that nothing can be done because the processes involve unavoidable and inevitable structural causes that are beyond the capacity of any political commitment and organising action; and (4) de-links the ethical and humanitarian imperatives (what *should* be done) from famine. In spite of these counter arguments, neo-Malthusianism persists, as for example in Brown (1995, 1996) and Ryan et al. (1999). Yet it would be foolish to be complacent about the implications of continuing high rates of population growth on food security in some regions of the world.

There may be a number of indirect 'neo-Malthusian' impacts of population growth on food production which have intuitive appeal but are also undermined by convincing counter-arguments. These indirect impacts of rapid population growth and 'pressure upon resources' occur in fragile and easily degraded environments. In these, food supplies may be disrupted by natural hazards that have been aggravated by human action. An example would be the reduction in the moisture-retention capacities of soil due to over-cultivation or overgrazing, such that a dry period becomes a drought. Human-aggravated drought leads to a downturn in food availability and increased variability of annual output. Some of the evidence that environmental degradation has negative consequences for food supply in the longer term is credible.[17]

But it is intellectually hazardous to link the development of famine conditions, back through unsafe conditions and economic and social pressures, to 'root causes' such as population growth. Any number of intervening processes and events may serve to modify or nullify the longer-term effects of environmental degradation and population pressure. We must understand that there can be human adaptation to the risk of famine, the use of compensating livelihood strategies and coping mechanisms when faced with hardship, and the application of tested measures to avert famine. All of these can, and should, stand between the dismal connection between population pressure-induced environmental degradation and famine, even though there are instances where a credible link between environmental degradation and reductions in the supply and yield variability of food can be made.

Environmental 'supply-side' explanations

There is a long tradition of blaming drought (and sometimes floods) as a significant cause of famine. Of the major famine disasters in the past 30 years (e.g. in the Sahel zone of west Africa, in Sudan and Ethiopia and in northeast Brazil), many have been explained principally in the popular media as being caused by drought. In recent years, global climate change has emerged as an additional factor in explanations of the reduction or disruption of food output, especially in relation to drought (Downing 1992). It is becoming a major focus in understanding the possible increase of extreme events, in which natural hazards are magnified in intensity and frequency. Yet it is almost certain that the reasons for this climate change are rooted in human activities generating increased levels of 'greenhouse gases' in the atmosphere.

There are also major climatic oscillations such as the El Niño Southern Oscillation (ENSO) which appears every three to seven years or so, and lasts 12–18 months. The severity and timing of ENSO may also be associated with longer-term climate change, although there is uncertainty about the strength and nature of this relationship, and whether or not ENSO is driven by global warming (McGuire et al. 2002; Glantz 2001). The ENSO events also involve El Niño's counterpart, La Niña, which is characterised by an exaggerated 'return to normal' from El Niño, and includes, for example, excessive rainfall returning to areas stricken by El Niño-related drought.

It is highly likely that these oscillations have been occurring for a very long time. Davis's account of *Late Victorian Holocausts* (2000) offers persuasive evidence that El Niño was associated with devastating droughts in northern China in 1875–1876; in Brazil, India and Morocco between 1877 and 1878; and in Russia, Korea and Ethiopia from 1888 to 1889. These acted as triggers for huge famines, which he argues were worsened as they combined with the emerging Victorian global economy in which the poverty-stricken were 'ground to bits'. Deliberate policies of colonial officers, heavy taxation of poor rural populations in newly conquered colonies and the exigencies of the marketplace (a version of neo-liberalism) turned ENSO events into disasters. Victorian colonialism ignored opportunities to reduce them, and mitigation by local rulers was no longer possible as it had been in the past (e.g. in China and in independent Ethiopia in 1899). Davis's book contains lessons for current global leaders who preside over policies which resemble those of the late nineteenth century.

The implications of longer-term, more general, climate change (often referred to as global warming) are discussed in general terms in Chapter 2. They include increased frequency and severity of tropical cyclones, with storm damage, landslides and flooding, raised sea levels and coastal inundation, and the shift of epizootics and epidemics to new locations. In relation to famine, climate change principally acts as a trigger through drought, the shifting of the timing of rainfall and its seasonal patterns (e.g. the Asian monsoon), floods which disrupt the production and distribution of food and the possible spread of disease to humans, livestock and crops. All of these increased risks are almost certainly caused by human action (in relation to greenhouse gases) and relate to social vulnerability and to the pre-existing 'normal' level of hazards. But with climate change, human action is responsible for both the generation of peoples' vulnerability *and* the increased level of hazard.

Economic theories of famine

Since the 1980s, academic literature on the explanation of famine has shifted from giving prominence to natural events towards an emphasis on the social causes and economic processes involved. Amartya Sen, perhaps the most influential writer of famine in the twentieth century, claims that both Marx and Adam Smith directed attention to the social causes of famine (Sen 1981: 13). This suggests that such explanations go beyond a simple divide between the political 'left' and 'right'. In a series of seminal writings in the 1980s, Sen focused on specific factors which (in the language of our Pressure and Release, or PAR, model) channel root causes into unsafe conditions and thereby to the 'point of pressure'. These specific causes at the 'point of pressure' require much more rigorous analysis than more general and under-theorised root causes because they must describe why specific people, at a particular place and time, cannot eat enough food to survive.

There are three main (and largely competing) economic explanations of famine, based on different sets of causal mechanisms. The *first*, which Sen seeks to identify as only one (and in his view, overstated) cause of famine, is termed Food Availability Decline (often abbreviated to FAD). This mechanism is simple and based on common sense. It tends to invoke natural triggers of famine (e.g. drought, flood, locusts) which directly cause crop failure, or heightened mortality of livestock, or both. This in turn reduces the aggregate amount of food available, so that famine is caused by there not being enough food to go around, and is termed a 'supply-side' explanation of famine. To identify drought or other natural events as the immediate cause of crop failure, and therefore of a decline in food supply, seems a common-sense deduction.

The major problem with such a simple causal connection between a fall in the aggregate food supply and a famine is the assumption that food availability is shared equitably among the population, and that its members have no source of food other than their own production. Both assumptions are usually unwarranted. While many explanations of famines start with a trigger event such as flood or drought, there are usually difficult anomalies between the relative severity of the famine and the decline in food supply in both space and time. For example, Currey (1981) uses cartographic data for the Bangladesh famine of 1974 to show that the spatial distribution of the number of famine deaths did not closely correlate with those areas of below average production of food staples. Indeed, Sen (1981: 92) claims that 1974 was 'a local peak year in terms of total output and *per capita* output of rice'. Likewise, Kumar (1987) shows that the number of districts reporting normal, above normal and below normal aggregate crop production for 1972–1973 in Ethiopia did not correspond well with the spatial distribution of famine-related mortality. He also states that the evidence points to a fall in food production *after* the main impact of the famine had been felt (ibid.: 13). Similar difficulties of finding precise and causal links that match, spatially and temporarily, famine mortality and a fall in aggregate food production confront almost every case of famine for which reasonable production and mortality data exist.

The conclusion to be drawn is that natural events may be implicated as a trigger to famine, but that the actual factors involved in reducing how much certain groups of people are able to eat are only very indirectly related to a drop in output. It may be much more strongly connected to economic factors (e.g. a relatively small drop in output can trigger a disproportionate increase in prices for those who need to purchase food). If falling aggregate food supply makes people more prone to famine, it is not because there is not enough to eat but because social processes are determining that those who need whatever food is available are unable to get it. The actual processes that translate the decline in output to peoples' hunger are not *causal* but parallel.

We must also keep in mind that famine may occur with no natural trigger involved at all. The most significant cases are the 'policy famines' (which result from the imposition of government policies or political interference in food production conditions) that occurred under forcible collectivisation in the Soviet Union in the 1930s, the Great Leap Forward famine in China and the current famine in North Korea. These have been responsible for the deaths of many millions of people, far more than others triggered by natural hazards.

There are two types of case which are crucial in demonstrating the inadequacy of FAD-type explanations. The first is where famines occur without any significant natural trigger, so that food availability *does* decline and people are prevented from getting enough to eat, although the causes are often directly political. The second is those cases where drought is supposedly causing hunger and perhaps famine while at the same time the country concerned has produced food which is being exported. Cannon gives several examples of countries experiencing famine, or with large numbers of malnourished people (including India in recent years), that have exported food:

> Over the four years 1999 to 2002, Ethiopia considered itself to have a surplus in national food production. In very controversial circumstances, Sudan has a number of times sought to sell foodgrain, including an attempt in 1992 to export sorghum to the European Union (EU) for use as animal feed, at the same time its government was requesting food aid.
>
> (Cannon 2003: 46)[18]

Such situations provide support for our emphasis on the need for 'sustainable livelihoods' and access to resources and entitlements for everybody, not for simplistic explanations that look mainly at natural factors or the supposed problem of population. In India the problem of hunger today is not one of famine of but widespread malnutrition amongst millions of people within an economy that has the capacity to export food to others who can afford to buy it.

For farmer households, hunger arises not only from an inability to purchase food, but mainly from the lack of resources to grow it (especially land and water), or from the reduction of production by the impact of natural hazards (especially drought or irregular rainfall). An analysis of famine in Ethiopia in the 1980s by Diriba (1991) has amply shown how long-term decline in cereal production per household and per hectare (as a result of population growth, land shortage, lack of economic development and environmental degradation) has made self-provisioning very difficult for many households.[19] Even during 'normal years', most households had to buy food or rely on aid to make up shortfalls in household production. Thus a long-term supply-side problem undoubtedly existed. However, it was the onerous taxation imposed by the state and the lack of alternative income-earning opportunities outside subsistence agriculture that made any enforced entry into the food market very risky for the majority. Thus, a decline in aggregate food supply within an area is an important factor, but, as always, the lack of alternative means of purchasing food turns this long-term decline into a potentially disastrous situation virtually every year. Goyder and Goyder (1988) and Kebbede (1992) agree that excessive taxation by the Ethiopian state has tended to push many marginal peasant producers over the edge, particularly when combined with other adverse conditions.

The *second* type of explanatory mechanism of famine is also on the supply side, but concerns the operation of markets, which in times of hunger may be unable to meet the demand for food by a set of people, even though effective demand (backed by the ability to purchase) exists (Seaman and Holt 1980). In other words, a functioning market which should be capable of supplying the food that is in demand is not able to operate. For example, Cutler (1984) and Devereux and Hay (1986) point to market failure as a major mechanism responsible for famine in Tigray and Wollo in the 1970s, where the market was not well developed, transport was poor and expensive and the proportion of food bought or sold in the household was very small. Poor transport is often put forward as one of the major reasons for this failure. Emerging from this is a related argument that modern transport can integrate markets and enable food to reach the areas where it is required. Conversely, some writers (e.g. Copans 1983) attribute a dangerous decline in self-provisioning to the spread of commercialisation, which is often aided by expansion of rail and road systems. Drèze considered the Indian situation under British rule in relation to the expansion of the railways:

> The fact that the expansion of the railways resulted in a greater tendency towards uniformity of prices there can be little doubt.... One may also generally expect a reduction of price disparities to reflect greater food movements towards famine-affected areas, and to result in an improvement of the food entitlements of vulnerable sections of the population in these regions. However, it is easy to think of counter-examples of which two are particularly important.
>
> (Drèze 1988: 19)

The first of these counter-examples is that price equalisation also went on at the international level, with large-scale exports of grain from India during famine periods. The second is that the equalisation of prices brought about by the introduction of railways is not the same as the equalisation of real wages or food entitlements. Thus there could be an area which exported grain but which also was a low wage area, with the result that labourers (who were unable to grow food for themselves) were unable to afford to purchase grain. Strictly speaking this is not a market failure, but an example of how an efficient, price-equalising market could precipitate a famine. In fact it is this situation that typifies the chronic malnutrition of the poor in India today, where hungry people watch as food passes them by; this type of analysis of markets is often as relevant in everyday life as it is in the 'exceptional' famine.

Other factors may impede the transfer of food from surplus to deficit areas, even when prices are higher in the latter and normally expected to become more equalised. For example Clay argues:

> If Bangladesh is regarded as a fully integrated production system with smooth inter-district flows of commodities, then there is no production problem. But to the extent that there are frictions and difficulties in moving commodities between districts, regional losses of production can have severe effects on food prices, intensifying the effects of loss of production, income and employment.
>
> (Clay 1985: 203)

Thus the problem is seen not as a fall in aggregate regional food production (as a FAD explanation would claim) but a market failure which is not enabling the satisfaction of effective demand. Also, the anticipation of a food shortage can lead to speculation and hoarding. It is possible that hoarding may be desirable given the expectation of future scarcity, though in many cases it is used by a minority of traders to influence current prices rather than to smooth out future price rises. In some cases, a lack of information about demand may also prevent the market mechanism from satisfying effective demand. For example, Ravallion (1985) found evidence for a systematic rise in rice prices during 1972–1975 in Bangladesh, caused by inaccurate and pessimistic information in the press about food prospects.

This led both traders and better-off consumers to hoard, causing further price rises. So according to this view, it is not a decline in food production which causes famine (after all, it could be imported), but that the market has failed to receive the correct signals to distribute food to where a demand exists.

The *third* economic mechanism of famine involves the decline of people's entitlement to food. In 1981 Sen brought together a number of disparate sources to provide a significant advance in the debate on famines. This was centred on the notion of entitlement failures, or food entitlement decline (FED), which involves a range of factors that include entitlements to production resources, the selling of labour for wages or selling produce for cash. The notion was challenging because it suggested that famines can be caused by the failure of a range of entitlements which had little connection with the overall aggregate availability of food. His work was provoked in part by his need to understand how starvation could occur when there is food available, as happened in the famine he witnessed as a boy in Bengal in 1943. He identified other famines in which there was no aggregate shortage of food, including several in different parts of Ethiopia in the 1970s (Sen 1981).

The argument here is that famine is a result of the lack of purchasing power, which itself may be caused by the failure of some groups of people's 'entitlements' to wages, or their inability to engage in other forms of exchange (e.g. trade) which normally enable them to satisfy their food needs. This is hunger caused by a reduction in people's effective demand (or 'pull failure'), rather than market failure (or 'response failure' as Sen [1985] terms it), or FAD. People may need to purchase food but lack the necessary cash or other exchangeable resources to purchase it. This approach distinguishes between aggregate availability or supply of food and an individual's access to, or ownership of, food.

People obtain food through five different types of 'entitlement relationships' in private-ownership market economies. These are adapted from Sen (1981: 2) and Drèze and Sen (1989: 10):

1   production-based entitlement, which is the right to own the food produced using one's own or hired resources;
2   trade-based entitlement, which describes the rights associated with ownership when they are transferred through commodity exchange;
3   own-labour entitlement, which is the trade-based and production-based entitlements derived from selling labour power;
4   inheritance and transfer entitlement, which is the right to own what is given by others (e.g. gifts) and transfers by the state, e.g. pensions;
5   extended entitlements, which are entitlements that exist outside legal rights (e.g. of ownership) and are based on legitimacy and expectations of access to resources.

Thus, entitlements are either 'owned' by a person (described by Sen as the person's 'endowment') or can be exchanged by that person for other commodities (termed 'exchange entitlement'). People are vulnerable to starvation if their endowment does not contain adequate food or the resources to produce food, and their capacity to exchange labour or other goods and services cannot be converted into sufficient food. This can occur without a decline in aggregate food supply, and without any disruption or malfunction of the market. Such events were possible in India even after the introduction of railways and their tendency to equalise prices for food staples.

With shifts in entitlements a person can suffer a fall in endowments (such as crop failure, or livestock deaths), a fall in exchange entitlements (in which food prices rise, wage income drops, or demand for one's own production falls so that the terms of trade with the market for food shifts against the individual) or a combination of these factors. Analysing exchange entitlements, as an additional mechanism for a person's failure to obtain enough food, permits explanation of famine occurring in situations where there is food available. This was the situation in the Irish famine of 1845–1848, and that in the Ukraine of 1930 (during which the Soviet government commandeered grain from there to be exported to other parts of the Soviet Union). However, academic debate became side-tracked (and, as is usual in these circumstances, heated) over whether Sen was claiming that there could *never* be a FAD-induced famine (Rangasami 1985; Bowbrick 1986). The passage of time and a certain weariness with arcane debates (carried out in well-fed university environments) seem to have left onlookers with the impression that Sen had not intended to mean that FAD could never be a contributory factor in famine.

The analysis of food entitlements decline (FED) was undoubtedly a great advance on theories of food availability decline (FAD), for a number of reasons. Firstly, FED acknowledges the importance of changes in purchasing power. Secondly, it disaggregates regional food production and availability, and follows the process by which food is distributed to individuals. It permits analysis of intra-household food allocation, although this is less well studied (Shepherd 1988). It explains why the rich never die in a famine, and shows indeed how some classes can become better-off while others die. Thirdly, it involves the regional, national and world economy in the analysis and draws attention to the possible prevention of famines by food import. In general terms it is an active model – governments *can* intervene to rescue endowment or entitlement failures, but are also scrutinised in terms of their contribution to causing the problem in the first place.

Food entitlements decline (FED) also exposes the shortcoming of the FAD hypothesis. Firstly, FAD deals only with supply factors while FED claims to deal with both supply and demand (and can therefore address the impact of rising prices of food staples). Secondly, FAD cannot deal with disaggregated populations and explain why some people starve while their neighbours do not. This point is well illustrated by a story from the 1970s Sahel famine recorded by Mamdani (1985). A fat man is reported to have said to a thin man: 'You should be ashamed of yourself. If someone visiting the country saw you before anyone else, he would think there was a famine here'. Replied the thin man: 'And if he saw you next, he would know the reason for the famine!'. Thirdly, the FAD hypothesis is ill-equipped to identify the social causes of vulnerability and poverty other than in general terms of factors such as low agricultural productivity, backward technology or perhaps environmental degradation, which do not lead to any analysis of the underlying social and political determinants.

On the other hand, there have also been criticisms of the FED approach, and particularly the way in which Sen has sought to defend it. Firstly, there is the scale and boundary problem. If the analysis is stretched to include a big enough area, of course there is enough food to avert a famine in any one part of it, and thus there is no aggregate food shortage. Secondly, some famines clearly have had their origins in FAD, whether it be because the government of the day engineered it, or because they neglected the warnings. Differences of opinion concerning the causes of the Wollo and Tigray famine of 1973–1974 in Ethiopia focused on food prices and market failure. Did prices rise and was it in the context of effective demand? Or did they not appreciably increase, in which case there would be support for Sen's 'pull failure' ('demand-side failure') hypothesis? The argument here revolved around 'price ripples' (or uneven spatial and temporal variations of prices which tended to arise in an area where there were serious food shortages), where effective demand was not met by supply because of high transportation costs and either a failure of information to reach traders or their unwillingness to risk distributing small lots of grain to scattered markets. These market imperfections in the presence of effective demand caused people to migrate in search of food, and this resulted in price rises in a series of ripples, corresponding to the movement of people away from the area, which brought with them an increase in effective demand.

Clearly, many famines have been preceded by FAD, and, although it may be incorrect (or arbitrary) to identify FAD as an ultimate or even the most important cause, it is an inescapable fact that a fall in the amount of locally produced food (because of war, drought or longer-term environmental decline) challenges the ability of people to find alternative sources. Sen's notion of FED initially tended to perceive endowments and entitlements as static and given. In this sense, this rather technical, economic approach to the causes of famine cannot explain *why*, in political-economic terms, entitlements fail. These structural reasons lie in the normal daily lives of vulnerable people. Also, there are political reasons why governments decline to take (or deliberately withhold) action to avoid or at least alleviate the collapse of entitlements to food, and it is to these that we turn next.

Complex emergencies, policy famines and human rights

The previous section claimed that the FED model and its entitlements approach has released famine studies from some significant theoretical and ideological constraints. It seems generally to have played a politically progressive role, because it exposed the ways in which entitlements and their failure are socially determined and are not a result of 'acts of God' and therefore unpredictable and unforeseen (and all the other descriptions in Hewitt's [1983b] 'disaster archipelago' we discussed in Chapters 1 and 2). However, some would claim that this economistic approach is too techno-cratic and conservative (Watts 1991; Devereux 1987). If famine is defined in terms of a failure of food systems (and not of displacement, dispossession and death), it becomes legitimate to limit the analysis to 'policy recommen-dations' in terms of technical tinkering (such as liberalising agricultural policies and the integration of markets), and to blame famine on poorly performing economies, 'weak' states or 'policy failure' rather than on delib-erate actions (Devereux 2000). In some cases too, there is the suggestion that famines in Africa have occurred because of 'incompetent' governments failing to follow the neo-liberal policies of the IMF and World Bank (von Braun et al. 1998), in spite of evidence that these policies have probably increased the vulnerability of those who are already poor.

That famine is a human rights issue arises from the premise that they are preventable (Kent 2003; FAO 2001). If this is coupled with the assumption that there is (or should be) a 'political contract' that safeguards citizens (between governments and their people, and between the international agen-cies and human beings, irrespective of their nationality), then a famine means that something avoidable has not been prevented properly[20]. If that contract is not honoured, then there is a breach of human rights, and a 'famine crime' (de Waal 1997, 1991) has been committed. The political conditions (the right to free speech, free circulation of information, the upholding of minority rights and accountability to an electorate) which favour such a contract are found in democratic regimes rather than totali-tarian ones (ibid., 1999, 2000). While this generalisation is broadly verifiable, there are some counter-examples which lead us (as in most discussions about famine) to the view that famines have multiple causes, and to explanations that are highly contextual. Still, the lack of political conditions for an anti-famine contract revolve around anti-democratic tendencies that abrogate any existing democratic rights, thereby hindering timely and effective action to prevent famine, and can therefore be said to involve a famine crime.[21]

Many African nations have failed to provide an anti-famine contract. Democratically elected governments have yet to gain legitimacy, and instead there is often reliance on ethnic and regional coalitions; there is a lack of information and absence of debate about anti-famine policy; human rights activism is weak, and there is often urban bias at the expense of rural food producers (in Chapter 9 we elaborate on such questions of 'governance'). Of course, the crime does not stop with these national governments. Other governments and international organisations are also at fault, sometimes turning a blind eye to regimes which are hostile to the West, or responding too late to crises. In some instances, the aid encounter associated with famine relief has led to demoralisation, dependency and the shifting of blame from national to international shoulders (de Waal 1997). There are other more acute reasons, which both magnify the root causes and dynamic pressures listed above and create more of their own: armed conflicts.

Leftwich and Harvie (1986) and Mafeje (1987) point to the disruptive effects of armed conflict on production and distribution of food. Ethiopia (1984 and 1990), Angola, Chad, the Sudan (throughout the 1990s), Mozambique (1984 and after) and Bangladesh (1974) are all well-documented examples. War tends to reduce both the area sown and to reduce the labour available for field preparation and harvesting. Occupying armies commandeer food and deliberately burn crops in rebel-held areas. Normal trading patterns are disrupted so that both FAD and FED can occur. The authorities often refuse safe passage to relief workers where hunger or even famine serve to put pressure on their opponents (Keen 1994; Macrae and Zwi 1994; Minear 1991; Messer 1991; Smith 1990; UNICEF 1989; Shindo 1985).

Armed conflict disrupts food production (farmers become soldiers and do not plant crops), armies feed off the land and requisition food, shoot at aid convoys and eat from lorries food destined for starving civilians; soldiers burn granaries, bomb houses, burn crops and lay mines to prevent cultivation. Significant areas – often of the best land – have been mined in many countries engaged in war or civil war in the last 30 years, especially Eritrea, Ethiopia, Cambodia, Afghanistan and Angola. Armies disrupt the flow of food (including food aid) from elsewhere; blockade provinces; refuse donors access to war-affected areas – in 1988 the Sudanese government blocked famine relief to Kordofan, where 7 per cent of the displaced people died every week (Devereux 2001: 137).

Refugees create cross-border tensions, and under-resourced refugee camps find it hard to cope with large numbers of displaced persons whose entitlements to food have virtually disappeared. There are military–commercial alliances where fortunes are made from protection rackets and profits from the illegal sale of looted goods, famine relief and medical supplies... the list goes on. The anti-famine contract and issues of human rights seem fragile in the face of the onslaught of war, and the reader may wonder how far the debates have come from the technical jousting of academic debating chambers where FAD, market imperfections and FED are discussed. It would be a mistake to dismiss these debates as arcane and irrelevant, since the whole issue of food security, food marketing and safety nets is rooted in theories such as these. But we also need to approach famine from the perspective of the holistic analysis of complex emergencies. This enables the political nature of famines to be identified and builds up a framework of experience on how to deal with famine and how to avoid the same problems in the future. Also, the dissemination of information about complex emergencies furthers the ends of activists by showing that famines may sometimes be promoted deliberately, but are always preventable, in addition to identifying the perpetrators of famine crimes.

The political character of some famines is much more evident in those that are directly a result of government policy, although they can vary considerably in the types of policies involved (Eide et al. 1984; Bread for the World 1991). These 'policy famines' could be considered a variant of FED, as they are dominated by shifts in people's entitlements that result from government interventions (or the deliberate decision not to intervene). Although they may sometimes be associated with natural hazards as triggers (as with the potato blight in the Irish famine), it appears that without the disruptive and damaging effects of the policy, the hazard itself is unlikely to have precipitated a disaster. Sen's early work on famine in 1981 analyses the Bengal famine of 1942 (at the time Bengal was united, and is today divided into Bangladesh, and West Bengal, a state of India). He demonstrated that there was no serious food shortage in the region, but that an estimated 1.5 to 3 million people starved because of disrupted access to work and own production, mainly as a result of the wartime policies of the British government.

Policy famines may have been responsible for more deaths in the twentieth century than the total for all other famines. For instance, the famine associated with the forced Soviet collectivisation of agriculture in the Ukraine in 1932–1933 is variously estimated to have killed between six and eight million (of a population of about 32 million).[22] In China, the introduction of rural People's Communes in 1958 led to a disruption of farm production (and, as in the Ukraine, the slaughter of livestock to prevent them from being taken over) for several years. Production was further compromised by extraction of labour for rapid rural industrialisation – the policy known as the Great Leap Forward. As local officials sought to satisfy the political objectives of the leadership, the real condition of the people in the countryside was concealed so that extractions for the urban population continued and harvests were declared good. The mortality figures are again highly contentious, and (as with the Soviet Union) information was withheld for many years. But it is generally accepted (since the mid-1980s by the Chinese authorities also) that at least 13 million people died, while some estimates suggest the total was 30 million or even more (Jowett 1990; Kane 1988; Yang 1996). Forced urban to rural displacement in Cambodia amounted to genocide (Kiljunen 1984).

There may also be 'policy famines' that result from inappropriate state interventions or deliberate absence of response. When the potato crop in Ireland failed because of the blight, the British government adhered to a laissez-faire policy of first no, and then limited, intervention. Limited food aid was eventually imported from the USA (in fact maize that could not be milled in Ireland) with no real reduction of misery. While there was mass starvation of the peasantry, cattle and wheat continued to be exported to Britain (Woodham-Smith 1962).

Causes, pressures, unsafe conditions and famine

This brief review of theories of famine indicates that even though they may have intellectual or ideological appeal, no single theory is dominant or capable of excluding the others. While each theory may have certain advantages over others (e.g. by being more pro-active or offering insights into public and political action), each famine in fact requires a unique explanation. In some cases, a hazard such as drought will impact upon vulnerable people and the 'best' explanation may be dominated by FAD or market failure approaches (in terms of hypothesis formation and empirical verification). In other situations, it may be chronic insecurity, the nature of the state and politics that will dominate, and the hazard trigger will either not exist or be unimportant. There are political considerations in the choice of explanation, which are governed both by ideological and discipline-based pre-dispositions, as well as by the nature of the famine itself. In most famines there will be an array of factors, and a narrative of the unfolding of the famine. For example, an Oxfam publicity booklet illustrates the problem of finding an explanation for one of the many famines in the Sudan, proposing the following reasons:

> The people who died did not do so because the rains failed in 1984. Despite hard work and calculation, they died because they could not grow enough food and were too poor to buy what they needed.
>
> (Cater 1986: 1)

There follows a brief description of some of the factors which combined to bring about a situation of 'not enough food' and 'too poor to buy'. These included drought, unsuitable technologies for providing water for humans and livestock, population pressure and a fragile ecosystem, deforestation and a fuel crisis, chronic uncertainties over land tenure, lack of credit, and monopolistic and monopsonistic power of merchants in rural areas. This booklet and other more academic literature (e.g. Shepherd 1988; Maxwell 1991; de Waal 1989b, 1997; de Waal and Amin 1986) provide further evidence of the importance of other larger scale and more deep-rooted causes. Curtis et al. (1988) add low income, poor import capacity and the disruption of the war to Oxfam's list (cf: Crossette 1992). Walker (1989) cites the over-cultivation of fragile sandy soils (*qoz*) in the worst affected areas of Sudan, along with drought, loss of access to land through expropriation by absentee merchants and landlords and widespread reliance on unreliable waged employment and thus on the market to buy food in situations of sky-rocketing prices. So far, FAD (with an environmental add-on factor), FED, market imperfections and the effects of war have all been suggested.

A later famine in the Sudan in 1998 (described by Deng 1999) is reviewed here using the PAR model (Figure 4.1). It illustrates both the organising power of the model, as well as the problem of allocating different factors (which interact dynamically) to the three levels (Root Causes, Dynamic Pressures, Unsafe Conditions) in the chain of explanation. Perhaps PAR can best be used as an aid for structuring a specific but theoretically informed narrative of a famine. The narrative has to be temporally and spatially situated, and so maps are used here. These signify root causes (shown in a general map of administrative regions), dynamic pressures (showing two examples: the civil war, and cattle raiding) and unsafe conditions (losses and disruptions of assets and livelihoods). However, the unfolding of (especially political) events through time, including war, armed incursions and so on, does not occur in a predictable way, as do seasonal shortages of food or secular declines in self-provisioning. Therefore the famine 'crunch' (between drought hazard on the right and the unsafe conditions) packs its destructive power both as a result of unique slow-onset factors (which are also tainted by history), but also as a result of sudden-onset, unpredictable events, which themselves modify, cancel out or magnify certain unsafe conditions or dynamic pressures.

This brings us to the second point about explanations of famine. If there are many combinations of factors and mechanisms that bring about famine, then each famine is unique. The task of building theories of famine is particularly difficult because of the complexity of each specific case. It will probably involve not only an understanding of the existing system of production, but also the distribution of food in terms of access to land and inputs as well as the operation of the market, the determination of prices and the behaviour of traders in food staples (Cannon 2003). Government policies with regard to food production and distribution and in famine relief itself may also have a profound effect. Then there is always a series of contextual events peculiar to each famine, a 'sequence of events' (Alamgir 1981), or using Currey's (1984) terminology, a 'concatenation'.

Therefore, the narrative of each event will be an important element in the explanation of particular famines, and at all times it is advisable to maintain a flexible analytical approach. For example, *Silent Violence* (Watts 1983a) is a book of over 600 pages which traces the changing contexts of famine in northern Nigeria. His research design is like a set of Chinese boxes or Russian dolls in which the local level of explanation is dependent on an understanding of the regional, national and international levels. So each instance of famine is a 'concatenation' involving factors that are related to each other by continuous changes in social relations of production, climatic fluctuations and the like. In this way, the impacts of changes in the political economy, or of climatic change, are linked to larger-scale and broader processes which can be characterised as root causes and dynamic pressures, as illustrated in the chain of explanation of the PAR model.[23]

Returning to Deng's (1999) description of the 1998 famine in southern Sudan, we see elements of FAD, FED, price ripple effects, complex emergency (including protracted war, systematic and sometimes gratuitous violence and genocide) and a natural hazard (an unusual distribution of rain-fall in 1997). One forceful conclusion to this account is that the causes of the famine were political, with long historical roots. The causes have their origins in a particular development of the (Sudanese) state from pre-colonial times, which continued through the colonial period (that unified diverse ethnic and religious groups into one country). The long-established practices of extracting tribute from weaker groups through enslavement, raiding and more organised war and destruction was perpetuated, except that by the late twentieth century it involved using aircraft, tanks and a motorised army. Donors were excluded from the war zone and thus inadequate information systems probably led to a failure to predict the crisis. The role of the natural hazard in this case probably cannot be described as a 'trigger' since the whole process would have detonated sooner or later anyway without the occurrence of unusual weather conditions. The result? Between 70,000 and 100,000 deaths.

Access and famines

A more specific and detailed development of the generalised Access model described in the last chapter, but here specifying the development of famine, is shown in Figures 4.2a and 4.2b. It is an explanatory model of 'normal life' before the operation of the more dynamic and acute processes of the PAR model, which potentially can lead to famine conditions.

In Figure 4.2a, Box 1 shows the broad political economy (social relations and flows of surplus) and provides a context in which individuals and households earn their livelihood. At the international level, certain aspects of the world economy impinge in an indirect manner on food consumption. For example, national foreign exchange reserves affect a country's ability to import grain to avert possible famines. The long-term decline in world food stocks from the 1970s onwards, combined with the entry of the then Soviet Union into the world market, has affected the ability of some countries with low foreign exchange reserves to import or hold adequate stocks of grain. At the national level, the nature of the state itself is extremely important. There must exist an ability and willingness to implement policies that reduce the impact of famine, and to having the political will to avert famine and respond to complex emergencies.

Box 1 also provides a caricatured and simplified diagram of the political economy of civil society, focusing on the social relations of production in the rural household. Other classes such as merchants, moneylenders and land-lords are usually important groups in a famine, although they seldom suffer the effects of it (on the contrary, they sometimes profit enormously from it). The social relations of production are important basic determinants of the distribution of access to the means of growing food or the income with which to buy it.

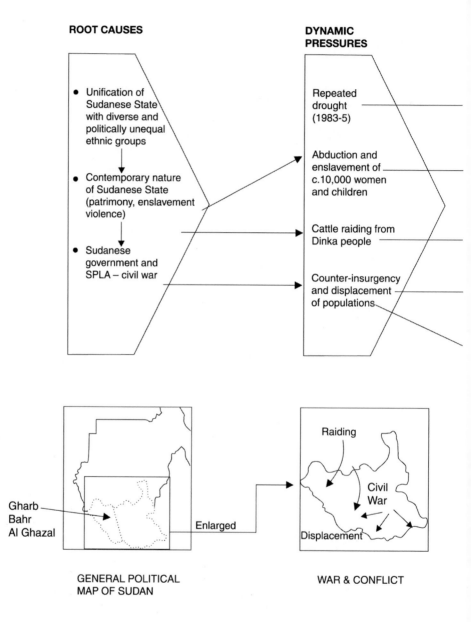

ROOT CAUSES

- Unification of Sudanese State with diverse and politically unequal ethnic groups
- Contemporary nature of Sudanese State (patrimony, enslavement violence)
- Sudanese government and SPLA – civil war

DYNAMIC PRESSURES

Repeated drought (1983-5)

Abduction and enslavement of c.10,000 women and children

Cattle raiding from Dinka people

Counter-insurgency and displacement of populations

Gharb Bahr Al Ghazal

Enlarged

GENERAL POLITICAL MAP OF SUDAN

Raiding

Civil War

Displacement

WAR & CONFLICT

Figure 4.1 Pressure and Release (PAR) model: famine in southern Sudan
Source of maps: www.reliefweb.int./w/map.nsf/home

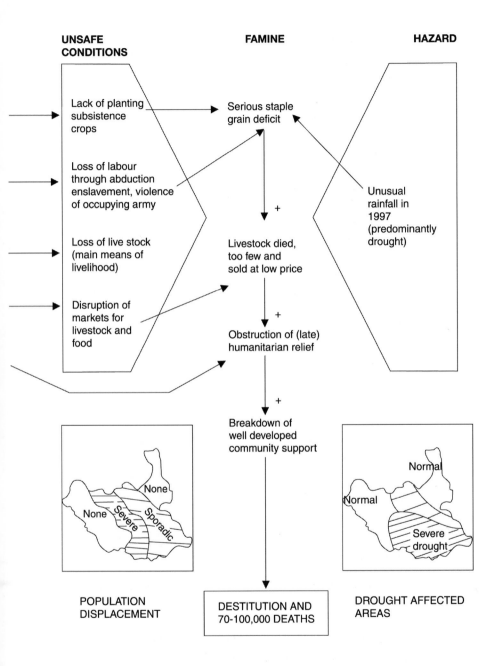

UNSAFE CONDITIONS

Lack of planting subsistence crops

Loss of labour through abduction enslavement, violence of occupying army

Loss of live stock (main means of livelihood)

Disruption of markets for livestock and food

FAMINE

Serious staple grain deficit

+

Livestock died, too few and sold at low price

+

Obstruction of (late) humanitarian relief

+

Breakdown of well developed community support

HAZARD

Unusual rainfall in 1997 (predominantly drought)

POPULATION DISPLACEMENT

None

None   Severe   Sporadic

DESTITUTION AND 70-100,000 DEATHS

DROUGHT AFFECTED AREAS

Normal

Normal

Severe drought

Figure 4.1 continued

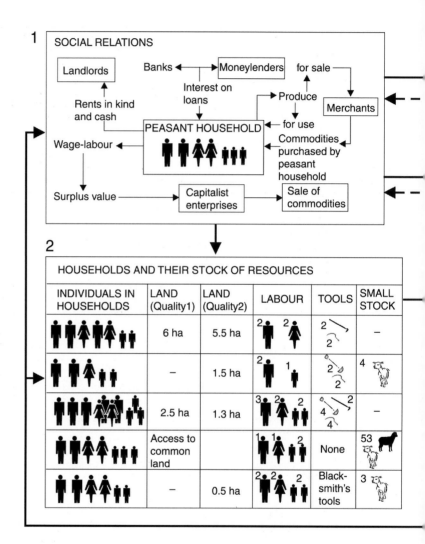

INDIVIDUALS IN HOUSEHOLDS	LAND (Quality1)	LAND (Quality2)	LABOUR	TOOLS	SMALL STOCK
	6 ha	5.5 ha	2 2	2 / 2	–
	–	1.5 ha	2 1	2 2	4
	2.5 ha	1.3 ha	3 2 2	4 2 4	–
	Access to common land		1 1 2	None	53
	–	0.5 ha	2 2 2	Black-smith's tools	3

Figure 4.2a 'Normal life' before a famine

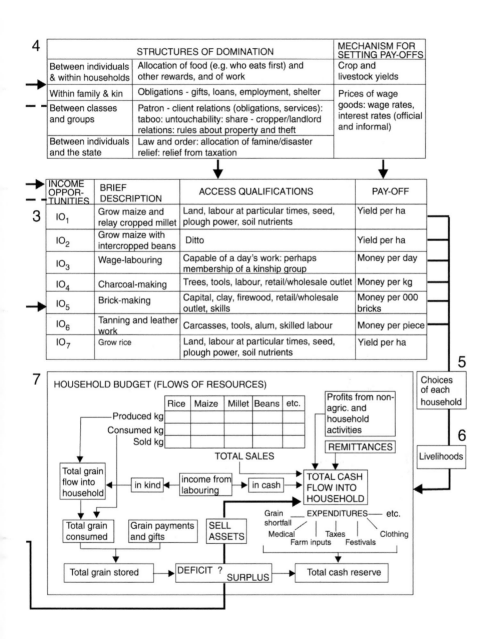

4

STRUCTURES OF DOMINATION		MECHANISM FOR SETTING PAY-OFFS
Between individuals & within households	Allocation of food (e.g. who eats first) and other rewards, and of work	Crop and livestock yields
Within family & kin	Obligations - gifts, loans, employment, shelter	Prices of wage goods: wage rates, interest rates (official and informal)
Between classes and groups	Patron - client relations (obligations, services): taboo: untouchability: share - cropper/landlord relations: rules about property and theft	
Between individuals and the state	Law and order: allocation of famine/disaster relief: relief from taxation	

3

INCOME OPPOR-TUNITIES	BRIEF DESCRIPTION	ACCESS QUALIFICATIONS	PAY-OFF
IO_1	Grow maize and relay cropped millet	Land, labour at particular times, seed, plough power, soil nutrients	Yield per ha
IO_2	Grow maize with intercropped beans	Ditto	Yield per ha
IO_3	Wage-labouring	Capable of a day's work: perhaps membership of a kinship group	Money per day
IO_4	Charcoal-making	Trees, tools, labour, retail/wholesale outlet	Money per kg
IO_5	Brick-making	Capital, clay, firewood, retail/wholesale outlet, skills	Money per 000 bricks
IO_6	Tanning and leather work	Carcasses, tools, alum, skilled labour	Money per piece
IO_7	Grow rice	Land, labour at particular times, seed, plough power, soil nutrients	Yield per ha

5 Choices of each household

6 Livelihoods

7 HOUSEHOLD BUDGET (FLOWS OF RESOURCES)

	Rice	Maize	Millet	Beans	etc.
Produced kg					
Consumed kg					
Sold kg					

TOTAL SALES

Profits from non-agric. and household activities

REMITTANCES

Total grain flow into household

in kind ← income from labouring → in cash

TOTAL CASH FLOW INTO HOUSEHOLD

Total grain consumed

Grain payments and gifts

SELL ASSETS

Grain shortfall — EXPENDITURES— etc.

Medical | Taxes | Clothing
Farm inputs | Festivals

Total grain stored → DEFICIT ? SURPLUS → Total cash reserve

Figure 4.2a continued

The individuals in households (Box 2) are therefore put in the context of the wider society and an illustrative sample of their assets and resources is listed here. The physical assets (similar, but not exactly equivalent to Sen's endowments) usually include access to land, seed, agricultural implements, etc. Other assets are social and refer to the membership of a tribe, segmented group or family as well as to rights to assistance in the form of loans, employment, food, etc. (and correspond to the social and human capital of the livelihoods approach mentioned in Chapter 3). As such, they are defined more widely than Sen's entitlements, and include various forms of expectations (or 'claims' as in Shepherd 1988) and the ability to mobilise resources commanded by others, such as labour and food in the form of gifts. In the livelihood literature, this would be described as social capital (Pelling 2002).

In Box 2 the first column shows men, women and children in pictorial form. It is important to distinguish between people of different gender and status within the household – something which Sen has taken up in later writings (1988, 1990). Sometimes 'older sisters will serve food to younger brothers ... and female etiquette may demand that women take less food and eat more slowly in the presence of the men and of guests' (Rahmato 1988: 237). Some survival strategies prioritise food intake (and ultimately survival) of men over women and adults over children and old people. It is important to establish whether these habits persist in times of actual famine. According to Rahmato they do not, but others report that there can be a considerable struggle between men and women over food resources during times of stress (Goheen 1991; Kerner and Cook 1991; Schoepf and Schoepf 1990). There is considerable evidence that females are more resilient than men to reductions in food intake. But there are some notable exceptions, as in China in the Great Leap Forward famine, where female children typically died first (Becker 1996: 3), or in the Ethiopian famine of 1972–1975, where girls under five years of age suffered much higher mortality than boys (Mariam 1986). Usually though, male mortality rates rise more than those for females.

Each of the people represented pictorially consumes a certain amount and type of food during 'normal' times. This provides nutrients that are absorbed and utilised by the body to the degree allowed by that person's state of health. Thus health and nutrition interact, and the resulting baseline nutritional status influences the ability of that person to survive food and health emergencies (see Chapter 5). 'Normal' food intake also influences the working capacity and productivity of household members as they utilise available resources.

Each household shown in Box 2 has a range of resources and assets at a point in time. The household then reviews a range of income opportunities, each with its access qualifications and payoffs as described in the general Access model (Box 3). This outcome reflects the contested interests within the household (between women and men, perhaps between generations, and between siblings). The idea of the household possessing a 'joint utility function' which subsumes individual (and often contested) utilities between males, females, children and the elderly within a household has recently been criticised. The payoff is a function of the relations of production, the technology used in production, current prices, when the output enters the market, rainfall, fertility of the soil, etc.

Box 4 shows structures of domination and allocation of resources, and are the economic, social and political expressions of the class and gender relations described in Box 1. These structures operate in different and complex ways: at the individual level within the household (usually revolving around gender and age but also genealogy); within family and kin; and between classes and groups (e.g. patron–client relations). It is also important to take account of the gender division of labour, and of expectations regarding crops and property, particularly those that normally safeguard private property (or which sanction theft under abnormal conditions). Finally, the state provides a broad framework in which rural households have to pay taxes, and are subjected to law and order (which may be exercised in favour of particular groups). The state may also engage in systematic abuse and harassment of some groups, especially on the basis of ethnicity and/or political allegiance. The structures of domination and allocation in any specific case have a major part to play in setting the payoffs for different income opportunities, and determine the level of people's access qualifications.

Each household possesses a particular package of endowments and entitlements (some as rights and expectations for future realisation). This determines the condition of the household budget, at the point in time outlined in Box 7. Here it is important to specify different foodstuffs and cash as it flows through the budget. This box has been added to the generalised Access model in order to trace the household's access to food. Apart from this, the reader will recognise this as the familiar model of a rural economy in normal times, corresponding to the model outlined in Figure 3.2.

Figure 4.2b shows the progression from unsafe conditions to a famine, as the process iterates through time to explain the slide from 'normal times' (together with unsafe conditions) into famine conditions (the 'disaster', in the more general terms of Chapter 3). For whatever reason (following trigger events or the combination of factors which activate unsafe conditions), a period of food shortages starts. Figure 4.2b illustrates the multiple changes in the access status of different households.

The components shown in Boxes 1 to 6, which had hitherto been changing relatively slowly over a matter of years or even decades, now start to change rapidly. Survival or coping strategies are put into action, usually in a sequence beginning with those that have least long-term impact on assets (Box 8). This may involve first protecting current food

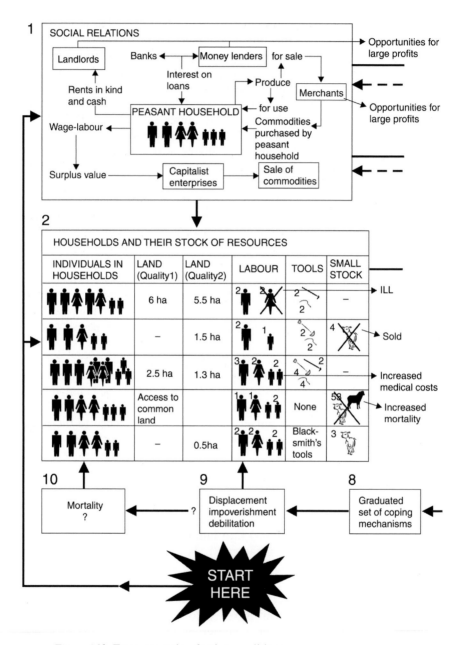

Figure 4.2b From normal to famine conditions

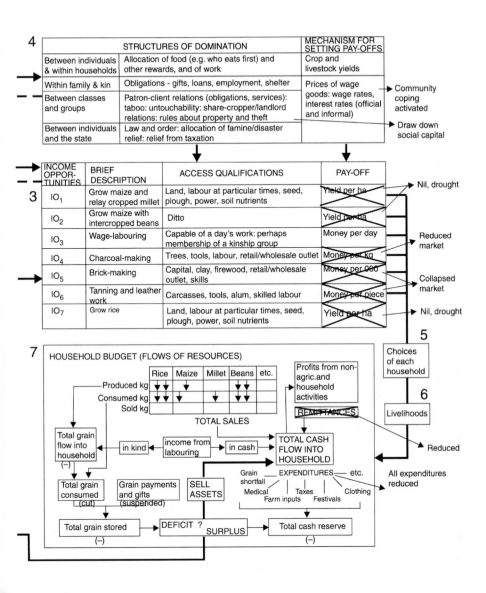

4

	STRUCTURES OF DOMINATION		MECHANISM FOR SETTING PAY-OFFS
Between individuals & within households	Allocation of food (e.g. who eats first) and other rewards, and of work		Crop and livestock yields
Within family & kin	Obligations - gifts, loans, employment, shelter		Prices of wage goods: wage rates, interest rates (official and informal)
Between classes and groups	Patron-client relations (obligations, services): taboo: untouchability: share-cropper/landlord relations: rules about property and theft		
Between individuals and the state	Law and order: allocation of famine/disaster relief: relief from taxation		

→ Community coping activated

→ Draw down social capital

3

INCOME OPPOR-TUNITIES	BRIEF DESCRIPTION	ACCESS QUALIFICATIONS	PAY-OFF
IO_1	Grow maize and relay cropped millet	Land, labour at particular times, seed, plough, power, soil nutrients	Yield per ha
IO_2	Grow maize with intercropped beans	Ditto	Yield per ha
IO_3	Wage-labouring	Capable of a day's work: perhaps membership of a kinship group	Money per day
IO_4	Charcoal-making	Trees, tools, labour, retail/wholesale outlet	Money per kg
IO_5	Brick-making	Capital, clay, firewood, retail/wholesale outlet, skills	Money per 000
IO_6	Tanning and leather work	Carcasses, tools, alum, skilled labour	Money per piece
IO_7	Grow rice	Land, labour at particular times, seed, plough, power, soil nutrients	Yield per ha

→ Nil, drought

→ Reduced market

→ Collapsed market

→ Nil, drought

5 Choices of each household

6 Livelihoods

→ Reduced

7

HOUSEHOLD BUDGET (FLOWS OF RESOURCES)

	Rice	Maize	Millet	Beans	etc.
Produced kg	↓↓	↓↓		↓↓	
Consumed kg	↓↓	↓		↓↓	
Sold kg					

TOTAL SALES

Profits from non-agric.and household activities

REMITTANCES

Total grain flow into household (−) ← in kind ← income from labouring → in cash → TOTAL CASH FLOW INTO HOUSEHOLD

Total grain consumed (cut) | Grain payments and gifts (suspended) | SELL ASSETS

Grain shortfall | EXPENDITURES — etc.
Medical | Taxes | Clothing
Farm inputs | Festivals

All expenditures reduced

Total grain stored (−) → DEFICIT ? SURPLUS → Total cash reserve (−)

Figure 4.2b continued

consumption but without endangering future food intake. But when conditions deteriorate, more radical strategies may be adopted that do undermine future access to food. In the meantime, coping strategies may allow the mobilisation of new resources in the short term. The household budget for those with a less well-developed access profile in normal times suffers, and food stocks in some households run down. Market relations also becomes crucial when those without the means of producing food have to support themselves through purchases. The economic debates about the operation of the market for food grains outlined above are relevant here. Also the 'anthropology of famine' becomes significant, especially how much the rules and institutions of society can adapt in abnormal times (Rangasami 1986).

For some authors, societal breakdown is both a marker and cause of famine (Alamgir 1980; Currey 1978; Cutler 1985). In the following example, however, Rangasami describes Firth's (1959) analysis of the Tikopia famine in Polynesia, arguing that there was no major 'societal breakdown':

> The skeleton of the social order was preserved though attenuated in content. He [Firth] offers interesting insights into famine: what the famine did was to reveal the solidarity of the elementary family. Even at the height of the famine, it appeared that within the elementary family full sharing of food continued to be the norm.
>
> (Rangasami 1986: 1597)

In other cases (as in the famine of southern Sudan, outlined earlier in Figure 4.1), Boxes 9 and 10 (Figure 4.2b) refer to changes in the ability of the individual and the household to continue to cope effectively with stress and to avail themselves of food by whatever means. If food intake falls for long enough, the capability of people to look for and perform work becomes impaired (Box 9). If food shortages persist, morbidity and, later, mortality increases (Box 10). After some months of severe food shortage, the stock of assets and resources of a household may be seriously impaired. Assets may have been disposed of to purchase food: perhaps cattle are sold, some land mortgaged, the seed corn eaten. Members of the household may themselves be sick (or have died), so that their capacity to earn through work is reduced.

Class relations and the associated structures of domination and allocation (Boxes 1 & 4) may also shift rapidly as a result of famine. While many households suffer, some merchants and moneylenders can amass fortunes by selling grain at inflated prices, foreclosing on mortgages (especially for land) and outstanding debts. Some may also gain bonded labour in return for food.

This approach to famine analyses its structures and processes in relation to ways of making a living in normal times. Even normal times have inherent unsafe conditions, but the risks from these are distributed differently between households. It is an iterative approach in which 'external' shocks and triggers (such as drought or war) have an impact on the structures and processes of political economy. Once again, these changes in the political economy can affect individuals and households differently. However, some external interventions may help people to survive, and these are illustrated in the next section.

Policy

Policy options can be examined in relation to the 'chain of explanation' which accounts for famine. This chain involves analysing the mechanisms which produce vulnerability by tracing back from the immediate and local causes (in space and time) to the less direct, longer-term and structural reasons (underlying dynamic pressures and root causes). The chain of explanation is paralleled by the PAR model) presented in Chapter 2, in which the root causes are linked to the more direct pressures that create vulnerability. When these overlie particularly adverse local conditions, the impact of a hazard event may act as a trigger for disaster.

Policy actions must address issues at the level that can be affected and altered. For example, the colonial history of the Sudan and the nature of the post-colonial state cannot be re-written. Instead, responsible outside agents may have to deal with how to ensure relief gets through war zones, how international agencies protect refugee camps from occupying armies or bands of insurrectionist troops, and so on. While an integrated explanation of a disaster may be intellectually fulfilling, a policy has to locate itself at the level it can make a significant impact. There must be effectiveness in the short run at least – human lives often depend on it. Yet the opposite pitfall also awaits the policy maker. As Kent warns:

> By dealing with disasters as isolated phenomena, we lose a sense of the real causes of vulnerability. Conceptually, it is a way of avoiding the full implications of the causes and solutions to disasters. One need not address the global inter-relationships between international trade, currency fluctuations, geo-political and commercial interests and a flooded delta in Bangladesh. In practical terms, by isolating disaster phenomena, one can demonstrate goodwill and test one's technological solutions without being 'mouse trapped' into more long-term commitments.
>
> (R. C. Kent 1987a: 174–175).

For each link in the chain of explanation of a famine there are a range of policy measures. At the international and national levels, achieving aggregate food security has been an important policy goal. The food security debate of the 1980s was heavily influenced by the work of Sen, and a number of plans emerged (especially among NGOs) that focused on access to food (through entitlements). But neo-liberalism was much more significant in driving policy, mainly through the Structural Adjustment Programmes (SAPs). These dominated the debate, and seriously curtailed scope for other forms of action.[24] But by the mid-1980s the dangerous implications of structural adjustment for food security and longer-term entitlement protection had become clear to many. Food security approaches continued their fitful career on the international stage from the 1990s up to the present time (Devereux and Maxwell 2001).

In the 1990s, significant changes in both the global political scene and thinking about famine once again shifted attention. The World Bank gave greater priority to poverty rather than food security issues. Also, the character of famine itself changed in terms of location and causes, with a shift from Asia to Africa, and from natural hazards to war and political instability. Where drought did occur and could potentially have caused hunger, it seemed that famines were more or less successfully averted. Attention therefore was directed to the management of complex emergencies in the short term, and diverted to more strategic issues.[25]

The Declaration and Plan of Action (FAO 1996) was a sort of balancing act between neo-Malthusian concerns of global food shortages, food self-sufficiency and human rights issues, which were now enshrined for the first time.[26] What this continuing flurry of grand international meetings and wish-lists actually means is open to discussion. To some extent, such rhetoric is important. It may not predict or explain famine, and therefore does little to reduce or prevent it, but it is a weapon that can be used to shape debates and stimulate more general international action.

Early warning systems

Famine Early Warning Systems (FEWS) are tools that have been developed particularly since the famines in the Sahel during the 1970s (although they can be traced back to nineteenth-century Indian Famine Codes). Information about present and future access to food is an essential, though not sufficient, resource in the prevention of famine. Even if early warning systems can provide useful information, it may never be used. Government structures are often poorly linked to those that produce the information. Data do not necessarily inform policy, and emergencies (rather than longer-term prevention measures) dominate the political concerns of most governments whose peoples are threatened by famine (Buchanan-Smith and Davies 1995). Institutional weaknesses in national and international policies can also allow famine conditions to emerge, in spite of adequate information. Complex emergencies (particularly involving war), and practical difficulties such as inadequate port facilities, poor or disrupted and damaged transport facilities such as roads and railways, can all contribute to famine even when there is adequate knowledge of the pre-conditions.

FEWS are tools that enable the measurement of a range of factors that affect people's food status (Hervio 1987; McIntire 1987; Downing 1991). There are many (some analysts think far too many) different approaches to such systems. For example, there are about forty FEWS developed for use in Africa, with at least six rival systems in Mali alone. There is the FAO's Global Information and Early Warning System developed in the 1970s, which predicted crop yields by estimating biomass from satellite imagery. New technologies such as satellite imagery, the internet, computer networking and established reporting systems have transformed the production and potential for dissemination of information. But it is not clear how far they will remain useful in relation to the shifting pattern of famines.

Warning systems such as these that focus on the supply-side suffer from the same problems as the FAD theories which underpin them. While useful, they cannot identify who actually will have access to food in sufficient quantity, and who will fail to do so. Such FEWS have a similar and inadequate intellectual ancestry as the food security programmes discussed above. Others work on the principle of a food balance sheet, where food production minus food needs equals imports plus food aid. This type of warning system, operating at the national level, fails to incorporate the essential element of *access*, and so has limited explanatory value. More sophisticated systems have since been developed, although the practicality of collecting the data needed for an ideal system is doubtful (Nichols 1988). Various types of indicators of famine have also been used as additional warnings (Cutler 1985). While some of these systems have become competent in predicting drought, they cannot warn of changing political situations, particularly in complex emergencies. Devereux suggests a number of improvements (Devereux 2001: 211–212). Most are linked directly to the directions suggested by both the PAR and Access models. These include: (1) broadening the focus from extreme events to food systems; (2) addressing chronic food insecurity and a failure of access to food; (3) improving information about market access to food, including the sale and purchase of food crops by the vulnerable; and (4) increasing the role of participatory food security assessments, in spite of scaling-up and consistency problems (see the following sections).

Strengthening livelihood systems

Moving to less direct measures for preventing famine, the strengthening of rural livelihoods is an obvious policy requirement. 'Sustainable livelihoods' is indeed the emblematic intellectual tool to guide policies with which the present British government hopes to halve world poverty by 2015. Of course, these measures have objectives that are not directly concerned with famine prevention, but have broader development and environmental objectives. Swift and Hamilton (2001) claim that a livelihoods approach (which follows our Access model in most important respects) allows a much more holistic view of food security issues. Livelihood diversification (especially involving urban–rural links, remittances and artisan activity) is an important process which is rapidly increasing in most areas of the developing world (Ellis 2000). It also draws attention away from a narrow 'food first' approach to a much broader spectrum of policy issues which address access to income opportunities in all sectors. This is taken up again in Chapter 9.

Response to famine from the grassroots level

Most of the policy options discussed so far are 'top down'. Complementary and sometimes conflicting responses to the trigger events that may bring famine can also be seen at the grassroots level. In addition to household- and community-level coping, discussed in Chapter 3, various NGOs have attempted many of the interventions just discussed 'from the bottom up'. Organisations close to grassroots (e.g. Save the Children Fund) are well placed to provide certain kinds of information for early warning systems to complement the coarse data provided by satellite. These systems are run by local people, who compile information on food and livestock prices and the sale of assets by households, among other indicators of a potential food emergency (York 1985; Cutler 1984; Beerlandt and Huysman 1999; SIDS 2003a, 2003b).

Many NGOs have found themselves providing famine relief, sometimes on behalf of the government or international donors, sometimes independently in isolated areas that do not receive official assistance. Some NGOs have tried to administer food relief in ways that strengthen local livelihoods in the long run (Scott 1987; Slim and Mitchell 1990). For instance, Operation Hunger in South Africa supported people in the so-called Bantustans who faced widespread hunger during the drought of 1991–1993. They supported women's irrigated gardening groups as well as providing conventional mass feeding.

Other grassroots initiatives aim to support livelihoods. For example, in Burkina Faso and other Sahelian countries, the Federation of Village Development (*naam*) Groups sponsored grain storage on a co-operative basis. This releases farmers from the cycle of indebtedness caused by having to sell at a low price every year to a trader, followed by purchase of grain from that same trader at a higher price later during the 'hungry season' (Adamson 1982; Uemura n.d.; Six-S n.d.; Woodrow 1989). The Grameen Bank in Bangladesh serves a similar function, providing credit for the 'non-creditworthy' in order to start up income-generating activities (see Chapter 9). Earlier we argued that failure of exchange entitlements are generally more important than decline in output as an explanation of famine. This suggests that extra income would be especially important in preventing famine, and that such schemes are capable of altering the pattern of access, enabling people to improve long-term entitlements and thereby reducing the risk of them facing famine.

Many other efforts to strengthen livelihoods have been directed towards rural people vulnerable to drought. One of the most interesting of these was undertaken by the British charity War on Want in Mauritania following the Sahel drought and famine 1967–1973. Rather than spend all the funds collected during the crisis on immediate relief, they launched a thorough study of the existing livelihood system of the people living in the Guidimaka region of Mauritania bordering the Senegal river (Bradley et al. 1977). On the basis of this study they proposed a long-term development programme centred on and controlled by village associations. The programme featured low-cost improvements of existing agricultural practices, which already included very sophisticated flood-retreat agriculture and careful livestock management. They attempted to diversify the livelihood system through the addition of fruit trees and craft industries. Theirs was one of the first programmes to introduce small bunds (embankments) to increase the infiltration of scarce rainfall into the soil – an innovation that was later widely promoted throughout the Sahel (Twose 1985; Harrison 1987; Murwira et al. 2000).

Conclusion

The causes of famine are too complex and varied for single theories to claim universal applicability, although the protagonists of some have done so, to the detriment of our understanding. The arcane debate in the hallowed groves of academe concerning FAD (food availability decline) versus FED (food entitlement decline) is a case in point. There is no single theory of famine but a number of theories of circumscribed relevance and ability to predict. In addition, it might be argued that some theories or propositions about famine are more politically helpful than others. For example, de Waal's notion of famine crimes throws a spotlight on the issue of who should bear responsibility, as in his view all famines are preventable. On the other hand, the FAD context may be helpful in understanding the processes which lead to some famines. But it still grants a significant role to 'extreme' natural events, which focuses attention on unpredictable nature. This means that it can avoid the analysis of how the history of vulnerability (involving impoverishment, marginalisation, persecution, dispossession and ethnic cleansing, etc.) operates to provide the context for the trigger event. The iterative access model allows a number of theories to operate, but under conditions relevant to each particular famine. Analysing famine in this way makes it possible to evolve a more flexible analytical method for explaining them, which in turn may assist the policy response to be more politically aware, relevant and flexible.

Notes

1 There is an enormous literature on hunger in history, including an overview of the evidence in pre-history (Cohen 1977). Recorded history is traced by De Castro (1977/1952), Sorokin (1975), Pankhurst (1974) and Arnold (1988); for the Greco-Roman world see Garnsey (1988); for ancient China, Mallory (1926). There is also a thorough interdisciplinary view edited by Newman (1990).
2 Reviews of the natural and physical science of drought can be found in collections edited by Wilhite and Easterling (1987) and Wilhite (2000).
3 As we observed in Chapters 1 and 2, war, and violent conflict more generally, have many links with disaster. The war in Iraq diverted some policy makers' attention away from the food crisis in southern Africa, and by April 2003 the WFP was $1 billion short in its funds needed for humanitarian operations in Africa (including southern Africa). In all, 40 million Africans were at risk, and the WFP executive director, James Morris, told the UN Security Council: 'As much as I don't like it, I cannot escape the thought that we have a double standard' (Carroll 2003: 1).
4 There were antecedents, especially in a critical understanding of hunger and famine: e.g. Gini and De Castro (1928). De Castro contributed more than half a century of famine studies that deviated from conventional wisdom by emphasising social relations. A Brazilian medical doctor and geographer, De Castro began documenting chronic under-nutrition and what we would today call 'vulnerability' in Latin America in the 1920s. In the early days of the United Nations he served the Food and Agricultural Organisation (FAO) in its nutrition division, but became disenchanted by international efforts to prevent famine. There is a bibliography of De Castro's major writing on famine and critical review in Wisner (1982); his best-known works include *Geopolitics of Hunger* (1977/1952) and *Death in the Northeast* (1966).
5 Data from World Food Programme (2002a); for regional context see MacGregor (2002), FAO (2002b).
6 Data from WHO (2002).
7 Data from United Nations Office for the Coordination of Humanitarian Affairs (OCHA 2002e).
8 Data from International Federation of Red Cross and Red Crescent Societies (IFRC 2002d).
9 Data from World Food Programme (2002b).
10 The following endnote appears in the original as note 4:

> Local estimates wrongly suggested that last year's maize shortfall of 600,000 MT could be partly offset by an increase in crop and tuber production. This resulted in complacency within the Malawi Government and the donor community. Nevertheless, the IMF and World Bank generate their own independent information data – which makes it all the more surprising that they and the European Commission (authors of the report that formed the basis of the IMF advice) could also get it so wrong. As early as July 2001, The Famine Early Warning System Network (FEWS NET), a USAID funded project, had noted the worrying signs. By September 2001, private traders were stockpiling maize to enable them to sell at higher prices later. The national maize stocks had run down to only 15,000 MT. Official warnings about serious food shortages circulated in October 2001, with reports from Save the Children (UK) and World Vision. A food security assessment conducted by the EU and the World Food Programme also reported that up to 25 per cent of households in 35 'food insecure extension planning areas' required relief assistance.
>
> (Owusu and Ng'ambi 2002: 49)

11 Southern African Development Co-ordinating Conference (SADCC), facilitated this logistical feat. SADCC was created in 1980 by Mozambique, Angola, Zimbabwe, Botswana, Swaziland, Zambia, Malawi, Lesotho, Tanzania and Namibia to provide a common front against apartheid South Africa. Since the end of apartheid, South Africa has joined the group, which renamed itself the Southern African Development Community (SADC). On their common food policy, see Morgan (1988); Prah (1988); Rau (1991:125–128); Wisner (1992a). See also SADC's Food Security Unit: http://www.sadc-fanr.org.zw/

12 Ninety per cent of the famine deaths during this period were in Mozambique and Angola, where civil wars had weakened the population and made transportation difficult (Green 1994: 44–45).

13 However, in an act reminiscent of the solidarity shown in 1991, neighbouring Tanzania did donate 2,000 tonnes of maize to Malawi (Agence France-Presse 2002e).

14 Agence France-Presse (2002f).

15 See *Economist* (2002a); FAO (2002).

16 Neo-Malthusianism is the term used for the significant resurrection of Malthus's ideas mainly in the period since the Second World War, when the colonial powers (and the USA) were confronting the independence of the emerging Third World. No longer directly responsible for their colonial dependents, Malthusian ideas resurfaced in the West as a means of blaming the people of the ex-colonies for the potential disruption of Western wealth, and as a way to explain the problems of the developing countries and their poverty.

17 For a general review and critical treatment of the 'pressure on resources' thesis, see Blaikie and Brookfield (1987).

18 Drèze (1988: 19) says that exports from India were common during famines in the nineteenth century.

19 'The underlying cause of famine is crop failure which undermines incomes of the already poor' (Mellor and Gavian 1987: 539). Platteau (1988) emphasises the importance of declining aggregate food production in Africa, but attributes it more to backward agricultural technology and pricing policies, as do many other authors. For the alternative emphasis, especially on the influence of government policy (e.g. pricing, credit, marketing, infrastructure, research and development) on African food production, see Bates (1981); Berry (1984); Glantz (1987); Raikes (1988); Akong'a (1988); Wisner (1988a); Odhiambo et al. (1988); Cheru (1989); Bernstein (1990); Achebe et al. (1990); Rau (1991) and Wisner (1992a, 1992b).

20 One could also argue in relation to landslides, earthquakes, etc. that people have a human right to protection from avoidable harm; see Wisner (2001e) and Chapter 9.

21 There is a more direct breach of contract when governments ignore the food needs of people by claiming there is no shortage, and when they deliberately withhold food from the hungry in order to punish them or force them into submission. Examples of this have been seen in recent civil conflicts in Ethiopia, Sudan and Zimbabwe. It has also occurred with international aid and the withholding of food aid by those who hold surpluses in order to achieve political goals (Cannon 2003).

22 Seventy years after the event, the Ukraine famine remains highly controversial, with heated disputes about its causes, and arguments about how many died. It is a highly politicised debate: some protagonists are sympathetic to the original communist cause (and suggest that high mortality figures are exaggerated right-wing propaganda); others are pro-Ukrainian nationalists who tend to maximise the numbers of deaths in order to blame both Russian domination and Soviet oppression. Some sources argue that many deaths were caused by epidemics, especially of typhus, and were not necessarily the result of hunger. A range of these arguments can be found easily by entering the terms 'Ukraine', 'famine' and '1930s' into an internet search engine such as Google.

23 Additional discussion of northern Nigeria is provided by Mortimore (1989). Similar treatments of famine's 'chain of explanation' are provided for Malawi by Vaughan (1987), for Kenya in 1971 by Wisner (1978b, 1980, 1988a), for Kenya in 1984 by Downing et al. (1989), for Ethiopia by Mariam (1986), Lemma 1985, and Kebbede (1992). Many authors have sought out 'root causes' of famine in Africa. For instance, Meillassoux (1973, 1974) and Copans (1975, 1983) are typical of the analysis at the time of famine in the west African Sahel region. Franke and Chasin (1980) reviewed and summarised much of the French literature of the period. Elsewhere in Africa, Wisner and Mbithi (1974), Wisner (1975, 1977, 1980) had much the same thing to say about famine in east Africa, while Bondestam (1974) and Kloos (1982) elaborated a 'political economy' explanation for the Awash Valley (in Ethiopia), and Hussein (1976) extended the analysis to Ethiopia as a whole. Similar analysis of the role of class relations in undermining traditional drought coping methods and in causing environmental destruction (which made the land more prone to the effect of drought) also appeared for southern Africa (Cliffe and Moorsom 1979; cf. Wilmsen 1989).

24 The Structural Adjustment Participatory Review International Network (SAPRIN) has produced an excellent research review of SAPs and their generally harmful effects on food security (SAPRIN 2003).See also Cannon 2003.

25 In dealing with the humanitarian crises of the 1990s, much attention was given to emergency feeding on a very large scale. The worldwide work of the WFP grew considerably. Much has been learned by field administrators and government officials over the last two decades. There is a body of experience on the management of famine refugee camps (Harrell-Bond 1986), provision of food aid (Drèze and Sen 1989), and early warning systems (Walker 1989; Borton 1994). See also Borton (1988) for an excellent review of recent British experiences in organising famine and emergency relief. On making food aid more effective, see Singer, Wood and Jennings (1987); Scott (1987); Cohen and Lewis (1987); Hopkins (1987) and Raikes (1988). Jackson (1982) and Crow (1990) are more critical of the aid process, especially the provision of food aid. R. C. Kent (1987) has summarised a large body of increasingly critical commentary on relief aid in many forms.

26 The document is available at: http://www.fao.org/docrep/003/w3613e/w3613e00.htm

5

BIOLOGICAL HAZARDS

Introduction

Human health, daily life and vulnerability

In Part I we introduced the notion of 'vulnerability' and its relations to a set of processes involving access to resources in the maintenance of livelihoods. In Chapter 4 (in Part II) we began to apply these concepts, and suggested that famine is seldom caused by extreme climatic conditions (such as drought) alone, and that a severe decline in food consumption which might be expected to result from a drought may either not occur at all, or may not be the prime cause of a disastrous famine. What is of prime importance is people's vulnerability, brought about by long-term or sudden disruptions to their access to resources of all kinds, both material and non-material, and people's ability to use them in the successful pursuit of a livelihood. In a similar way, biological hazards may be both a trigger to a disaster and exacerbate its consequences, or follow on from other socio-economic root causes and unsafe conditions, once the disaster process is under way.

The present chapter focuses on hazards that originate in the life processes and conditions of other living things that affect humans and their livelihoods and assets. On the macro-scale, humans and their built environments are often surrounded by vegetation. Forests, for example, can provide protection from high winds and a reserve of survival food as a buffer against famine. However, they can also burn catastrophically.[1] Macrofauna (organisms visible to the naked eye) can be beneficial to humans, but some animals, such as rats, some birds and also insects, can damage crops or stored food. On the microbiological scale, minute organisms can also benefit or harm humans with 'biological disasters' that can affect both people (epidemics) and their animals or crops (epizootics, explosive plant disease and pest infestations).

In Part I of this book we emphasised how the conditions of daily life, to a large extent, account for the vulnerability of individuals, households and social groups. Daily life is, above all, *biological*, so it is important to recall this theme of the 'everyday' or the 'normal' (Lefebvre 1991; Wisner 1993;

Harvey 1996). We all eat, breathe and drink water. We are one type of organism among other living things. However much our 'embodiment' (Weiss and Haber 1999; Kruks 2002) may be culturally or socially 'constructed' and interpreted, our existence and our daily lives revolve around the care and use of bodies.

The emergence of the 'risk society' (see Chapter 1) was predicated on the growing complexity of modern technologies. It also relates to the view that science had become increasingly necessary to inform the public and public policy, and yet remained inadequate for that task. In richer countries, biological risks are increasingly connected in people's minds with very large, sometimes unimaginable risks (such as a major nuclear accident), or with risks that seem even to threaten our biological identity (maybe this is one reason why there has been so much controversy over the patenting of life forms and the buying and selling of human body parts; see Andrews and Nelkin 2001).

The fact of our embodiment and biological dependency has also meant that development studies has turned to health as one of the main indicators of 'human development', as in the UNDP Human Development Index (HDI) which includes measures of health. The World Health Organisation (WHO) prefers to measure the average number of DALYs (Disability Adjusted Life Years) lived by a population group, and not simply the simple life expectancy.[2] Philosophers Doyal and Gough (1991) propose a theory of human need in which health and autonomy are the basic elements.[3]

Under the 'magnifying glass' of the ordinary daily life of people in Chapter 2, we saw a series of 'unsafe conditions' lying dormant, to be activated by the trigger of an extreme natural event. 'Normal' health status is clearly among these conditions of daily life. Although planning and policy making require us to step back and generalise, it is important that we remember also the fine-grained detail of the biological in daily life, 'the visceral, mortal and, above all, interconnected rhythms of living people, animals, plants and places' (Whatmore 1999: 159).

What are biological hazards?

Biological hazards include micro-organisms such as those responsible for epidemic human disease, 'epidemics' (epizootics) such as rinderpest, foot/hoof and mouth and swine fever that affect livestock, and diseases of plants (especially crops). Insects (mosquitoes, rats, lice, fleas) and other animals (dioch birds, locusts, army worms, grasshoppers) can transmit disease or can destroy crops.

To varying degrees, humans have developed social (as well as biological) resistance to such hazards. Human culture has also developed ways of tolerating crop and animal losses up to certain levels. Indeed, people may be ambivalent towards certain pests (the grasshoppers in south-east Nigeria that constitute the major crop hazard are seen as a food source by women and children) (Richards 1985). Elaborate adjustments of techno-social systems have developed in the face of diseases of plants and animals, and of crop losses due to pests and vermin (Mascarenhas 1971) and post-harvest losses (Bates 1986). In most cases the existence of a 'normal surplus' is enough simply to absorb such losses. That is, during 'normal' times people's subsistence systems tend to *overproduce* beyond subsistence needs to ensure that needs will still be met in all but the very worst times (Allan 1965; Porter 1979). What has been common practice among peasant farmers and pastoralists for centuries has recently been rediscovered in the context of European and North American agriculture as 'integrated pest management'. A degree of loss is tolerated until it exceeds the marginal cost of action against the pest (Altieri 1987).

Apart from specialised public health writing, the disaster literature has tended to neglect biological hazards.[4] Epidemics and other biological hazards were also excluded from the programme of the UN International Decade for Natural Disaster Reduction (IDNDR). With increased public awareness of the potential for biological terrorism and awareness of high-profile epidemics such as HIV-AIDS and SARS,[5] both specialist and more general writing on the subject is increasing. However, an understanding of biological hazards has even wider application to the field of disaster management. It permits better comprehension of how health problems contribute to people's vulnerability to other hazards, and also allows under-standing of the health impacts of other hazards, for example where floods expose people to new health risks.

In this chapter, biological hazards are understood in several ways. They can be seen to be partly independent of and prior to other types of hazards (that is, they develop into, contribute to, or exacerbate vulnerability leading to disasters). The HIV-AIDS pandemic in southern Africa, discussed later, is such an example. There are also biological hazards that result from disas-trous events. War and drought can lead to malnutrition and famine conditions, and these, in turn, make outbreaks of disease more likely and reduce resistance to these diseases, leading to much higher rates of morbidity and mortality and precipitating a more rapid and profound descent from 'normal life', as explained in Chapters 3 and 4. Finally, biolog-ical hazards can act as the trigger of disasters in their own right, as with serious pest infestations, the potato fungal blight in the Irish famine of 1845–1848 (see below), the worldwide influenza pandemic of 1917–1919 or the outbreak of a previously unknown virus responsible for severe acute respiratory syndrome (SARS) in 2003.

Limitations to our treatment of biological hazards

In Chapter 1 we cautioned that differential vulnerability may not be a major determinant of the distribution of impact for all disasters, or if it is, then it is sometimes in latent and complex forms for which there may be no clear-cut policy responses. Differential vulnerability across populations (which is usually also variable across space and time) is relevant in most disasters. But there may be 'limiting cases', where vulnerability resulting from social processes are of little significance for the outcome of the disaster. In such cases, the issue of how a hazard's impact is distributed among a particular population would be irrelevant. The Black Death in fourteenth-century Europe may have been such a case.

In the past, a disease outbreak over a large area, affecting a high proportion of the population, may not have respected the social class or other differential characteristics of the population. These pandemics contrast strongly with more limited disease outbreaks (e.g. plague in London in the mid-1660s or yellow fever in Memphis, Tennessee in the 1870s[6]) when the wealthier residents escaped into the countryside to wait out the epidemics. In the present situation, with HIV-AIDS in developing countries, infection is clearly determined by social factors, including gender attitudes and sexual practice, social class, geographical location and access to costly medical care. The extension of life and the likelihood of dying are socially determined: a relatively high mortality or reduced life span are concentrated among people too poor to afford therapeutic drugs or in countries which have a poorly developed distribution system of health care.[7] This chapter will explore such limits to the applicability of 'vulnerability' as a concept in relation to biological disasters. Severe and widespread plagues, pestilence and infestations, as well as biological terrorism, serve as an extreme test of the vulnerability concept.

Bujra, in judging the appropriateness of the 'risk society' (see Chapter 1) framework for understanding HIV-AIDS in Africa, also explores that boundary.

> Whilst the patterns of transmission and spread of AIDS in Africa may be linked to the changes wrought by capitalist penetration, the disease itself is not a product of 'science', that is, equivalent to pollution, toxicity, the nuclear threat or genetic engineering.
>
> (Bujra 2000: 62)

But what of the case of arsenic poisoning in Bangladesh? There, an estimated 77 million people are suffering chronic arsenic poisoning because an ambitious, donor-funded well-digging project in the 1980s failed to test groundwater for this substance.[8] The project has become an enormous controversy involving UNICEF and the British Geological Survey (BGS), the agency involved that failed to test the water for arsenic.[9] This is a case of how 'science', in the form of humanitarian and 'top down' projects in less developed countries (LDCs), can have unforeseen consequences for human health, just as 'science' in industrial countries has created 'diseases of modernity'.

We also exclude from treatment in this chapter the threat of biological terrorism. The 'science' devoted to building military stockpiles and technologies of dispersion during the Cold War can be credited with creating this hazard.[10] The anthrax spores sent in the US mail that killed several people and became infamous for shutting down the US Congressional office buildings in 2001 had their origins in US Army laboratories. This risk is a classic example of the construction of the 'risk society' (Beck 1992). Although it could be argued that in this case it was predominantly black, working-class postal employees who were more vulnerable to harm from letters containing anthrax, it is more complex than that. We have to set the subject of bio-terror aside due to these complications and to limits on the size of this book.

However, in one important respect the concern about biological terrorism and other forms of domestic attack in the USA, and elsewhere, has a direct bearing on our book. Firstly, budget priorities in the US have been diverted from routine public health activities (occupational health, environmental health) and extra money ($6 billion) provided for laboratory identification of terror pathogens and rapid communication of information about outbreaks. The running-down of 'routine' public health activities is possibly a much greater risk than missing the start of an epidemic created by terrorists (Stolberg 2002a: A20). Although a counter-argument has been made, namely that the rapid response to the SARS epidemic of the US Centers for Disease Control and Prevention and its improved links with state and municipal health authorities are the result of enhancements of the public health system brought about by investments in preparedness for biological terrorism (Goldstein 2003).

The inward-looking American preoccupation with 'homeland security' may also cause reductions in foreign aid in the fight against 'natural' biological hazards that far outweigh the risks from bio-terrorism. When UN Secretary General Kofi Annan requested more funding from the US President for the global campaign against HIV-AIDS, tuberculosis and malaria, the response was that the needs of the war against terrorism came first and limited the ability of the US to contribute.[11] It must be said that public health expenditure in any country is determined by many complex political and economic factors. However, in already poor countries with large numbers of vulnerable people, reductions in state-financed health care often greatly increase the level of biological risks. The result is likely to be increased morbidity, disability and mortality, and the deepening of vulnerability over time.

The US decision to ignore the 1997 Kyoto Treaty on global climate change and to introduce, instead, a system of voluntary increases in energy efficiency based on tax incentives may also have a large impact on biological hazards worldwide. Whatever effect the US faith in market-based solutions, it cannot possibly reduce its carbon emissions to the 1990 level by 2010, which is the Kyoto goal. Current international scientific consensus is that even if this goal were reached, profound changes in the distribution of agro-ecological zones, disease vector habitats and extreme climate hazards will occur. All of these changes will bring with them health consequences (as well as interconnected social, economic and political consequences) for millions of people.[12]

Biological links with other hazards

Many other hazards such as floods, drought and tropical cyclones can have potentially severe biological and epidemiological consequences as secondary effects (Noji 1997; de Boer and Dubouloz 2000). Therefore this chapter shares some concerns with other chapters in Part II. For instance, flooding in LDCs is commonly followed by epidemics of diarrhoea due to contamination of drinking water. Volcanic eruptions are often followed by outbreaks of respiratory problems due to the ash fall and emission of gases (Baxter 1997). There is no doubt that people who survive the immediate impact of some hazards can then suffer additionally from a consequent health crisis.

Conversely, vulnerability to the impacts of many hazards (e.g. drought, heat waves, severe winter storms) is increased in populations with chronically bad health. Hence an understanding of disease helps us grasp the implications of access (to adequate resources and livelihoods) in understanding the full impact of flood, drought and other extreme natural events. A vivid and tragic example of this can be seen in events in a remote part of Afghanistan in 2002. Despite a pledge of some $4.5 billion to rebuild the country, life in the villages in the so-called 'hunger belt' is unlikely to change soon. Four years of the impact of drought, exacerbated by war and isolation, meant that tuberculosis as well as starvation was rampant. In the words of an eye witness:

> Here, as in nearly all the 380-odd villages of Jawand, hunger and disease ravage the population, culling babies, women, and the elderly. The living stagger on, coughing their lungs and their lives out with tuberculosis. People are so weakened by hunger that even flu can kill.
>
> (Goldenberg 2002: 3)[13]

There has been a protracted debate on the causes of death during famines, and recently this has focused more on the role of disease. Until the 1980s deaths were assumed to be caused mainly by starvation. More recent studies argue that hunger reduces people's resistance to disease, and often dysentery and gastro-enteritis are the cause of death. This view was questioned by de Waal (1989b), who suggested that higher mortality was caused rather by increased exposure to infection, especially in camps for displaced persons and refugees. Indeed, in some Sudanese refugee camps in 1985 de Waal found no correlation between wealth and mortality. The rich had the money to buy food, but perished along with their poorer neighbours because of similar levels of exposure to biological hazards.

Although the crisis was a result of conflict and not drought/famine, de Waal's point about crowded camps and exposure to disease agents is supported by what happened when large numbers of people fled the genocide in Rwanda in 1994. Refugees numbering 800,000 were accommodated in camps near Goma (eastern Zaire, now Congo). There was an epidemic of cholera (a water- and food-borne disease) during the first month of the exodus from Rwanda which may have killed as many as 30,000 people in the Goma camps (Borton et al. 1996).[14] Undoubtedly there has been a degree of generalisation from one type of famine to another. It seems reasonable to take account of the varying conditions of each famine, and to understand the linkages between mortality and under-nutrition, malnutrition, resistance to disease and exposure to disease.

It is not just in large-scale famine relief and war-time refuges for displaced civilians that density and sanitary conditions can compromise health. In other, smaller evacuations of people from approaching cyclones or likely volcanic eruptions there have also been cases of poor hygiene and health hazards. In some of these cases, the consequences of the hazard itself may be insignificant compared with those of removal to shelters or camps, in which there are opportunities for the increased spread of communicable disease and where sanitation is usually rudimentary (PAHO 1982: 3–12; Simmonds et al. 1983: 125–165; Médecins Sans Frontières 1997).

In extreme cases of the destruction of infrastructure in cities there may be significant water and sanitation hazards. Examples include the volcanic eruption near Goma in 2002 and the earthquake in the Indian state of Gujarat (2001) (see Chapter 8 for more detailed discussion of both occurrences). After both of these events, the survivors were deprived of adequate water and sanitation, and there were outbreaks of cholera. Other extreme events (floods, tropical cyclones, tsunamis) are more likely to lead to increased water- and vector-borne diseases such as typhoid, cholera, leptospirosis and typhus, malaria and encephalitis.[15] In such situations environmental engineering and surveillance are generally advised rather than mass immunisation (Woodruff et al. 1990; PAHO 1982: 13–21, 53–60, 2000b: 47–49; see Chapter 6: 220–221).

In other hazard events, it is likely that moderate earthquakes, landslides and tornadoes have few serious secondary epidemiological threats.[16] In fact, the disease consequences of natural hazard events such as moderate earthquakes have often been overestimated or misinterpreted (Cuny 1983: 44–49). The mass burial of victims of moderate earthquakes is still common, despite repeated statements by the WHO and other health authorities that corpses do not present an acute public health risk.

Livelihoods, resources and disease

As with the other hazards treated in this book, understanding normal daily life tells us important things that may be hidden by an extreme situation such as a cholera outbreak in a refugee camp. The Access model developed in Chapter 3 can be applied to the way in which normal 'access to health' is affected by social, economic and political processes.

The role of access

The Access model in Figure 3.2 (p. 99) shows the factors and processes that determine household and individual livelihood resources and income opportunities (Boxes 1 to 4). Households have differential access to production resources such as land, labour, tools, knowledge, skill, livestock and to social networks and employment opportunities with which to provide a livelihood (Box 2b). 'Health' (nutritional level, housing and sanitation, health care, etc.) and well-being are a function of this 'livelihood' – the more secure and adequate the livelihood, the lower the level of the health component of vulnerability. Well-being and health are attained both by accessing adequate public services (if they are not privatised), and by using household income and assets to provide adequate nutrition, and to treat illness or injury (Boxes 2b and 7).

Some of the livelihood activities may in themselves be hazardous in terms of health. Pesticides may be applied by people without adequate protection such as gloves, mask or boots. Many 'development' projects, including irrigation schemes, have associated health risks, including malaria and schistosomiasis (bilharzia) (Hughes and Hunter 1970; Wisner 1976a; Bradley 1977; Spielman and D'Antonio 2001: 172–8). Migrants may expose other family members to disease when they return home (see the classic studies by Forde [1972] on contact with tsetse flies, Prothero [1965] on anopheline mosquitoes) or sources of infection not found at home.[17] In this way, access to resources that are meant to sustain family health may be achieved only at the cost of exposing them to other risks.

'Normal' levels of morbidity and mortality are influenced by these processes. The household's reproductive fertility is determined by a complex chain of cultural, psychological and physiological events (Boxes 8 and 9), which in turn influence the health of individual members of the household through the sharing of food, the ante-natal health of children and the well-being of mothers after childbirth.

Even the condition of 'normality' (often not 'healthy') can potentially contribute to biological disaster in three ways. Firstly, there may be sudden changes in class relations or structures of domination (Boxes 1 and 4) that reduce access to the resources essential to maintain even minimum 'normal' levels of health. Expropriation of land is an example of such change. Secondly, a breakdown in law and order may disrupt access to resources, increase the uncertainty of future income or propel people into headlong flight. When displaced people come into contact with disease vectors or agents new to them, widespread mortality can occur (Hansen and Oliver-Smith 1982). In these examples, biological disaster occurs as a sequel to social disruptions such as war, pogroms, dispossession and famine. One needs only to think, in the early twenty-first century, of the suffering endured by the displaced people of Colombia, Congo, Burundi, Angola, Sudan, Afghanistan and Iraq, to mention just a few of the most severe conflicts.

A third set of issues in the disruption of normality involves the breakdown in animal health, which can result in a deterioration in human health. In such cases a 'vicious cycle' of vulnerability is apparent. For instance, in the late 1970s an epizootic of foot/hoof and mouth disease among cattle of the pastoral Maasai in Kenya led the government to prohibit the sale of Maasai animals (because of a fear that Kenyan beef would be banned from European markets). The quarantine disrupted livelihoods, and income with which to buy food was severely curtailed. The consequence was increased human morbidity and mortality (Campbell 1987). The Maasai's increased vulnerability thus arose from a sudden decline in nutrition originally triggered by a biological hazard which affected their main source of livelihood and the economic shock of government policy, further complicated by endemic tuberculosis and measles. This series of events is a good example of the role of 'food entitlement decline' (FED) rather than 'food availability decline' (FAD) as the main cause of hunger (see Chapter 4). The chain of causation of the Maasai crisis has root causes that include the historic loss of two-thirds of their grazing lands under British colonial rule and their isolation and neglect during colonial and post-colonial periods. Dynamic pressures include the lack of diversified income opportunities for pastoral people in contemporary Kenya, and the country's acute external debt crisis that led to the national government's urgent desire to maintain its share of the European market for beef.

Vulnerability-creating processes

Since vulnerability analysis should be useful to planners and other development workers, we need to specify as carefully as possible how causal chains of social and economic actions may affect people's vulnerability, and the role of biological hazards in exacerbating this. We do so by discussing how these causal chains manifest themselves in different social and natural environments.

The micro-environment

Diet, shelter, sanitation and water supply work together at the household level to determine vulnerability to biological disaster. The synergism linking disease resistance and nutrition has already been mentioned. During the drought in the Sahel (1967–1973), most of the 100,000 lives lost were due to the interaction of starvation and measles in children (Morris and Sheets 1974). Baseline nutrition as a factor in famine has already been mentioned in Chapter 4 and is discussed earlier in this chapter. Previously well-nourished populations (such as the Dutch in the latter months of the Second World War) are able to survive famine conditions that would kill many more people from less well-nourished groups.[18]

Vulnerability may be specifically affected by the type and location of housing. Location, especially in urban areas, is constrained by law (such as zoning legislation), land prices, distance to livelihoods and availability of building materials. There are a million or so living as inhabitants of Cairo's 'City of the Dead' (originally a cemetery) (Rodenbeck 2000) and thousands who inhabit the garbage dumps of Manila, Guatemala City and other cities in LDCs. They did not choose these locations on account of their pleasantness, but for the resources and income opportunities they make possible. But this also determines the quality of housing, water and sanitation. In later chapters we show how some groups of people are constrained in their locational decisions within a specific urban political ecology, so increasing their vulnerability to mudslides, floods, storms and earthquakes. Biological hazards are clearly worsened by various social and economic factors, especially those affecting housing type and location.

Migration and biological hazards

The impact of migration on people also plays an important role in determining vulnerability (Earickson and Meade 2000). Roundy (1983) has shown that movements of Ethiopians, even over short distances, can introduce people to health threats if a large change in altitude or a shift to another ecosystem is involved. Seasonal wage migration of highland dwellers to the coasts of Central and Andean South America produces similar effects. Long-distance migrants can become victims of malaria, sleeping sickness and other diseases. Migrant workers are also a significant factor in spreading HIV-AIDS (see below).

When whole families migrate to very different environments (willingly or with various degrees of state coercion), severe health problems have been recorded. Migrants from the North-east and South of Brazil suffered high morbidity and mortality in newly settled Amazonian habitats (Schmink and Wood 1992). Migrants from the northern parts of Ethiopia who moved to the south-west (usually under powerful state coercion) also suffered greatly (Kebbede 1992; Clay et al. 1988). Similarly, in the 1920s Soviet settlers in Siberia endured a range of viral diseases that were transmitted via the wild rodent population (Pavlovsky n.d.). Longitudinal studies have revealed that several decades after being resettled from the valley flooded for the Kariba dam on the Zambia/Zimbabwe border, people were still suffering measurable health consequences (Scudder 1980, 1989; cf. Hansen and Oliver-Smith 1982). Forced displacement has had even more severe health consequences than migration.[19]

All such displacements have resulted in significant levels of epidemic disease. Even the best-run refugee camps have problems with sanitation (Khan and Shahidullah 1982; PAHO 1982, 2000b; Morris et al. 1982; Harrell-Bond 1986; Médecins Sans Frontières 1997; Wisner and Adams 2003). Besides the cholera epidemic in the Rwandan refugee camps mentioned earlier, outbreaks of epidemic diarrhoeal disease have occurred in Latin American camps (Isaza et al. 1980) and elsewhere in Africa (Rivers et al. 1974) as well as Asia (Anton et al. 1981; Temcharoen et al. 1979). Crowding produces ideal conditions for the spread of air-borne infections and other diseases transmitted by personal contact. Until it was controlled, smallpox could spread rapidly through non-immunised refugee populations (Mazumder and Chakrabarty 1973; McNeil 1979). Higher population densities also make the transmission of malaria easier, especially where there is drug resistance (Reacher et al. 1980) or acute nutritional stress (Murray et al. 1978).

Regional physical environment

Soil erosion, desertification and alkalinisation have been labelled 'pervasive' or 'slow-onset' hazards (Pryor 1982; Blaikie 1985b). Degradation of environments can so reduce livelihood resources that people suffer increased vulnerability to a number of hazards, including biological ones. But it can also affect vulnerability in other ways, by reducing or damaging the earth's genetic materials. Air- or water-borne industrial pollution can also have this effect. As well as rendering fisheries and farmland barren, it can cause the death of trees, wildlife and mangroves (Eckholm 1976; Maltby 1986), thus reducing livelihood resources and at the same time damaging biodiversity.

There has been much discussion of worldwide threats to biodiversity (E. O. Wilson 1988, 1989; Juma 1989; Fowler and Mooney 1990). Efforts to protect genetic resources include the Convention on Biodiversity which was agreed as part of the Rio Earth Summit in 1992, the Seed Treaty, agreed at FAO headquarters in 2001, and an appeal by NGOs from 50 countries to establish a worldwide 'genetic commons'.[20] In the long run, declining biodiversity could provoke or at least amplify biological disasters, especially as the wild ancestors of our food crops are lost or abandoned. Since roughly one-quarter of all pharmaceutical products have plant or animal origins, diminished biodiversity could undermine the scope of therapy in the face of new diseases. Depletion of regional genetic resources has already contributed (as 'trigger') to disasters, and should be considered a pervasive hazard, as illustrated by the Irish potato famine (Box 5.1).

Box 5.1: The Irish Potato Famine, 1845–1848

This disaster, like all others, involved unsafe conditions with origins in history.[21] The reliance of the impoverished rural population of Ireland for 90 per cent of their food energy on potatoes meant that these people were highly vulnerable to the blight caused by the organism *Phytophora infestans*. The potatoes used by the Irish peasantry were all descended from a few hundred tubers brought from the Andes in the early 1600s. By the 1840s the potato gene pool in Ireland was extremely homogeneous, and thus vulnerable to widespread damage by disease. Genetic diversity would have conferred at least some protection, by increasing the chance that some potato varieties were resistant to blight.

The root causes, dynamic pressures and unsafe conditions at work in this biological disaster are illustrated in Figure 5.1, a version of the Pressure and Release (PAR) model.

The root causes can be traced back to the 1650s, when Cromwell's conquest of Ireland began systematic discrimination against the Catholic peasant majority. Firstly, a large number were forced to resettle in the west of the country. Two centuries later this had produced a population distribution in western Ireland with high rural densities and small farms. It also encouraged reliance on potatoes because they could yield up to 2.5 tons per acre. The Penal Laws (1695) reinforced this pattern by making it illegal for most Catholics to own land. By then, English absentee landlords had been granted ownership of most of the land in Ireland. Protected by the Corn Laws, landlords profited from the export of grain and meat to England without facing any competition from cheaper foreign imports.

Contributing pressures that provided the background for specific vulnerability effects included the increase of the Irish Catholic population. In the many non-blighted years the potato had provided increased dietary energy, leading to higher child survival rates. The pressure remained to produce large families both to work the land and, above all, to seek paid employment in England and America (a part of the wage would be remitted to the rural home in Ireland). Population growth exacerbated pressure on land, particularly in the west, and maintained the trend towards very small farms. Export of grain and other food to England further reinforced a reliance on potatoes by the majority. Ireland's economic dependence on England produced additional pressure on subsistence potato farming because the resulting low rate of domestic (Irish) saving meant that few non-farm jobs were available locally as an alternative.

Specific vulnerability effects included a diet heavily dependent on potatoes, with reports that an adult would consume up to six kilograms a day, an amount that would provide 6,000 Kcal (Aykroyd 1974: 32). At

Box 5.1 continued

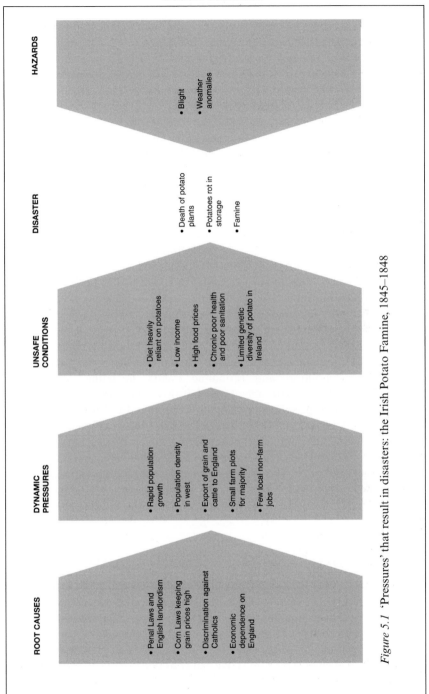

Figure 5.1 'Pressures' that result in disasters: the Irish Potato Famine, 1845–1848

ROOT CAUSES
- Penal Laws and English landlordism
- Corn Laws keeping grain prices high
- Discrimination against Catholics
- Economic dependence on England

DYNAMIC PRESSURES
- Rapid population growth
- Population density in west
- Export of grain and cattle to England
- Small farm plots for majority
- Few local non-farm jobs

UNSAFE CONDITIONS
- Diet heavily reliant on potatoes
- Low income
- High food prices
- Chronic poor health and poor sanitation
- Limited genetic diversity of potato in Ireland

DISASTER
- Death of potato plants
- Potatoes rot in storage
- Famine

HAZARDS
- Blight
- Weather anomalies

Box 5.1 continued

the time that the 'Great Starvation' became fully manifest, one-half of the Irish population was dependent on the potato (Regan 1983: 114). Besides a hog, fattened on kitchen scraps for sale to meet taxes and other monetary expenses, potatoes and peat to burn for heat were the mainstays of the peasant economy. Vulnerability was further exacerbated by the low wages paid by English landlords, making the purchase of alternative foods very difficult. Because of the Corn Laws (which forbade the import of cheaper grains from the New World), prices for alternative foodstuffs such as grain were always high, and of course very much higher during the famine. Crowding, poor housing and sanitation added to vulnerability by undermining people's health status, making them less resistant to the physiological effects of hunger and cold.

Figure 5.1 shows that the physical hazard side must include not only the disease agent *Phytophora infestans*, but also the exceptionally warm weather in 1845 that triggered the explosion of blight, and the unusually cold winter weather of 1846–1847. The cold is said to have killed many who had been weakened by hunger during the previous year. Such sequential impacts of different hazards – drought followed by flood, tornado followed by flood, earthquake followed by freezing weather – are often the triggers of serious disasters.

Widespread rotting of the primary food crop following blight intensified the chronic hunger into famine. In addition there was a series of contributory political and economic events, including prolonged debates in the British parliament on what the scope and nature of relief should be, and predatory pricing and lending by Irish merchants (Middleton and O'Keefe 1998, citing Poirteir 1995 and Whelan 1996). The impact of this biological hazard on the vulnerable was the death of at least 1.5 million between 1845 and 1848, and the forced emigration of another 1.5 million.[22] The 1881 census revealed an Irish population still three million people fewer than the pre-famine total. The political fallout from this disaster has had impacts that last to this day.

Pressures affecting defences against biological hazards

To understand how occasional biological disasters occur, it is clear that we need to analyse many factors that may increase people's vulnerability to the impact of diseases of humans, plants and animals, or to infestations of pests. In addition we need to evaluate environmentally damaging processes that diminish the availability of genetic material, thereby undermining food sources and medicinal repositories. We can consider these in terms of genetic, environmental and cultural defences that represent examples of the coping mechanisms discussed in Chapter 3 in terms specific to biological hazards.

Genetic defences

Genetic polymorphism (diversity within a species) confers a degree of resilience in a population of plants, animals or humans (Anderson and May 1982; Ruffié 1987; McKeown 1988; Ewald 1993). These benefits can be defeated in a number of ways. Firstly, an unknown, new and virulent organism can be introduced to an existing biological community and do a great deal of damage, as with the great loss of life due to smallpox when it was introduced into the New World.[23] Similarly, pigs (hogs), horses, cattle and the black rat thrived after they were taken to the New World, and pushed competitors out of ecological niches (Crosby 1991, 1986). In East Africa, zebu cattle (*Bos indicus*) had no resistance to the disease rinderpest that was carried by donkeys used by German colonial military convoys. The resulting mortality among cattle was as high as 90 per cent. Their hapless owners were weakened by the economic stress of losing so many animals. The people were also suffering social disruption due to German and British colonial land occupation and attempts to tax rural populations and limit their movements. They had to confront new disease organisms brought from West Africa: guinea worm and jiggers. The resulting mortality was probably one of the major reasons for the collapse of militant resistance to colonialism in the region (Kjekshus 1977; Maddox et al. 1996).

Even when there is some resistance to a disease agent in a population, sufficient time may have elapsed since the last outbreak to produce a large number of non-immune individuals. European history is punctuated with catastrophic epidemics of bubonic plague (transmitted by fleas living on rats) even though it was endemic for very long periods of time.[24] In addition to a non-immune population of sufficient size, well-established and busy communications networks are necessary to channel the infection and support such epidemics. Thus, the Plague of Justinian apparently originated in Egypt and spread along trade routes. These routes were very busy with soldiers and refugees of the wars then being fought to regain parts of the Roman Empire lost to the Vandals and Ostrogoths. Bubonic plague appeared in Byzantium (capital of the Eastern Roman Empire), in AD542, and by the end of that century this city had lost half its population. Following a second route along the Mediterranean, plague appeared in France in AD543. Depopulation was so great that much land fell idle, taxes were not paid and estates were widely replaced in the seventh century by a pattern of small freeholders (Russell 1968).

A sudden and severe economic loss of crops through disease or pests can be avoided for a time by the heavy use of protective agro-chemicals, which many LDCs have to import, with negative impacts on their foreign exchange. But pests and diseases often develop resistance to the chemical agent, and competing or beneficial organisms may be killed. This produces the well-known pesticide 'treadmill' of resistance and resurgence of the pest

(Debach 1974; Altieri 1987). For example, the cultivation of cotton in the vast irrigated Gezira scheme in Sudan began in the 1920s. In the early 1950s the crop was sprayed with pesticides once or twice a year. By the early 1980s, cotton was sprayed nineteen times during the growing season in order to control a growing number of pests. Widespread use of agro-chemicals (especially DDT to combat malarial mosquitoes) has produced pesticide-resistant mosquitoes and malaria is resurgent in many parts of the world.[25]

The expansion of export crops into forest, fallow land and plots previously devoted to subsistence crops has caused the extinction of many local varieties of legumes and other food crops, as well as many gatherable forest products (Juma 1989). These genetic resources are lost to future generations who use them to increase diversity and strengthen livelihoods and to reduce disaster vulnerability. The urban market for red kidney beans in Kenya caused farmers to stop using many indigenous varieties of legume and concentrate on *Phaseolus vulgaris* because they needed cash. The potential for a collapse in cash incomes through pest and disease infestations of a much narrower genetically based range of crops are clear.

Environmental and cultural defences

A very important environmental defence against catastrophic disease and infestation is dispersed settlement. Rapid urbanisation is also associated with increased vulnerability to epidemics in cities. This seems to be true whether urban growth is the result of rural insecurity (as during the decline of Rome and the early Middle Ages) or due to the rise of capitalism (from the fifteenth century), or because of rural depression and relative wage differentials between town and country during the last 50 years.

Potable water and sanitation systems are also a factor. There was probably little significant difference in sanitation between rich and poor in Europe until the nineteenth century.[26] Today, in cities of the LDCs, the more affluent enjoy indoor plumbing while the relatively poor are lucky if they have a communal water standpipe within a few hundred metres of their front door (Feachem et al. 1978; Agarwal et al. 1989; Cairncross et al. 1990a; McGranahan et al. 2001). Many mega-cities (e.g. Calcutta, Lagos, Mexico City) have sanitation systems based on drains and water mains that are at least 100 years old. Some, such as Howrah (a city of two million lying across the Hooghly River from Calcutta), until quite recently had no sewers at all (KMDA 2003; Wright 1997).

Different classes of people were also more or less able to get access to safe havens and flee from the Black Death. While it was not guaranteed protection, escape to country houses from affected cities was an option enjoyed only by the rich (Ziegler 1970). The spatial organisation of residence in the mega-cities of Africa, Asia and Latin America is likewise significant in producing differential vulnerability to disease.

The dangers are increased with declining budgets for maintenance of even minimal urban infrastructure in many countries. Reduced public expenditure is often due to International Monetary Fund (IMF) and World Bank insistence on austerity programmes in the face of foreign debt (Hardoy and Satterthwaite 1989; Cairncross et al. 1990a). In Chapter 2 we introduced this as one of the major 'dynamic pressures' that create 'unsafe conditions'. This pressure is made worse where the municipal organisation of waste management cannot recover more than a small fraction of the cost of sanitation services. One study in Africa found that the best waste management system (Abidjan, Côte d'Ivoire) recovered 30 per cent of its costs, while the rate was less than 5 per cent in Ibadan (Nigeria), Dar es Salaam (Tanzania) and Johannesburg (South Africa) (Onibokun 1999).

We discussed earlier the diverse ways in which people tolerate crop and livestock losses. One of the most important of these is to combine, if possible, an assortment of livelihood activities (see the section on 'coping' in Chapter 3). Poor farmers not only try to grow a variety of crops, usually intercropped, but they may engage in a variety of non-farm activities, including trade, craft production and services, fishing, small-scale mining and forestry, etc., in an effort to diversify their livelihood portfolios (Chambers 1983; Guyer 1981; Wisner 1988a; Ellis 2000).[27] Current 'development' efforts often incorporate rural households into commodity production in a manner that reduces the diversity of rural livelihood opportunities and increases the risk of sudden biological hazards (Bernstein 1977, 1990; Wisner 1988a: 187–197; Bryceson 1999).

Root causes and pressures

Biological hazards and vulnerability in Africa

In the PAR model, 'root causes' of vulnerability are often found in global economic and political processes, while 'dynamic pressures' are to be found in the structure of particular societies. This section examines some of the linkages between people's vulnerability to biological hazards and the political and economic root causes and pressures that explain such vulnerability. The illustrations are drawn from Africa, and a case study of HIV-AIDS is given in Box 5.2.

Since the era of independence in the 1960s, which prompted such high expectations for the people of most sub-Saharan African countries, the Four Horsemen of the Apocalypse have ridden over nearly every part of the continent. War and famine have affected many of these countries during the decades of the 1970s, 1980s and 1990s, and have interacted in numerous ways with the other two biblical threats, disease and pestilence.[28] Many of the more than 100 armed conflicts in the world (1989–2000) took place in Africa (Wallensteen and Sollenberg 2001). During this period a possible 40 million people have been displaced. The resulting disruption of livelihoods and resort to refugee camps have contributed to high levels of civilian deaths. In early 2002, nearly one-third of the entire population of Angola was internally displaced, and about 300,000 people were threatened by famine and disease.[29] Internally displaced persons (IDPs) who flee into the forest and isolated rangeland margins are vulnerable to such deadly diseases as Ebola virus and meningitis. With the rainy season can come devastating epidemics of pneumonia, diarrhoea and measles. Meanwhile drought, flood and cyclone may affect an already weakened population that has also had to cope, perhaps, with locusts, army worms, Sudan dioch birds or live-stock diseases. Cholera, measles, tuberculosis, meningitis and, above all, HIV-AIDS may also take their toll (Timberlake 1988; Wisner 1992a; Mekendamp et al. 1999; Barnett and Whiteside 2001).

The broader context of these tragedies is a failure of social and economic development (Wisner 1988a; Seidman and Anang 1992; Toulmin and Wisner 2003). Many African nations lost ground during the 1980s compared to their position in the 1960s or 1970s in terms of infrastructure development (including water supply; see Thompson et al. 2002), services and per capita productivity (Cornia et al. 1987; Whitaker 1988; Rau 1991; Cheru 2002). Mamdani (2002; UNDP 2003b) reminds us that:

> [t]he first two decades of independence were decades of moderate progress. Between 1967 and 1980 more than a dozen African countries registered a growth rate of 6%. This included not only mineral-rich countries such as Gabon, Congo, Nigeria and Botswana but also countries such as Egypt, Kenya and Ivory Coast.

Foreign indebtedness in Africa is also very high when compared with the ability to service debt (George 1988; Onimode 1989; Mkandawire and Soludo 1999). Dependency on foreign aid is as high as it has ever been, and despite much campaigning and paper commitments to debt relief for the Highly Indebted Poor Countries (HIPCs), there has actually been very little debt cancelled (Hanlon 1996; Cheru 2002) (see also Chapter 2, section on 'Global economic pressures').

African governments have cut their health budgets due to reduced export revenues (world market prices for many LDC exports fell consistently from the 1970s) and financial austerity programmes mandated by the IMF and World Bank (Wisner 1992a). Maintenance of infrastructure, procurement of medicines, training and plans to improve primary health care have all suffered. Lack of road maintenance and shortage of foreign exchange to import fuel and spare parts for vehicles has meant that mobile services to isolated villages have been interrupted. Such services had in the past been effective in both providing an early warning of famine and epidemic disease hazards, as well as a contribution to their treatment, or at least palliation (see Box 4.1 on Malawi in Chapter 4).

Nigeria provides a useful illustration of the links back from unsafe conditions, through dynamic pressures to root causes. The petroleum boom of the 1970s weakened agricultural and other rural livelihoods in Nigeria, especially in terms of self-provisioning (Watts 1986). Economic crisis in the 1980s eroded the purchasing power of cash wages and many people lost employment. The decreasing resilience of livelihoods to these shocks was structured by long-standing gender, ethnic, religious and class biases. Rural–urban differences in access to health care and income differences between classes had always been great. The IMF Structural Adjustment Programme (SAP) increased both of these differences (Nafziger 1988: 123–124; Wisner 1992a: 152). Women have very heavy work burdens and poor health. In the north, maternal mortality (an index of both health care and 'normal' health) was 1,500 per 100,000, compared with 150 in Zimbabwe and 5 in Europe (Wisner 1992a: 161).

In 1991, half of world deaths from cholera (7,200) occurred in Nigeria. Ten years later, cholera was still endemic, and a serious outbreak with more than 2,000 cases occurred in the northern Nigerian state of Kano (WHO 2001b). The deteriorating health care system combined with increasing individual vulnerability required only the presence of the cholera organism (the biophysical 'hazard') for a disaster to occur. The spread of cholera reflected spatial structures (infrastructure, patterns of urbanisation) and spatial process (frequency of public markets, festivals, etc.) (cf. Stock 1976).

Box 5.2: HIV-AIDS in Africa

HIV-AIDS is a disease of the immune system caused by the human immuno-deficiency virus (HIV). People may be infected (HIV-positive) for many years before full AIDS (acquired immuno-deficiency syndrome) develops, and they may be unaware of their status. The disease is transmitted through sexual intercourse, or by blood from an infected person being introduced into another through shared syringes or use of contaminated blood products (including transfusions). Mothers can also pass on the virus to their unborn children. Inadequate precautions when sterilising needles and surgical instruments, as well as when screening blood from blood donor services, all increase the risk of contracting HIV-AIDS by means other than sexual intercourse. For these reasons, people in countries with poor medical infrastructure are at greater risk of contracting HIV-AIDS.

There are a number of co-factors which are thought to be associated with increased rates of transmission of HIV-AIDS through sex. These include genital sores, other active sexually transmitted diseases and cervical erosion. They are responsible in part for higher rates of infection, since the virus enters the body more easily through lesions or unhealed wounds. Poor medical facilities mean that these conditions

---

Box 5.2 continued

remain untreated and increase the rates of infection of HIV-AIDS. Once an individual becomes HIV-positive, the progress of the disease is profoundly affected by the pre-existing status of the immune system. Those with a damaged or stressed immune system may develop symptoms from opportunistic infections earlier than they otherwise would (Packard and Epstein 1987).

These were precisely the epidemiological conditions in Uganda and elsewhere in eastern and central Africa when the HIV-AIDS epidemic began in the late 1970s and 1980s. Endemic malaria, filariasis, war and refugee movements had affected Uganda and neighbouring parts of Tanzania, Kenya, Rwanda, Burundi and Zaire (now Congo). Some of these areas had also been affected by human trypanosomiasis (Langlands 1968; Forde 1972; Wisner 1976a). Disrupted livelihoods, especially in Uganda (caused by a wide variety of factors including inept and dictatorial rule, the collapse of most of the trading networks following the expulsion of Asians, civil war and a war with Tanzania), had resulted in reduced levels of general health, including parasitic diseases such as malaria which stress the immune system. Typically one-third or more of the children suffered chronic malnutrition (UNICEF 1985; Wisner 1988a). Very little has changed. In spite of a remarkable recovery in the economic fortunes of Uganda, in 2002 half the population of Uganda still lacked access to safe water, and only about 30 per cent had access to adequate sanitation (OCHA 2002d).

Using a vulnerability framework to analyse the early days of the African HIV-AIDS pandemic reveals a number of social shifts that encouraged the spread of HIV-AIDS. War, economic crisis and upheaval of family life in Uganda led to a greater spatial mixing of populations and a relaxation in men's sexual control over women. *Magendo* (smuggling) was rife and involved the movement of illicit goods from Mombasa, via Lake Victoria, to Rwanda and Zaire (now Congo). Roving bands of traders, often armed, were away from home for months at a time. Spare cash was spent on casual sex. Indeed, both smuggling along roads and across Lake Victoria, together with legal motorised traffic, gave rise to overnight stopovers and hotels, and a rise in prostitution. Many women had a precarious economic existence, often being *de facto* barred from owning or renting land. Thus marriage or temporary liaisons with men provided the only 'meal-ticket'. Selling beer and occasional prostitution offered ready cash in an economic climate that offered little in the way of permanent and independent livelihoods for unattached women. Unfortunately, these assertions, however well supported by hearsay and observation, are quite difficult

Box 5.2 continued

to corroborate through detailed sociological study (Barnett and Blaikie 1992, 1994).

The unstable economic conditions of the 1970s and 1980s added to women's deteriorating economic security, and this may have led to the spread of the virus. However, economic insecurity alone is not a sufficient factor to explain the cause of the HIV-AIDS pandemic. Every society develops an HIV-AIDS epidemic that reflects the sexual practices of its population. Behind the structuring of this pattern of sexual practice lie relations of gender inequality (Barnett and Blaikie 1992).

In the 1990s, however, the rate of increase of HIV infection began to decrease in Uganda due to vigorous health education campaigns and condom distribution. But by the beginning of the twenty-first century, many parts of eastern, central and southern Africa were affected (UN Economic Commission on Africa 2001). The centre of the epidemic had shifted to southern Africa, where one in four adults in Botswana and Zimbabwe are thought to be infected. In another ten African countries 1 in 10 adults carry the HIV virus.[30] The *World Development Report* for 2000/2001 reflected upon this situation:

The effect on life expectancy will be devastating. Had AIDS not affected these countries, life expectancy would have reached 64 years by 2010–15. Instead, it will have regressed to 47 years, reversing the gains of the past 30 years. The impact on child mortality is also enormous. In Zambia and Zimbabwe 25 per cent more infants are dying than would have without HIV.

(World Bank 2001: 139)

Although varying in detail because of differences in local sexual practices, HIV-AIDS transmission in the 1990s in southern Africa depended on many of the same socio-economic and spatial processes as it had during the 1980s in eastern and central Africa. Long-distance labour migration was deeply rooted in the economic history of southern Africa (Murray 1981; First 1983). Hundreds of thousands of men had been leaving rural homes in the surrounding countries to work in South African mines and factories. They lived in all male hostels. They visited prostitutes. They carried the virus home with them to Botswana, Lesotho, Swaziland, Namibia, Zambia, Malawi, Mozambique and Zimbabwe.

In 2001 and 2002 there were debates, campaigns and law suits over monopoly pricing of the anti-retroviral drugs that can prolong the life of those living with HIV-AIDS. People in LDCs (especially in Africa where the largest number of sufferers live) cannot afford these drugs.

Box 5.2 continued

In a landmark case in South Africa, the major international pharmaceutical companies agreed to let poor countries import or manufacture less expensive generic equivalents of these patented drugs. However, in a tragic twist, South African President Thabo Mbeki blocked their use, even that of a drug, Nevirapine, that can prevent the infection crossing from mothers to infants (AIDS ACTION 2001). In 2002 some provincial health departments in South Africa began to distribute the drug in defiance of the central government, and former president and elder statesman Nelson Mandela criticised Mbeki for his irrational position (*Economist* 2002b). In 2003, the South African government was finally forced by legal action taken by the NGO, Treatment Action Campaign (TAC), to provide medication to pregnant women living with HIV-AIDS. However, as of early May 2003, the government was still refusing to provide the anti-retroviral drugs that would prolong the lives of five million South Africans who are infected – this despite the fact that the annual cost of treatment had fallen to less than $2,000 per patient, from over $10,000, due to the availability of generic drugs. With a death rate of 600 per day, this biological disaster is the moral equivalent of the crash of two large airliners or a catastrophic flood *every day* (Thompson 2003).

The UN AIDS Programme compares HIV-AIDS in Africa to war: 'HIV is deadlier than war itself: in 1998, 200 000 Africans died in war but more than 2 million died of AIDS. AIDS has become a full blown development crisis' (UNAIDS 2000: 21). Barnett and Whiteside are even more vehement about the disastrous consequences for human development. They argue that despite the large number of studies, the impact of HIV-AIDS is still underestimated:

> It is our contention that serious though HIV-AIDS is for those indicators that can measure its impact they do not really show the full effects of this disease. There are a number of reasons for this.
> • The indicators are based on the demographic event of death! The problem with AIDS is that death is preceded by a period of long, debilitating and unpleasant illness. This is not picked up in these indicators.
> • The complexity of the disease is such that it may have unexpected impacts. For example we believe that enrolment in primary education is likely to decrease because parents cannot afford to send their children to school, or they need the child labour at home, or the teachers have been sick and died so there is no school.
> One of the most significant features of AIDS is its link with poverty and the fact that it pushes households and individuals into a downward cycle.
> (Barnett and Whiteside 2001: 7–8)

Box 5.2 continued

Some aspects of this 'downward cycle' can be illustrated by using our 'Access' model, as in Figure 5.2. Household labour power, livelihood options, remitted income, household income levels, purchases and levels of welfare will all be affected (FAO n.d.; Barnett and Blaikie 1989; Collins and Rau 2000; de Waal 2002). In the diagram, Box 2 now has a new input derived from the selective mortality of adults, which reduces the number of active people in households. The sick need nursing, and resources are used up in caring and for medicine. This, combined with the reduced labour force, means that less food can be grown and earning opportunities are fewer (Box 3). In some areas, the deaths of parents has led to a crisis in caring for orphans, with displaced and asset-less children growing up in a world in which their livelihoods are disrupted. This shows another side to the AIDS disaster that goes far beyond its medical effects. The implications of HIV-AIDS for a reduction in livelihoods are suggested in Box 3, bringing about radical choices to safeguard a minimum income, and care for the chronically ill and dying (Box 5). A sample of the impacts upon the household budget are shown in Box 7.

While Figure 5.2 identifies the critical pressure points which the disease brings to bear on livelihoods, Figure 5.3 charts some of the coping responses and indicates some of the ways in which outside agencies may be able to assist and palliate the economics and social impacts. Figure 5.3 is reproduced and redrawn from Barnett and Blaikie (1992, 1994) and reports empirical data on household coping mechanisms in the face of HIV-AIDS, and also outlines in further detail responses to the impact of the disease on individual family members.

The UN FAO has identified the impact of HIV-AIDS on rural production as a major threat to food security (FAO n.d; IFRC 2000c: 56–57). Yamano et al. (2002) found that in Kenyan households which suffered the death of the adult household head, crop production fell by 60 per cent.[31] In Chapter 4 we showed that the food emergency in Malawi in 2002 was due, in part, to the high incidence of HIV-AIDS in the population.

De Waal believes that the HIV-AIDS pandemic will 'change Africa as we know it'.[32] There will be profound economic effects, for example South Africa's economy will be 22 per cent smaller in 2010 than it would have been without HIV-AIDS (de Waal 2001). He believes that an 'AIDS related national crisis' will 'fasten onto the weak points of governance or socio-political relations that already occur in society' and produce profound social and political changes, as well as economic ones (ibid.: 3). This view is compatible with our PAR model, if HIV-AIDS is considered the 'root cause' in the 1980s and 1990s, the effects of which will ripple through into the future, transmitted and shaped by 'dynamic pressures' such as war, weak governance, corruption and indebtedness.

Box 5.2 continued

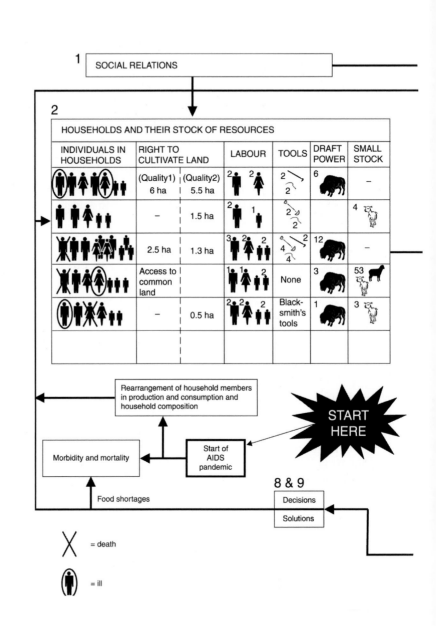

Figure 5.2 Access to resources to maintain livelihoods: the impact of AIDS

Box 5.2 continued

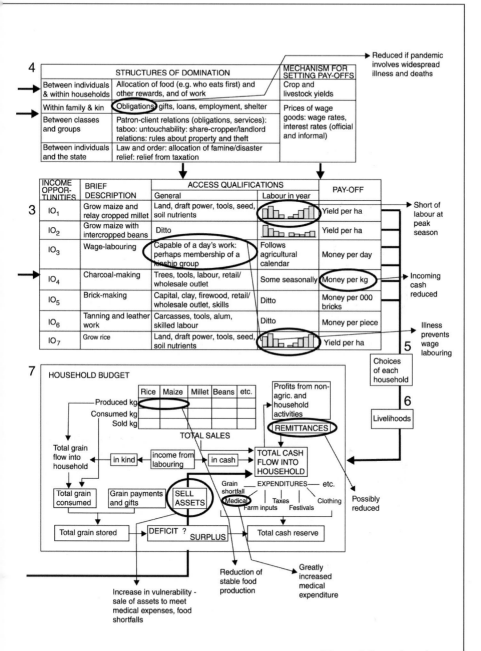

Figure 5.2 continued

Box 5.2 continued

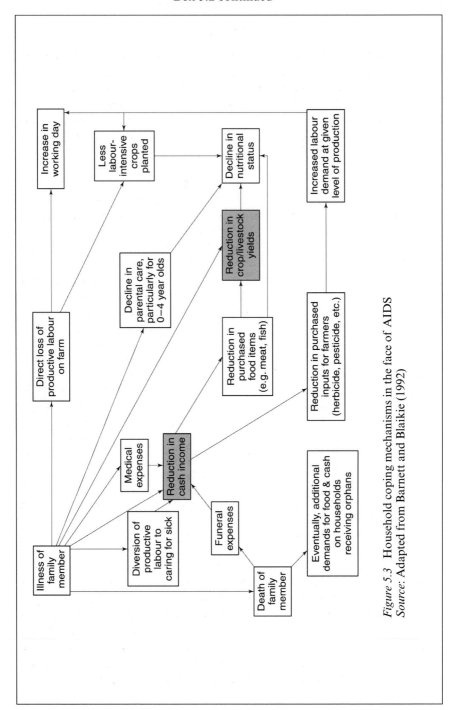

Figure 5.3 Household coping mechanisms in the face of AIDS
Source: Adapted from Barnett and Blaikie (1992)

Steps towards risk reduction

Earlier successes

Successful campaigns against certain illnesses have been recorded that have operated both from 'the top down' as well as 'from the bottom up'. Control of yaws just after the Second World War and the more recent eradication of smallpox were successes in large-scale biomedical administration. The disease agents and mode of transmission in both cases were straightforward. Treatment with penicillin in the first case,[33] vaccination in the second, were uncomplicated, and follow-up was not necessary. No insect vectors with complicated life cycles were involved.

UNICEF is currently in the midst of a more ambitious worldwide effort to immunise all children against tuberculosis, tetanus, measles, diphtheria, pertussis and polio. Some of the vaccines (measles, for instance) must be kept cool until injected, and so require a 'cold chain' of refrigerated facilities in storage and transport. Others require follow-up immunisation (polio, for instance). There is some debate as to whether malnourished children with depressed immune systems can form antibodies in response to immunisation in some cases. (This did not affect the anti-smallpox campaign because the immune reaction against live cowpox antigen is universally strong.) Another partial 'top down' success against a biological hazard involves locusts in Africa and the Middle East. Aerial surveillance (conditional on international agreements with regard to air space) and massive aerial spraying of locust breeding sites with insecticide, seem to have been moderately successful.[34] On the other hand, the biggest increase in human longevity came as the result of improved food supply, housing and sanitation in nineteenth-century Europe and North America, and not through biomedical intervention at all (McKeown 1988). In other words, human development is intimately bound up with disaster risk reduction and vice versa, a general conclusion to which we will return in Chapter 9.

A number of the potential health disasters facing us are not suited to the 'top down' approach. HIV-AIDS, cholera, plague, malaria, as well as other complex biological threats such as deforestation, desertification and the loss of species, cannot be tackled by 'top down' problem solving alone. The development of broad policy measures on a country or regional basis is essential as a framework for intervention in areas of public health, public education, work in schools with pupils and teachers, and so on. However, this is not enough. In addition, detailed knowledge of highly variable social and environmental situations is needed to address local problems. This can be provided by responsive, community-based, flexible 'action research' and interaction between governments, NGOs and community-based organisations. In Chapter 4 we saw that civil society has a role to play in famine early warning systems; in the same way, a wide variety of people, including teachers, community leaders and prostitutes, can become HIV-AIDS educators (Schoepf 1992), and villagers can serve as water and sanitation engineers (White 1981; Wright 1997). Using a network of volunteer counsellors, Uganda's AIDS Information Centre and other AIDS-focused NGOs have made a good deal of progress and stand as an example for other African countries (IFRC 2000c: 65–66).

The experience of the 'bottom up' success of some LDCs in the implementation of primary health care (PHC) is possibly more relevant than 'top down' successes. From the 1960s to the early 1980s, China was able to improve health significantly through health education campaigns and by mobilising labour for the improvement of water supplies, drainage, sanitation and housing. Schistosomiasis, sexually transmitted diseases and tuberculosis were reduced significantly (Horn 1965; Sidel and Sidel 1982). Unfortunately much of this rural social infrastructure in China has disappeared during the past decade, which is one of the reasons why the Chinese authorities are alarmed by the prospect that SARS might escape the cities and become established in the countryside (Pomfret 2003).

Also in the 1960s and 1970s much progress was made in establishing a PHC network in several African countries including Sudan, Tanzania, Kenya, Mozambique and Guinea-Bissau (Wisner 1976a, 1992a). In Cuba also there has been notable success in preventing dengue fever outbreaks through the use of neighbourhood Committees for the Defence of the Revolution which have carried out case finding and environmental management at the micro-level – seeking out and removing mosquito breeding sites (Susman 2001). Later, in Chapter 9, we will argue that the knowledge and skill of local people are often neglected in planning to mitigate loss and to prevent disasters. Participatory and inclusive planning has already been used in the field of public health for some time (PAHO 1994; Macaulay et al. 1999), and should be used more widely by disaster planners.

Policy directions

The first steps toward reducing the risk of biological disaster should be the extension and strengthening of PHC networks. This was the goal of the WHO's campaign 'Health for All by the Year 2000' (Wisner 1988b, 1992a). Left purely to 'market forces', health services are not accessible to the poor. In fact, observing the distribution of care in the UK more than 30 years ago, Hart (1971) proposed an 'inverse care law' which states that the greater the need for care, the less that is provided. NGOs, especially grassroots people's development associations, can improve access to health care in a variety of ways, even for the poorest households (Packard et al. 1989; Cairncross et al. 1990b). The effects of such work in promoting safer conditions (reducing vulnerability) can be traced in several of the boxes in our Access model shown in Figure 3.1.[35] Dr John Lindsay, a medical officer with the government of Manitoba, Canada, goes even further by arguing that the means and the ends of population health policy are identical with those of disaster risk mitigation (Lindsay 2002).

Second, a large effort should be put into meeting the Millennium Goal of providing improved access to water supply and sanitation to some 800 million people by the year 2015. In this chapter the close connection between routine or 'normal' access to such essential services and resilience in the face of the health consequences of disaster has been pointed out several times, as well as the obvious point that vulnerability to such biological hazards as cholera is reduced when water and sanitation are improved.

Third, vulnerable groups should receive special help in improving their nutritional status. Good nutrition provides resilience. But the steps needed to improve nutrition may involve measures that are difficult to implement. Land reform, rural credit to women, access to livelihoods in the urban 'informal' sector, food pricing policy, targeted nutritional supplementation and urban gardening have all proven effective in providing improved nutritional status.[36]

Fourth, agricultural research and extension should greatly increase resources devoted to genetic diversity, identifying and preserving local genetic diversity and resource management techniques. Integrated pest management (IPM) and integrated vector management (IVM) should substitute for the import-intensive agro-chemical approach to plant and animal hygiene. However, there are strong pressures, both commercial (from seed companies and the manufacturers of pesticides and herbicides) and bureaucratic (science-led agencies chasing targets, quotas for innovation adopters), which work against this strategy. Also, it cannot become an exclusive strategy, but one in which risks of reducing biodiversity and adopting modern techniques of agriculture and conservation are more carefully researched and understood.

Lastly, health education and agricultural extension should pay much more attention to the real constraints on people's lives. In this way, the 'messages' of educators will be more relevant, and the 'educators' should be more open to what they can learn from civil society (Barth-Eide 1978; Kent 1988; Turner and Ingle 1985; Wisner 1987a).

Precautionary science

During the last two centuries, biodiversity has suffered irreparable damage. Consumption of fossil fuels has begun to change the earth's climate, with a whole series of consequences for food security and health. Destruction of the ozone layer in the atmosphere could produce a significant rise in new cases of skin cancer (Benedick 1991: 21; CIESIN 2003). The effect of additional ultra-violet radiation on phytoplankton, zooplankton and other life forms at the bottom of the earth's food chain is not known. In addition, large doses of ultra-violet radiation may harm the immune systems of higher animals, including humans. Such a possibility brings us around full circle, as this chapter began with a discussion of human biological adaptation and immunity.

In view of the massive changes urban-industrial civilisation has brought about in a short span of time, the current enthusiasm for biotechnology should be reconsidered carefully. A healthy scepticism and cautious attitude would seem appropriate before the next round of humankind's heroic environmental modification goes too far. No one foresaw the health consequences of releasing tens of thousands of newly synthesised chemicals into the environment until Rachel Carson wrote *Silent Spring* (1962). What will the deliberate release (and accidental escape) of genetically engineered life forms do? There are certainly some beneficial uses of recombinant DNA technology in medicine and agriculture. But who should decide which applications are worth the environmental risks (Walgate 1990; Harremoes et al. 2002)? If we have learned anything about human health since Hippocrates wrote *Airs, Waters and Places* approximately 2,400 years ago, it is that each person's health is inseparable from the health of others and the health of the environment. A reduction in vulnerability to biological hazard will therefore require both social justice (to ensure the health of others) and technological humility (to restore the health of the environment).

Notes

1 During the 1990s there were serious wildfires on the urban–wildland interface affecting Cape Town, Sydney, Los Angeles and San Diego. Large forest fires occurred often as well in Brazil, Mexico, the US and the south of France. Most memorable, perhaps, and an event with significant health consequences, was the series of fires in Indonesia in 1997. Tens of millions of people were affected by the resulting air pollution in many downwind countries including Malaysia, Singapore, as well as parts of Thailand and the Philippines, and elsewhere in Indonesia (Brauer and Hisham-Hashim 1998). In 1871 a forest fire in Peshtigo, Wisconsin, in the US, killed more than 2,200 people who lived in the midst of a large area devoted to the timber industry (Foote 1997: 101; Gess and Lutz 2002).

2 WHO http://www.who.int/msa/mnh/ems/dalys/intro.htm.

3 The literature on human needs is large and diverse. Some believe that 'need discourse' is politically and ideologically manipulated, and that it does not make sense to assert the existence of universal, 'basic' human needs (Sopher 1981). Others assert that some needs are not only real and universal, but imply the human right to their satisfaction (Bay 1988), a view that is opposed by Streeten (1984). Nobel Laureate Sen links rights and needs: 'The exercise of basic political rights makes it more likely not only that there would be a policy response to economic needs, but also that the conceptualization – including comprehension – of "economic needs" itself may require the exercise of such rights' (2000: 153).

4 On the epidemiology of disasters, see Chen (1973); de Ville and Lechat (1976); PAHO (1982, 2000b); UNDRO (1982a); Seaman et al. (1984); Alexander (1985); Sapir and Lechat (1986); WHO (1990); Noji (1997); Wisner and Adams (2003).

5 Severe Acute Respiratory Syndrome, a highly infectious viral disease that first appeared in southern China in November 2002, and was declared a worldwide threat in March 2003. By 9 May 2003 there were 7,000 cases worldwide with 500 deaths (WHO 2003). SARS was officially contained by July 2003.

6 Yellow fever outbreaks were common in Memphis and other US cities in the nineteenth century. See Yellow Fever Collection (2003).

7 South Africa is also experiencing socially determined mortality for a different reason, as the government has opposed drug treatments, for ideological rather than financial reasons. As this book goes to press some provincial governors have defied national policy and are providing HIV-AIDS medication to expectant mothers, and there are discussions that might lead to a change in national policy. A factory or mine worker whose employer provides HIV-AIDS medication (so that the workforce does not die off) is less vulnerable than an urban worker who does not have such access to treatment.

8 West Bengal and Bangladesh Arsenic Crisis Information Centre http://bicn.com/acic/

9 The BGS may be sued for being negligent in failing to test for arsenic (Clark 2001). An internet search on the terms BGS and arsenic will provide a great deal of material on the controversy, including statements from critics of UNICEF and the BGS, and responses from those organisations, e.g. UNICEF at http://www.unicef.org/arsenic/

10 US Centers for Disease Control and Prevention bioterrorism site http://www.bt.cdc.gov/

11 US Secretary of Health and Human Services testifying to the US Senate Foreign Relations Committee said: '[U]nder the circumstances that we are facing right now, I think [$200 million toward the UN goal of $7–10 billion] is a tremendously generous contribution'. Stolberg [2002b] comments that 'under the circumstances' is an apparent reference to the war on terrorism.

12 On global climate change and human health, see McMichaels et al. (1996); Watson et al. (1999); Houghton et al. (2001); McCarthy et al. (2001); Glantz (2001); World Bank (2002).

13 A frightening and regrettable urban mirror image of this situation seemed to be developing at the time of writing in Iraq under US occupation. On 8 May 2003, the WHO gave notice of cases of cholera in Basra and warned of the possibility of a large epidemic. The population's baseline health and nutritional situation is poor, and the water and sanitation infrastructure was not yet functioning properly due to damage sustained during the first and second Gulf Wars as well as years of sanctions (Wright 2003).

14 The US Centers for Disease Control and Prevention reported a crude mortality rate (CMR) among refugees in Goma in July 1994 of 34–54 per 10,000 per day. Normally, the expected CMR in an LDC is below 0.5 per 10,000 per day. Much of the excess mortality is explained by epidemic cholera. See: http://www.cdc.gov/epo/mmwr/preview/mmwrhtml/00052860.htm

15 For example, after hurricane Mitch (1998) PAHO (1999) summarised its health impacts as follows

> Cholera ... jumped from a total of 2,836 cases in 1998, before Hurricane Mitch, to 3,544 cases in the nine weeks after Mitch, with the average number of weekly cases rocketing from 16.4 cases to 78.7 cases. After Mitch, 708 leptospirosis cases were reported, while only one case was reported in 1998 before the hurricane struck. Malaria jumped from an average of 433 cases a week per country before Mitch to 642 cases, a 48 percent increase, with Belize reporting the lowest (27 cases a week) and Guatemala the highest

(1,528 malaria cases per week). Dengue also rose after Mitch struck Central America, from an average of 141 cases a week to 162 cases a week per country.

16 Mental health consequences, on the other hand, can be long lasting and may accompany any disaster or other trauma; see, for example, WHO (1997). In this book we deal with mental health only insofar as it affects well-being with regard to household access and livelihoods as presented in Chapter 3. This is, admittedly, another self-imposed limitation, and is not meant to suggest that people in LDCs (the focus of most of our case studies) suffer any less from grief and other longer lasting post-traumatic stress. We simply cannot accommodate a review of this large and growing literature. For an introduction one might consult Gerrity and Flynn (1997).

17 Here we would include exposure of the migrant to sexually transmitted diseases (STDs) – possibly including HIV-AIDS – and tuberculosis. These are known to have been spread to rural areas that provide contract labour to the South African mines (de Beer 1986; Packard 1989). The spread of HIV infection from urban centres to villages is also documented (Barnett and Blaikie 1992; Collins and Rau 2000).

18 Dutch agricultural production fell by a half from 1938 to 1944/1945 due to war disruption. Nevertheless the Nazis first thoroughly integrated the Dutch economy into their war efforts, exporting large quantities of food to Germany. Then, as a reprisal for the Dutch railroad strike, they cut off all imports of food, fuel and electricity into the Netherlands during the last eight months of the war. Finally, the occupiers intentionally flooded 8 per cent of the country to impede the Allied advance, further weakening agricultural production. The result was famine, with urban rations as low as 500 Kcal in January 1945 (one-third of adult subsistence level). Twenty soup kitchens were run by Amsterdam city officials, feeding up to 160,000 people daily. People subsisted on beet-sugar and food obtained from the black market, where a loaf of bread could cost $27. In view of the severity of the famine it is remarkable that only 15–18,000 Dutch perished out of a population of about 9 million (Mass 1970; Warmbrunn 1972).

19 Some information on health crises is given in Chen (1973) for the Bangladesh war of 1970, and in CIMADE (1986), Kibreab (1985) and Rekacewicz (2000) on refugees in Africa.

20 FAO (2002a); World Social Forum (2002).

21 General discussions of the Irish Potato Famine and its causes include Walford (1879); Woodham-Smith (1962); O'Brien and O'Brien (1972); Aykroyd (1974); Sen (1981); Regan (1983).

22 In 1998, the 150th anniversary of the Irish famine, descendants of the survivors who immigrated to the US established a Famine Institute in Boston that raises money to respond to other food emergencies in the world, beginning with aid to post-hurricane Mitch Nicaragua (see Chapter 7). The institute also supports research into the causes and prevention of famine, and has provided donations to CONCERN, the largest relief and development non-governmental organisation in Eire, as well as to the University College Cork's International Famine Centre. See: http://www.boston.com/famine/irishmore.stm

23 The impact often equalled or surpassed that of the Black Death in Europe. Thomas Jefferson, for instance, wrote in 1781 of the Indians of the Powhatan confederacy in Virginia:

What would be the melancholy sequel of their history, may however be augured from the census of 1669; by which we discover that the tribes therein enumerated were, in the space of 62 years, reduced to about one-third of their former numbers. Spirituous liquors, the small-pox, war, and an abridgement of territory, to a people who lived principally on the spontaneous productions of nature, had committed terrible havoc among them.

(Quoted in Peterson 1977: 135)

McNeil (1979) also records the devastating impact of smallpox in 1530–1531 on the Indian ability to withstand the military invasion of the conquistadors.

24 Major pandemics of bubonic plague occurred in the mid-sixth to mid-seventh centuries (Plague of Justinian and its aftermath in England and Ireland), the latter half of the fourteenth century (Black Death) and the mid-seventeenth to mid-eighteenth centuries, including the Great Plague of London in 1665 (Marks and Beatty 1976).

25 On the resurgence and persistence of malaria, see Chapin and Wasserstrom (1981); Sharma and Mehrotra (1986); Learmonth (1988: 208–211); Matthiessen (1992); Malaria Foundation International (1999); Spielman and D'Antonio (2001: 141–178); Whitfield (2002).

26 Extensive sewer systems were first dug in Hamburg in 1844–1888 and in London in 1854–1865. The association between cholera and water used for drinking and cooking was suggested only in 1849 by Dr John Snow (Read 1970).

27 The intense concentration on potato cultivation in Ireland was an anomaly compared with accounts of peasant farming in other parts of Europe during the same period (Shanin 1971).

28 By the mid-1980s more than four million people had died directly or indirectly in violent conflict in Africa since the era of independence began in 1958. The largest losses were: more than two million in Biafra (Nigeria), 600,000 in Ethiopia, 550,000 in Uganda, 500,000 in Mozambique, 300,000 in Sudan and more than 100,000 each in Zaire, Burundi, Rwanda and Angola (Barnaby 1988). This was, of course, before the complete breakdown of the state in Somalia from 1991 onwards, the genocide in Rwanda and deaths from cholera in refugee camps (1994–1996), the war between Ethiopia and Eritrea, the horrific wars in Zaire/Congo in the late 1990s and civil wars in West Africa. In addition, conflict continued into the twenty-first century in Angola and Sudan. Therefore, an educated guess might put the total number of lives lost by 2002 at double Barnaby's earlier estimate: 8 million.

29 The global IDP Project at http://www.idpproject.org/weekly_news/weekly_news. htm#4 has information on Angola and IDPs worldwide.

30 See also Balter (1998), Cohen (2000a, 2000b).

31 This figure was derived from a two-year study of 1,422 households. The study also found that grain production was more affected by the death of an adult female head of household, while a male death had a greater effect on 'cash' crops. In addition this research revealed that it was very difficult for households to recover from the death of an adult, and that the result of such a loss was often loss of assets such as goats, and a decrease in the overall size of the household. Both these effects could well be part of the household 'coping' strategy discussed in Chapter 3.

32 See also Africa Justice web site at http://www.justiceafrica.org/

33 The reader may be less familiar with yaws than with smallpox. Yaws is a skin and bone disease caused by a bacterium similar to that responsible for syphilis, and is thus one of a family of non-venereal treponematoses (diseases caused by *Treponema spp.*). Until the early 1950s yaws was a very widespread tropical disease causing much disability and even death. The WHO began treating it with penicillin in the 1950s, treating over 50 million cases in 46 countries. The response to a single dose of the antibiotic was curative, and today yaws is of minor importance. See: http://www.intelihealth.com/IH/ihtIH/WSIHOOD/9339/23810.html

34 Of course there is the question of what these large quantities of insecticide may do in the environment, but in general this book cannot deal with the enormous area of technological hazards.

35 Some NGOs have challenged PHC as 'not cost effective'. This challenge arises out of the debate that has accompanied a retreat from 'basic needs' as a fundamental goal (Wisner 1988b; Newell 1988). Some, including the World Bank, have questioned the whole idea of publicly financed health care. Privatisation schemes of various kinds grew up in many countries during the 1980s and 1990s. The long-term impact of such privatisation of essential services, as well as of lifeline infrastructure such as water supply, sanitation and drainage, has not yet been studied systematically.

36 The literature on policies for improving nutrition includes Pinstrup-Anderson (1985); Cornia et al. (1987); Pinstrup-Anderson (1988); Hall et al. (1996); Neefjes (2000); Francis (2000).

6

FLOODS

Introduction

Flood disasters, challenges and changes in thinking

The past decade has been a very significant period in relation to floods around the world, for several reasons.[1] Firstly, some of the most extensive, damaging and costly floods have occurred in developed, wealthy countries: for example, in 1993 in the Mississippi basin (including its major tributaries, the Missouri and Red River) and on the Rhine and its tributaries; in 1998 and 2000 in England; in 1990 in Australia, where an area twice the size of Texas was under water; and in parts of eastern and central Europe in 1997, 1998, 1999, 2000 and worst of all in 2002. While these countries have never been exempt from floods, the severity of these disasters seemed to shock not only the victims, but also governments, planners and insurers. It was as if wealth, infrastructure and order were being unfairly challenged by nature, in societies that considered themselves immune or robust, unlike the less developed countries (LDCs).

Secondly, flooding in LDCs has appeared to be increasingly frequent and serious, to the extent that supposedly 100-year floods were occurring almost yearly in Bangladesh and China, and severe floods afflicted south-east Asia over several years (especially 1996, 1998, 1999 and 2000) and Africa, including Mozambique and Malawi (2000), Ethiopia and Somalia (1997). Thirdly, such floods (rightly or wrongly) have become increasingly associated with climate change: the popular and media perception has been of an increased frequency of floods and storms supposedly resulting from global warming. This general outlook has been linked to a more specific belief that the El Niño Southern Oscillation (ENSO) is increasing in frequency and intensity as a part of climate change. Indeed, such is the popular adoption of El Niño as a climate change phenomenon that it has become a part of television weather forecast in the USA and Europe in the past ten years.[2] This perception has developed alongside increased scientific understanding of ENSO, which has shown it to be not only a regional phenomenon of the Pacific, but part of a system that circumscribes the world, with patterns of drought-desiccation alternating with floods and storms (Glantz 2001, 2002).[3]

Fourthly, the significant impact of floods on wealthy countries (which often had histories of engineering works intended to prevent floods) meant that a new debate opened up (allied loosely to environmentalism and the Green Movement) which permitted new thinking about the need to allow rivers to run unconstrained by earthworks, embankments, artificial levees, concrete and walls. Instead, a parallel popular consciousness arose that rivers should be allowed to flow freely in their valleys, enabling the flood plains to be restored to exactly that: flood plains (Smith 2000).

This shift in thinking also affected the types of policies that developed countries could advocate in LDCs. It has become difficult to advocate engineering solutions to the Bangladesh flood problem, as had been proposed in the 1989 Bangladesh Flood Action Plan (BFAP) (Box 6.3). Crucial to this shift in thinking was also a growing sense that flood disasters are caused by *people* and not just water. The media and popular conceptions of floods shifted significantly to suggestions that the disasters were happening because people and buildings were in the wrong places on flood-prone land. This ties up neatly with the notion that rivers should remain untamed and that people should beat an organised retreat to higher and more appropriate land. The policy fits with the recognition of the needs of nature and the inappropriate behaviour of people.

Despite all this new thinking and increased awareness of floods, we should not lose sight of the fact that floods have long been considered to affect more people and cause more economic losses than any other hazard (although the data often fail to separate out those floods associated with tropical cyclones; see note 1). Parker (2000: 5) cites data that suggest floods were the most common type of disaster trigger in the second half of the twentieth century. The reinsurance company Munich Re has reported that for most years in the late 1990s the largest share of economic losses from all natural hazards were due to floods (Munich Re 1997, 1998), and that 55 per cent of deaths in the period 1986–1995 were caused by flooding, making it the leading cause of disaster mortality in that period (IFRC 1999a: 5).

Yet in many parts of the world, floods are also a normal and an essential component of agricultural and ecological systems, because they provide the basis for the regeneration of crops, plant and aquatic life, and of livelihoods derived from them. In surveys of people who live and cope with floods in Bangladesh, a majority of villagers generally disapprove of attempts to stop floods entirely (Leaf 1997; Haque 1997: 303–304). Floods there are even given different names to allow a distinction to be made between those that are beneficial and those that are destructive. This ambiguous character of floods is important, and is discussed later in the context of flood control measures (since it is difficult to control destructive flooding without also preventing beneficial flooding). If there really is a shift towards allowing rivers to be restored to their 'natural' flow regimes, then the corollary is that the benefits of floods will be reinstated. Paradoxically, floods are also likely to affect areas that at other times are prone to drought. Even in Bangladesh, farming in the winter months requires irrigation in much of the country.

The enormous floods in developed countries over the last decade caused relatively few deaths but tens of billions of dollars of damage, much of it uninsured losses. Along with the damage from water itself (and sewage) came health hazards from contamination by chemical leaks, fuel and other pollutants leaking from damaged industrial plants, and enormous amounts of garbage and debris. These floods had a significant impact on flood policies, not only for Europe and North America, but also for other parts of the world, such as Bangladesh (see Box 6.3). In particular, they severely reduced the trust in conventional 'engineered' flood control measures. This is evidenced in a commentary on the US floods in *Engineering News Record*:

> In large measure, these increases in flood damages have been self-inflicted. Development of our flood plains has continued as a result of engineering hubris, disaster-denial mentality and a willingness to pursue short-term profit in the face of long-term risk. Integral to this problem has been an unhealthy over-reliance on levees too close to rivers and 100-year flood-plain zoning.
>
> (Mount 1998: 59)

The collapse of confidence in engineered flood prevention has allowed an increased interest in a 'living with floods' approach to emerge, which also recognises that rivers, their banks and flood plains, provide valuable 'ecological services' (which can include the absorption of some flood water). This shift in outlook led to an acceptance of the need to understand the function of rivers and their flow regimes in relation to the wider environment, and a new interest in the idea of returning rivers to their 'natural' state. This approach is coupled with a wider interest in river 'restoration', which (especially in the USA) has led to the dismantling of dams, and the renewal of flood regimes through planned release from dams (e.g. as has happened on the Colorado River through the Grand Canyon). Underlying this shift in thinking is a general acceptance that rivers and their floods have considerable benefits that have been lost through damming and flood control.

Although we understand all too well the damage floods do, we have not, until recently, understood very well the many beneficial aspects of flooding. Floods are critical for maintaining and restoring many of the important services provided to humans by riparian ecosystems. Among other things, flooding provides critical habitat for fish, waterfowl and wildlife, and helps maintain high levels of plant and animal diversity. Floodwaters also replenish agricultural soils with nutrients and transport sediment that is necessary to maintain downstream delta and coastal areas. Indeed, recent attempts in the United States to restore riverine ecosystems have increasingly turned to 'managed' floods – the manipulation of water flows from dams and other impoundments – to achieve the benefits of flooding (Haeuber and Michener 1998: 74).

These complex ecological relationships are clear in the case of the lower Mississippi. Before elaborate flood control systems were built, the city of New Orleans was partly protected from coastal storms by marshes and barrier islands (Figure 7.1). However, after decades of coastal erosion without the replenishment of the silt provided in the past by those floods, the 'natural defences' for the city are disappearing. Given this , coupled with the low-lying topography of much of the city and its demographic growth, a very large hurricane-flood disaster is possible – one that could kill up to 45,000 people (Nordheimer 2002: D1&D4; see also Chapter 7).

The new attitude to rivers and floods has also meant a wider acceptance that flood plains might be inappropriate locations for some human economic activity. Policies are therefore being explored for the withdrawal of some economic functions, along with the removal of flood control measures. The floods in the USA even raised doubts about the viability of some flood prevention measures, and it is worth quoting Mount again at length:

> Most of the flooding that occurred last year [1997] was associated with structural failure of levees, rather than with over-topping. In most cases, levees failed due to prolonged high flood stages associated with unusually large runoff in a system divorced from its ancestral flood plain. Other failures resulted from poor levee maintenance. Floods in 1997 in California's Central Valley revealed a disorganized and underfunded maintenance and inspection system, and a significant number of orphaned levees. But several spectacular failures also occurred on levees that were well-engineered, well-maintained, recently inspected and operating within design capacity. It's easy, but politically and economically naive, to state that the silver-bullet solution to preventing levee failures is to simply do a better job of design and maintenance. It's more difficult to acknowledge the most worrisome aspect of levees: No matter how rigorous the engineering and maintenance, even the best levees will fail occasionally.
>
> (Mount 1998: 59)

The implications of this for poor countries (where maintenance is an even greater problem than for the wealthy USA, and reliance on levees puts millions of people, and not just property at risk) are clearly serious.

However, this awareness of the significance of the flood plain does not mean that pressure for use of its 'cheap' flat land has diminished everywhere. In the UK, the severe housing shortage in some regions, coupled with a reduction in the planning powers of local authorities (initiated under Prime Ministers Thatcher and Major, and not redressed under Blair), has led to increasing use of flood plains for housing development. In Britain, insured flood losses have risen enormously in the past decade, with unprecedented flooding almost every year. The insurance industry is now withdrawing its cover from such areas. Often the people buying houses are unaware that they will not be able to get cover, while others are having insurance renewals refused (Crichton 2001: 188; Blackstock 2002).

Floods as known risks

Because of their repetitive behaviour patterns, most types of floods are 'known risks' (White 1942). Riverine floods are of course normally restricted to flood plains, where events over thousands of years have deposited silt and levelled the land so that it is good for agriculture and cheap for modern construction. Flash floods also occur in the hilly upper reaches of river basins, when heavy rain over a limited area drains rapidly into a main channel. This makes warnings and avoidance difficult, although areas at risk of such sudden events are themselves predictable, so that with given rainfall conditions warnings should be possible. Floods affect some low-lying inland areas as a result of rainfall, and some coastlines are liable both to rain flooding and sea invasion (especially under storm surge or unusual tidal conditions). Coastal areas are also at risk of *tsunami*, the waves triggered by earthquakes, volcanoes or undersea landslides that can cross oceans at 800 km/h and rise to enormous heights (often 10 m and sometimes up to 30 m) when striking the shore (see Chapter 8).

As a result of earlier events, and by conforming to predictable patterns, the places affected by flood hazards are generally known. This means that both self and social protection measures should be possible for most types of flood hazards. However, this basic 'expectedness' is complicated by the wide ranges of intensity and duration of floods that can affect the same area at different times, and variation in return periods (the average number of years between floods of a given magnitude recurring). This makes some people gamble against self-protection, on the basis that, over time, the risks are outweighed by the costs of protection. The social protection offered by those who advocate 'technical-fix' engineering interventions are usually costly, and often conflict with the benefits of floods.

Self- and social-protection measures to prepare for floods are made all the more complex and contradictory by the gains that floods provide. Agriculture benefits from the use of flood-plain alluvium (usually with more fertile soil and better moisture retention). The floods themselves provide a form of natural irrigation. As the waters recede, people in many parts of the world practise flood-retreat agriculture, sowing their seeds in the wet soil. In Bangladesh, the regular annual floods that affect much of the country help to restore the soil's fertility with a new layer of productive silt. This process is essential to life and livelihoods, and supports the dense population of farmers and fishers. Here, the absence of the flood may be considered a drought (Schmuck-Widmann 1996, 2001; Paul 1984). On the *char* (the silt islands that litter the channels of the main rivers), inundation is also welcomed by the inhabitants as a means of cleansing the land of pests, especially rats (ibid.). Less often considered are the enhanced fishing opportunities derived from the nutrient-rich waters brought to ponds, lakes and rivers by freshwater inundation.

Areas that are at risk of floods have attracted farming communities and towns for thousands of years, as well as industrial and commercial development in the past two hundred, precisely because of flood-associated benefits. For centuries people have engaged in complex (and sometimes disastrous) 'trade-offs' between coping with floods and using these benefits. But in many countries, those people who have few alternative livelihoods or low income are forced to put themselves at risk because they have no option but to try and survive in flood-prone locations. In effect, vulnerability to floods is determined by their position in society, not by the existence of the flood hazard.

The paradox of flood control

Floods may also be associated with dams, and not simply their catastrophic collapse. Dams can be specifically designed to help reduce flood risk, but be inadequate or inappropriate for the task, or provide a false sense of security (see Box 6.1 for examples in China). The most tragic examples are of dams built to inadequate standards or capacities (or on unsafe sites) which suddenly fail and release a flash flood. Ironically, such dams may have been built to control floods. In other circumstances, managers of dams that are designed to control flooding (usually in addition to generating hydro-electric power [HEP]) have inappropriately released water from reservoirs. This is done to protect the dam from being damaged by high water or overspill, with the result that this water has exacerbated downstream floods.

There have been many dam failures in industrialised countries as well as LDCs. For instance, in 1975 the Grand Teton dam in Idaho (USA) failed, causing $2 billion worth of damage in a number of downstream towns. Fourteen people died, a low number in the circumstances thanks to a timely warning and an evacuation, but 8,000 homes were destroyed along with 350 businesses (US–BR 2003; Christensen 2002).[4] One estimate suggests that 'on average ten significant dam failures have occurred somewhere in the world each decade, in addition to damaging near-failures' (Veltrop 1990: 10; cf: Le Moigne et al. 1990).

Pearce (2001) provides a catalogue of dam overtopping, failures and collapses, and instances of harmful releases of water from dams when their reservoirs were full. In a number of cases he offers evidence that existing floods were made much worse by such releases (for instance in Central America during hurricane Mitch in 1998).[5] Dam failures are often associated with periods of heavy rainfall and runoff that exceed design expectations. Probably the worst examples of dam failures and subsequent deaths are from China, where the official *China Daily* newspaper has admitted that 322 dams have failed since 1949 (Pearce 2001: 5). The worst case was the 1975 collapse of the Banqiao dam in Henan province after a typhoon deposited more than 1.5 m rainfall in three days. The disaster was covered up for many years, and the total number of deaths is a controversial issue. Senior Chinese scientists who oppose the Three Gorges dam (and therefore use the Banqiao collapse as evidence of the risks of the new dam) have claimed that 230,000 people died; an official figure circulated internally gives 86,000 (Human Rights Watch 1995; see also Yi 1998). More recently, a *China Daily* (21 February 2002) article reported the death toll as 20,000 (see also World Commission on Dams 2000a: 7). But the dangers of dam collapse are significant even in rich countries such as the USA, where official reports have shown that a high percentage of dams are unsafe (McCully 1996: 126).

It is also worth recognising that some dams shift the flood water problem elsewhere, as with the Ganges's peak flows which are diverted through Bangladesh by India's Farakka Barrage (Monan 1989: 27). Other dams built for HEP and irrigation (or even flood control) result in permanent flooding, often of inhabited areas, as with the large dams on the Narmada in India, which will mean the forcible displacement of a million people, and the Three Gorges (Sanxia) dam on the Yangtze which requires the expulsion of 1.5 million people (Human Rights Watch 1995; Dai Qing 1998). The Three Gorges dam is specifically claimed to be primarily for flood protection (although it is also a major generator of HEP, and is also intended for irrigation and improved river transport). However, after the extreme 1998 floods in the Yangtze valley (see Box 6.1), there has been no published report on how the dam would have prevented or reduced the damage.

The reservoirs created by dams eventually silt up so that the dam becomes ineffective for power generation, irrigation or flood control, while the submerged land cannot be recovered. There is considerable controversy over the rates of silting, and design capacity and duration have often been very optimistic compared with actual performance. In many cases, a life span of forty years or less is normal for a reservoir. This seems a relatively short period in which to enjoy the supposed benefits, given the enormous financial costs, the high level of borrowing (and the resulting strains on the economy) and the disruption of millions of people's lives. The controversies in the winning of contracts for dams, and the irregularities in working out the ratio of benefits to costs, are well known (Pearce 1992; McCully 2001), and do little to disprove the view that dams are being built more for the benefit of the contractors and politicians than people.

Box 6.1: Floods in China, 1998

The year 1998 brought serious floods to most countries in east and south-east Asia. The rainfall in many countries was much above average, and was linked to an intense El Niño. In China meteorologists were aware of the ENSO event unfolding, and in late 1997 predicted that there would be severe rainfall the following spring and summer. There were attempts to prepare in response to these warnings, in addition to conventional and long-standing flood defence systems that have been in use for hundreds of years. But warnings could not have prepared China for the extent and duration of the floods. They affected all but one province, including the oilfields of Daqing in the north-east to Xinjiang in the far north-west. In all, it was claimed that around 240 million people were affected, equivalent to almost the entire population of the USA. Of these, 14 million were forced to abandon their homes, and official figures claimed 11 million hectares of cropland was 'destroyed' (although it is unclear what this meant) (Disaster Relief 1998a). The estimated costs of the damage had reached $24 billion before the end of August.

The worst hit region was the Yangtze basin, and especially the middle reaches where a number of rivers converge with the Yangtze. This area is a low-lying plain with myriad waterways and lakes that itself received record amounts of rainfall. The city of Jingdezhen recorded 16 inches (406 mm) during a 10-day period in July, after having received 9 inches (229 mm) in just one day in June (NCDC 1998). Floods in this region are common, and have killed many thousands over the last century. In 1954 an estimated 30,000 perished, many drowning as dikes and protective measures failed. In 1931 floods and subsequent hunger and disease killed as many as 3.7 million along the Yangtze (Clark 1982). The 1998 event was considered as serious in its extent and the damage it caused as 1954, but the actual rainfall and peak river flows were lower. The number of deaths was a fraction, although the official death toll of around 4,000 is believed by some observers to be much lower than the real figure. One report suggested that 20,000 were missing from Jiujiang, a city of over 4 million in Jiangxi Province, swept away when a dike collapsed on the Yangtze (Disaster Relief 1998b).

The 1998 disaster was worse in terms of extent and damage mainly as a result of human activity over the nearly fifty years since 1954. Environmental damage and mismanagement have exacerbated the impacts of rainfall and river flooding, reducing the effectiveness of precautionary measures. In particular, the river basin has lost much of its capacity to absorb and store flood waters in lakes and detention basins because of the pressure on peasant households that have lost the

Box 6.1 continued

livelihood options once provided by communal farms to find land to settle and farm. As a result, while a massive (and in many ways effective) civil defence exercise went into action during the floods, it was having to cope with a situation that was not amenable to short-term solutions. The flood precautions and responses – the 'hardware' and the civil defence response – were structurally flawed and inadequate because of the build-up of failures in human and environmental management. In particular, civil defence measures that involve the expulsion of people from their homes and land, with no serious alternatives being provided, are not an effective way of reducing vulnerability (Disaster Relief 1998c).

The main human factors which increased the vulnerability of the population include deforestation, inappropriate precautions and settlements encroaching into flood protection zones. Most of these problems are consequences of central government policies since the Communist Party came to power in 1949 (and so include both central planning up to 1979 and subsequent 'market socialism'). Since 1985 it is estimated that forest cover in the upper Yangtze basin has fallen by 30 per cent (BBC 1999). In the aftermath of the 1998 disaster, Chinese officials up to the highest level cited deforestation as the major cause of the disaster, and an immediate ban was imposed on logging. It was claimed that a million workers in the industry would be 'relocated', and programmes of reforestation put into effect urgently (ibid.). Part of the problem is that a lot of the logging has been carried out by local governments, which in effect 'mine' the environment to raise local revenues and satisfy local leaders' aspirations – both for personal gain and local development (Cannon 2000a). Reforesting means not only losing out on earnings, but incurring expenses; the transition is going to be difficult.

The effect of deforestation is to speed up the runoff in the catchment areas and alter the flow regime of the Yangtze and other rivers so that peak rainfall is not retained and arrives in the stream flow in higher crests. Deforested slopes are also more quickly eroded, and the downstream silting of river beds (and detention basins) has increased rapidly, so that river channels can hold less water (Brush et al. 1989). In many places rivers already run several metres above the level of the surrounding countryside, as the embankments containing the channels have been raised higher and higher over many decades. This highlights the second type of human-induced problem: inappropriate control measures. China's main riverine flood defences are embankments that run the length of the rivers, along with diversion basins and lakes. In many cases these embankments run close to the channel, and there is a continuous need to

Box 6.1 continued

maintain the flow capacity by raising them to compensate for the silt deposited on the channel bottom.

This policy is costly in terms of labour and requires careful and constant maintenance of the banks. During heavy rain and high flows, hundreds of thousands of local people and army personnel have to be posted along the banks to ensure that waterlogged and weakened portions are reinforced, and breaches quickly repaired (in 1998 this included sinking large barges in some gaps and using bags of rice and soya to swell and block others).

A key issue is that the dikes create potentially catastrophic disasters both for those who take refuge on them, and for those who are not evacuated from land likely to be inundated when one fails. However, their existence creates a sense of security and stability, so encouraging the growth of settlements and economic activities in sites along the river valleys which would be at risk of catastrophic dike failure. Continuing to raise them higher and higher increases this danger, and makes main-tenance and repair all the more imperative. Unfortunately, under the economic reforms of the past 20 years or so, the incentives for local people to engage in maintenance are reduced, and the desire of local governments to invest in them has declined. There are other activities that bring profit and revenue so, at the same time, local governments have powerful incentives to build in inappropriate places, and to defy or avoid central government planning guidelines for land use.

One potential way to improve matters is to restore and expand the lakes, diversion basins and wetlands that have been relied on for hundreds of years as flood retention measures. These flood protection resources have suffered considerable reduction in number, size and effectiveness in recent decades. Silting has reduced their capacity; some lakes that could previously absorb peak flows have been drained and lost to this function. The number of lakes in Hubei Province alone (in the middle reaches of the Yangtze) has been reduced from 1,332 in the 1950s to 843 in the 1980s (*Washington Post* 1998). These lakes, along with other retention basins (areas set aside for flood waters diverted from the main river channels), have been encroached on for farmland and settlement by farmers desiring better land and livelihood opportunities under the 'go-it-alone' ethos of the economic reforms. However, since the 1998 floods, the government has acted to remove people from such areas, and claims to have removed 1.8 million (out of a target of 2.45 million by early 2002) (CND 2002). While this may improve flood preparedness, it has in effect meant that millions of (mainly) farming families have lost their livelihoods and are highly likely to become impoverished.

Box 6.1 continued

Dongting Lake, one of the largest of the flood basins, has been squatted over the past 50 years by hundreds of thousands of people desperate for farmland. It has shrunk from 4,350 sq km in 1949 to 2,820 sq km in 1980 (US Embassy 1998). Attempts to remove the people and restore such basins to their flood function are fraught with difficulty. Higher-level authorities evict the people, but the local authorities are unwilling to take responsibility for this 'unofficial' population. Since 1998 almost half a million people have been removed from around Poyang lake in Jianxi province, and embankments breached to return it to its natural size (Gittings 2002). This is the uncomfortable downside of letting rivers return to their natural flow regime.

To substitute for the loss of conventional retention areas, deliberate breaches had to be made in the dikes along the Yangtze in 1998, and the affected people forcibly removed from their land and villages (another reason for the army and police presence). To protect Wuhan, a key industrial city of seven million inhabitants, levees were to be breached to flood farmland in Jianli county. The 50,000 people resisted forcible removal (Disaster Relief 1998c). In the most significant reported case, 330,000 people were to be moved from an area in central Hubei province, although in the end this was not necessary. Those forced from their homes by floods were unable to move back for over a month.

Silting is a factor common to both river bed accumulation and reduced effectiveness of flood retention. But it is too simple to blame deforestation for the flood problem, and it can become a visible but misleading target for policy. Deforestation may be a significant factor in the flooding, but the causal link has been widely accepted in recent decades in many parts of the world as a rather simplistic way of bringing human interference into the picture. It is likely that it is only one factor, and possibly not even the main cause. If so, even successful reforestation will not necessarily solve the flood problem. It is all too clear that there were severe, disastrous floods in China long before the most serious deforestation of the past 20 years during economic liberalisation. The farmers who want land to cultivate are twice victims: of the floods themselves, and of the official attempts to reduce flood disasters by removing them from the flood plain. What needs to be addressed is the vulnerability of rural people in the free-for-all commercialised economy, not just deforestation.

Another key issue in regard to Yangtze flooding is the role of the highly controversial Three Gorges dam (Edmonds 2000), promoted officially as a flood control measure. However, much of the flood problem on the Yangtze is downstream of the Three Gorges dam in low-lying terrain that is subject to rainfall inundation and riverine floods on the

Box 6.1 continued

tributaries that enter the Yangtze below the dam. The dam's functions are contradictory: to maximise HEP requires the dam operators to keep the reservoir as full as possible, while flood protection needs it to be left partly empty in anticipation of summer rains. Practice in other countries, including in the Mozambique floods of 2000 (Pearce 2001), demonstrates that dam operators are often reluctant (or even forbidden) to release water in preparation for possible flood peaks, and only release water in desperation when they fear damage to the dam from the incoming water. Another problem is that the Three Gorges reservoir includes mountains and steep slopes and myriad tributaries, some with their own rocky gorges. Some Chinese scientists have expressed grave concern in a public warning about the dangers of landslides affecting people, especially some of those resettled on riverbanks and hillsides around the reservoir (Pearce 2002b). Such landslides might be triggered by earthquakes, by rainfall or by settlement of unsuitable land. Any major slide into the reservoir would trigger an enormous wave (in effect a local tsunami) which could overtop or damage the dam.

Natural dams

There is another source of floods – the collapse of 'natural dams'. These dams are formed by landslides blocking a valley, or glaciers holding back lakes, creating a reservoir of water that can race down a valley when the natural blockage is eroded or melts. Because the subsequent flow is rapid, it creates a flash flood (similar to a dam failure) which is difficult for people to escape. Yet in many cases such natural dams are detected, and preventive measures are possible. A flood of this type occurred in 2000 after a landslide (which happened in April and was reportedly the largest ever in Asia at 60 m high) blocked the river Tsangpo in Tibet (which becomes the Brahmaputra in India and Bangladesh). The dam burst in July, and the resulting flood surge downstream into north-east India left perhaps 2,000 people dead or missing, and 50,000 people homeless in one of the poorest and most remote parts of India (*Indian Express* 2000). The lack of any warning for this entirely predictable event arose partly from a lack of proper communication between the Chinese and Indian authorities, and demonstrates that (as with the Farakka Barrage) international politics and mistrust can generate vulnerability factors affecting ordinary people.

Flash floods such as these, and those associated with glacier dams and outburst floods, are likely to increase with climate change (Pearce 2002c). Snow- and ice-covered mountain slopes are becoming more unstable with rising temperatures, leading to a possible increase in the frequency of avalanches and landslides that can create natural dams. More rapid melting of glaciers may increase the number of outburst hazards from impounded water. In the Himalayas, local people, governments and a number of NGOs are working on projects to prevent devastation of villagers by attempting to siphon water out of dangerous lakes.[6]

In central Asia there is serious concern about the risk from Lake Sarez in Tajikistan. This 60 km lake was formed after an enormous landslide in 1911 created a natural dam which is 600 m high and 4 km across. If it collapses – it has already been damaged by frequent earthquakes in the region – a flash flood wave of an estimated 150 m initial height will devastate the homes and lives of up to five million people downstream (IFRC 1999a: 39). A warning system was set up (but only in 1984) by the Soviet Union, but it is regarded as very poor, and in any case alert procedures and evacuation plans are inadequate. The anticipated disaster will not only affect Tajikistan, but also parts of Afghanistan, Uzbekistan and Turkmenistan. Co-ordination and co-operation between these now independent countries is much more difficult than under the Soviet regime, and has made it difficult for foreign assistance to be organised. Rivalry and suspicion between some of the governments does not improve the chances of strategies being drawn up to plan for the dam's collapse (through warning systems or by strengthening the dam) or of implementing measures to reduce the amount of water through siphons or channels.

Another political source of vulnerability is associated with a debris dam which impounded a lake in the crater of Mount Pinatubo in the Philippines.[7] This crater was enlarged by the eruption of 1991, and was filled with rainwater. On one side it was held in place by a natural dam of mud and loose debris, and if this failed it would have sent an avalanche of water, mud and rubble into the towns below. In 2000 local people and Oxfam highlighted the risk of this debris collapsing and releasing the lake in such a flash flood (Marchant 2001; Aglionby 2001). There was a suggestion to dig a channel through the blockage, or set up a siphon to drain water harmlessly from the lake. But there was delay, confusion and reluctance on the part of some of the authorities to take appropriate action before the water reached the rim (ibid.). Fortunately a catastrophic collapse did not happen, and the lake level was successfully lowered by engineering means in 2002 (Punongbayan 2002, 2003).

Flash floods and landslides

Typically, flash floods are small-scale events, but often with a high mortality rate. By their very nature it is often difficult to give a timely warning of flash floods, although possible danger areas should be relatively easy to identify. Extreme events with long or unknown return periods are understandably difficult to anticipate. But for many upland or hilly regions of the world, where there is the risk of flash floods with much shorter return periods, precautions ought to be possible.[8] When warnings are not given, the numbers of casualties can be high. The force of the flow of water and the debris carried in it can be extremely damaging to homes and to farmland, and it is common for thousands of people to be displaced from their ruined settlements for months. Recent events include the deaths of 156 in Nepal and Tibet (August 1998); and in the space of just one month (August 1999), 40 dead in Vietnam, 63 killed in a suburb of Khartoum (Sudan) and 77 dead and 120,000 displaced in Chenzhou, China.[9]

Heavy rainfall and flash floods are often associated with landslides, which are frequent events in hilly and mountainous areas around the world. Usually the numbers killed are regarded as insignificant, and therefore such events do not normally reach the media but, as with flash floods, total mortality due to landslides is much higher annually than from some other disaster triggers. However some landslides can have enormous impact, such as the one in 1920 in Gansu province (China), when 200,000 were reported killed (USGS 2003). Much more recently, a whole series of flash floods and associated mudslides and landslides affected Venezuela in December 1999, leaving an estimated 30,000 dead (many people were buried or washed away, and so there is no accurate mortality figure), and around 400,000 homeless (Wieczorek et al. 2001: 2; BBC NEWS 1999; PAHO 2002a). In general, landslides are thought to have relatively low casualty statistics in comparison with other hazards. Red Cross data for the decade 1991–2000 suggests that 'only' 10,000 people were killed in 173 landslides (IFRC 2000a, 2001b). However, the data are misleading, since landslides often occur as a secondary consequence of another type of hazard, such as flooding (as in the Venezuela disaster already mentioned), storms or as a result of an earthquake. So landslide casualties are often added to the total of deaths and injuries attributable to these other events, while those specifically linked to landslides are probably under-reported.

Landslides involve material that can vary considerably in its character, including rock, debris, mud, soil, snow, ice or several of these in combination (Alexander 1989: 157). Alexander includes landslides that are generated by a wide variety of 'agents': the failure of a coal-mining waste tip in Aberfan village (Wales) in 1966 (triggered by heavy rain); a volcanic eruption in Colombia (Nevado del Ruiz) in 1985 which set off a *lahar* (mud and debris landslide) which is estimated to have killed 23,000 people (see Chapter 8); an earthquake in Peru (Mt Huascaran); or flooding in Rio de Janeiro (Brazil). Applying vulnerability analysis to the case of landslides, we need to move beyond the physical hazard to inquire about human activities that might act as 'triggers' for the physical event. We must also understand the 'mechanism' that translates exposure to the risk differently for various categories of people. Differential ability to recover after a landslide is also important since it can cause people to be more exposed to future hazards, as we have seen in previous chapters.

Consider four typical examples of landslides that took place between 1985 and 1988:

- Mameyes, near Ponce (Puerto Rico), 8 October 1985, killing 180, 260 homes destroyed (Wisner 1985; Doerner 1985).
- Rio de Janeiro (Brazil), February 1988, where 277 were killed, 735 injured and more than 22,000 displaced in shanty towns (Allen 1994; Byrne 1988; Margolis 1988; Michaels 1988; Munasinghe et al. 1991: 28–31).
- Catak (Turkey), 23 June 1988, killing approximately 75 (Gurdilek 1988).
- Hat Yai (Thailand) in November 1988, killing 400 (*Economist* 1989; Nuguid 1990; West 1989).

The analysis of the likely causes (or 'root causes' and 'dynamic pressures' in terms of the PAR model outlined in Chapter 2) of these landslides shows a number of interesting similarities. The first commonly cited cause is deforestation. In Thailand there was an outcry against logging following landslides there. West notes that this protest:

> did not come from bearded ecologists and trendy 'green' politicians, but from the local farmers and townspeople, those in fact, who had suffered. The anger comes from below, and is aimed especially at the greedy loggers, frequently Chinese businessmen in partnership with senior officials in the police and army.
>
> (West 1989: 18)

The Prime Minister of Thailand visited the site of the disaster and announced that logging operations would be banned. Forty years ago, 70 per cent of the country was covered by forests, but by 1989 this had dropped to just 12 per cent (*Economist* 1989).

In Rio, the authorities were criticised for not taking effective action to tackle the problems of the denuded hills where all the *favelas* (squatter settlements) had been constructed. These housed a million of the eight million people in the city. There had been extensive deforestation in these areas to make way for dwellings as well as to providing fuelwood. Socio-economic factors are an obvious 'pressure' that both forces squatters to inhabit unsafe locations and forces them to cut vegetation for fuel or building materials, since alternatives are too expensive (Allen 1994; Smyth and Royle 2000).

Road building of poor quality or in the wrong location is also commonly mentioned as a cause. Turkish authorities commented that the roads in Catak should have been cut into the contours, rather than running parallel with them. Frequently, roads are cut into steep slopes with minimal understanding of the geomorphology of the setting, and can interrupt drainage patterns. The actual 'cut and fill technique' of road building on steep slopes can contribute to landslide risk (Smith 1992: 165).

Environmental damage to sub-soil stability is also frequently cited as a cause. Changes in the water table can occur due to leaking tube wells, stand pipes and septic tanks, and appear to have been a contributory cause to the landslides in Puerto Rico and Rio de Janeiro. Unsafe, unauthorised building on dangerously steep slopes is very often cited as a cause of landslide disasters. The location of squatter settlements themselves may have been a contributory cause to the landslides in Puerto Rico and in Rio. Many cities also lack warning systems for predicting water flow and arranging the evacuation of communities at risk. This also appears to have contributed to the landslide disasters in the aforementioned places. There had been extensive rain for a number of days prior to the mudslides, but no monitoring or planning for such contingencies.

Disastrous outcomes for vulnerable people

The chain of causation of flood disasters is shown in an adapted PAR diagram (Figure 6.1). Unsafe conditions in floods may be a combination of factors involving the physical environment, the local economy and the performance of public actions and institutions. In the physical environment, people with high levels of vulnerability will include those who are unable to protect themselves. They may be unable to afford a house site on raised, safer land, or a house with inadequate construction quality. Inadequate action by public institutions may mean that social protection is lacking in flood protection measures or warning systems.

Specific dynamic pressures that lead to these vulnerable conditions will include class relations and income distribution (affecting assets, livelihood opportunities, level of self-protection). Gender relations may be highly significant in some cases, e.g. where women are more likely to be undernourished and less resilient in the aftermath of flooding. The type of state, and its willingness to act impartially on behalf of all its citizens, is likely to influence the distribution of assets and livelihood opportunities, as well as the scope and efficiency of social protection measures. In the sections below, and in the boxed case studies, examples of factors and processes influencing the different levels of that chain are explored in relation to particular countries and flood disasters.

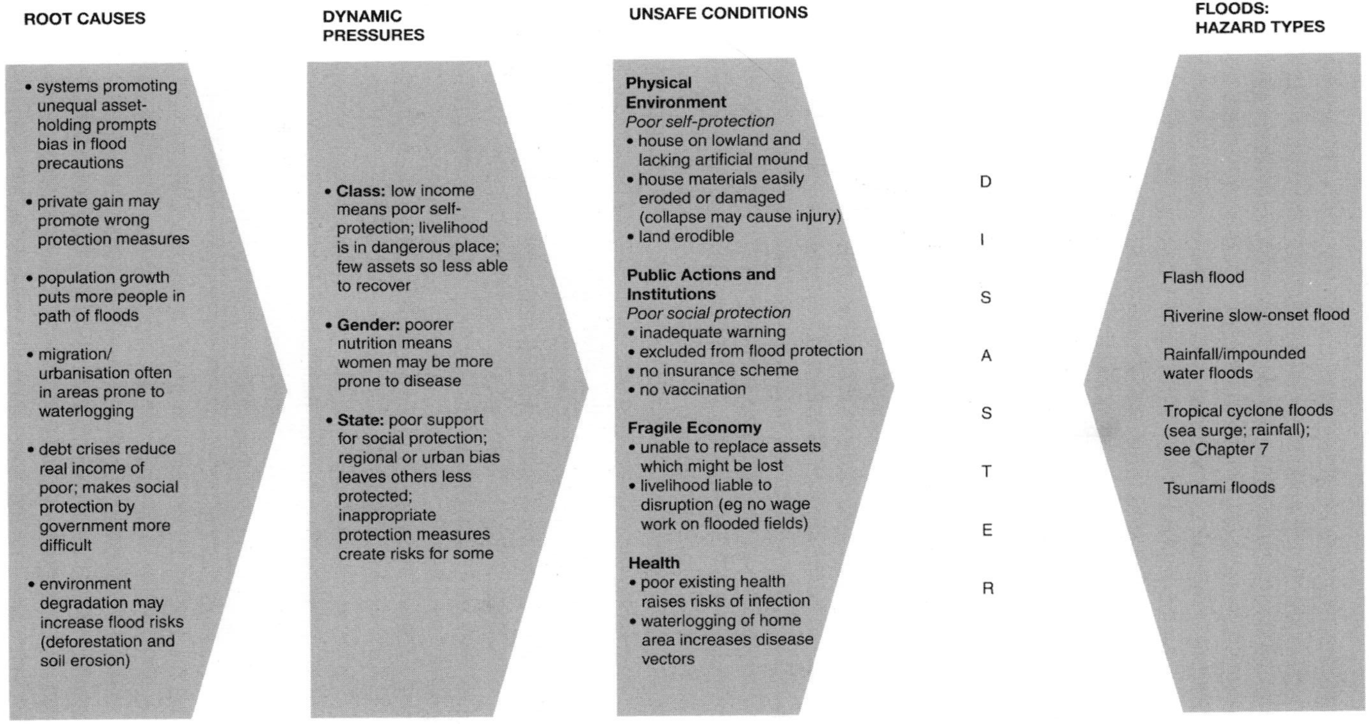

ROOT CAUSES	DYNAMIC PRESSURES	UNSAFE CONDITIONS		FLOODS: HAZARD TYPES

ROOT CAUSES

- systems promoting unequal asset-holding prompts bias in flood precautions

- private gain may promote wrong protection measures

- population growth puts more people in path of floods

- migration/urbanisation often in areas prone to waterlogging

- debt crises reduce real income of poor; makes social protection by government more difficult

- environment degradation may increase flood risks (deforestation and soil erosion)

DYNAMIC PRESSURES

- **Class:** low income means poor self-protection; livelihood is in dangerous place; few assets so less able to recover

- **Gender:** poorer nutrition means women may be more prone to disease

- **State:** poor support for social protection; regional or urban bias leaves others less protected; inappropriate protection measures create risks for some

UNSAFE CONDITIONS

Physical Environment
Poor self-protection
- house on lowland and lacking artificial mound
- house materials easily eroded or damaged (collapse may cause injury)
- land erodible

Public Actions and Institutions
Poor social protection
- inadequate warning
- excluded from flood protection
- no insurance scheme
- no vaccination

Fragile Economy
- unable to replace assets which might be lost
- livelihood liable to disruption (eg no wage work on flooded fields)

Health
- poor existing health raises risks of infection
- waterlogging of home area increases disease vectors

D
I
S
A
S
T
E
R

FLOODS: HAZARD TYPES

Flash flood

Riverine slow-onset flood

Rainfall/impounded water floods

Tropical cyclone floods (sea surge: rainfall); see Chapter 7

Tsunami floods

Figure 6.1 'Pressures' that result in disasters: flood hazards

Mortality, morbidity and injury

As we have seen in this chapter, there are many types of primary flood hazards, including riverine and rainfall, flash floods and dam collapses, plus those produced by tropical cyclones (see Chapter 7), and secondary hazards produced by associated land- and mudslides. Floods and their associated impacts are not only one of the most widespread of natural hazards; they also lead to the greatest loss of life, immediately through drowning and fatal injury and later through illness and sometimes famine. It is difficult with the available data to separate the impacts of different types of floods. However, although deaths associated with cyclones have sometimes been very high (in Bangladesh possibly 300,000 in 1970 and over 138,000 in 1991, and an estimated 10,000 in Orissa, India in 1999), other types of floods are far more frequent and widespread, and range in scale from those in small watersheds to some which cover large areas of an entire country. A few single events have been responsible for millions of deaths: for instance an estimated 2 million in the Huanghe (Yellow) River flood in China, 1887 (Smith 2000). However, rainfall floods and slow-onset river inundation of flood plains generally result in lower direct casualties. These are less likely to be a result of drowning and are due more often to building collapse (people may remain in upper storeys or on roofs), other injuries and snake bites. The risks of disease and malnutrition are also increased by slow-onset (and slow-retreat) floods, with the effects lasting months or even years.

As with other disasters, there is a tendency for events that cause the most deaths and damage to receive the most attention (both in the media, and more seriously in terms of international relief efforts). But as well as the large-scale events, it is essential to consider the myriad small and medium disasters. Not only do these add up to an impact which globally may well exceed the major events, but they may affect the same people and places repeatedly. And as we have seen above, mortality from flash floods – usually regarded as small events – probably exceeds deaths from slower-onset river floods in most years. The 'hidden' problem of these smaller floods is considered in Box 6.2.

Box 6.2: 'Small' floods – a hidden problem

Researchers and activists in the Latin American network La RED argue that not enough attention is given to small and medium-sized disasters. These events are far more common than the dramatic 'super cyclones' or 'catastrophic earthquakes' that win media attention. Such 'small' disasters are a frequent experience for excluded, marginal groups (Lavell 1994), and their economic, social and human costs accumulate.[10]

A number of floods in early 2002 in Bolivia, Indonesia and Senegal demonstrate the policy challenge of small disasters. On 19 February, La Paz (Bolivia) experienced an unusually intense rain and hail storm,

---

Box 6.2 continued

at the rate of 41 mm/h. Although the storm lasted just 45 minutes, flood waters up to 2 m high surged through the streets. Urban drainage was unable to cope, especially in low-income sections of the city. The resulting flash floods killed 63 people, another 13 were listed as missing, 146 were injured and 5,000 displaced. Thirty-two schools were damaged; 126 homes were damaged and 28 destroyed (OCHA 2002b). Some might say that no city in a semi-arid climate (average rainfall 540 mm per year) should or could be prepared for such an unusual event. But this kind of urban flash flooding is very well known in the region. It is also known that generally the lower the rainfall the more variable it is, with greater potential intensity during precipitation. A week later, there were other floods in Bolivia, at Santa Cruz and Cochabamba, that affected 1,300 families (OCHA 2002a). On 5 March, some 150 km from La Paz, flooding on the Yara affected 100 families and displaced 225 people (World Vision 2002c). As Lavell and his colleagues in La RED suggest, wherever one looks, there are many 'smaller' or 'minor' events that disrupt lives and livelihoods.

In Jakarta (Indonesia) torrential rain fell for five days in January 2002. Twenty-nine of Jakarta's 37 districts were affected, with hundreds of thousands of residents displaced and suffering loss of property and interrupted livelihoods. Twenty-two persons were killed. Meanwhile, there were floods elsewhere in Indonesia, including north and east Java, where floods and landslides killed 75 people (Agence France-Presse 2002a). Flooding in Indonesia at the beginning of 2002 killed at least 150 people and destroyed or damaged $177 million worth of infrastructure: roads, school buildings, dikes and drainage works. In Jakarta 100,000 homes were damaged (Agence France-Presse 2002b).

Beyond the aggregate losses there are further complexities. For example, Indonesia's Mental Guidance and Social Affairs Agency found that infant food was generally not available in the shelters set up for flood victims, putting 20,000 of 26,000 flood-victim babies at risk of malnutrition (World Vision 2002b). Recalling the discussion in Chapter 5, the reader will not be surprised that the flooding was followed by coughs, skin rashes and diarrhoea. The City Health Department reported that 18,000 of 384,256 flood victims in Jakarta were suffering from diarrhoea, and that eight infants died from dehydration due to chronic diarrhoea (World Vision 2002a).

This example underscores the vulnerability of people in rapidly-growing cities in a period of economic stringency. The Indonesian economy was among those most badly affected by the 1997 Asian financial crisis. Many of the canals and floodgates used for flood control in Jakarta are very old, dating from the Dutch colonial period.

Box 6.2 continued

Indonesian environmentalists said that the cause of the floods was not the rainfall by itself, but 'uncontrolled development of green spaces and of natural water catchment areas, along with broken or refuse-choked drains' (Agence France-Presse 2002a).

Finally, there is an example from Senegal in the same month. The International Red Cross described the situation as follows:

> Heavy off-season rains swept through Senegal between 9–11 January, killing 28 people and affecting over 100,000 others. The situation was made worse by an unusual cold spell with temperatures plunging from 40 to 16 degrees Celsius. A damage assessment carried out by government authorities and the Senegalese Red Cross revealed that an estimated 105,471 head of livestock had perished while heavy rains demolished 13,993 homes, washing away another 581 hectares of crops. Approximately 1,537 tons of rice was also destroyed.
>
> (IFRC 2002c)

On the scale of the great floods in Mozambique (see Chapter 7), Bangladesh or China, this was not a major disaster. But our approach suggests that many hundreds of small and medium-sized disasters can have the same long-term impacts on livelihoods and well-being as a large event, especially where rural households are already stressed by chronic disease, poverty and isolation. The key point about a disaster is not its scale, but the impact of a hazard of whatever intensity on a vulnerable population. If disasters are defined only in terms of the large events that warrant media coverage or outside assistance, we are almost certainly missing a larger affected population whose disasters are hardly ever noticed by outsiders. The boundary between disaster and everyday life can be very thin, as vulnerable people are at risk from normality as well as the exceptional.

Flood waters bring increased risk of diseases such as cholera and dysentery, arising from sewage contamination of drinking water (see Chapter 5). There may be rapid growth in the incidence of malaria and yellow fever because of the multiplication of insect vectors in the stagnant water which remains ponded after a flood. In some regions (this is common in Bangladesh and India), water is held back by raised structures such as roads and railways, and these cause local floods that are sometimes severe.[11] Stagnant water may be trapped in such places for months. These structures should have ducts and culverts that permit the return flow of water back to river channels, but often these are inadequate or badly maintained. In some floods, they may also be blocked by silt. In Bangladesh, plastic bags were banned in 2002 because they frequently block urban drainage conduits. But there has been fierce opposition from the businesses that profit from making them. An environmental campaigner has received death threats as a result of his 13-year campaign to have the bags outlawed (Mahmud 2002).

Respiratory illnesses often become more prevalent in the aftermath of slow-onset floods, and take a toll, especially among very young children, babies and the elderly. Illness or injuries caused in the flood are important factors that increase existing vulnerability and extend it to new groups of people (see Chapter 5). The sick and injured usually cannot work, and a family's loss of their labour, especially during attempts to recover after a hazard event, can be a significant element of the disaster.

There are few studies of what actually happens to people after floods strike. A survey was conducted in Bangladesh after the 1998 floods (see Box 6.3). People were asked whether they had experienced diarrhoea, respiratory illness or fever in the previous two weeks. All flood-affected people suffered more than those in the non-flooded control group. On average, 31 per cent of flood-affected people were ill, with rates for those exposed to very severe flooding as high as 40 per cent (del Ninno et al. 2001: 73). However, it is worth noting that 23 per cent of the non-flooded people were also ill, suggesting that there is prevalent and pervasive illness throughout much of the rural population under 'normal' conditions.

Such health problems were highlighted in studies of floods on the west coast of South America brought about by El Niño in 1982–1983. In that southern summer, El Niño struck badly, principally affecting Peru and Ecuador. In parts of Peru a state of emergency was declared: rainfall in the first six months of 1983 was many times more than the total rainfall of the previous ten years (Gueri et al. 1986). Mortality caused directly by floods does not seem to have been high, but disease and health problems were made much worse, and people's livelihoods suffered enormously as will be seen later. In Ecuador, faced with the failure of the waters to subside, many rural people fled to towns and cities, more optimistic about conditions there. They brought malaria infection with them, leading to the reinfection of urban areas previously cleared of the disease. The floods greatly increased the number of cases of malaria anyway. Despite a massive increase in insecticide sprayings, the number of cases rose in 1983 and 1984 to levels ten to twenty times (depending on location) those of previous years (Cedeno 1986). In neighbouring north Peru, a study of government health centres showed morbidity rates up by 75 per cent for respiratory and 150 per cent for gastro-intestinal illnesses in the first six months of 1983 (compared with the same period the previous year) (Gueri et al. 1986). These illnesses led to a large increase in deaths: 6,327 compared with 3,226 for the same period (the centres surveyed covered a population of 630,000).

Livelihood disruption

While death, illness and disablement lead to a reduced capacity for work in affected families, there are other impacts on people's livelihoods that make some more vulnerable and yet which enrich others. Not all groups in flood areas are necessarily disaster victims. The impact on different social and economic groups may be more or less severe. In floods it is of course true that much property (including livestock) is damaged, destroyed or swept away. But even flooded land or remaining animals can be sold by a destitute farming family to buy food, despite the low prices arising from many others making such 'distress sales' at the same time. So there are beneficiaries of the disaster who can accumulate land or other assets at depressed prices. Others may benefit from selling food stocks at higher prices. Still others may have saleable goods or services on which they can thrive, perhaps trading in drinking water by virtue of owning a boat in which to carry it around.

Flood impact therefore must also be understood in terms of the disruption and destruction it can cause to livelihoods, and the changes in the access profiles of affected people (see Chapter 3). Losing or being forced to sell land and other assets as a result of floods may shift people into poverty or worsen their existing poverty. The loss of assets or ability to work, of land and animals, or suffering due to injury and illness, may still be affecting people when another flood arrives a year later, and possibly even for years afterwards. A recent IFPRI study of the 1998 floods in Bangladesh gives rare details of these effects from household surveys (del Ninno et al. 2001: xvi–xviii, Chs 4 and 5). Any deaths from illness or injury which occur after a time-lag are unlikely to be linked officially with the flood, so mortality data are often underestimated. As with famine and biological disasters, our Access model (Chapter 3) indicates how vulnerability to *future* floods (or other hazards) can be increased by the longer-term impacts of floods on household assets, labour power and social networks.

Each household's 'bundle' of property and assets (including land and animals for farmers, or boats and nets for fishers) and economic links to others, may be lost, enhanced, disrupted or reinforced in a number of permutations. This sort of disaggregated approach to the impact of hazards shows that although it is possible for a large majority of people to be made worse off, floods may not be a disaster for everyone. The existing social and economic system involves 'rules' and patterns which govern how a particular flood hazard makes its impact on existing patterns of vulnerability, moving some people down in terms of assets, others sideways, and leaving a few with enhanced endowments and livelihoods.

Even where a flood disaster does not create hunger, the impact of the deluge on many people's livelihoods causes at least medium-term disruption and will entail hunger for some groups of people. Follow-up studies of such impacts are extremely rare (or fairly crude, e.g. the evaluations done by governments and aid agencies to determine food aid needs in flood-affected areas). The International Food Policy Research Institute (IFPRI) study provides evidence that people in lower income groups reduce their food intake considerably, and that overall consumption goes down (ibid.: 56–63). This study found that Bangladeshi food markets had been reformed as a result of liberalisation policies in the decade preceding the floods, and are now mainly operated by the private sector. This, combined with government action and food aid, seems to have ensured that there was no significant increase in food prices at the time of the floods. IFPRI claims that, as a result, serious hunger was averted because a food price rise like that which caused famine in 1974 was avoided. It is difficult to know if this success can be replicated at other times or in other places. Normally food prices rise after floods, and may lead to hunger for some people even when adequate food supplies exist (ibid.; Sen 1981; Crow 1984). Our discussion of Sen's theory of exchange entitlements (of FAD versus FED) in Chapter 4 explained why this is a possible outcome.

The majority of people in poor countries are unlikely to be insured. The loss of the home is a major setback to the family's livelihood, not because it is necessary for earning a living (although it often is), but because providing a replacement places an excessive burden on limited finances. The cost is to be measured not simply in terms of cash outlay, but frequently includes lost time that would otherwise be used in direct livelihood-earning activities. As well as the building, many simple household items may also need replacing, such as cooking pots and water vessels. Uninsured people with no savings lose twice in a flood disaster: they lose the goods which are essential to life and they lose the time which they have to spend working to replace them. Having reserves or insurance means being able to return quicker to normal livelihood activities.

There are other losses which may directly disrupt household livelihood. Standing crops are a loss for the farmers who own them (and for the poorer families this is perhaps the most serious aspect of flooding). In many areas of the world, there is an unhappy coincidence of the season in which floods are most likely and the time when crops are ripening for harvest. An added 'ratchet effect' arises because this pre-harvest season is also often the 'hungry season' when household food stores and income are low and the physiological reserves of people are depleted (Chambers et al. 1981). Crops in some parts of the world are well adapted to expected levels of flood. For instance, thousands of indigenous varieties of rice have been developed in south and south-east Asia. They include the floating varieties which are planted in many areas to be grown alongside floods. It is estimated that over 15,000 varieties of rice suited to all types of conditions have been developed by farmers in Bangladesh (Shore 1999). But even these will succumb to inundation under some circumstances, along with non-adapted crops.

Large land-owners do not need labourers when their fields are flooded. This consequent loss of paid employment may be disastrous for families that rely for a large part of their livelihood on such income-earning opportunities. This was one of the main causes of hardship and hunger in Bangladesh following the 1974 floods, when millions of people found that falling incomes and rising prices meant they could not buy food (Sen 1981; Clay 1985). Estimates of those who starved to death at that time range from the official total of 26,000 up to 100,000 in just one district (Sen 1981: 134).

The length of time that water remains on the land can also affect the subsequent normal planting, or the planting of a 'catch crop' aimed at recovering some of the losses. Paradoxically, it often happens that the overall annual harvest in countries affected (even by serious floods) is not lower, and is sometimes higher than normal. While heavy rainfall brings floods to some areas, in others it increases yield. Even in flooded areas, the soil retains more moisture that can be taken up in plant growth later during the dry season. In Bangladesh, the cereal output in both 1987 and 1988 (years of serious floods) was almost identical to that of 1986 (at around 24 million tonnes), and in 1989 increased to 27 million tonnes. In 1998, with floods regarded as being even worse than those of the 1980s, Bangladesh produced 31.58 million tonnes (up 2 million from 1997), and the following year output was 36.5 million tonnes.[12] The same was true of China: in the major flood year of 1998, production was 458.4 million tonnes, up 13 million from the previous year. In 1999, output was down by only 3 million tonnes.

These figures are perhaps rather surprising, but of course are only relevant at the aggregate national level. At the other end of the scale, a household whose water buffalo has died in a flood, or which for other reasons cannot take advantage of these moist soil conditions, will not be able to plant in a timely way and 'catch up'. Animals may be swept away and drowned or injured, and their loss to those families which use livestock produce for subsistence or sale is comparable to the loss of those reliant on crops. Animals are often the main source of draught power and transport for significant sectors of the rural population in many poor countries. In Sikander's (1983) study of floods from that period in Pakistan, the surveyed villagers reported losses of 35 per cent among their animals.

Flooded rivers are part of a physical process whereby land is destroyed by erosion, and recreated in areas where silt is deposited as the sediment-laden waters are slowed down. Flooded rivers are by definition flowing beyond their usual banks. Their route across the countryside, if unconstrained by human interventions, will be via new channels through the lowest-lying land. New channels carved in this way are often miles away from their previous course. People who lose land in this process are unlikely to have access to other land to replace it. Yet others may find that fortuitously the river has abandoned a channel near them, making it possible in time to colonise that land. However, it is more likely that the powerful and already better-off households will gain control of such new land, as has been shown in Bangladesh (Elahi 1989; Rogge and Elahi 1989).

Land is 'lost' in other ways too. Depending on the speed of the flood waters, in some places the soil itself may be carried away. Generally though, as flood waters spread out across the land, they slow to a pace at which they can no longer carry their suspended load of silt and sand. This sediment is then deposited on top of the earth. In some regions the sediment usually has a beneficial effect, replenishing minerals which improve fertility. In such cases the silt deposition is benign, but this is not always so. The size of deposited particles may be much larger, and may cover extensive areas with infertile sand or gravel that inhibits plant growth. When the Kosi River floods in north Bihar (north India), it also normally deposits a layer of sand over agricultural land, rendering it useless for up to 50 years (Lyngdoh 1988). The mineral content of the sediment may also be too saline or alkaline, rendering the ground toxic to plants. Depending on the combination of these different factors, the land left behind can be enriched and newly fertilised by the layer of deposited silt, or made more barren and less productive. Flash floods in Rajasthan (west India) are likely to produce the latter situation (Seth et al. 1981), in a region which usually has to face problems of drought rather than inundation.

In Bangladesh it is more likely that flooded land is enriched by a new layer of silt cast on it by the floods. Further downstream in Bangladesh, the waters of the Ganges and Brahmaputra arrive at the Bay of Bengal laden with silt, are stilled by the sea and add new material which expands the delta. This new land, often in the form of islands called *char* in the middle of the many channels of this complex river system, is quickly occupied by poor and landless peasants from elsewhere, who otherwise have no means of subsistence (Schmuck-Widmann 1996; Zaman 1991). Their precarious existence on the edge of this watery boundary-zone is discussed in Chapter 7.

Box 6.3: Bangladesh – reducing vulnerability to floods is not the same as stopping floods

Bangladesh attracted much international concern to 'solve its flood problem' when it experienced a supposed '100-year flood' in 1987, followed by another – even worse – in 1988 (British Overseas Development 1990; Parker 1992; Westcoat 1992; Hossain et al. 1992; Brammer 1990a). A decade later, in 1998, the country endured floods that were considered the worst ever. As in 1988 they covered around 60 per cent of the country, but although in general the waters were no deeper, they lasted two months longer, destroying more of the summer harvest and delaying opportunities for replanting and recovery.

Box 6.3 continued

What happened in the intervening decade, during which costly studies had been undertaken to find solutions to 'the flood problem'? Whose definition of the problem had been listened to, and what changes had occurred in understanding how flooding in the country should be perceived? And why did a significant foreign donor-supported initiative – BFAP – not provide a comprehensive solution?

In 1988 the Government of Bangladesh (GoB) in co-operation with the UN Development Programme (UNDP), plus official teams from Japan, France and the United States, began studies of the 'flood problem' which cost around $200 million (Brammer 2000; for a summary of the plan, see World Bank 1990: 25–30). These studies differed considerably in their prescriptions for dealing with the hazard, ranging from a capital-intensive 'high-tech' intervention (France) to a version of a 'living-with-floods' approach. The World Bank took on the co-ordination of foreign and GoB proposals. It produced the Bangladesh Flood Action Plan (BFAP) in conjunction with the GoB, which amalgamates some of the ideas from the various other studies (World Bank 1990). Initial budgets of $146 million went into 26 studies, with some capital going into the repair of existing projects. The World Bank in effect became a broker between the different interests represented in the G7 and the GoB, trying to keep on board all potential donors, and mediating between the different flood plan proposals. This was a difficult task, since many of the BFAP components conflicted with each other (both in terms of their technical requirements and their objectives). Also, some proposals were of very doubtful financial viability, and if judged by normal World Bank auditing criteria would not be approved (see Boyce 1990: 422–423).

A further problem was that the so-called action plan was really a series of studies, some of them designed to research the possibilities of various approaches to flood control (Brammer 2000), but was not a complete set of projects to solve flooding. And while it is certain that major interest groups in the GoB strongly favoured an approach that would 'stop' flooding (Leaf 1997), the BFAP was not the means by which that could be directly achieved. Moreover, after some of the studies had been underway for a few years, the immense and extremely costly 1993 floods on the Mississippi, Red River and Missouri in the USA severely undermined confidence in (or the appropriateness of) the river training (engineering) methods being proposed by powerful interest groups for Bangladesh.

The key issue in regard to the 'flood problem' is what type of approach is favoured and implemented. One model calls for high

Box 6.3 continued

levels of investment in large-scale engineering works that will supposedly contain the rivers between high embankments. Such
embankment methods for containing rivers in spate were not new for
Bangladesh, and they have been recorded for over 200 years
(Brammer 2000). However, the BFAP claimed that this 'river training'
approach would also help by promoting agriculture in protected
enclosed areas (like polders), for which dry-season irrigation could be
provided. Proponents of the plan believed it to be flexible and not an
overly 'technical fix' (e.g. Brammer 1993, 2000). Yet there were
widespread and serious criticisms emerging from many in
Bangladesh, including affected people in the countryside, academics
and engineers, and NGOs (Brammer 2000 provides a summary of the
main objections). Some Bangladeshi organisations and experts
considered the approach to be top down, and insensitive to the causes
of vulnerability (Adnan 1993; Farooque 1993; Leaf 1997).

As a result of the extraordinary coincidence of two very exceptional floods in 1987 and 1988, it seems that an excuse had been
found to try to end all floods, including those that most rural people
consider to be beneficial and essential. This point is crucial, since for
the majority of the rural population in most years there is no 'flood
problem'; indeed, when there is no flooding, it is considered to be a
disastrous 'drought'. It is highly unlikely that many among the rural
population would be willing to give up the normal annual summer
flooding which has enormous value (mainly for agriculture and fisheries) in order to reap dubious and costly benefits of schemes which
stop floods altogether. As Leaf says of the claims that flood protection would improve agriculture (in his discussion of the Flood
Response Study in which he participated as part of the BFAP):

[T]he implication was that such benefits [of improved farming]
might eventually outweigh the costs. However, there are many
reasons to doubt this. The existing agricultural system in
Bangladesh is closely adapted to the cycle of inundations and is
unquestionably productive, supporting one of the highest rural
population densities in the world. No rain-fed system and few
irrigation systems could achieve as much.

(Leaf 1997: 181)

The BFAP failed to take adequate account of the fact that flood
prevention will produce its own set of victims: those who are going
to be made worse off by the proposed projects. It is even possible
that more people will suffer longer-term damage to their livelihoods

Box 6.3 continued

as a result of the flood prevention projects than the number who suffer during flooding. Leaf states that:

[T]he clear historical pattern in Bangladesh has been that the construction of flood control projects has consistently resulted in an immediate net decline in production in the project areas, evidently because of poor advance information and the lack of co-ordination between agencies. This has typically extended five to seven years, until farmers can discover and make the necessary compensating adjustments.

(Ibid.: 196–197)

The principal 'hard' projects for flood prevention suggested in the BFAP can be summarised as the repair, consolidation or construction of high embankments along the length of the main rivers (Ganges, Brahmaputra, Meghna: see Figure 6.2), combined with protection of land for High Yielding Varieties (HYV), 'Green Revolution' agriculture. In addition, some new or improved anti-storm surge bunds were to be constructed or repaired on low land at the mouths of the delta distributaries. These are designed to protect farmland from saltwater incursion during cyclones (but were not located in the worst-hit areas of the 1991 cyclone, on the east coast). The 'soft' components include the development and improvement of flood warning systems. (Summaries of the projects are to be found in Boyce 1990; Brammer 1990b, 2000; Dalal-Clayton 1990).

The basic conception of the BFAP was that water would be contained and moved downstream between large, long embankments, despite the fact that there is little knowledge of the consequences downstream (including possible flooding in other areas). The newly protected 'HYV lands' behind these banks were to be separated into 'compartments', surrounded by their own embankments. It was proposed that these compartments could be deliberately allowed to flood, in a version of the concept of 'living-with-floods'. The idea is that water can be controlled to allow optimum levels for irrigation in the wet season, but with the ability to keep out flood peaks. Some such Flood Control Drainage and Irrigation (FCDI) projects have been in operation for more than a decade, but have neither solved poverty nor even removed all vulnerability to flooding.

The issue of land tenure and the distribution of the assets and income generated by such schemes was not to be dealt with within the projects. Even if such FCDI schemes were to work properly and increase agricultural output, the social factors which prevent poor

Box 6.3 continued

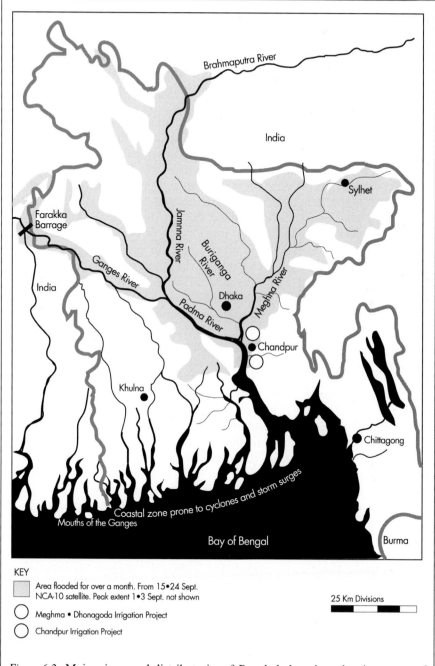

KEY

▨ Area flooded for over a month. From 15●24 Sept.
NCA-10 satellite. Peak extent 1●3 Sept. not shown

◯ Meghma ● Dhonagoda Irrigation Project

◯ Chandpur Irrigation Project

25 Km Divisions

Figure 6.2 Major rivers and distributaries of Bangladesh region, showing extent of flooding in September 1988

Box 6.3 continued

people from receiving adequate nutrition were not addressed. Increased output is not a sufficient nor a necessary condition for solving hunger, and there is no guarantee that better production under flood prevention will actually help, since the problem of hunger is not a lack of food, but the inability of people either to grow their own (land and asset poverty) or to buy it (income poverty) (see Chapter 4 and BRAC 1983a, 1983b; Hartmann and Boyce 1983).

Supporters argued that the BFAP was not monolithic and could be adapted to the most appropriate policies, in sympathy with the needs of the people. Critics argued that alternatives which might work better and be much cheaper were being ignored, and that the existing knowledge of many Bangladeshi and international NGOs was not being properly considered. To understand the gap between these perspectives, we need to see how they fit into existing power systems. Disaster reduction is not isolated from other aspects of life in Bangladesh. It operates in a hierarchy which connects the vulnerable and poor in the villages to national and international interests. It involves either ignoring or failing to understand the factors which generate people's vulnerability to floods. The existing distribution of power, income and assets is a major component of that vulnerability. Numerous NGOs and people's organisations in Bangladesh are concerned that the proposed projects (and even the pilot studies) reflect both a continuation of existing processes of generating vulnerability, and a failure to deal with the needs of the majority (Chowdhury 1991; Adnan 1993; Farooque 1993). One British member of the panel of experts advising the plan conceded that consultation and public participation 'is not an easy task in a strongly hierarchical society, and it would be unrealistic to expect overnight success' (Brammer 1993: 9).

Two interlinked factors were involved in setting the BFAP on its course towards a 'tech-fix' approach, factors which are unlikely to change very significantly. The first is the availability of the technical fix itself, as a set of expensive methods involving major engineering works. It reinforces the benefits of the power system for those already in control, both in Bangladesh and internationally. Because it involves large-scale engineering contracts, foreign donors or lenders get back a very substantial share of the spending, through consultancy fees and the purchase of equipment. Likewise, politicians and others in the local elite benefit from kickbacks on the contracts, consultancy and brokerage fees for the donor agencies in arranging local projects. Other ways of doing things are not nearly so attractive to either group because they would involve less spending. The only study in the BFAP which really sought to find

Box 6.3 continued

out what vulnerable people themselves wanted was strongly opposed by some GoB officials at the highest level because it showed that the technical fix was neither wanted nor needed (Leaf 1997).

Secondly, a preference for the tech-fix approach is also linked to the desire of local elites to protect their own land and property, especially in towns and cities, and in areas of existing Green Revolution projects. One researcher recorded the following opinion:

'Some would argue', says Dalal-Clayton, that the flood plan amounts to 'a political response to the clamour for flood protection from wealthier, influential, urban-based groups ... a great many people who presently live along the main rivers and on islands in the river channels will be exposed to increased risks from flooding.

(Quoted in Pearce 1991: 40)

The national priority is to raise agricultural production (without asset or income redistribution). Paradoxically, the worst floods (e.g. in 1987 and 1988) led to the largest harvests, as moisture remained in the soil in the dry season and non-flooded areas received more rain. The Flood Response Study found that the areas with the worst rice production included non-flooded areas (Leaf 1997). But the real problem for those who focus on raising output is that both low-rainfall monsoons in summer and the normally dry winters lead to low yields, except in areas with irrigation. It is extremely difficult to calculate, but it is conceivable that the losses from low water availability in winter, and from summers with poor rainfall, are greater than those which accompany floods. In a sense the need is more to deal with water shortages and *drought control* than to prevent floods. Flood projects tend to benefit middle peasants and wealthier land owners. It is to be expected that new projects created under the 'flood protection' works will benefit the same interest groups. Poorer farmers and the landless may derive some benefit through higher demands for labour and a general rise in output, but this seems to be a secondary consideration in the prospective designs. Similar FCDI projects already exist, for example around Chandpur, south-east of Dhaka. Some sections of the poor, including those vulnerable to floods, benefit from the protection in this area.

The issue is whether it is the best way to protect the vulnerable, and whether the impact of protection and flood prevention on other groups will be less beneficial. In rural areas, many of those already vulnerable to flooding may not be relocated safely when embankments are built. The impact on transfer of water by the river-training measures is also unpredictable, and it is possible that other areas will become more prone to flooding. In addition, it is likely that asset and income positions, and livelihood opportunities (especially of those who depend on common property resources), will deteriorate. There is widespread concern that flood protection works may eliminate the ponds that remain after floods and in which fish and shrimps spawn. These 'common property' fisheries are of immense value to poor people, both for nutrition and as a source of income for those who catch and sell on to others. It has been estimated that inland fisheries account for 77 per cent of the national harvest (Boyce 1990: 423).

The prevention of flooding may also seriously diminish the renewal of soil fertility by the effects of inundation, resultant decomposition of plant residues and an associated bloom of nitrogen-fixing algae (Brammer 1990b: 164). There is a further potential and significant loss of 'free' groundwater replenishment. In place of the natural recharging of the water table by normal floods, the impact of compartmentalised agriculture will be to require expensive pumping from rivers and channels to provide for both winter and summer agriculture.

Maintenance of extensive and large flood control constructions (like those suggested by the BFAP) would be both difficult and expensive. Commentators on the BFAP pointed out that past experience of repairs to embankments shows that the tech-fix is unworkable. This is an important argument, because any failure of a major embankment will have appalling consequences, especially as river beds are likely to rise above the level of the surrounding countryside when they become deposition zones for river silt. Even the impact on the compartmentalised FCDI agriculture could be serious if the banks fail. One similar FCDI scheme in existence at Meghna-Dhonagoda (to the north of the Chandpur project mentioned above) suffered serious losses in 1987 because of a breached embankment (Thompson and Penning-Rowsell 1994: 6). Another likely maintenance problem, widely observed with existing flood prevention banks, is blocked drainage conduits preventing the return to rivers of impounded flood and rainwater. Roads and railways are normally elevated on embankments to avoid inundation, but these structures commonly cause flooding by not allowing proper return drainage.

Alternatives to the plan: a people's needs-based approach?

The BFAP was intended to diagnose problems and seek appropriate solutions. But there was widespread concern both in the country and outside that it would automatically prescribe a 'tech-fix' approach, a

Box 6.3 continued

fear that Leaf (1997) highlights in the conflicts that arose from the Flood Response Plan with which he was involved (and which was the only component of the BFAP that actually discussed with affected people their wants and needs with regard to the floods). The main criticisms concern the costs involved (in relation to the uncertainties of the claimed benefits), the uncertainties of success, the increased dangers of failure (a breached embankment may cause a worse disaster than a bad 'natural' flood), the ignoring of alternatives and the failure to deal with the problem by first finding out how the majority of affected people themselves define it. It is also essential to calculate the impacts on the poor and vulnerable (for example, through taxation) of any increased financial burdens the country may face as a result of the borrowing required for construction projects.

The emphasis of the USAID report (see Rogers et al. 1989) was on managing the flood, or living with flood, rather than attempting expensive prevention. This approach was undoubtedly influenced by the many Bangladeshi scientists, social scientists and NGOs that advocated modified 'living with flood' strategies. The Flood Response Study found that affected people had their own priorities for flooding, and very little interest in stopping floods altogether. What is really crucial to understanding the problem is to ascertain what causes people to be vulnerable to the negative aspects of flooding – and especially to 'extreme' floods which are disastrous for some – and therefore to assess what measures can reduce vulnerability. These may be far removed from the need to prevent floods through measures the costs of which the poor will have to bear disproportionately anyway. Figure 6.3 maps out some of these components of vulnerability, and it is there that the beginnings of river and rainfall flood disaster avoidance should be found.

To protect people and their homes, there needs to be investment in flood shelters (including elevated schools and health centres) and raised house sites. To ensure their health (and thereby indirectly their livelihoods) there needs to be elevated or protected drinking water sources and primary health care (some authorities also advocate safe flood-proof latrines). To protect their livelihoods there needs to be access to land, replacements for land lost through erosion, compensation for losses of animals and other production assets. Where necessary, there must, of course, also be flood prevention measures, but based on an assessment of what is actually needed to reduce vulnerability rather than some grand design which supposedly prevents all flooding without considering who benefits and who loses, who pays for it and whether it is needed anyway.

Box 6.3 continued

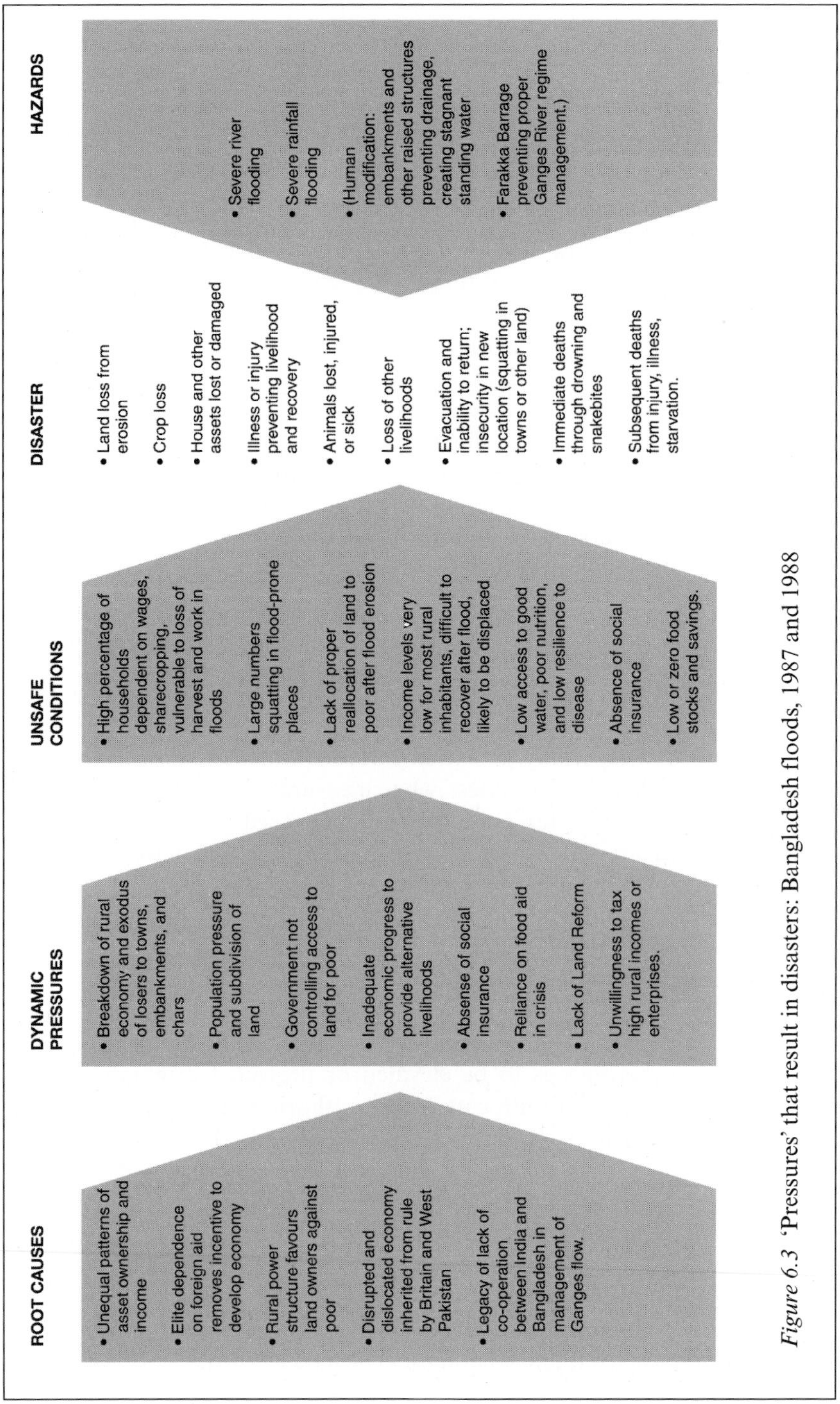

ROOT CAUSES	DYNAMIC PRESSURES	UNSAFE CONDITIONS	DISASTER	HAZARDS
• Unequal patterns of asset ownership and income	• Breakdown of rural economy and exodus of losers to towns, embankments, and chars	• High percentage of households dependent on wages, sharecropping, vulnerable to loss of harvest and work in floods	• Land loss from erosion	• Severe river flooding
• Elite dependence on foreign aid removes incentive to develop economy	• Population pressure and subdivision of land	• Large numbers squatting in flood-prone places	• Crop loss	• Severe rainfall flooding
• Rural power structure favours land owners against poor	• Government not controlling access to land for poor	• Lack of proper reallocation of land to poor after flood erosion	• House and other assets lost or damaged	• (Human modification: embankments and other raised structures preventing drainage, creating stagnant standing water
• Disrupted and dislocated economy inherited from rule by Britain and West Pakistan	• Inadequate economic progress to provide alternative livelihoods	• Income levels very low for most rural inhabitants, difficult to recover after flood, likely to be displaced	• Illness or injury preventing livelihood and recovery	• Farakka Barrage preventing proper Ganges River regime management.)
• Legacy of lack of co-operation between India and Bangladesh in management of Ganges flow.	• Absence of social insurance	• Low access to good water, poor nutrition, and low resilience to disease	• Animals lost, injured, or sick	
	• Reliance on food aid in crisis	• Absence of social insurance	• Loss of other livelihoods	
	• Lack of Land Reform	• Low or zero food stocks and savings.	• Evacuation and inability to return; insecurity in new location (squatting in towns or other land)	
	• Unwillingness to tax high rural incomes or enterprises.		• Immediate deaths through drowning and snakebites	
			• Subsequent deaths from injury, illness, starvation.	

Figure 6.3 'Pressures' that result in disasters: Bangladesh floods, 1987 and 1988

Box 6.3 continued

In 1991 tens of thousands died as the immediate consequence of flooding brought about by a tropical cyclone on the south-east coast of Bangladesh (see Chapter 7). By contrast, the worst-ever river and rainfall floods of 1987, 1988 and 1998 reportedly each caused less than 2,000 immediate deaths (Brammer 1993; World Bank 1990: 40; del Ninno et al. 2001). Although one component of the BFAP deals with cyclone protection, at the time it seemed extraordinary that there appeared to be much less concern for measures which provide coastal villagers with effective cyclone shelters and other forms of protection. While the desire to reduce the harmful effects of floods is laudable, and tens of millions of people endured intense suffering in 1987, 1988 and 1998, few lives were lost, and the hardship remains for many of the poor during non-flood years. Basic vulnerability issues need to be addressed not through prevention of floods, but through changes in the processes that create the unsafe conditions.

Water itself is an important part of the resource or livelihood rights of many people likely to be affected by floods. Rivers are crucial for livelihoods based not just in agriculture but also on transport, trading and fishing. River channel shifting may disrupt these livelihoods too, creating havoc among whole sections of a population.

In some circumstances, the normal flood regime of a river is used to good advantage by farmers. On some rivers 'flood-retreat' agriculture (also known as recession agriculture) is practised, where the receding waters reveal moist soil primed for planting with food crops. This was the case on the Nile before the irrigation schemes fed from the Aswan dam were installed. Such a system has existed for centuries on many rivers in Africa, including the Senegal on its route through Mali, Mauritania and Senegal in west Africa, as well as in south Asia. Livelihoods based on farming and fishing (in the ponds which remain as the flood goes down) are severely undermined when 'development' projects attempt to control such a river. For example, a dam has been constructed at Manantali on the upper Senegal, mainly for the generation of HEP and to regulate the river's flow to permit year-round barge traffic up to Kayes in Mali. It was also intended to irrigate farmland, but for large-scale projects that will grow crops which local people do not eat. This will not compensate those who lose out in the transfer of resources and loss of natural floods in the valley. Another of the costs of such a project is that the new regulated river flow will not allow traditional flood-retreat agriculture to take place (Horowitz 1989; Horowitz and Salem-Murdoch 1990). Similar losses to small-scale traditional irrigation were the result of damming a river to the south of Kano (Nigeria) so that large-scale irrigated wheat production could supply urban bakeries (Andrae and Beckman 1985).[13]

Box 6.4: Flooding and deforestation – the causation controversy

There is a widespread assumption by those concerned about flooding that it may be increased (in frequency and/or intensity) as a result of land in the rainfall catchment being cleared of vegetation and forest cover. This causation is frequently mentioned in relation to floods in China (after the severe 1998 floods, the government banned all logging in several provinces in the catchment basins of the Yangtze), in Bangladesh and north India (in relation to deforestation in the foothills of the Himalayas) and Thailand (where the government has also supposedly banned logging to stop floods). In this chapter, we have also mentioned the problem of deforestation, especially in rela-tion to landslides (where there may be a much more immediate connection between cause and effect). But although the link between deforestation and flooding has become almost general knowledge and is widely taken for granted, it may vary in its significance from place to place and river to river: the issues are extremely complex. The most significant challenge to the simple idea of causation comes from work on the Himalayas in relation to flooding in Bangladesh. Since the issue is of considerable relevance to flooding, it is discussed here briefly.

There are many causes of deforestation and de-vegetation of river catchments, and it is potentially misleading to put them all together as if they were similar. There are also significant differences in the effects of different patterns of land-use and associated deforestation. In some upland or sloping land areas, logging for timber is practised by national and foreign companies and by governments, and is often considered to lead to more rapid rainfall runoff, silting of rivers and, sometimes, landslides (which also block rivers with debris and sedi-ment). Associated roads may be constructed inappropriately, causing slope instability and vegetation loss. Land shortage in upland areas may increase the rate of deforestation, as people clear more land for agriculture, or damage trees for fuel and fodder. The people who do this may be new arrivals who clear new areas, or local people who have to expand their cultivated area or reduce the fallow period in upland swidden (slash and burn) agriculture.

These factors may all generate local flooding (through stream-damming) and increase the sediment load of rivers, contributing to a rise in the level of river beds downstream and increased flood hazard. However there are disputes among scientists about the significance of the different factors in this process, especially concerning the Himalayas (Thompson and Warburton 1988). One disagreement is about whether or not there has been an increased incidence of flooding during recent decades when, it is supposed, rapid deforestation has occurred. Some argue that the evidence for a strong connection between deforestation

Box 6.4 continued

and increased flooding is uncertain, and that hydrological data do not demonstrate that good vegetative cover in large river basins is necessarily a factor in preventing rapid runoff of storm water (Ross 1984: 224–225).

Others suggest that flooding of equivalent severity and frequency to current times has been apparent in river basins for centuries, long before recent increases in deforestation. For example, discussing the situation in Sichuan province, Ross presents arguments by one Chinese engineer that 'historical records show a high incidence of flooding even before modern increases in population and logging' (ibid.: 223). Ives and Messerli (1989) argue likewise for the Himalayas, that there is no convincing evidence of an increase in runoff during the preceding forty years, despite the supposed increased incidence of flood disasters. The rivers of the Ganges–Brahmaputra basin have been contributing immense amounts of sediment to the Ganges plain and Bengal delta for thousands of years, owing to climatic and tectonic factors in the mass wasting of Himalayan slopes, rather than recent human action. They ascribe the common perception of an increase in flood disasters not to greater amounts of water in the drainage system, but to human systems having put more people in more risk-prone places.

The significance of this controversy for a discussion of vulnerability to floods is two-fold. Firstly, if we are to accept that vulnerability is a condition deriving from economic and social systems, it is not certain whether the deforestation process should be included as a significant contributor to rising vulnerability. Secondly, arising from this, reducing vulnerability in areas that are downstream of significant deforestation may not necessarily be achieved by reducing that deforestation (or by reforestation). There may be other very good reasons to reduce such damage, but a reduction in flooding may not occur.

The key issue is that after the ground has become saturated by sufficient rainfall, it does not really matter how much vegetation there is covering it, since that water will flow and potentially cause flooding anyway. There may be an issue of silting and the reduced flow capacity of rivers resulting from the deposition of more silt after slopes become denuded and barren, but even here it is difficult to ascribe the shares in the amount of silt deposited from deforestation as compared with natural wasting processes (at least in the tectonically active mountains of the Pacific rim). In short, it may be harmful to the livelihoods of already poor, vulnerable people if they are blamed for downstream flooding. There may be other very good reasons to reduce deforestation (especially by logging enterprises where local people are displaced and impoverished anyway), but the inappropriate blaming of deforestation for supposed increases in floods may divert attention from the need to develop other policies to reduce peoples' vulnerability to floods.

Summary: floods and vulnerability

Flood hazards have a variable impact on people according to vulnerability patterns generated by the socio-economic system in which they live. Class relations and other structures of domination are crucial for explaining vulnerability to floods as they determine levels of ownership and control over assets and livelihood opportunities that may already be inadequate to providing basic needs for the household (Pelling 1997, 1999). So the initial 'pre-hazard' conditions of people are largely determined by patterns of vulnerability that result from the economic and social system in which they exist.

Our Access model explains many specific mechanisms that turn flood hazards into disaster. These include the location of homes (and their proneness to inundation) and the structure and type of housing and workplaces (and their resistance to floods). Both of these are a function of household income, legal or social limitations on land-use, availability or cost of building materials, and the location of livelihood activities. The Access model also describes the daily and yearly pattern of work and other activities, which in turn interact with the temporal patterns of flood hazard occurrence. These variables not only affect the risk of death and injury but also the potential for the destruction of assets, and of livelihood opportunities. All these can be summarised in the PAR model for floods, which shows how the more remote root causes of flood vulnerability are translated into unsafe conditions (see Figure 6.3). As we shall also see in the case of coastal storms (Chapter 7), distant pressures and factors from the past can influence vulnerability to flooding.

Ethnic divisions are often superimposed on class patterns, and may become the dominant factor determining vulnerability. This can be seen in differential access to, or possession of, resources, or inequalities of participation in different livelihoods, according to imposed racial or ethnic distinctions. However, very few studies of flood disasters seem to take up the issue of ethnicity as a vulnerability factor. One very significant case is that of the Venezuela floods/landslides of 1999 (see above), when Afro-Venezuelans were disproportionately affected. Another example comes from the exceptional flooding around Alice Springs in central Australia in 1985. Aboriginal people did not receive flood warnings and lived in flimsy accommodation on low-lying land. The radio broadcasts that alerted the white people were not on channels which were customarily used by the Aborigines (Keen et al. 1988).

It is also crucial to understand the differential vulnerability which is dependent on gender. There is usually inequality between women and men in their ownership and access to resources. Economic and cultural systems are generally male-dominated, and allocate power and resources in favour of men. Even the effort put into disaster recovery may be disproportionately carried by women, who in most 'normal' situations have to work harder in paid and unpaid work than men. In addition, women may be more prone to post-flood disease, largely as a result of their poorer initial well-being (nutritional condition and physical susceptibility). Men's and women's time and place patterns of daily and seasonal activities also differ, and this may produce inequalities in their exposure to flood hazards. To the extent that young children are more likely to be with women than men, this also affects their relative vulnerability.

Flood prevention and mitigation

Precautionary measures and policies for dealing with floods are aimed at modifying or predicting the hazard involved in triggering disasters, rather than at other causes of vulnerability. The measures include strategies intended to reduce the intensity of the hazard, the mitigation of flood effects, prediction and preparedness. Policies need to go beyond these aspect and look at the implications of vulnerability analysis in the development of different means of disaster avoidance.

Local-level mitigation

Local-level, indigenous responses include people's own strategies for dealing with flood risks. These entail a combination of self-protection and social protection by communities or NGOs. These responses have been developed by people in many places, often over hundreds of years, especially where people have had to colonise and cultivate new lands in flood plains. In some regions, for instance in parts of rural India and Bangladesh, houses are usually built on artificial mounds that raise them above normal flood levels. But this is not always possible for all people. For instance, in the Gangetic plain of north India, villages in flood-prone areas often have higher ground at their centres. The more substantially built houses of the wealthier groups are often near these village centres. Poorer classes, including the lower castes and untouchables, are to be found mainly round the edges of the settlements, on low-lying sites.[14]

Social protection and flood precautions

Precautions against riverine flood hazards often involve attempts to *prevent* them, through measures that modify the stream flow by using embankments, barrages, dams, retention basins and other engineering installations. These nearly always involve a high level of technical (and therefore capital) investment. Such attempts at 'social protection' (that is, measures carried out by institutions within society at levels above that of the household) are increasingly being called into question as inappropriate (as discussed above), and sometimes seem to be carried out for the benefit of the higher-level institutions (including the contractors, politicians and other interest groups) rather than (or as well as) the people who are affected by floods. Dams and barrages constructed for flood prevention are often controversial. A problem of preventive projects is that they can induce a false sense of security, leading people to settle in supposedly protected areas that may be completely devastated if the construction measures fail.

Channel control methods in developing countries often involve employing thousands of workers along lengthy stretches of river. The most common approach is to constrain river channels within artificial embankments or dikes (river training), or to use dikes (called bunds in some countries) to prevent the river from spilling into its natural flood plain. In addition, embankments may be used to encircle areas or places (e.g. ring bunds around towns, cities or heritage sites) which are deemed to need special protection. Such methods have a long history in some parts of the world, including India and China. In China the channel of the Yellow River (Huanghe) has been repeatedly enclosed within dikes for much of its course across the North China Plain for hundreds of years. As a result it now flows for some of its journey in impounded channels 5 m above the level of the surrounding countryside. It is a policy that requires massive expenditure, and the dikes must constantly be raised higher and higher. Breaches of these dikes are akin to a dam bursting and can devastate settlements and farmland across extensive areas of the surrounding flood plain.

Other channel control methods are used (often in conjunction with river training) to provide emergency storage for flood water (called detention or retention basins). These may be existing lakes which adjoin the river channel, or artificial depressions or naturally low-lying areas. The embankment between river and lake can be breached deliberately, and water from the peak flow is then stored to prevent the river reaching danger levels further downstream. Such basins have been crucial to flood management in China for decades, but as discussed in Box 6.1, various economic pressures have led people to settle there, and they no longer function as a means of flood prevention.

Where there are known river flood hazards, land-zoning measures can be effective in preventing disaster by literally avoiding the flood, or giving priority in land-use on flood-prone areas to certain types of users. But it is common (especially in the cities of developing countries) for many people to evade such restrictions because they are forced by high rents elsewhere to seek such 'free' land to use for housing. Similar economic pressures force the poor to squat in shanty towns, often on unstable hill slopes which can collapse in heavy rain.

Flood mitigation and preparedness

Mitigation policies can save lives and protect property even though the flood itself cannot be prevented, contained or avoided. The most conventional of such preparatory methods is a flood warning system, the effectiveness of which has been shown in a range of countries. The value of warnings depends greatly on their timeliness (people are very reluctant to leave their dwellings and assets unguarded, and so delay evacuation until the last possible moment) and accuracy (this affects their credibility), the lead-time available to permit preparedness and evacuation and the effectiveness of the message delivery system (Zschau and Kueppers 2002; see also Chapter 7).

Notes

1 In this chapter we concentrate on river and rainfall flooding, with some reference to inundation associated with tropical storms when relevant. Although tropical storms (especially cyclones or hurricanes) are associated with floods in disaster research and policy, they are often treated separately from other types of floods because they also cause damage through high winds (and saltwater incursions on coasts). We have therefore dealt with tropical storms separately in Chapter 7 to mirror this common distinction. As we have stressed, the focus on particular hazards in this book is to reflect common usage, even though we consider that it is people's vulnerability which is the key focus.

2 The British supermarket Sainsbury's displayed a notice in its vegetable section in Autumn 1998 that said: 'Because of the effects of El Niño in Peru, we are unable to supply asparagus'.

3 As yet, modelling efforts have not revealed precise connections between climate change and ENSO, but research continues into these complexities (McGuire et al. 2002).

4 By contrast, a dam failure in Johnstown, Pennsylvania, in 1889 killed 2,000 people and is credited with having ushered in the modern period of national flood control in the USA (Foote 1997).

5 In the Mozambique floods of 2000, there was some release of water from dams that may have made the floods worse downstream (Christie and Hanlon 2001: 112–117; see Chapter 7).

6 Examples include projects in Nepal on the Imja glacier lake as well as the country's largest lake, Tsho Rolpa. Information from NepalNet at: www.panasia.org.sg/nepanet/water/khumbu.htm, and Bridges: Rowaling Project at http://namche.net/ooo/bn-tsho-rolpa-c.html

7 This is a good example of how a supposedly sudden impact hazard can continue to have potentially devastating effects years afterwards. As well as the threat from this lake, there have been many casualties since the eruption as *lahars* have swept down the volcano's slopes (see also Chapter 8).

8 Cannon became aware of the potential for warnings when on holiday in Corsica. Relaxing by a river in the hills with several hundred other tourists, all were saved from drowning by the local fire brigade, which travelled up the valley to evacuate the riverside. Aware of heavy storms over the upper catchments in the mountains above, Cannon had predicted the flood, but the experience of the fire brigade in knowing the local patterns and moving people out was vital in ensuring people's safety. Within fifteen minutes, a flash flood with a 5 m wave filled the canyon where everyone had been enjoying themselves.

9 Flash Flood Lab, n.d, http://www.cira.colostate.edu/fflab/recentff.htm

10 La RED and others have developed an accounting system for compiling information on such 'small' disasters from local records, provincial news media and oral history. The system is known as DESINVENTAR, and has been used in Guatemala, Ecuador, El Salvador, Peru, Colombia, Panama, Argentina, Mexico, Costa Rica, Nicaragua and Honduras. It is available on the internet at http://www.desinventar.org/desinventar.html (in English and Spanish).

11 Leaf (1997) found that the respondents in his survey were often very concerned about the impact of existing and possible new structures on creating these floods. Where they prevent rain runoff entering rivers they actually create floods where none would otherwise exist.

12 All output figures from FAO website database at http://apps.fao.org/page/collections?subset=agriculture

13 Other examples of losses of previous agriculture and fishing resources in African rivers with flood-recession capability are given in Fiselier (1990) and Scudder (1989).

14 Field observations by Cannon in Haryana and Uttar Pradesh, north India, in 1976 and 1979.

7

COASTAL STORMS

Introduction

People have lived along coasts since antiquity. The most recent phase of colonial expansion (since the middle of the nineteenth century) and the establishment of a world market have greatly increased the numbers of urban settlements, plantations, ports and naval bases and other centres of population in coastal areas. More recently, tourism and the global expansion of export-oriented industries have added to the attraction of coastal locations. The 1999 Hangzhou Declaration of representatives of large coastal cities noted the following:

- More than half of the world's population ... lives in coastal areas and is expected to develop uninterruptedly in the following decades.
- This process has been associated with population growth and tourist pressure increase ....
- Coastal mega-cities (cities with eight million inhabitants or more) have increased in number in such a way as to become the key component of coastal areas.
- Small coastal cities (three to eight million inhabitants) are also proliferating.
- The most populated mega-cities are located in the developing world, and this spatial process is expected to accelerate in the twenty-first century.
- Coastal urbanisation has increased coastal erosion, and the rise of new resource uses have produced increasingly complicated coastal use patterns.
- Coastal cities have become key spatial elements of globalisation processes.
- These spatial processes ... have provoked acceleration in human pressure on the local ecosystems and natural resources.

Several of the fastest growing cities, all projected to have 20–30 million inhabitants each by the year 2025, have long histories of exposure to severe tropical storms. These include Karachi, Jakarta, Calcutta and Dhaka (Davis 1986: 279), as well as several coastal Chinese cities including Shanghai and Hong Kong,[1] Manila, Tokyo, Miami, New Orleans and Darwin.

The 'attractiveness' of coastal location is often taken for granted. For instance, Griggs and Gilchrist (1983: 274) review numerous high-risk situations on the Gulf and Atlantic coasts of the USA and conclude that 'people want to live in the sun and be able to look at the ocean; realtors and developers want to make money, and local governments want more tax dollars' (cf: Dean 1999: 92–119). Burton, Kates and White (1978: 4–17) emphasise the attractiveness of rich alluvial soils in the case of coastal Bangladesh, a factor that is significant in many other regions.

In this chapter we examine the reasons why people inhabit coasts and the implications for their vulnerability in the face of hazards.[2] In seeking the causes of vulnerability to cyclones, typhoons and hurricanes (these being different regional names for the same meteorological phenomenon), common assumptions why people live in coastal regions need to be investigated more specifically. Cyclonic storms do not affect all coasts equally (White 1974; Southern 1978). Furthermore, where such storms are frequent, not all people suffer equally. In instances where people suffer from coastal storms, not all people are able to reconstruct their lives rapidly or equally well. Such differences in vulnerability can be seen in the case of storms affecting more developed countries (MDCs) such as Australia, Japan and the USA as well as less developed countries (LDCs) such as Fiji, Mozambique, Nicaragua, Bangladesh and the Philippines. For every 'voluntary' resident in a high-risk coastal location (those seeking 'sun and surf', for instance) there are very many who have no alternative because their livelihoods are tied to jobs in oil refineries or export enclaves, to jobs in the service sector spun off from the tourist trade, to jobs on fishing boats or to employment on coastal farms and plantations.

Patterns of death and damage due to these storms and the ability of people to reconstruct their livelihoods show variation according to national wealth, history and sociopolitical organisation. Recovery following hurricanes in the Caribbean in 1988 and 1989 demonstrated such contrasts. Nicaragua mobilised a nationwide effort to help victims of hurricane Joan on that country's Atlantic coast (Nicaraguan Ecumenical Group 1988). In part, this was an opportunity for the Sandinista government to build support among sections of the people that had not benefited substantially from the 1979 revolution, and in some cases had supported the opposition Contras.

By contrast, relief efforts in Jamaica following hurricane Gilbert in 1988 were rife with partisan politics and corruption; so much so that mismanagement was one of the factors that led to a change of government in the elections that followed. In the USA, the major obstacle to relief and reconstruction on the South Carolina coast following hurricane Hugo in 1989 was bureaucratic blindness to the needs of low-income people who lacked insurance and other support systems (Miller and Simile 1992). Many of these marginal people, who may never recover from hurricane Hugo, are African Americans. Two decades earlier, when hurricane Camille devastated the Mississippi delta, the US Senate investigated charges that relief assistance had been racially biased (Popkin 1990: 124). Similar problems have been identified by Laird (1992) among poor, Spanish-speaking people near the epicentre of the 1989 earthquake in northern California, and by Bolin and Stanford (1998a, 1998b) in a study of the recovery by Hispanics after the Northridge earthquake in southern California in 1994 (see Chapter 8).

The extraordinary differences in mortality from similar physical events should alert planners, citizen activists and development agencies to significant differences in preparedness, response and vulnerability. Australia suffered two very similar cyclones shortly after the catastrophic 1970 storm in the Bay of Bengal that killed at least 300,000 in Bangladesh (Carter 1987: 490). Yet the death toll in Australia was less than 100 (Stark and Walker 1979; Western and Milne 1979), and this is not simply because of the lower population densities.

Sociopolitical organisation can be as significant as national wealth in disaster preparedness. For example, in 1971 North Vietnam survived a combination of coastal storm surge and torrential rain in the Red River delta that could have cost as many lives as in Bangladesh in 1970. But only a few hundred lives were lost in North Vietnam, largely because of highly efficient war-time village-level organisation that allowed rapid evacuation and provision of first aid (Wisner 1978a).

Again, in 1974, when cyclonic storms delivered equivalent amounts of rain and wind in two parts of the world, 49 were killed in Darwin, Australia (Western and Milne 1979: 488) and 8,000 in Honduras (CIIR 1975: 1; Pulwarty and Riebsame 1997: 195–196). The major difference seems to have been the pattern of rural land ownership in Honduras, where 63 per cent of the farmers had access to only 6 per cent of the arable land (CIIR 1975: 13). Large-scale beef ranches and banana plantations had displaced peasants over several decades into isolated valleys and steep hillside farms. Here they received little warning and were at risk from mudslides that accompanied the hurricane's rainfall. Deforestation by these peasants seeking to carve out subsistence farms had made the hillsides unstable. In the northern town of Choloma, 2,300 people were killed when a dam created by landslides into a nearby river burst, sending masses of black mud into the streets (ibid.: 3).

Hurricane Mitch brought similar death and destruction in 1998, for much the same reasons. Honduras was again badly affected, and again it was the poor, living on steep slopes and on flood plains, who died; in Nicaragua, landslides and flooding killed more than 3,000 people (Comfort et al. 1999). By contrast, in 2001 hurricane Michelle traversed Cuba across its most populated region. Although considerable economic damage was done, only five people perished. One has to ask 'What did Cuba do right?', and indeed, this chapter takes up that question (Wisner 2001c).

The physical hazard

The tropical cyclone is one of the most powerful atmospheric phenomena.[3] A fully developed hurricane releases energy equivalent to many Hiroshima-sized atom bombs (Cuny 1983; Milne 1986: 71; Alvarez 1999; Gray 2000). These storms arise during the summer over the oceans in a belt north and south of the Equator. In addition to the wind damage and flooding caused by cyclones, there are a wide variety of possible physical effects involving a web of social and natural linkages. Water piled up by wind against coasts along shallow ocean shelves is called a storm surge. When these occur at high tide, they can reach up to 6 m in height. The extreme low pressure that is characteristic of these storms also adds to the height of storm surges because the ocean surface is actually lifted up to 1 cm/millibar reduction in air pressure (McGuire et al. 2002: 60). Storm surges accounted for many of the dead in such infamous cyclones as those that affected Bangladesh in 1970 and 1991, the Indian coastal states of Andhra Pradesh in 1977 and Orissa in 1999, and the Indian state of Gujarat along its coast facing the Arabian sea in 1998.

Wind and wave action have immediate impacts, but the effects of the erosion and saltwater incursion can handicap an economy for months or even years (Campbell 1984). Damage to roads, telecommunications and power facilities can have both short- and long-term effects, complicating relief efforts and the challenge of economic recovery. Even in areas remote from the coast, associated heavy rainfall can provoke mudslides and other mass movements such as the one associated with hurricane Mitch in Nicaragua. In 1998 up to 2 m of rainfall from that storm fell over a six-day period, triggering the partial collapse of the summit crater of Las Casitas volcano. This unleashed a landslide 16 km long and up to 8 km wide that killed over 2,000 people and destroyed ten villages and the town of Posoltega (McGuire et al. 2002: 78; IFRC 1999a: 45). Later in this chapter we will discuss the differential vulnerability of various social groups to these kinds of physical damage.

Tropical storms are seasonal yet highly unpredictable. Year-to-year severity and frequency of storms may be related to factors working at the global atmospheric level, such as ocean temperature and current changes, including some correlation with El Niño and La Niña cycles (McGuire et al. 2002: 45–48). As we noted in Chapters 2 and 4, global climate change is characterised above all by increasing variability (Adger 1999). It is a distinct possibility that the severity of cyclones (hurricanes and typhoons) will increase as the atmosphere continues to warm. Having weighed up all of the uncertainties and difficulties of modelling cyclonogenesis within the context of models of global climate change (challenging in their own right), McGuire, Mason and Kilburn conclude that 'it seems possible that a warmer world will experience hurricanes of a somewhat greater average intensity' (McGuire et al. 2002: 48). This is also the conclusion of a large

US-based team of scientists (Easterling et al. 2000; cf: Wilson and Rachman 1989).

There is additional uncertainty because the direction, speed and growth dynamics of such storms have not yet been fully understood, despite attempts at computer modeling. Accurate and timely knowledge of the track, and thus the specific point of landfall, is the most vital information needed to provide effective warnings, and although there has been progress in this area of meteorology, there are still gaps (Elsberry 2002). As a result, warnings broadcast over the media sometimes result in needless evacuations, making it more difficult to convince the public on later occasions. For instance, in the years following the devastating 1970 cyclone in Bangladesh (then East Pakistan), many warnings – some of them false – were issued over the radio. Some authors claim that the losses in the 1985 cyclone might have been lower if the public had been less cynical about broadcast warnings (Milne 1986: 73–74). On the other hand, millions of people did respond to warnings of the great cyclone that hit Bangladesh in 1991, and a timely warning and response saved many lives in cyclonic storms affecting Tonga, Mauritius[4] and Cuba during 2001.

Patterns of vulnerability

Coastal locations were frequently the first point of contact between indigenous people and colonial powers. These footholds, established from the sixteenth to the nineteenth centuries (first in Latin America and Asia, later in Africa) often grew into major urban centres. Livelihoods all along the new corridors of administration and extraction became dependent on the needs of the coloniser (e.g. the market for labour power, groundnuts, cotton, beef) (Franke and Chasin 1980). Squatter urbanisation began around the newly established coastal cities (Hardoy and Sattertwaite 1989). Migration to coastal cities has become commonplace in the post-colonial world. For example, many millions of Chinese have moved to coastal cities that have been integrated into the global economy within the past 20 years (Eckholm 2003). Within the more general process of urban growth we identified in Chapter 2 as a 'dynamic pressure', it is specifically the growth of coastal cities in the tropics and sub-tropics that constitutes the background to increased vulnerability to these powerful storms (International Centre 1989).

Much of this urban population lives in crowded areas, many residing in low-lying, flood-prone areas, in flimsy housing and with a lack of infrastructure. The millions who have caused these former colonial cities to swell into today's coastal mega-cities are part of the patterns of extractive, export-oriented economic activity established a century or more ago.

Contemporary coastal settlement and the cyclone hazard

In both wealthy countries, such as Australia and the USA, and in the poorest former colonies such as Bangladesh, Mozambique and Jamaica, there are large towns and cities on hurricane-prone coasts. Potential economic losses are a function of this pattern of urbanisation.

Hurricane Andrew, a storm that hit the US city of Miami in 1992 caused $30 billion in losses (Munich Re 2002). Hurricane Gilbert (1988) travelled directly over Kingston, capital of Jamaica, and traversed the full length of that island. The death toll was 45 and damage was estimated at $1 billion (Barker and Miller 1990).[5] Cyclone Tracy (1974) destroyed 50–60 per cent of the houses in Darwin (Australia) and did A$3.2 billion damage to the city (Wilkie and Neal 1979: 473). If it had struck the coast just 100 km either side of that city, the losses would have been negligible (Stark and Walker 1979: 191). There have been near misses that highlight the potential damage. In 1969 hurricane Camille missed the major US city of New Orleans by about 100 km. Even so, 262 people died and losses totalled $1.4 billion (Petak and Atkisson 1982: 332). Given its coastal situation and its location between a large lake and the Mississippi River, a direct hit on New Orleans would cost hundreds of billions of dollars and probably take thousands of lives (Figure 7.1) (see Chapter 6: 204).

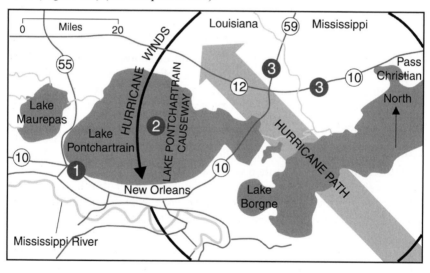

EVACUATION WORRIES
There are three main routes out of the city, all of them problematic

1 Interstate 10 is prone to flooding where it passes over a corner of the lake.

2 The 24-mile Lake Pontchartrain Causeway is closed when winds exceed 50 m.p.h.

3 Interstate Highways 10 and 59 would very likely be clogged with traffic from vulnerable areas of the Mississippi and Gulf coast.

Figure 7.1 New Orleans: hurricane evacuation worries

Source: Adapted from an article by Jon Nordheimer, *New York Times*, 30 April 2002

Rural areas in the tropics may contain capital-intensive plantations and other agro-industrial facilities. Jamaica lost 30 per cent of its sugar acreage, 54 per cent of its coffee and more than 90 per cent of its bananas and cocoa to hurricane Gilbert. This deprived Jamaica of more than $27 million in exports in 1988/1989 alone (Barker and Miller 1990: 111). Mozambique lost two cashew processing factories during the cyclone that struck in 1979, in addition to many thousands of economically important trees (coconut and cashew) (Wisner 1979).

In countries with high rural population densities and considerable inequalities in income and access to land, rural concentrations of people in the high-risk coastal zones can be very great (for example in the Philippines, parts of Indonesia, the Sundarbans and islands such as Sandwip in Bangladesh, and India). In these cases, as we show in detail below, local or even distant pressure on land forces the poor to remove protective vegetation, destroying buffer zones and increasing their vulnerability to storms.

But there are also vulnerable groups in richer countries such as the USA. Since the 1960s there are coastal Florida counties that have consistently grown by at least 50 per cent each decade. A significant proportion of the newcomers to the coast of Florida are older retired people (Graff and Wiseman 1978).[6] During hurricane Agnes in 1972 it was realised that the elderly require special assistance during evacuations (Briggs 1973: 134). Retirement homes, tourist and recreational centres have grown rapidly near and even quite literally on these coasts. It has been predicted that an extreme Camille-sized hurricane hitting the major Florida recreation and retirement centre of Dade County (near Miami) would destroy an estimated 18 per cent of the structures directly on the coast, and up to 7 per cent as far away as 100 miles (Cochrane 1975: 19). When hurricane Andrew passed through southern Dade county in 1992 these calculations proved to be underestimates. Post disaster surveys found that the low-income population, including retirees, as well as women, African Americans, Haitians and Hispanics experienced the most difficulty recovering (Peacock et al. 1997).

Skilled workers in the USA are sold the 'American dream' of retirement in the sun, in cheaply-built properties in high-risk locations. This is another facet of vulnerability created by corporate production, sales and investment strategies (Steinberg 2000). This local business activity has similar results to the activities of US corporations in many parts of the world such as the Philippines, where their acquisition of land for growing pineapple, banana and other export commodities increased the land pressure on the local people. The small farmers who are dispossessed are thus forced to live in dangerous urban coastal locations through the actions (direct or indirect) of a transnational corporation (Boyce 1992). The skilled working class and middle class in the USA are tempted into similarly hazardous zones on the other side of the globe as the value extracted abroad passes through the financial circuits of the system in the form of coastal land development.

Small island nations finding themselves in the path of frequent tropical storms should be considered special cases. Their population growth rates have often been lower than in other exposed areas. Yet the ravages of a tropical storm can encompass an entire island and its people (Lewis 1981, 1984a, 1984b, 1999), and the effect of global warming on the frequency and power of storms, and sea level rise, magnify the risk (Lewis 1989, 1990). Increasing their vulnerability, these islands also tend to rely on one or two main export crops, or on tourism, as an economic base, and they have fewer reserves for recovery (Pelling and Uitto 2002). For instance, in 1989, hurricane Hugo destroyed 85 per cent of the housing on the Caribbean island of St Croix, and in 1972, hurricane Bebe left 20 per cent of Fiji's population homeless. The economy of such islands is based on a small number of tree crops such as bananas and nutmeg in many Caribbean islands and coconuts in Oceania (Shakow and O'Keefe 1981). The small-scale fishing sector is often also seriously damaged, and although more easily replaced than a coconut plantation, a poor fisher may find it very difficult to replace a lost boat or equipment. Access to credit may speed recovery, but a vicious spiral of indebtedness can result.[7]

Coastal livelihoods

People's livelihoods have already been mentioned in the preceding chapters. In most rural coastal areas, the political economy of access to resources is complex and fragmented. Even where large landowners, companies or the state have rights to coastal plantations or fisheries, various kinds of formal and informal, legal and illegal access may be available to workers, the seasonally employed and neighbouring poor people. Small-scale fishing tends to be controlled by those who own boats and who employ a large percentage of fishers, as in the coastal waters of north-eastern Brazil, Sri Lanka and Andhra Pradesh (India). There is a trend toward concentration of ownership, the effects of which can be seen in US coastal fisheries, where (as with the 'family farm' in the US) high interest rates and changing production technology have put pressure on small, self-employed units. The latter tend to underinsure their equipment due to cash-flow pressures and are highly vulnerable to the economic effects of production being interrupted by a hurricane.

Coastal land and marine resources worldwide increasingly tend to be under the control of absentee interests. This is true of both LDCs and MDCs An economic calculus applied from a distance often leads to land-use decisions that put people and the sustainability of local ecosystems at risk. This is the case where remote financial interests continue to drive speculative residential development in Florida. The effect of similarly 'distanced' economic rationality can be seen in Haiti and coastal Kenya. There, urban commercial interests encourage conversion of timber from the remaining mangrove wetlands to charcoal for sale as urban fuel (World Resources

Institute 1986: 146–149). In Jamaica, there has been a proposal to mine large portions of coastal wetlands for peat as a substitute for imported energy (Maltby 1986: 135–144). Such a process would remove vegetative barriers that anchor soil and absorb some of the force of storm winds and waves. In the case of Jamaica it has been predicted that illegal cannabis production would be displaced from the Nigril Morass (wetland) to mountains further east. Deforestation of those mountains by cannabis producers would increase storm runoff and erosion (Maltby 1985) and could put the growers at risk from mudslides during severe storms.[8]

It is useful to distinguish between patterns of livelihood in the immediate vicinity of large coastal cities and those in more remote coastal areas. In the former case, vulnerable rural people may have relatives and livelihood options open to them in the city as a refuge when a storm hits. But the disruptive influence of the city over the immediate rural coastal area (e.g. loss of access to resources because of tourism, industrial pollution or commercial market gardening) has to be weighed against the advantages of a possible urban 'fall back' position.

The petty-commodity producer and semi-subsistence farmer in many tropical coastal areas utilises a variety of complementary livelihood options. Examples include 'community forest' farming in Sri Lanka and the combination of fishing and farming in Fiji and other parts of the world. However, these multiple options have been undermined by government and commercial encouragement of monocropping, the growing importance of waged employment and competition for resources with absentee commercial interests. Thus in coastal India, Thailand and the Philippines small-scale fishing for domestic consumption is declining due to competition with, and overfishing by, commercial trawlers in the same coastal waters (G. Kent 1987). In some areas it has been noted that the diversity of locally adapted food crops, including those most hurricane resistant, has decreased (Campbell 1984; Pacific Islands Development Program n.d.).

The livelihood systems of the rural poor in exposed coastal areas are heavily influenced by spatial and temporal constraints and opportunities. There is not only a connection between seasonality and rural poverty (Chambers et al. 1981; Chambers 1983), but vulnerability has its own temporal rhythms (as noted in Chapter 3). The timing of certain farming operations is critical to yields, and the plants and animals themselves are more vulnerable to damage from flooding or wind at some times than at others.

Fishing is also seasonal, and this can sometimes lead to complex risk calculations. If fish yields are better at greater distances from the shore during the hurricane season, where should the 'rational' fisher place nets? The fisher has a greater chance of saving the nets within the usual storm warning time if they are set closer to shore, but will catch more by gambling on the more distant fishery (Davenport 1960; Gould 1969). In fact, over-fishing by commercial competitors may be forcing small-scale fishers to go further out to sea even during the stormy season, thus adding to the risk of fishing as a livelihood option.

Waged work is also more readily available in certain seasons (planting, harvest, the run of a particular kind of fish, for example). Typically, the harvest season sees large numbers of rural poor engaged in temporary work in fields and plantations. Tropical storms striking at this time of the year may mean large economic losses for the owners and drastic reduction in income to workers who can no longer find employment. There is a parallel here with drought; in many cases it not only denies the small farmer or herder their subsistence, but it also deprives them and the rural landless of employment on larger farms (see Chapter 4). Riverine flooding in Bangladesh denies the landless and land-poor employment in the rice and jute fields of the more prosperous (Currey 1984; Clay 1985; see also Chapter 6).

We can summarise the relationship of vulnerability, livelihood and cyclones by applying the Access model in relation to hurricanes (as in Figure 3.2). Depending on their level of vulnerability, households will experience storm impacts differently on their assets and livelihoods (Boxes 2b and 6). Family members die and are injured, but more so among the poorer social groups. According to one account of the 1977 Andhra Pradesh cyclone, 75 per cent of the 8,000 deaths were in small farming, fishing or marginal farming families (Winchester 1986). In Bangladesh in 1991 the wealthy survived in houses constructed of concrete-blocks (Khan 1991).

Loss of labour power in poorer households makes it harder for them to recover economically. Illness arising after a cyclone, or that was already affecting the poor and malnourished, can only mean lower productivity. The poorest households may also use a high proportion of their funds on health care when it could have gone towards recovery of assets (Winchester 1992: 120 and 184).

Assets are also lost. The wealthy lose more in absolute terms, but less in relation to their total property. In particular, farm buildings, livestock and perennial (tree) crops are lost, as well as the standing annual crop. Saltwater flooding of fields may further complicate recovery (Campbell 1984). All of this modifies the household's income opportunities (Box 3a). They may be able to add a quick 'catch crop' if seed and other inputs are available, but otherwise tree crops may be missing from their list of options for some years. On the plus side, income from waged labour in reconstruction may be available as a new option for some households.

Social relations and structures of domination (Boxes 1 and 4) influence the new array of livelihood options. Poorer households will not be able to pay off loans or meet rent and other obligations. Credit provided by government and other sources may not be so easily obtained by the poorer social groups. This was true in Winchester's longitudinal study in Andhra Pradesh. Ten years after banks opened branches in the cyclone-affected area, the small farmers, landless and fishers still resorted to traditional money lenders

(Winchester 1992: 120).[9] Indebtedness and dependency will tend to reinforce the allocation of social power, and hence the structures of domination. In this, the storm Access model differs somewhat from that applicable to a disaster such as famine. In cases of famine, the legitimacy of structures of domination have sometimes been called into question by the majority of the population.

The household budget (Box 7) is radically changed. Indeed, for a period during relief and recovery the poorest households may have lost everything and not have any money, and will be dependent largely on social networks and aid from outside (as in Bangladesh following cyclones in 1970, 1985 and 1991, the cyclones in Andhra Pradesh in 1977, Orissa in 1999, and hurricane Mitch in 1998). If aid is adequate, malnutrition and disease should be avoided, but there is a chance that a weakened population can fall prey to epidemic disease owing to poor sanitary conditions, especially when there is a lack of safe drinking water (see Chapters 5 and 6).

Large cyclones are common on India's eastern coastline. In 1864 an enormous storm surge killed an estimated 30–35,000 people. Since then, severe cyclones have occurred in 1927 and 1949, with two in 1969 (Cohen and Raghavulu 1979: 9). What made the 1977 situation unusual was the coincidence of the cyclone with a high tide. The storm also made a sudden change of course that confused meteorological forecasting and meant there was no warning.[10] The worst-hit population lived in Divi Taluk, which lies at the mouth of the Krishna river between its two branches and the Bay of Bengal. Winchester (1986) mapped the deaths and found a high correlation between percentage deaths and the height of the storm surge (Figure 7.2). But his detailed analysis of topography showed that even within this relatively flat, alluvial delta, there were higher sites that were more secure from the storm surge (which was 3–7 m in height). For the most part, the wealthy farmers and petty officials who lived on these sites in *pucca* ('good') concrete houses survived. The rest of the community lacked public shelters of this quality, and those buildings perceived as safe (schools, temples, administrative headquarters) were located in low-lying sites.

Ten years later, when a cyclone hit this same area in 1987, cyclone shelters had been built, and although some people were excluded because of caste differences and overcrowding, most people were able to take refuge and thus survive (Winchester 1992: 121). However, after both the 1977 and 1987 cyclones the poorer social groups found themselves in a weaker position in terms of rebuilding their livelihood systems, and their recovery was very difficult. They lost tools, draft animals, milk cows and labour power.

Why do people expose themselves to such risks? The reasons are illustrated by the 'Pressure' model for the cyclone shown in Figure 7.3. The pressures leading to vulnerability included land hunger and the desire by low-caste farm labourers and fishing folk to live near their employment. Roads and transport in and out of this administrative area (*taluk*) were

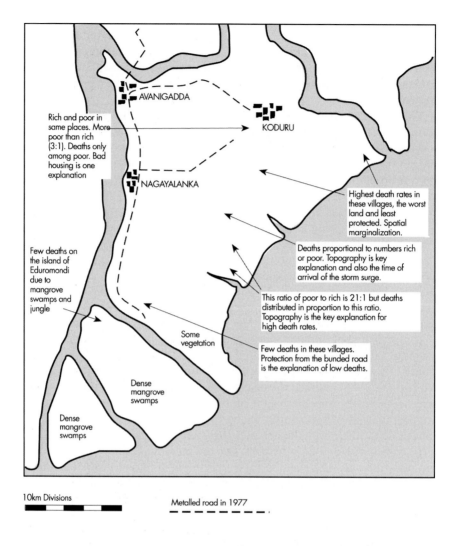

Figure 7.2 Explanation for deaths caused in Andhra Pradesh cyclone of 1977

poor, so labourers could not easily commute to work from higher land further inland.[11] Finally, the root causes of these pressures included the caste system, land tenure and the nature of government. These three deeply embedded systems combined to allocate most resources to the wealthy farmers and petty officials. They also denied the area adequate investment in public shelters, all-weather transport to inland areas and rural development assistance for the poorer social groups that might enable them to find livelihood options that did not demand residence in exposed sites[12]

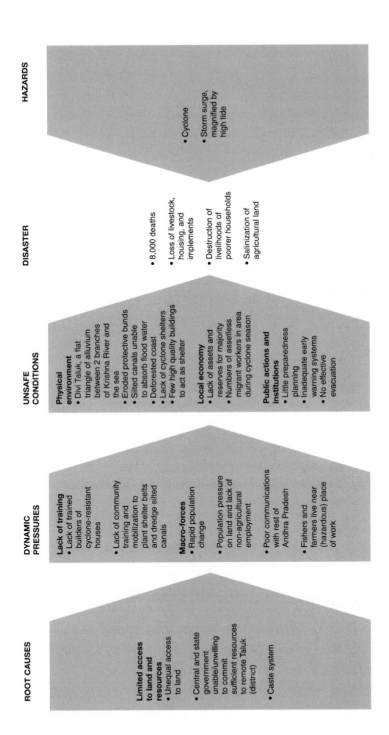

ROOT CAUSES

Limited access to land and resources
- Unequal access to land
- Central and state government unable/unwilling to commit sufficient resources to remote Taluk (district)
- Caste system

DYNAMIC PRESSURES

Lack of training
- Lack of trained builders of cyclone-resistant houses
- Lack of community training and mobilization to plant shelter belts and dredge silted canals

Macro-forces
- Rapid population change
- Population pressure on land and lack of non-agricultural employment
- Poor communications with rest of Andhra Pradesh
- Fishers and farmers live near (hazardous) place of work

UNSAFE CONDITIONS

Physical environment
- Divi Taluk, a flat triangle of alluvium between 2 branches of Krishna River and the sea
- Eroded protective bunds
- Silted canals unable to absorb flood water
- Deforested coast
- Lack of cyclone shelters
- Few high quality buildings to act as shelter

Local economy
- Lack of assets and reserves for majority
- Numbers of assetless migrant workers in area during cyclone season

Public actions and institutions
- Little preparedness planning
- Inadequate early warning systems
- No effective evacuation

DISASTER
- 8,000 deaths
- Loss of livestock, housing, and implements
- Destruction of livelihoods of poorer households
- Salinization of agricultural land

HAZARDS
- Cyclone
- Storm surge, magnified by high tide

Figure 7.3 'Pressures' that result in disasters: Divi Taluk, Krishna Delta, Andhra Pradesh cyclone of 1977

Some lessons do seem to have been learned since then. Reddy (1991) reports that in May 1990 there was a remarkable example of successful evacuation of more than 500 villages as a result also of a better warning system and improvements in practical arrangements made by the government. The process involved more than 2,000 evacuation teams and more than a thousand temporary relief camps. In all, 650,000 people were evacuated from the path of the cyclone, and the number of casualties was much lower than in the same area in 1977.

It is a shame that there is such variability in state capacity and political will in various Indian states, for the lessons of Andhra Pradesh do not seem to have been applied in neighbouring Orissa. In October 1999, two cyclones made landfall, the second one very powerful. It is considered to be the worst tropical cyclone in Indian meteorological history, with sustained winds of 160 mph (256 km/h) that peaked at 223 mph (359 km/h). This storm drenched the coastal plains with torrential rainfall for a full 24 hours and 30 ft (9 m) high storm surges inundated much low-lying land. At least 10,000 people died. A million hectares of cropland were lost under saltwater, and farmers lost more than 400,000 head of livestock – vital for their income, nutrition and traction power (Kriner 2000; Government of India 1999). The World Health Organisation's New Delhi office estimated that 15 million people were affected out of the state's population of 35 million (WHO 2001a).

Pre-cyclone conditions in Orissa were much like those Winchester found in Andhra Pradesh: 90 per cent rural and 60 per cent of the population living below the Indian poverty line (WHO 2001a). Recovery efforts there have been slow and characterised by much criticism of the state government, which was in fact voted out of office.

Case studies

Less densely populated coasts

Mozambique has a coastline of more than 2,000 km along the Indian Ocean, all of it subject to tropical storms including some of cyclone force. The colonial pattern of urbanisation was highly centralised, and Lourenço Marques (present day Maputo), the capital on the coast in the extreme south, dominated all other centres. Thus very long stretches of coast and the surrounding hinterlands were, and remain today, relatively remote from the three major coastal cities (Maputo, Quelimane, Beira), except for small administrative towns with few economic or social functions. These are connected by very poor roads.

Livelihoods on these more remote stretches of coast revolve around small-scale fishing, subsistence farming, limited livestock herding, remittance of wages from South Africa, income from limited sale of cashews, cotton, groundnuts and food crops, and casual employment in sugar and coconut plantations. The most important agro-industries in the coastal zone are coconut, cashew processing and sugar cane, but these industries declined dramatically during 20 years of civil war and 25 years of a centralised, command economy.

After independence, some state farms were developed that concentrated on basic food crops (maize, rice, potatoes), sunflower and cotton. Wages in plantations were low, and much lower than migrant wages in South Africa. However, since Mozambican independence South Africa cut back on the numbers of Mozambican men it allowed in to work there. Furthermore, producer prices for crops sold to the government were kept low and, more importantly, commodities for purchase in the countryside have become increasingly scarce and expensive since the early 1980s. As a result of all these factors, much of the activity of the people in these more remote zones is focused on their own needs. There is limited local trade and barter, but little on the national market.

During the period of recovery following a peace agreement in 1992 some limited market activity began again in Mozambique's coastal hinterland, but this has been slow as fully one-third of the total national population had been uprooted from their farms and fishing grounds by the war (United Nations 1995). In addition, although donor-imposed economic policies have replaced Mozambique's central planning, there is still little consideration of the impact of the export-oriented policy and large-scale development projects on the livelihoods and conditions of ordinary Mozambicans (Hanlon 1991, 1996). Mozambique faces even more serious challenges to its development in this post-war and post-socialist situation (Ferraz and Munslow 1999).

The shift from socialist to capitalist policies in Mozambique allows a comparison of the state's response to three cyclones: one in 1979 (before the war), one in 1984 (at the height of the war and the disruption caused by sabotage and terrorism) and another in 2000 (after some years of peace, but with the end of socialism). In 1979 the central and northern coast was battered by a cyclone that destroyed two of three cashew processing factories and uprooted 50,000 cashew trees (Wisner 1979: 304). At the time cashews were a major agricultural export. Apart from the cost of repairing these processing facilities, losses to the cotton and maize crops and to the system of feeder roads amounted to $5 million.

Loss of life, however, was negligible. The impact of floods in 1976, 1977 and 1978 (the first three years of independence) made the government accelerate its programme of planned, voluntary resettlement in communal villages. The policy concentrated on people living in flood-prone locations (ibid.: 202, Wisner 1984). As many as 500,000 people had been successfully resettled in more secure locations, many of whom were safe in 1979 as a result. Fortunately, there was also considerable foreign assistance already available in the country because of a drought that had affected some parts of the country for four years. Feeding camps had also been set up to respond to the needs of the internally displaced population seeking refuge from the initial phase of attacks by RENAMO (Resistencia Nacional Mocambicana; a South African-backed group of mercenaries and disaffected Mozambicans which launched a war of destabilisation on the government). Thus the existing aid and relief infrastructure was able to help storm victims immediately.

In 1979, the government response to the cyclone also showed the optimism and energy of a newly independent nation, and relief was organised swiftly. The president and cabinet ministers toured the affected areas within days, and a nationwide collection of donations came from factories and schools, with 'nation-building' as the positive message to overcome the tragedy. The cashew industry was rebuilt; peasants were rehoused in a massive effort to accelerate the programme of communal villages in the affected areas; public health and nutritional standards were maintained; and livelihoods were rebuilt. Here we provide a 'Release' version of the PAR model (Figure 7.4), which is designed to show how hazard vulnerability can be reduced. It is a counterpart to the 'Pressure' model applied to the Andhra cyclone (see above).

In 1984, by contrast, cyclone Daimone slammed into the Mozambican coast when the people and government had been trying to cope with drought for two years, while also dealing with increasing sabotage and massacres by RENAMO. The storm itself was similar to the one in 1979, and the cashew and coconut industries suffered again. It also flooded the largest sugar plantation in the country, Sena Sugar Estates, whose system of protective dikes had fallen into disrepair. The government's difficulty in responding in 1984 compared with the success of 1979 can be explained by a number of factors. By 1983–1984 RENAMO attacks were responsible for numerous communication breakdowns, making response to the cyclone more difficult. Also, economic disruption by RENAMO in the 1980s had been so great that the government had fewer resources with which to respond.[13]

Four cyclones made landfall in Mozambique between February and April, 2000. Rainfall from these storms was added to above average rain that had been falling since October 1999. There was widespread flooding of rural and urban areas in the south and centre of the country. About 800 people died or were reported missing. Ten per cent of the country's cultivated land and 100,000 hectares of subsistence and commercial crops were affected, as were 4.5 million persons, and the estimated loss to the country was nearly $700 million (Martin et al. 2001; Agence France-Presse 2000).

The government and people had learned many lessons from earlier cyclones such as those in 1979 and 1984. There had been early warning of possible flooding as early as September 1999, and the government acti-

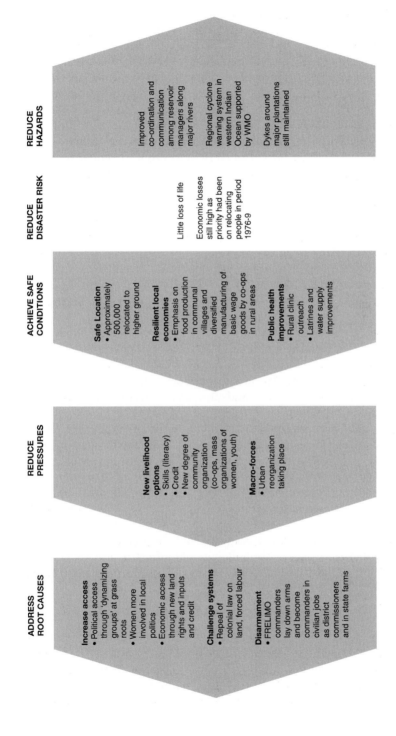

Figure 7.4 The 'release' of pressures to reduce disasters: the Mozambique cyclone of 1979

vated its disaster management plans (Christie and Hanlon 2001). However, the successive cyclones and the sheer scale of the flooding overcame dikes protecting such cities as Chokwe, and caused the response and relief system to spread itself thinly. Nevertheless, the death toll was low considering the power of these storms and magnitude of the flooding.

It seems to us that 20 years after the cyclone that is represented in the 'Release' model (Figure 7.4), much that had accounted for the earlier 'release' of pressure was still in place. In particular, the growth of an efficient national disaster planning bureaucracy and the growth of local development non-governmental organisations (NGOs) should be added to Figure 7.4. There is also evidence of acute awareness by Mozambican policy makers of hazard vulnerability as a key element in poverty reduction. The Mozambican *Action Plan for the Reduction of Absolute Poverty 2001–05* specifically deals with natural hazards and notes that they 'constitute an obstacle to a definitive break with certain degrees and patterns of poverty' (cited in World Bank 2002: 27).

On the other hand, a troubling sign for future preparedness and mitigation is the polarisation of rich and poor that has accompanied the adoption of a neo-liberal model of development in the country (Hanlon 1996). Mozambique's experience suggests that the vulnerability of much of the population may have been increased by the privatisation of public services (health care, water and sanitation, public works) and the emphasis on export production.

Earlier in this chapter we noted that social mobilisation of the kind seen in Mozambique in 1979 had also been effective in saving lives in Vietnam and Cuba (Wisner 1978a, 2001c). Maskrey (1989: 79–86) reviews what he calls community-based mitigation programmes. He discusses other positive cases, including the response to floods in western Nicaragua in 1982 by neighbourhood 'committees for Sandinista defence' (citing Bommer 1985). We will return to the issue of community-based mitigation later in this chapter, when we take up a case study from Cuba, and again in Chapter 9.

Nicaragua faced a challenge similar to that of Mozambique in 1988, when hurricane Joan virtually destroyed Bluefields, the largest city on the Atlantic. Eighty per cent of the buildings and infrastructure were destroyed, as well as 95 per cent of all structures on nearby Corn Island. The storm flooded land for hundreds of miles along Nicaragua's river border with Costa Rica. This affected important beef-producing areas and caused flooding and landslides due to torrential rains in the rich coffee-producing central mountains. Losses were estimated to have been more than those of the 1972 earthquake (Nicaraguan Ecumenical Group 1988).

Until the revolution in 1979, the Atlantic coast of Nicaragua had never been developed as part of the country. The population was a mixture of indigenous people and migrants from the Caribbean islands. English was more common than Spanish, population density was low and livelihoods centred around fishing and extraction of products from the rainforest that covers much of the region (Ballard 1984).

The hurricane came at a time when Nicaragua's left-wing Sandinista government was being forced to fight a war against the Contras (right-wing rebels who were armed and encouraged by the USA). But unlike Mozambique in 1984, the war was shifting in favour of the government, and it was possible to mobilise the nation and provide considerable aid for the stricken region. It was a fall in the world market price of coffee (by about 30 per cent) in 1989 that ultimately did more harm to the national economy, already weakened by a trade embargo imposed by the USA and the long war against the Contras. The dual economic shock of hurricane Joan and the crash of coffee prices were important in turning voters against the Sandinistas and eventually led to a change in government. Remembering the 'global pressures' introduced in Chapter 2, we can reflect that even though production losses due to the hurricane were very large, the politics and economics of US opposition to the Nicaraguan government did more damage to the economy.

Later in this chapter we will look at hurricane Mitch, which devastated Nicaragua (and Honduras) in 1998. By then Nicaragua had abandoned any vestiges of socialism. Bowing to the demands of the World Bank, it had drastically cut public investment in health care, road and bridge mainte-nance, drainage and local government. Since many elected municipal governments were in the control of pro-Sandinista officials, the national government had systematically tried to isolate and starve them of resources. The result (perhaps a cautionary tale for Mozambique, which seems to have set off on a parallel neo-liberal path) was that the hurricane Mitch disaster was worse than it should have been, with more than 3,000 deaths (Comfort et al. 1999; Buvinic et al. 1999).

Rural hinterlands

Although high winds at the coast and storm surges are dramatic manifesta-tions of the tropical cyclone's destructive power, intense rainfall in the hinterland is also a major hazard. Hurricane Mitch, a storm that killed 27,000 people,[14] produced rainfall of this kind when it stalled over Honduras and Nicaragua and dropped as much as four feet of rain in as many days. This intense precipitation falling on steep, deforested slopes caused an estimated one million landslides in the two countries (Lavell 2002). The largest of these occurred when Las Casitas volcano sent mud and debris sliding down to bury more than 2,000 people in ten villages and the town of Posoltega in Nicaragua (IFRC 1999a: 45; McGuire et al. 2002: 78). In both Nicaragua and Honduras, much forest cover had been lost since 1950 as commercial crops (coffee, bananas, sugar, cotton), beef ranching and timber extraction grew (Comfort et al.1999). Successive waves of expansion of these commercial activities left small farmers expropriated and forced them further toward the agricultural frontier, where they cleared land for subsistence crops, often on steep slopes (Wisner 2000b).

In the case of Honduras, forest cover had decreased from 63 per cent in 1960 to 37 per cent in 1998. The hills around Choluteca had been denuded during the 1990s by the timber industry, cattle ranchers, the establishment of export plantations and by small farmers. It was in this zone that some of the deadliest floods took place (IFRC 1999a: 47) By contrast with the Himalaya case (Box 6.4), here the role of deforestation is very clear.

Densely populated coasts

India

Many lives have been lost in the coastal zone of south Asia (Sri Lanka, India, Pakistan and Bangladesh). Table 7.1 lists the most intense tropical cyclones in this region from 1970 to 2000.

Returning briefly to a case we discussed earlier in this chapter, the cyclone that hit the Indian coastal state of Andhra Pradesh in 1977 exemplifies some of the complex interactions between the physical hazard and human vulnerability (Winchester 1986, 1992). The storm surge killed between 8 and 12,000 people (accounts vary), with a gigantic storm surge that raised a mass of water between 9 and 20 ft high (3 to 6 m), 50 miles wide and 15 miles deep (Cohen and Raghavulu 1979: 1). Who were the people killed? Winchester (1986: 184) shows death rates of between 23 and 27 per cent for small farmers, fishers and marginal farmers, and only 3 to 4 per cent for large farmers and petty officials. The ability to evacuate by road, the type

Table 7.1 Most intense cyclones in north Indian ocean, 1970–2000

Date	Location	Deaths
1970	Bangladesh	300,000
1971	Orissa, India	10,000
1977	Andhra Pradesh, India	8,547
1978	Sri Lanka	1,000
1988	Indo-Bangladesh border	8,000
1990	Andhra Pradesh, India	967
1991	Bangladesh	138,882
1998	Gujarat, India	3,000
1999	Sind, Pakistan	454
1999	Orissa, India	9,887

Source: India Meteorological Department, New Delhi.
http://agricoop.nic.in/statistics/history.htm

and strength of house, and small variations in topography within villages, were reasons why the wealthier farmers and petty officials survived to a much greater degree. In fact the wealthier lost more property (housing, stalls, cattle, carts and standing crops) than the poor, but lost a smaller proportion of their total property. They and the petty officials were able to evacuate and take their valuables because they had access to motor vehicles.

Another major aspect of differential vulnerability was the capacity of different households to recover from repeated shocks. Winchester (ibid.: 228) investigated rates of economic recovery from the 1977 cyclone among six occupational groups up to 1981. After four years large farmers, who had lost most, recovered more quickly and ended up better off. Small farmers recovered more slowly and were also somewhat better off. Petty officials were more prosperous in 1981. By contrast, marginal farmers, fisher people and landless labourers barely regained pre-cyclone levels of prosperity. Repeating his study of the same sample of different socio-economic groups again ten years after the cyclone, Winchester concluded:

[T]here had been a return to the *status quo* of 1977 manifested mainly by a decline in wage-labour opportunities resulting from the alternative land uses and increase in minimally labour intensive cash-crop activities. The economic and spatial marginalisation of the landless and small and marginal farmers appeared to have continued ... [because of] the influx of entrepreneurs taking over the previously uncultivable land (i.e., common grazing and forestry land) for cash crops (fish farms and coconut plantations).

(Winchester 1992: 119)

Physical reconstruction of shelter cut deeply into the limited resources of all but the wealthiest groups. Winchester (1986) criticises credit facilities, suggesting that government plans for building more secure housing were likely to make matters worse by emphasising concrete block construction that the poor could not afford, rather than more feasible lower-cost improvements.

Bangladesh

Bangladesh unfortunately is also very familiar with severe cyclones (Islam 1974; C. Haque 1997). The storm that struck the coast in 1970 brought a surge that coincided with high tides. Many thousands of poor migrants were sleeping in the fields they had come to harvest in coastal areas. Most at risk were those in fields on low-lying silt islands (*char*) in the gigantic delta of the Ganges and Brahmaputra rivers (see Figure 6.2). Mortality was estimated at 600,000–700,000 by Burton et al. (1978) and 300,000 by Carter (1987). The country also suffered great losses to severe storms in 1961, 1963 and 1965. The 1963 cyclone swept the coast from Cox's Bazaar to Feni, killing 80,000;

while the death toll in 1965 was 18,000 (Hanson 1967: 12–18). In 1985 another killed 10,000 people, destroyed 17,000 homes and damaged a further 122,000, swept away some 140,000 cattle and destroyed 500,000 acres of rice and jute (Milne 1986: 74). Another large cyclone hit Bangladesh in December 1988, displacing millions of people only months after the largest riverine floods ever recorded (see Chapter 6).

On 30 April 1991 a cyclone larger than the one in 1970 crossed the coast near Chittagong and killed tens of thousands of people (Haque and Blair 1992). Residents of islands were most vulnerable, along with coastal dwellers from Cox's Bazaar north to opposite Sandwip Island. At least half of the 10,000 people on Sandwip Island were killed. While the magnitude of the human loss and destruction in 1991 is hard to imagine, it is considerably less than the toll in 1970. Since then, thousands of Red Crescent volunteers with radios have been trained to listen for and to spread evacuation warnings (Fazlul n.d.). The government had spent $3 million to improve its warning system since 1985 (Mahmud 1988), and many hundreds of thousands of people did make it to safety. Likewise, public storm shelters built since 1970 fared well, although the problem lies in there being only 302 of them: 10,000 such shelters (costing $5,000 each) are needed to protect the exposed population. Public works programmes have also systematically been building embankments, but these are generally too low to protect against the greatest cyclones. Raising embankments to the required height of 18–25 ft (6–8 m) costs $25,000 for each 100 ft (30 m) in length (Kristof 1991).

Such engineering efforts are aimed mainly at achieving improved safety for those at risk in the face of the hazard itself, and to some extent at modifying the intensity of the wind in specific places. But although welcome and supported (or even initiated) at the grassroots level, they are inadequate compared with the numbers of people who need protection. Nothing is being done further back in the links that create those unsafe conditions. The pressures and root causes (which are very similar to those shown in Figure 6.3) are not being addressed (see Figure 7.5). Without the political will and significant changes in the national and international factors that affect Bangladesh, efforts at the local level are likely to remain inadequate.

The Philippines

The Philippines is a nation of more than 7,000 islands which are home to some 76.5 million people (NSO 2003). Much of it is at risk from tropical cyclones (Delica 2002: 41), with the capital, Manila (11 million people), and the heavily populated island of Luzon to the north frequently affected. Within one week in October 1989, the Philippines was hit twice by typhoons (Angela and Dan), killing 159 people and leaving 429,000 homeless (Newman 1989). Similar suffering was recorded nearly every year during the 1980s. For instance, in 1988 there were also two typhoons (Ruby and Skip) that caused

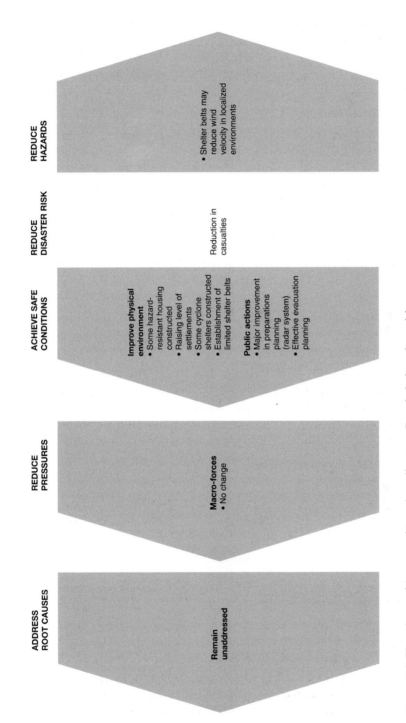

ADDRESS ROOT CAUSES

Remain unaddressed

REDUCE PRESSURES

Macro-forces
• No change

ACHIEVE SAFE CONDITIONS

Improve physical environment
• Some hazard-resistant housing constructed
• Raising level of settlements
• Some cyclone shelters constructed
• Establishment of limited shelter belts

Public actions
• Major improvement in preparations planning (radar system)
• Effective evacuation planning

REDUCE DISASTER RISK

Reduction in casualties

REDUCE HAZARDS

• Shelter belts may reduce wind velocity in localized environments

Figure 7.5 The release of 'pressures' to reduce disasters: Bangladesh cyclone risk

extensive flooding in central Luzon and made 200,000 people homeless on Leyte Island.

The cascade of mutually reinforcing causes for the people's vulnerability in the Philippines begins with land hunger, which is a result of highly unequal land tenure. Despite a nominal commitment to land reform by a succession of governments, the land shortage of poorer people has changed little (Collins 1989; Boyce 1992; Putzel 1992, 2000). Landlessness leads to a desperate search for livelihood alternatives in the growing cities, where migrants live in poorly constructed squatter settlements (some of them on stilts at the water's edge), in low-lying flood plains and on wasteland, rubbish dumps and on steep slopes. These are the areas where urban land owners cannot command high rents and where commercial alternatives to shanty development are not profitable.

In 1981, 1.7 million people in greater Manila lived in 415 slum colonies, amounting to 23 per cent of the city's then 7.5 million inhabitants (Ramos-Jimenez et al. 1986: 11–13). These slum dwellers were highly vulnerable to the impacts of wind and flooding due to typhoons. Twenty years on, the situation has not changed very much, and in 2002 an estimated 30 per cent of a much larger Manila (the city having grown from 7.5 million in 1981 to 11 million) were living in slums (Wisner 2002a).

In parts of rural coastal Philippines, economic development involved entrepreneurs seeking out raw materials and sites for fish farming and industrial salt production, destroying mangroves in the process (World Resources Institute 1986: 146–150). Land pressure also meant that small farmers and indigenous people such as the Palaw'ans (who live on one of the more isolated islands to the west of the archipelago) retreat further into steeper mountain areas. They are forced to do this as government-sponsored migrants and companies take over land at lower elevations for rice farming (Lopez 1987). Clearing land and attempting to farm on such slopes puts these people at risk from the same typhoons that wash away the urban squatter's home. Torrential rains cause landslides such as the one reported in 1988 (Union News 1988). The resulting erosion caused silting of rivers downstream, adding to the potential for flooding at coastal locations, including the sites of urban slums.

Mexico

On the other side of the Pacific, the Mexican tourist centre of Acapulco was hit by hurricane Pauline in 1997, and the patterns of vulnerability were abundantly clear. Not a single tourist was killed, but more than 100 of the low-wage service workers who lived in the ravines above the beach resorts died, and 5,000 lost their homes (Puente 1999). Many cities and towns along the Pacific and Caribbean coasts of Mexico and Central America have grown rapidly as centres for beach-oriented tourism. In most of these sites the stark discrepancy between living conditions of the tourism service population and the tourists is bound to produce other, similar tragedies.

Islands and small island states

Those people living and earning their livelihoods on islands, including small island states, are particularly vulnerable to tropical cyclones, and, more generally, to climate change (Pelling and Uitto 2002).

Fiji

Fiji lost approximately 30 per cent of its agricultural production capacity in 1985 when four hurricanes swept these 361 islands over two months. Tree crops suffered losses of up to 80 per cent, and sugar cane production was heavily affected, reducing employment. In addition, food crops and livestock losses were high (Chung 1987). Fields growing subsistence crops were flooded with brackish water. Those root crops that had proved more resistant to storm damage over previous decades had declined in popularity due to rises in income and the purchase of store-bought food.[15]

Recovery from the 1985 hurricane raised many questions about the dependency of the island on a single export crop and the dependence of the people on imported food (Campbell 1984; Pacific Islands Development Program n.d.). Despite appeals for 'modern recovery measures [that] should merge ... with traditional and time-proven practices' (Chung 1987: 48), there is little evidence that things are moving in this direction. Fiji has continued to suffer damaging cyclones (most recently in 1993, 1997, 1998, 2002 and 2003); however, patterns of personal and social protection do not seem to have improved.

Cuba

Cuba is an island nation that has experienced many hurricanes. In the past death tolls were considerably higher, but more recently the number of people killed has decreased dramatically, although economic losses are still high. As a result, the relationship between deaths and economic losses in Cuba is similar to that in richer countries.

In 2001, hurricane Michelle made a direct hit on Cuba. Hurricane Michelle was a dangerous category 3 storm (on the five-point Saffir–Simpson scale). It made landfall at the Bay of Pigs on Cuba's southern coast with winds of 216 km/hr. The storm travelled north across the island, where it eventually caused heavy damage to homes (22,400 damaged; 2,800 destroyed), agriculture, industry and infrastructure in much of the western half of the island, including Havana.[16] This was the worst hurricane to hit Cuba since 1944. In all, however, only five deaths were reported. By contrast, when Michelle passed across Central America, 10 people were killed and 26 listed as missing (Cawthorne 2001: 2). Previously we have discussed the terrible loss of life in Central America during hurricane Mitch. How did Cuba save lives?

The most important factor seems to have been a timely evacuation. Roughly 700,000 people were evacuated out of Cuba's 11 million population.[17] This was quite a feat given Cuba's dilapidated fleet of vehicles, fuel shortage and poor roads. It was possible only because of:

- advance preparations, training and planning;
- a cadre of local personnel;
- confidence in warnings on the part of the population;
- co-operation with the Cuban Red Cross.

In addition, the electricity was turned off in Havana to avoid deaths or injuries from electrocution, as were the tap water supplies in anticipation of possible contamination. Reports say that Havana's population was advised to store water and food, and that on the whole they complied. The population at large in Havana also participated in clearing debris from the streets that could have become dangerous in strong winds, and was involved in tying down loose roofing.[18] Cuban state television broadcasts included references to the 1932 hurricane that had killed more than 3,000 people (Cawthorne 2001).[19]

These preparations also illustrate:

- an effective risk communication system;
- the use of the memory of past disasters being actively encouraged by the authorities;
- neighbourhood-based organisations capable of mobilising labour;[20]
- trust on the part of the general population.

Havana is a large city of two million inhabitants, with a long history of deaths due to passing hurricanes. In 1844 some 500 people lost their lives there. In 1866 the death toll in the city was 600, and in 1944 there were 330 fatalities and 269 collapsed buildings.[21] However, 2001 was not the first time that preparations and emergency response have saved lives in Cuba. In 1996 some historic buildings were destroyed by hurricane Lili, but no one died (Longshore 2000: 81).

Policy response

Coastal storms demonstrate the dynamics of vulnerability discussed in other chapters: the intimate link between vulnerability and livelihood strategies, and the complex and shifting pattern of differential class, ethnicity, caste, gender and age-specific vulnerability. Severe storms also demonstrate vulnerability in more specific ways. Unprotected, poor-quality urbanisation is basic to creating vulnerability to wind, flood and mudslide hazards associated with severe storms. The relationship between town and countryside can exacerbate the hazard when coastal vegetation is removed or disturbed. The role of economic dependency in creating vulnerability emerged clearly from the case studies of islands dependent on one or a small number of export crops. We have also seen the negative role of absentee landlords in both urban and rural settings.

There are, in short, many things that donors and national policy makers can do to reduce vulnerability to storms – and which will simultaneously reduce other risks – providing better access to livelihood resources, establishing better governance and social protection, engaging more vigorously in comprehensive urban and regional planning. We return to these more general policy changes in Chapter 9.

There are also some very specific, more focused policies and actions that have shown promise in reducing vulnerability to storms. Warning systems can be developed (Lee and Davis 1998; Zschau and Kueppers 2002) and will be effective if there is state capacity to ensure that people trust the authorities, that they receive the message and that communities then know what to do. In 2001 the low death tolls for storms in Cuba, Tonga and Mauritius showed the value of early warnings and preparedness. Unfortunately, some of the most vulnerable people may not hear or believe such warnings (Twigg 2002).

A striking example of this weakness is provided by the cyclone that struck the western coast of Gujarat in June 1998, affecting extremely poor, isolated people who make their living producing salt (Disaster Relief 1998d; Kalsi and Gupta 2002).[22] Some 3,000 people lost their lives, 200,000 houses were damaged and the economy suffered losses of $700 million. However, there had been reasonably accurate identification and tracking of this storm for five days. The problem seems to have been that cyclones are much less frequent on this western coast of India than they are in the east. People were therefore much less familiar with them, and preparedness was also insufficient. Some labourers and isolated low-income workers such as those who worked in the salt pans either did not receive, did not believe or did not understand the warning (Kalsi and Gupta 2002). There also seems to be a connection between the failure of the warning system and what we would call 'root causes' and 'dynamic pressures' in the terminology of the vulnerability approach:

> Due to privatisation of ports and the development of coastal areas for commercial and tourism interests, the fisherfolk have been involuntarily displaced from their places of residence to areas that are viewed as more disaster-prone. These fisherfolk were affected by a storm surge (up to 3 m high) and entire settlements were washed away. Being new to areas where they had been displaced, and without good communication lines, they were highly vulnerable
>
> (Kalsi and Gupta 2002: 202).

Another possible policy response is the construction of coastal shelters and meeting points. These have undoubtedly proved effective in saving lives, but more are needed, and access needs to be guaranteed for all those affected. In a study cited earlier there was evidence that people in Andhra

Pradesh had been excluded from shelters because of their caste. Winchester's consecutive studies of Andhra Pradesh disasters also showed that the major infrastructure improvements, and the new economic opportunities and credit facilities made available after the 1977 cyclone, went to the rich and to entrepreneurs who were attracted from the outside.

Severe storms underline the futility of adopting only engineering-based and administrative approaches: these are necessary but not sufficient. The threat of storm hazard must be placed in the context of 'normal' development policy. This was clear in the case of Andhra Pradesh, and Winchester concludes:

> As long as technical experts perceive the solution to vulnerability in terms of technical adjustments in isolation from political realities, then solutions will only treat symptoms and not causes, and spirals of dependency and expectation will inevitably be created which in time may create their own political cyclones.
>
> (Winchester 1992: 194)

In the case of Andhra Pradesh, roads and means of transportation, alternative livelihood possibilities, availability of credit and affordable health care were highlighted as crucial factors in reducing vulnerability. These are elements of sustainable human development, and are not specifically policies for dealing with disasters. Giving hurricane preparedness in high-risk zones a role as part of mainstream development can also be as straightforward as building community shelters as multiple-purpose structures (e.g. schools or community centres), as some NGOs are doing now in Bangladesh (Sattaur 1991; Schmuck-Widmann 2002; cf: Ullah 1988).

Drought, river flood and epidemic disease hazards raise similar issues of more general development policy (see Chapters 4, 5 and 6), especially as global warming may intensify all these hazards and spread their impact to new areas. When considering such hazards, it is essential that the livelihood strategies of all the different social groups affected are understood in some detail. Social profiling in this way is an essential complement to hazard mapping.

Since the coping patterns of ordinary people strongly influence the level of loss suffered, their chances of recovery, as well as their survival, it is essential that evaluation of vulnerability to hurricanes includes detailed accounts of how people cope. The striking differences in recovery rate seen in the case of groups affected by the Andhra cyclone can be avoided if aid is provided in appropriate form to the most vulnerable. This requires an understanding of their coping mechanisms as well as their needs.

As emphasised so often in this book, there is no substitute for painstaking work *with* the survivors as co-designers and co-directors of the recovery process (Maskrey 1989; Lewis 1999). In this way, it is more likely that vulnerability to the next coastal storm will be reduced. Since rebuilding of houses, other structures and facilities is usually involved in storm recovery, many of these same lessons will be seen in the next chapter, which deals with earthquakes; we will take up the question of recovery in Chapter 9.

Notes

1 China's 3,000 km coast line has experienced 319 typhoons and storm surges between 1915 and 1999, which have killed 82,551 people (Longshore 2000: 69). These numbers include the approximately 60,000 people who died in just one event in 1922. Hong Kong experienced deadly storms in 1841, 1906 (at least 10,000 killed), 1925, 1947 (more than 2,000 deaths), 1957, 1960, 1964 (twice), 1971 and 1995. The southern port of Canton (Guangzhou) had 37,000 reported fatalities in a storm in1862. Shanghai suffered typhoons in 1926 and 1949 (when most of the city was destroyed). In 1927, a storm hitting the central coastal port of Kwangtuang claimed 5,000 lives (Longshore 2000: 69–73).

2 We should also note that, as with floods and some volcanic impacts, there are positive attributes of cyclones for some societies. They can redistribute moisture to drought-prone areas and are looked on as a positive reduction of drought risk in the Philippines. In Mauritius, the authorities regard cyclones as vital to the islands' ecosystem, although of course there are significant dangers involved as well.

3 Destructive windstorms include tropical and extra-tropical cyclones, tornadoes, *derechos* (severe downdraughts of wind within a thunderstorm), dust storms and large thunderstorms with associated wind, hail and lightning (McGuire et al. 2002: Ch. 2). We focus our attention here on tropical cyclones because of their deadly and physically destructive impact on the coasts of many LDCs. Extra-tropical cyclones can also be destructive of lives and property. For instance, in 1953 the wind and storm surge associated with a cyclone claimed 2,000 lives in Netherlands and England (ibid.: 60). Tornadoes are created during intense thunderstorms, and some also have been deadly, killing several hundred people at a time; examples are those that struck in the USA in 1974, the Soviet Union in 1984 and Bangladesh in 1996 (ibid.: 48). A vulnerability approach is well suited to the analysis of tornadoes in the USA, since such a large proportion of their victims are low-income people living in flimsy mobile homes ('manufactured housing') (Steinberg 2000: 228, endnote 24, found that in 1981–1997, 35 per cent of the 296 tornado deaths in the USA were people living in such accommodation.

4 The islands comprising Mauritius have been the focus of attempts to improve tropical cyclone warning and preparedness, and this seems to have paid off in recent years, except that there are still disparities in the coverage of warning systems between richer and poorer islands, and the needs of some low-income groups such as fishers have not yet been taken into account fully (Vaghjee and Lee Man Yan 2002; Parker and Budgen 1998: 1.52).

5 It is important to note that the valuations of such economic losses may have little relationship with the real losses to livelihood and assets of different types of people. Moreover, comparisons between countries do not take into account the fact that in a poor country the monetary value of damaged property may be lower, and yet represent a much greater loss in terms of the values of local people.

6 In 1990 there were 12 coastal counties in Florida with more than 24 per cent of their populations aged 65 years or more and eight coastal countries with more than 7 per cent aged 85 years or more (FREAC 2001a, 2001b).

7 Such indebtedness to procure boats is a problem facing many small-scale fishing communities, not just those on small islands. Chambers (1983) and Winchester (1986, 1992) have assessed this problem in south India.

8 In part because of growing awareness of this destructive chain of events and, more importantly, the growing importance of tourist income to the Jamaican economy, the peatlands of Jamaica were not, in fact, developed. By comparison with other nations in the region, Jamaica's peat energy potential is actually quite small. Measured in thousands of hectares, the World Energy Council (2003: Table 8.1) shows Jamaica as possessing 12, while Costa Rica has 37, Haiti 48, Belize 90, Nicaragua 371, Honduras 453 and Cuba 658. Nevertheless, future economic and policy changes could see the rebirth of interest in mangrove and coastal peat areas in fragile ecosystems in Jamaica and elsewhere.

9 Such empirical results cast some doubt on the future of 'micro-credit' as a way of financing household-level self-protection, which seems to be the vision of the World Bank and the ProVention Consortium: http://www.proventionconsortium.org/

10 On the complexities of forecasting, see Jan-Hwa (1998) and Holland (2000).

11 After the 1977 cyclone, the government employed one-quarter of the male workforce of the area in public works. This included considerable road repairs and upgrading. However, in 1986 bus transportation was privatised, and the bus companies decided they would use only properly surfaced roads. The net effect was that transportation was not improved and the ability to evacuate this hazardous zone remained problematic (Winchester 1992: 120).

12 The Andhra cyclone has been analysed in considerable detail in Winchester (1992); much of this information is derived from his previous work (Winchester 1986, 1990). Also see Chapter 3: 103–110.

13 Attacks by RENAMO began in 1979 and grew steadily until by 1988 5.9 million people had been displaced and were dependent on food aid from a population of 13 million (Kruks and Wisner 1989: 166; D'Souza 1988).

14 The official figure used by the national governments of the affected countries was 9,000 (Consultative Group for the Reconstruction and Transformation of Central America 1999), however Gass (2002) provides detailed data from Honduras, Nicaragua and El Salvador that alone total 26,919 (even without data from Guatemala and Belize) when one adds official deaths and 'disappeared' or 'missing' together. The vast majority of deaths occurred in Honduras and Nicaragua.

15 Ironically, Fiji also suffers from drought, and the 1997 drought caused losses of $175 million; $18 million worth of food and water had to be distributed (World Bank 2002: 15). These costs were added to a decade of storm losses in Fiji, where cyclones cost $0.8 billion during the 1990s (ibid.).

16 Cuban Civil Defense authorities reported 'about 30 factories, 45 livestock and agricultural installations' destroyed and 'over 80 hospitals and clinics' damaged, according to the Deutsche Presse Agentur 2002 (accessed on ReliefWeb). On the positive side, more than three-quarters of a million animals were herded to higher ground (OCHA 2001).

17 Another report mentions 'nearly a million' evacuees (Deutsche Presse Agentur 2002).

18 OCHA Situation Report No. 7 (2001) and Reuters (Cawthorne 2001: 2–3) (accessed on ReliefWeb).

19 Cuba's communication of risks has been described as being very effective. Nevertheless, in a recent review, the Cuban authorities themselves note that they need to do a better job of educating people at all levels of society (Glantz 2001: 37).

20 Paul Susman, Bucknell University, has argued (in an unpublished study) that the Committees for the Defense of the Revolution played an important role in reducing dengue fever outbreaks because they were able to mobilise labour for highly localised, house-by-house monitoring and control of mosquito breeding sites. As a result of this, as well as prompt case finding and treatment, Cuba has the lowest prevalence of dengue in the Caribbean and Central America (personal communication; susman@bucknell.edu).

21 Longshore (2000: 79–81); in Havana 1,550 buildings are reported to have been damaged, 26 schools and 10 day care centers (OCHA 2001: 1).

22 Assessments of other warning systems considered to be generally affective, such as those in Mauritius, China, the Philippines and Vietnam, have all included the caveat that there are still problems reaching isolated or 'marginal' populations (see Parker and Budgen 1998: 1.10; Wills et al. 1998: esp. 4.50).

8

EARTHQUAKES AND VOLCANOES

Introduction

In this chapter attention is centred on earthquakes and volcanic eruptions. These are highly energetic natural events that occur irrespective and independently of social action and any modification of the environment. We mention the significance of human action in relation to these natural trigger events in order to highlight not the insignificance of humanity in relation to these geological processes, but to underscore the fact that human action and inaction can nevertheless impact upon the *outcomes* of earthquakes and volcanic eruptions.

To illustrate this fact, we begin by comparing two hazard events which occurred 100 years apart. The first was on the island of Martinique (in the Caribbean), the second affected the city of Catas in the southern coastal region of Peru. At the most general level of root causes, there are familiar processes involved in determining unsafe conditions: the interests of the powerful, bureaucratic incompetence and ignorance. However, in the example of Catas, evacuation orders were given and there was some evidence of lessons having been learnt from experience.

The first eruption was in 1902, when an irresponsible and opportunistic political leader refused to order the evacuation of Martinique because of an impending election (in which he expected to benefit), despite the immediate threat of a volcanic eruption on the island. The consequence was the worst loss of life in a volcanic eruption in the entire twentieth century. One hundred years later, our second example relates to seismic hazard mapping in Peru, in the aftermath of an earthquake that led to an evacuation order to relocate a community, supposedly for their protection. But was public safety the reason for the order, or was there another motive to remove the community – of the discovery of oil on their land?

During the twentieth century around 98,000 people were reportedly killed in volcanic eruptions (CRED 2002). More than a third of this total occurred in just one event: the catastrophic 1902 eruption of Mount Pelée in Martinique (Caribbean), when the entire population of 29,000 in the town of St Pierre was killed in less than two minutes (Scarth 1997: 138–141). This disaster is often blamed on the ferocity of the volcanic explosion and the blast of superheated gases (estimated at 700–1,200° Celsius) which struck the town. However, the massive loss of life is an early example of politically induced vulnerability. Three months before the eruption nauseous smells, similar to hydrogen sulphide, were experienced in St Pierre. Sixteen days before the eruption, animals began to die on the slopes of the Mount Pelée volcano due to gas emissions; then three days before there was a major pyroclastic flow, with associated ash falls. Despite this mounting evidence of an impending eruption, the Governor issued soothing messages and suggestions for an evacuation were rejected (Ferguson 2002: 14–19).

> [I]t seems that there were political motives for this ill-informed optimism. Elections were due on 9 May and the Governor was determined that they should proceed smoothly. A mass evacuation of St. Pierre would have created an enormous administrative headache, and both the leading political parties were adamant that their supporters would not leave and surrender their votes. Was there a conspiracy to down-play the possibility of a disaster? It appears more likely that ignorance lay behind official inaction.
>
> (Frampton 1998: 59)

The Governor's refusal to order an evacuation became his own death warrant, as when the volcano exploded on 8 May he and his family perished with the rest of the townspeople (Bishop 1998: 58; Zebroski 2002).

During the twentieth century 1.87 million people are estimated to have been killed in earthquakes. Thus for every person killed in a volcanic eruption, 19 have died in earthquakes (CRED 2002). We tend only to know about the large-scale earthquake disasters reported in the media. However, there are many small events in which tens or hundreds of people die, with hardly a mention in international news coverage. One such event involved the indirect deaths from an earthquake of 8.1 Richter that occurred on 23 June 2001 in the area of Arequipa and La Punta (southern coastal region of Peru). A hundred people died, half of them drowned, as a tsunami hit their villages. In the subsequent months the authorities (with funding partly from United Nations Development Programme [UNDP]) decided to map the seismic hazard in the area. One of the areas being mapped was the village of Catas, where three died and 63 of 71 families lost their homes in the earthquake. In 2002, most of the survivors lived on official welfare provision, occupying tents and being fed by a charity food kitchen. To make matters worse, the wells were contaminated by salt as a result of the earthquake and all water had to be transported into the area by road.

The seismic hazard mapping team advised the villagers to relocate, since the water table had risen as a result of the earthquake to just 70 cm below the ground surface, posing a threat of severe ground shaking in future

earthquakes. Having identified the area as a 'high-risk zone' the municipality refused to carry out repairs in Catas, and earmarked new safer land
for the relocation of the community. However the villagers were told they
would need to pay $40,000 for the land, which they could not afford.
Therefore they found themselves stranded 'between the devil and the deep
blue sea': highly vulnerable, dependent on external welfare, living in tents
without the money to move and with no government support for essential
repairs. The head of the village committee, Fernando Herrera, was suspi-
cious as to why his village had been the only one scheduled for relocation on
account of the seismic risk. He suspected that there was another motive for
the proposed relocation: 'The real trouble is the survey people found oil
here, that is what they told us, and many people here think that is why the
authorities want us to go' (quoted in Pearce 2002a: 34).

These contrasting examples of official action in response to different
hazards, with completely opposite scales of impact, in different geographical
regions, and separated by a century, have at least one feature in common.
They are reminders that people are not only at risk from earthquakes and
volcanic eruptions; they are also at risk from the action, or inaction, of politicians. And especially from those leaders with agendas other than the safety of
the public they are supposed to serve. This theme of political vulnerability
will recur as one of many levels of public exposure to earthquakes and volcanoes that are described in this chapter in a series of case studies.

Almost two million people killed by earthquakes in the last century represents a bleak death toll. Yet for all their dramatic impact, earthquakes and
volcanic eruptions do not remotely match the scale of casualties that result
from droughts and floods (Sapir and Lechat 1986; CRED 2002), and of the
millions of earthquakes that take place every year, few are actually deadly
(Hays 1990). About half of all earthquake deaths have occurred in China
(Noji 1997: 136), which also suffered the most devastating single event (the
1976 Tangshan earthquake) which resulted in between 242,000 (Coburn and
Spence 1993) and 290,000 (Munich Re 1999) deaths.

Determinants of vulnerability to earthquakes

Understanding the vulnerability of people to an earthquake in a particular
area, both *ex ante* (potential) and *ex post* (what happens after the initial
shock and in the process of recovery), involves two related tasks, outlined in
the Access model given in Chapter 3. Broadly, these are to understand the
space- and time-related characteristics of the earthquake, which is 'cross-
tabulated' with the socio-economic characteristics of the population at risk.
The space–time characteristics of the earthquake are then traced to the
space–time characteristics of human activities, which help to explain 'who
was where when' as the earthquake struck, and how they were affected by it.
Thereafter, the structure of access to resources, and the capability of

applying knowledge to the seeking out and utilisation of these resources, is crucial for the recovery of affected people after the earthquake.

The location of the earthquake has to be specified using a number of different scales. It is self-evident that its location is of prime importance. The probability of earthquakes can be mapped for the large scale and seismic belts (with higher chances of both the severity and number of events) can be identified. Secondly, at the meso-scale location is of great importance. In eastern Turkey, for example, villages and small towns are located on the sites of springs, which themselves occur along fault lines and are therefore more susceptible to earthquakes. Finally, the greatest share of mortality and loss of property from earthquakes occur as a result of damage to, and collapse of, buildings and from falling masonry. Therefore the micro-geography of the settlement and its cross-tabulation with the spatial configuration of the earthquake (and the characteristics of underlying rocks and soils) are of prime importance.

The temporal characteristics of earthquakes are also crucial. Firstly, the frequency of earthquake events over months, years, decades or longer is important, since it can condition the level of risk awareness (for institutions such as the state, as well as for civil society and individuals). For instance, this affects whether people and institutions adapt their building design and city planning to take account of memories of earthquake in the past. Secondly, the occurrence of the earthquake by time of year and day of the week can be critical. There are festivals, market days and national holidays that concentrate large numbers of people in particular places on specific dates. Also the season is important. For example, in the harsh winter conditions of the Armenia earthquake of 1988 few survived, while in warmer climates or seasons, trapped people stand a greater chance of surviving until rescued (Noji 1997: 148). Finally, the time of impact is of critical importance. If the earthquake occurs at night (as in both the 1976 Guatemala and 1993 Latur (India) earthquakes), casualties are always higher. People are likely to sleep through the fore-shocks that in daytime can provide sufficient warning to escape from buildings. Also, when lying flat in bed inside a building, people are much more exposed to injury from falling debris (Alexander 1985).

Characteristics of buildings and unsafe structures also have a strong influence on vulnerability to earthquake hazard (Cuny 1983; Coburn and Spence 2002). It is clear that many communities in seismic areas, particularly in LDCs, have many dangerous structures that will collapse under extreme seismic forces (French 1989). Over 95 per cent of all deaths in earthquakes result from building failures (Alexander 1985). Many fewer deaths and injuries are caused by the secondary impact of an earthquake, such as induced landslide, fire or drowning from a *tsunami* (Seaman et al. 1984: 10–11; Noji 1997: 155–157) (see Chapter 6: 105).

Since buildings are such a critical factor in seismic risk, concerned officials need simply to look very hard at the elements that relate to building safety, with specific consideration given to the shape, siting, building materials and constructional details of buildings, to find answers to three vital questions. Firstly, where are buildings likely to fail? Secondly, and perhaps even more significantly, what are the root causes of this dangerous situation? Thirdly, what action can be taken to reduce this underlying pressure? If the first question is tackled while the second is ignored, then the lives and property of the population will remain at risk even though certain individual buildings are made safe, since symptoms (unsafe building design) will have been addressed rather than the root causes (Noji 1997: 156–157).

Protective measures include structural and non-structural regulations and, in more general terms, policies that officials have taken, in advance of any disaster, to reduce seismic risks. What has been done, where, how, why, by whom and for whose benefit? Actions will include detailed seismic risk assessment, enforcement of land-use planning controls, building safety codes, developing a detailed disaster plan, having an effective search and rescue capacity, etc. This subject is addressed in detail in Chapter 9. After the earthquake, relief measures become important in determining the recovery of different people from the event. These measures relate to the types of injury sustained, the availability of medical assistance and the timing of whether, or when, a person is able to receive medical attention, and are all critical factors in the vulnerability of persons to earthquake impact (WHO 1997).

Access to resources in normal life and transition to disaster

Throughout this book it will be clear that political economy causes people to be differentiated by degrees of vulnerability. In the case of earthquakes, housing design, quality of building and the degree of maintenance are of crucial importance. In turn, these characteristics are shaped by the pattern of ownership of buildings, the level of rents and the distribution of income among urban dwellers, which interact to determine where people live and the degree of hazard associated with those locations. Later in this chapter a study of the 1985 earthquake in Mexico City will be used to illustrate the social construction of vulnerability, its spatial variation throughout the city, the importance of class, occupation, and physical type and ownership of residence (and the political conditions under which municipal building regulations were applied or circumvented). Overall, there is an urban ecology which is expressed in terms of access to resources before and after an earthquake.

Earthquakes can be viewed with the help of our Pressure and Release (PAR) model, as well as by seeking out information on vulnerability factors that derive from our Access model. Both models will be illustrated in the following case studies. The PAR model investigates the root causes and dynamic pressures that lead to unsafe conditions. In order to investigate these conditions of vulnerability with respect to earthquakes, we use a comparative analysis of four widely differing earthquake-related disasters that have occurred in the past quarter of a century: Guatemala (1976), Mexico City (1985), Kobe, Japan (1995) and Gujarat, India (2001).

Classic case studies: Guatemala and Mexico

The complexity of the social, political, economic and technical factors we have so far introduced were made apparent by two very destructive earthquakes: in Guatemala in 1976 and Mexico City in 1985. Both events have shaped subsequent international policies of seismic risk reduction.

The Guatemala earthquake, 02:58 hours, 4 February 1976

This earthquake proved to be a crucial experience (both negative and positive) for many agencies involved in disaster assistance. Some major 'aid blunders' were committed and some innovative ideas in educating small builders in the construction of earthquake-resistant (aseismic) houses were pioneered in its aftermath (Cuny 1983: 164–193). The disaster also focused attention on the vulnerability of the urban and rural poor to exploitation by landlords. It also highlighted the difficulties of certain aid agencies, which pursued assistance policies they probably now regret, or accept as important 'learning experiences'.

The earthquake killed 22,000 people living in unsafe housing in the rural highlands of Guatemala as well as within dangerous squatter settlements in Guatemala City. It left the upper and middle classes virtually unscathed. This was the first major earthquake widely recognised as having such a markedly selective impact, hence its common designation by people on the street as a 'class-quake'.

Vulnerability variations can be clearly detected in the Guatemalan case. Firstly, there was a strong ethnic as well as a class factor at work. The highland rural people who died were not only poor, but were indigenous Mayans. The dead in Guatemala City (some 1,200 people) and the 90,000 made homeless were almost exclusively concentrated in the city's slums (*Latin America* 1976). Secondly, it was exceedingly difficult for both the indigenous Mayans and urban squatters to obtain post-disaster assistance from the government.

The socio-economic forces that led to so many people living in unsafe conditions, and the political forces that controlled post-disaster aid, mirrored Guatemalan society at large (Plant 1978; EPOCA 1990). What makes Guatemala unusual is the high degree of awareness of these social weaknesses on the part of a large proportion of the population, so that post-disaster relief and rehabilitation became a political battleground. In the words of a contemporary journalist:

In this well-known fault zone the houses of the rich have been built to costly anti-earthquake specifications. Most of the poorest housing, on the other hand, is in the ravines or gorges which are highly susceptible to landslides whenever earth movements occur. The city received proportionately little aid largely because it is governed by the most radical opposition tolerated in Guatemala, the *Frente Unido de la Revolucion*, a social democratic coalition. Its leader Manual Colon Arguetta, was wounded by unknown gunmen on 29 March, the latest victim of a wave of terror attacks that has claimed 40 lives since the earthquake. One city official, Rolando Andrade Peña, was shot down two weeks after the earthquake after suggesting that homeless people should be encouraged to rebuild on unoccupied private land.

(*Latin America* 1976: 115)

In 1989, 13 years after the earthquake, one of the present authors visited Guatemala City to determine whether the people were any less vulnerable than they had been in 1976. In many ways there seemed to be more positive signs. While there were still houses on the steep slopes, they were certainly not as crowded or precarious. Many of the urban poor who lost their homes yet managed to survive immediately vacated the most dangerous slopes for flat or gently sloping sites a short distance away. This illegal 'invasion' had taken place from the day of the earthquake, and ever since the barrio has been known as '4th of February'. When survivors first 'invaded' safer sites, there were a large numbers of visiting newsmen in the city to report on the disaster and the authorities had to turn a blind eye to the influx of displaced families. Eventually (perhaps due to the sheer force of numbers linked to sustained political pressure), the occupiers were granted legal titles to the land by the government.

However there is no evidence that the builders of these houses had any knowledge of earthquake-resistant construction. So, although their sites are safer from earthquake-induced landslides, flash floods and eviction orders, their dwellings remain as dangerous. In fact, the risk of their houses collapsing may have increased significantly. When they were illegally sited they were generally built from lightweight materials, including corrugated iron sheet roofing. But when they were legalised many families began to build using heavy materials such as reinforced concrete that is potentially more harmful.

Oxfam America was one of the many NGOs closely involved in the reconstruction programmes based on co-operative activity. In 1982 they published an account of the reign of terror that ensued, including a series of interviews with local leaders.

[T]he earthquake tore open many holes in the social fabric which had already been stretched thin. The rich and those in power came out richer and the poor came out poorer, and differences and inequalities became more visible. More protest led to more repression to contain the forces of change. Those in power do not want to share the wealth ... I think this region has become the target for increased repression and violence against the population ... [since] many people in this area were very active in reconstruction efforts after the earthquake.

(Quoted in Davis and Hodson 1982: 15)

Miculax and Schramm wrote a case study of the long-term consequences of one of the Housing Education programmes in 1989, 13 years after the earthquake:

A terribly unfortunate negative consequence of these improvements in community organisation should be noted. During the 'violence' (*la violencia*) of the 1980's, individuals who had developed their personal capacities during the post-disaster relief project were seen as 'troublemakers'. Many were killed by the army and others sought exile in neighbouring countries.

(Quoted in Anderson and Woodrow 1989: 237)

Thus in Guatemala 'political vulnerability' expanded as a direct consequence of community development and leadership training specifically intended to reduce vulnerability to economic factors or seismic hazards.

The Mexico City earthquake, 07:00 hours, 19 September 1985

The impact of this disaster was very different to that in Guatemala. While there are millions of people living in the 'informal settlements' of Mexico City, in conditions very similar to the steep ravines of Guatemala City, they were not the main victims of this earthquake. Those who suffered in Mexico City were not the poorest residents, and this case reminds us that vulnerability is *not* synonymous with poverty, although the two are often strongly associated. We will need to analyse the situation in several ways in order to understand all the relationships that determined vulnerability in this complicated case.

Within greater Mexico City, a mega-city with perhaps 18–20 million residents, the long-suffering population has learned to cope with a host of natural and technological hazards. Earthquake remains a major threat and was the trigger for the disaster that led to as many as 17,000 deaths in 1985. Yet this is not the highest risk on Mexico's political agenda: in the 1990s it was air pollution, and in the early years of the twenty-first century traffic, a dispute over a new airport, water supply and drainage.

Historical influences on the physical environment

The seismic hazards facing Mexico City can be mapped with some accuracy in terms of their frequency, severity of impact, damage patterns, type of ground motion and location in relation to topography and soil conditions.

However, even such physical factors can be affected by human intervention. Tobriner has written an account of the way Mexico City has developed over the past six centuries on a site that could hardly be more dangerous. Over this period it has been at risk from floods, soil shrinkage, volcanic activity and earthquake impact. He observes that it is 'one of the great ironies of urban history that Mexico City, perhaps the largest city in the world, stands on one of the planet's most unstable soils' (Tobriner 1988: 469–479). Figure 8.1 shows the relationship between building damage and the bed of the old Lake Texcoco, with its legacy of these dangerous conditions. The explanation of why this occurred is rooted in the history of the site from the thirteenth century, when the Aztecs made it their capital, Tenochtitlan. As Oliver-Smith points out in his account of the '500 hundred year earthquake' in Peru, the root causes of vulnerability sometimes lie in the very remote past.[1] When the Spanish arrived they found it a well-adapted settlement. But conquest and its symbolic needs required them to destroy it and substitute a new city that required the draining of Lake Texcoco.

However the site of Tenochtitlan suffered from four natural hazards: volcanic eruptions, earthquakes, drought and severe flooding. The historic centre of Mexico City now sits on a lake bed, with an alluvial subsoil up to 60 m deep. In the 1985 earthquake this soil behaved like a liquid, with massive ground shaking causing damage almost exclusively within the area of the original lake bed. This is the tragic legacy of a major city set on an unstable site that has its roots in the power of Aztec kings and colonial rulers a full four and a half centuries ago. In Figure 8.2, this pre-conditioning factor is depicted as a 'root cause' of vulnerability.

In the years following the 1985 disaster the government of Mexico City took a bold decision to stop any future high-rise buildings from being constructed in the part of the city situated over the old lake bed, based on the evidence from the 1985 severe amplification of earthquake shaking in this area (Adams and Spence 1999: 54). There are, however, many remaining buildings at risk in this unstable zone.

Buildings at risk

In an assessment of the exposure of buildings to seismic risk, a survey was made of about 20,000 buildings in the historic centre. Seventeen factors were considered in the survey, including such matters as levels of maintenance, the shape of buildings, the 'hammer effect' of one building knocking into another during an earthquake, building height and type of construction (Aysan et al. 1989). A large number of these buildings could collapse in any future large earthquake.

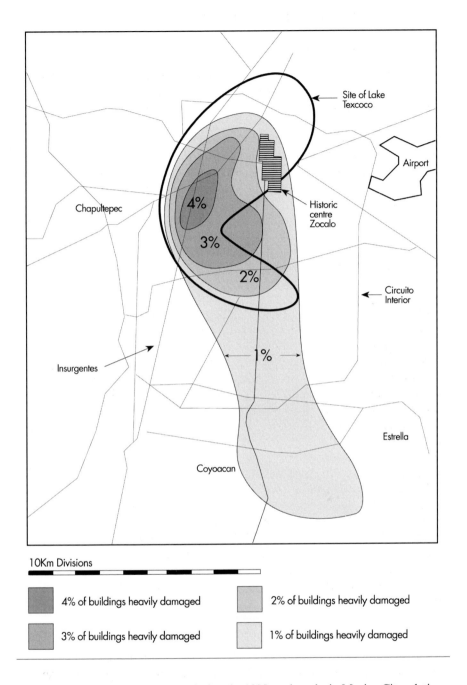

Figure 8.1 Distribution of damage during the 1985 earthquake in Mexico City relative
to the lake bed

Source: Adapted from Aysan et al. (1989) and Coburn and Spence (1993), p.211, (2002), p.256,
Figure 7.12

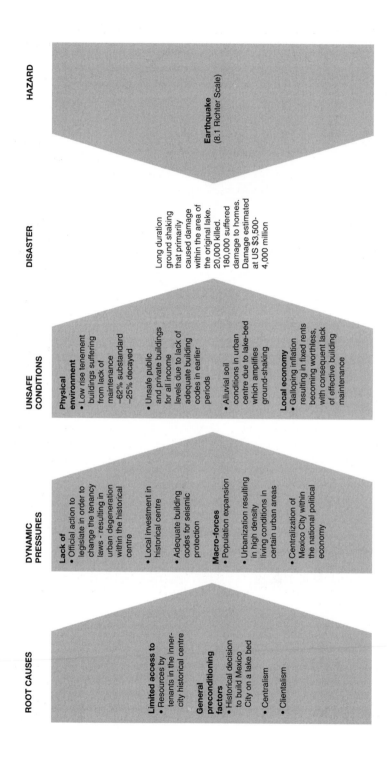

ROOT CAUSES

Limited access to
• Resources by tenants in the inner-city historical centre

General preconditioning factors
• Historical decision to build Mexico City on a lake bed
• Centralism
• Clientalism

DYNAMIC PRESSURES

Lack of
• Official action to legislate in order to change the tenancy laws - resulting in urban degeneration within the historical centre
• Local investment in historical centre
• Adequate building codes for seismic protection

Macro-forces
• Population expansion
• Urbanization resulting in high density living conditions in certain urban areas
• Centralization of Mexico City within the national political economy

UNSAFE CONDITIONS

Physical environment
• Low rise tenement buildings suffering from lack of maintenance
  −62% substandard
  −25% decayed
• Unsafe public and private buildings for all income levels due to lack of adequate building codes in earlier periods
• Alluvial soil conditions in urban centre due to lake-bed which amplifies ground-shaking

Local economy
• Galloping inflation resulting in fixed rents becoming worthless, with consequent lack of effective building maintenance

DISASTER

Long duration ground shaking that primarily caused damage within the area of the original lake. 20,000 killed. 180,000 suffered damage to homes. Damage estimated at US $3,500-4,000 million

HAZARD

Earthquake
(8.1 Richter Scale)

Figure 8.2 'Pressures' that result in disasters: Mexico City earthquake, 19 September 1985

Detailed analysis of earthquake damage has revealed that the primary factors causing damage to buildings related to their siting and their height. It was found that on sites within the lake bed, rigid structures (such as stone masonry buildings) generally performed better than flexible ones (such as reinforced concrete structures). The height of the building was an even more significant factor in creating vulnerability for people living in, or in close proximity to, dangerous buildings. Medium- to high-rise buildings between 6 and 20 stories were worst affected, with particularly severe damage occurring to buildings between 9 and 11 stories high. At this height they were prone to resonate in harmony with the low frequency waves that emanated from the epicentre 230 miles away (Degg 1989).

One major factor that has still to be considered in mapping seismic risk concerns the age of reinforced concrete buildings. Those erected between 1925 and 1942 were built to very high standards. Then, from 1942 to 1964 quality fell, the result of a building boom and a consequent lack of supervision of construction methods. After 1964, seismic codes were applied and the quality improved (Ambraseys 1988).

A very serious characteristic of the 1985 disaster was the destruction and damage to public buildings, including those with high and constant levels of occupancy such as hotels and hospitals. Five hospitals collapsed and 22 were seriously damaged, which resulted in a 28 per cent loss of public hospital capacity at precisely the time when it was most needed – 1,619 people lost their lives in the collapse of just six buildings (Kreimer and Echeverria 1991).

In addition to the failure of large high-rises, another group of buildings suffered severe damage, but this was not as well publicised. These were smaller structures, with high levels of occupancy, of mixed commercial and domestic use. Analysis of casualty statistics indicates the rather obvious fact that during day-time earthquakes people manage to escape from low-rise buildings more easily than from high-rise ones (Aysan et al. 1989). In the Mexico City case it is possible to see some relationship between underlying causes and unsafe conditions in the older tenements, where there had been a failure to maintain what tend to be severely overcrowded buildings. In some instances there was evidence of a failure to supervise building construction adequately. The question of the level of building maintenance relates to patterns of ownership and occupation as well as to the role of the state in enforcing maintenance standards. These issues provide a link with the next layer of the study, that of human vulnerability.

People at risk

The discussion above involves issues that are well-established subjects for those who engage in the physical mapping of hazards. In contrast to physical hazard mapping, human vulnerability analysis focuses on social patterns and institutions (termed 'social relations' and 'structures of domination' in the Access model, Figure 3.1), society-wide and intra-household social relations, and economic activity (gender and age relations are particularly important as noted earlier), and the psychology of risk.

Given the distribution in time and space of the earthquake event (hazard) itself and the distribution just described of historically remote 'causes' and unsafe buildings, what specific mechanisms were at work in 1985 that placed certain people in those unsafe buildings at the critical moment? These include density of population (a function of location factors such as site of employment, land prices and rents); the ownership of buildings relative to their maintenance; patterns of building use (seen in terms of both space and time); people's own perception of risk; cultural values such as a desire to remain in one's natal neighbourhood; and the existence of local institutions that could play a key role in post-disaster recovery.

The local economy at risk

The effect of the earthquake can be considered in two overlapping ways: the destruction of property and the impact on lives. In the Mexico City earthquake, two distinct categories of buildings collapsed or were damaged. Both were both constructed on alluvial soils that formed the bed of a long-vanished lake. The first category involved people who died or were injured in high-rise buildings, including a hotel and a number of hospitals. These casualties came from all levels of Mexican society. In contrast, the second group was the predominately low-income residents of nineteenth-century low-rise tenements. As we pointed out in Part I, we are concerned to define and to underscore the relevance and practicability of vulnerability analysis by pointing out how it often must go beyond simple criteria such as income and status. The losses in the first category included some victims who would not be considered vulnerable in terms of their income. However, vulnerability is closely related to the low-income status of tenement residents.

The earthquake occurred at 07:00 hours on 19 September 1985 when most people were on their way to work. For those on foot, the hazard was falling masonry crashing onto pavements, but for the thousands inside metro trains or motor vehicles the immediate environment was protective. Estimates of the number killed range from 5,000 to 17,000. The vast majority of those who died were inside medium- to high-rise buildings, in the central area of the city. Some 12,700 buildings were affected, 65 per cent of which were residential. The housing for 180,000 people was damaged, and 50,000 needed temporary accommodation (Kreimer and Echeverria 1991). Since the damage affected some high-investment buildings, the financial losses were enormous, estimated at $4 billion (ibid.). The reinsurance industry at that time assessed the earthquake as one of the three most disastrous of this century, the others being the San Francisco and Tokyo earthquakes of 1906 and 1923 respectively (Degg 1989).

Such financial loss dwarfs the dollar value of dislocated livelihoods. Yet for those who relied on work in the 1,200 small industrial workshops destroyed, the cost was great. Once more it is clear that issues of recovery and rehabilitation cannot be separated from the profile of vulnerability. Are those workers still unemployed, or have they found an alternative? Did the earthquake begin a spiral into poverty for those households? The number of casualties varied in relation to their location (i.e. in a building on the alluvial soils) and height of the building. The timing of the earthquake was even more critical in determining the number who died or were injured. Had it occurred just three hours earlier, when people were asleep, there would have been much higher casualties (but the same property losses).

An annotated Access model of this earthquake is provided in Figure 8.3, which graphically summarises the commentary above, and provides a holistic and causally coherent explanatory model of the earthquake.

Starting in the top left-hand corner of the Access model, the earthquake is mapped in space and time. The proportion of damaged buildings (Box B), along with the distribution of different types of building with a varying resistance to collapse (Box C), together constitute the main determinants of damage. Box A shows the severity and timing of previous earthquakes of the twentieth century (the 1985 quake was the most severe since 1908). The decline in their frequency may have affected people's attitude to the need to prepare for them. Box D shows the distribution of damaged buildings in relation to the sub-soil of the drained lake bed of Lake Texcoco. The 'net susceptibility' to earthquakes (in terms of building design to resist earthquakes) is illustrated in three different zones in Box D, while the actual impact of the earthquake (in terms of the percentage of buildings damaged) is shown in Box E (a simplified version of Figure 8.1), to the right of the zigzag line that represents the disaster. One of the key questions was 'who was where when?'. As the quake occurred at 0700 many people were going to work, and fortunately not asleep in buildings which subsequently collapsed. Also, the spatial structure and urban ecology of the city shaped the pattern of commuting to work and of travel and location of different groups of people. Many areas were characterised by very high residential densities which put large numbers at high risk during an earthquake, and this contributed to the very high death toll.

The impact of the earthquake on social relations (Box F) and structures of domination (Box G) was significant both immediately and subsequently, but probably much less so than for an earthquake in a rural area. It occurred in the capital city itself, which, as a classic example of a primate city, dominated the rest of the country politically, economically and demographically. The key social relations determining earthquake risk were technical, political and financial: these all had a role in shaping of the construction of public and private buildings. As will be seen later in the Access model, these were modified by public action, although they were not radically altered. Also,

287

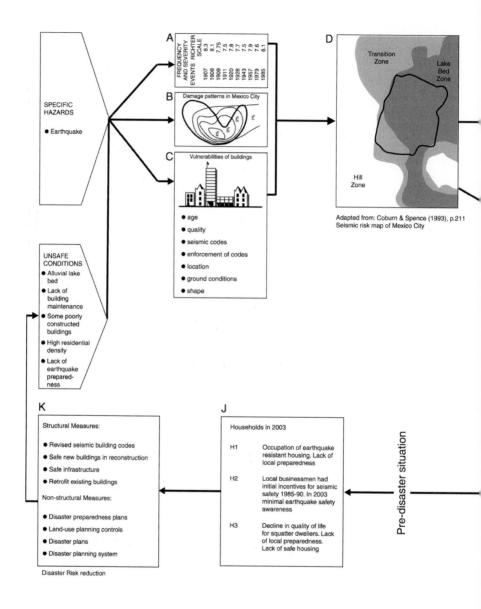

A

FREQUENCY AND SEVERITY	EVENTS	RICHTER SCALE
	1907	8.3
	1908	8.1
	1909	7.75
	1911	7.8
	1920	7.7
	1928	7.5
	1943	7.9
	1957	7.6
	1979	7.6
	1985	8.1

B
Damage patterns in Mexico City

4% 3% 2% 1%

C
Vulnerabilities of buildings

- age
- quality
- seismic codes
- enforcement of codes
- location
- ground conditions
- shape

D
Transition Zone
Lake Bed Zone
Hill Zone

Adapted from: Coburn & Spence (1993), p.211
Seismic risk map of Mexico City

SPECIFIC HAZARDS

- Earthquake

UNSAFE CONDITIONS

- Alluvial lake bed
- Lack of building maintenance
- Some poorly constructed buildings
- High residential density
- Lack of earthquake prepared-ness

K

Structural Measures:

- Revised seismic building codes
- Safe new buildings in reconstruction
- Safe infrastructure
- Retrofit existing buildings

Non-structural Measures:

- Disaster preparedness plans
- Land-use planning controls
- Disaster plans
- Disaster planning system

Disaster Risk reduction

J

Households in 2003

H1   Occupation of earthquake resistant housing. Lack of local preparedness

H2   Local businessmen had initial incentives for seismic safety 1985-90. In 2003 minimal earthquake safety awareness

H3   Decline in quality of life for squatter dwellers. Lack of local preparedness. Lack of safe housing

Pre-disaster situation

Figure 8.3 Access, vulnerability and recovery from the Mexico City earthquake

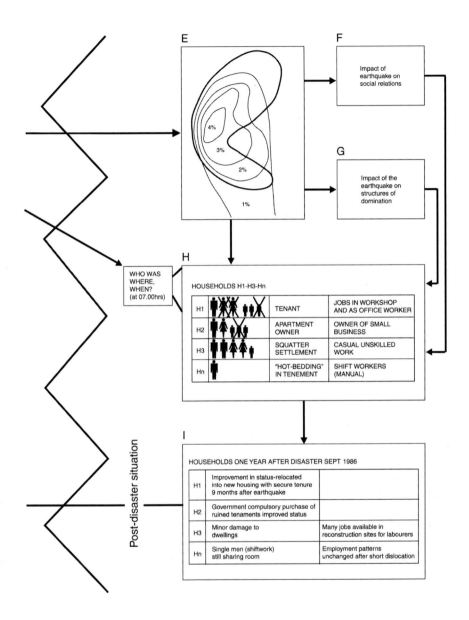

E

F

Impact of
earthquake on
social relations

G

Impact of the
earthquake on
structures of
domination

WHO WAS
WHERE,
WHEN?
(at 07.00hrs)

H

HOUSEHOLDS H1-H3-Hn

H1		TENANT	JOBS IN WORKSHOP AND AS OFFICE WORKER
H2		APARTMENT OWNER	OWNER OF SMALL BUSINESS
H3		SQUATTER SETTLEMENT	CASUAL UNSKILLED WORK
Hn		"HOT-BEDDING" IN TENEMENT	SHIFT WORKERS (MANUAL)

I

HOUSEHOLDS ONE YEAR AFTER DISASTER SEPT 1986

H1	Improvement in status-relocated into new housing with secure tenure 9 months after earthquake	
H2	Government compulsory purchase of ruined tenaments improved status	
H3	Minor damage to dwellings	Many jobs available in reconstruction sites for labourers
Hn	Single men (shiftwork) still sharing room	Employment patterns unchanged after short dislocation

Post-disaster situation

Figure 8.3 continued

Mexico City is highly inegalitarian, with very poorly maintained and high-density residential areas. Hyperinflation meant that in properties with fixed rents, landlords' incomes had become virtually worthless; they therefore neglected maintenance, particularly in the historic city centre. Squatter settlements, 'hot bedding' (multiple occupation of a single bed in shifts) and general overcrowding were (and still are) commonplace.

The impacts on three stereotypical households are illustrated in Box H (the pre-earthquake situation) and Box I (after the earthquake). The first household is a tenant in a highly damage-prone tenement, whose main income is as an office cleaner; the second. a small business owner living in a privately owned apartment in a high-rise building; the third lives in a squatter settlement; while the fourth is not really a 'household' at all but a 'hot bedder' in very cheap and overcrowded accommodation. In all cases, there was a lack of individual and community preparedness in the event of a disaster. The impacts included mortality and temporary disruption of liveli-hoods (office cleaning and casual and unskilled work stopped in the cases of the poorer households). However, as Box I shows, the longer-term impacts on livelihoods were by no means universally adverse, and some job opportu-nities increased. Moving from Box I to Box J the model shifts from the post-disaster situation and recovery to the pre-disaster conditions of the next disaster. Box J shows the notional situation of households 1, 2 and 3 in 2003, and what has (and has not) happened in their living conditions. Box K then summarises the official measures which were taken to reduce vulnera-bility, which in turn affect the initial 'unsafe conditions' that form the basis for the next iteration of the model, and for future disasters.

Figure 8.4 is an attempt to identify the factors which could be adopted by the Mexican authorities to release the pressures that have created unsafe conditions in Mexico City in the past (Coburn and Spence 1993: 130; Gomez 1991: 56–57; Echeverria 1991: 60–1). The model addresses vulnera-bility indirectly and directly. Indirect measures include reducing the size of Mexico City. Urbanisation is addressed by improving rural economic oppor-tunities, lessening the migration to the city, and by decentralising some of the federal government's functions (and hence employment opportunities) to other cities in Mexico. In this way one of the most important dynamic pres-sures that translates root causes into unsafe conditions is relieved.

In addition, unsafe conditions are addressed directly by improved aseismic building codes and their enforcement, strengthening existing struc-tures, and reducing the densities in certain weaker structures by changing patterns of use. These steps, none of them unthinkable in contemporary Mexico – although some more difficult politically and more expensive than others – complement a programme of improved disaster preparedness plan-ning. However, by the mid-1990s, a mere 10 years after this major disaster, a marked decline in 'political will' was already evident, with concerns for seismic safety having been displaced from the political agenda by other more

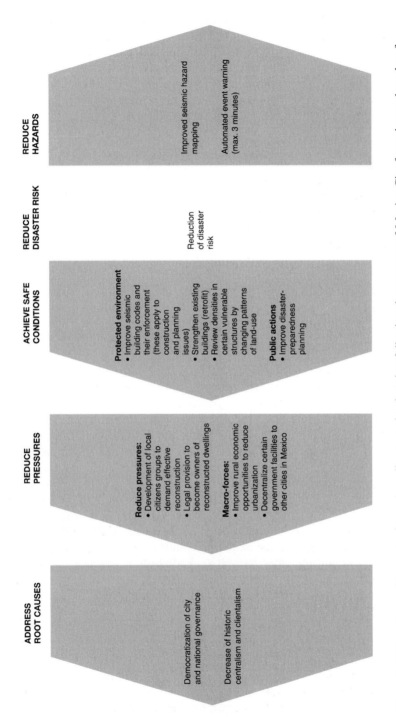

Figure 8.4 The release of 'pressures' to reduce disasters: the situation following the recovery of Mexico City from the earthquake of 19 September 1985

ADDRESS ROOT CAUSES

Democratization of city and national governance

Decrease of historic centralism and clientalism

REDUCE PRESSURES

Reduce pressures:
- Development of local citizens groups to demand effective reconstruction
- Legal provision to become owners of reconstructed dwellings

Macro-forces:
- Improve rural economic opportunities to reduce urbanization
- Decentralize certain government facilities to other cities in Mexico

ACHIEVE SAFE CONDITIONS

Protected environment
- Improve seismic building codes and their enforcement (these apply to construction and planning issues)
- Strengthen existing buildings (retrofit)
- Review densities in certain vulnerable structures by changing patterns of land-use

Public actions
- Improve disaster-preparedness planning

REDUCE DISASTER RISK

Reduction of disaster risk

REDUCE HAZARDS

Improved seismic hazard mapping

Automated event warning (max. 3 minutes)

pressing political and environmental problems.[2] In the early years of the twenty-first century, Mexico City was governed for the first time by officials from a party in opposition to the PRI that had dominated national and local politics for nearly 80 years. However, despite some changes in rhetoric and considerable improvements in hazard mapping, engineering knowledge and communication technology, Mexico's huge mega-city may not be any better prepared for its next earthquake (Puente 1999).

Recent case studies

We have selected two additional case studies from the many disasters which have occurred since Mexico: the Kobe and Gujarat earthquakes. The choice was not easy, since other earthquake disasters also raise important issues and bring insights to the analysis of social vulnerability (Davidson et al. 2000; Yamakazi et al. 1996). One such example is the 1994 Northridge earthquake in a suburb of Los Angeles. This was one of the costliest disasters in the history of the USA and resulted in 681,000 people applying for financial assistance. Despite the record physical and financial losses, only 57 people died, a testimony to the effectiveness of seismic risk-reduction measures in California in protecting lives (Bolin and Stanford 1998a, 1998b). Other useful data derived from the Northridge case concern how difficult it was for low-income Hispanic people to gain access to the resources needed for recovery (Bolin and Stanford 1998b). We will not treat this case in depth in this chapter however, but will return to it when we address the question of recovery in Chapter 9.

Another such example is the Latur (India) earthquake of 1993. This raises the point that primary victims can be the wealthier people, who in this case were highly vulnerable because they lived in non-reinforced heavy masonry dwellings whilst the majority of the rural poor occupied safer, lightweight structures (IFRC 1994: 123).

Other lessons can be learnt from the experience of Turkey, where the people suffered two devastating earthquakes in 1999 that resulted in the highest death toll in any earthquake in Turkey since 1939. In this case the principle lessons concern the acute hazardousness of the building stock in the rapidly urbanising cities, as well as inadequate emergency management by the authorities (IFRC 2000b: 26–27; ISDR 2001). Finally, the 2001 El Salvador earthquake reveals extensive data concerning the 'root causes' for our PAR model. These involve the economic and political marginality of the victims as well as the role of environmental degradation and the long-term legacy of the civil war in creating vulnerability (Wisner 2001f).

Despite the obvious value of a more detailed examination of any of these examples, or of others important because of scale and the wide significance of the issues raised, we have chosen here to include as new case studies the earthquake disasters in Kobe (1995) and Gujarat (2001).

The Kobe earthquake, 05:46 hours, 17 January 1995

The Kobe (Japan) earthquake is of great importance because it exposed complacency and a false sense of preparedness in the minds, policies and practices of public officials. For many in the field it was an important lesson about the world's seismic zones: before this disaster, officials in many highly-developed urban environments regarded major earthquakes as a problem mainly for developing societies, and assumed that their own seismic protection measures would minimise their losses. How could an earthquake of 7.2 Richter scale, occurring in a prosperous city in the second largest economy in the world, kill 6,279 persons, injure 34,900 and destroy or damage 136,000 buildings? In the end, 310,000 persons lost their homes, and there were financial losses of $147 billion (not including the economic effects from loss of lives, business interruption and loss of production). This made it the most severe economic loss for any disaster in world history. It came as a very big surprise for planners and policy makers: if this could happen in Kobe, what could be expected elsewhere (Mitchell 1999b: 1; Swiss Re 1995; EQE 1995a)?

This critical question, which has been asked repeatedly since January 1995, concerns the fragility of urban systems and infrastructure in spite of the existence and enforcement of disaster planning and building codes. How secure could other similar cities be – Tokyo, Vancouver, San Francisco, Lisbon, Wellington and other major conurbations in the seismic zones of wealthy countries? In Kobe, vulnerability was clearly evident in four critical areas: social, physical, economic and in governmental disaster protection measures.

Social vulnerability: high-risk social groups

The disaster revealed important information concerning the creation of particular patterns of social vulnerability, with those over 60 accounting for 58 per cent of all deaths. Females were also at greater risk, constituting 59.3 per cent of deaths, with more killed in every age group.[3] Although temporary housing was provided in a lavish and efficient manner, there is evidence of deep social distress on the part of the elderly population of lonely and possibly traumatised people. One citizen group claims that 2,900 deaths were attributable either to suicide or neglect within temporary housing (Bishop 1998: 50).

However, there was another key vulnerability factor that failed to surface in official reporting by the government; it was also ignored in most NGO reports. In Kobe (as in Japan's other major cities) there is a minority underclass that is particularly vulnerable. It is subject to official neglect and economic deprivation (its workers receive lower wages, and enjoy fewer rights). This minority is composed of Korean-Japanese workers and foreign workers (legal or illegal migrants from the Philippines, North Korea and

China). Within the most severely affected wards of Kobe City there were 130,000 foreign and migrant workers. Most were paid low wages in small businesses that were damaged or destroyed by the earthquake (IFRC 1996). This closure of so many such businesses made recovery even more difficult for them.

In addition, concentrated in the Osaka and Kobe regions, there is a group of people known as the *Burakumin* ('village people') who are ethnically and genealogically Japanese but who have been discriminated against for centuries. Their status, and the treatment they receive, resembles that of the outcastes in India: the *Dalits* or 'untouchables'.[4] Ancestors of the *Buraku* had been low-caste butchers and leather workers, but centuries later the associations with 'unclean' professions of the medieval caste system have persisted (Laidlaw 2001). Some Japanese parents even hire specialist investigation firms to make sure a prospective son- or daughter-in-law has no *Buraku* ancestors. Laidlaw (2002) has noted the effect of this social stigma and economic deprivation on their vulnerability in earlier disasters:

> The *Buraku* were generally situated on river banks, mountainsides and undesirable areas. Because of this they were the first to suffer during natural disasters, and also the most likely to be subject to heavy damage.

In Kobe City there was a concentration of *Buraku* in Nagata Ward, one of the poorest in the city, where they were employed in leather dyeing industries and in plastic shoe manufacture (Velasquez 1995; Ukai 1996). It was here that the most devastating fires occurred after the earthquake (EQE 1995b). In Nagata Ward alone there were 919 deaths, 15,521 collapsed houses and 4,759 burned houses (Kobe City 1995). The other two wards with the highest numbers of deaths, Nada (933) and Higashi Nada (1,471), also have concentrations of *Burakumin*, as can be seen by the presence of *Buraku* cultural centres in Figure 8.5. More than half the deaths (53 per cent) in this earthquake and the subsequent fires took place in these three wards. The intensity and rapid spread of the fire was due to the presence of inflammable chemicals used in dyeing and shoe making, in combination with the concentration of sub-standard timber buildings. Laidlaw (2002) writes of these three wards:

> The entire wards themselves were not actually *Buraku* areas, but there were indeed *Buraku* areas within them. The *Buraku* areas also appeared to be in the areas that suffered the most damage, which was probably partly due to the fact that *Buraku* housing, on average, is sub-standard when compared to non-*Buraku* housing and was therefore more likely to collapse.

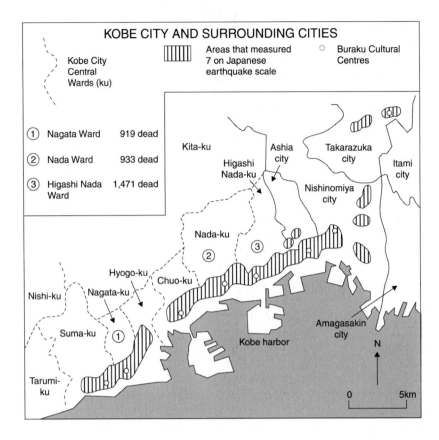

Figure 8.5 Zones of greatest damage in Kobe, 1995

Source: Adapted from Rinpokan Renraku Kyogikai (1995)

Unsafe dwellings

As often noted in this chapter, the vulnerability of people to earthquakes is inseparable from the stability of the buildings they occupy. The carefully documented statistics of damage and casualties in Kobe provides further convincing evidence. Of the victims, 86 per cent died in their homes; 73 per cent of these deaths were due to suffocation or being crushed when 54,949 buildings collapsed, while 9 per cent died from burns (a further 7,061 buildings were destroyed by post-earthquake fires). In addition to those destroyed, approximately 88,000 buildings were damaged. The majority of the buildings that collapsed were two-storey wooden dwellings with heavy tiled roofs (IFRC1996: 69; Nagle 1998: 19, Adams and Spence 1999: 53)

Professor Murosaki, (Kobe University Architecture Department) has described the dangers of these wooden dwellings in relation to a major gap in higher education syllabuses on earthquake engineering:

295

There is something seriously wrong with the way we approach earthquake engineering. About 100 universities in Japan teach earthquake engineering and architecture, but I know of only one or two that teach anything about wooden house design. Yet 80 per cent of Japanese people live in wooden houses. We teach our students how to build office blocks and hotels, where the prestige and money is. Maybe we will see some change now but I am not hopeful.

(Quoted in IFRC 1996: 69)

Most of these wooden-framed houses were built as part of a rapid reconstruction programme after the Second World War, when the pressure for cheap, rapid construction caused standards to suffer and meant inadequate earthquake-resistance measures. They also incorporated a design fault of very heavy tiled roofs, a highly undesirable feature given the severe horizontal shaking common during earthquakes (Swiss Re 1995).

Similar ill-conceived recovery actions after other disasters have shown how they can inadvertently sow the seeds of the next. After the San Francisco earthquake of 1906, rubble from the clearing up was used as landfill in the bay to expand the city.[5] This unstable reclaimed land was built on by opportunistic developers who capitalised on the demand for housing after the earthquake. Most of the damage to dwellings in the Loma Prieta earthquake of 1989 occurred because of severe ground-shaking of these unstable soils (Hewitt 1997: 216). Much of Yokohama has also been built on landfill from debris following the 1923 Tokyo earthquake, and experts predict severe damage in these areas when the next Tokyo Bay earthquake occurs.

In Kobe a surprising 85 per cent of all schools were damaged. Had the earthquake occurred three hours later, when the children were inside the school buildings, the earthquake impact would have been devastating, with the probable loss of a large number of children.

Economic vulnerability

The Kobe earthquake resulted in larger financial and economic losses than any disaster in world history, and only 7 per cent of house owners or businesses had earthquake damage insurance cover. Its economic impact was felt throughout Japan and across the world. The viability and reliability of the 'just-in time' system of logistical delivery to industry and commerce was thrown into question by the interruption in the supply of computer and automobile parts as a result of the collapse of key road and rail networks in Kobe (Munich Re 1999).

Livelihoods were lost or severely disrupted as a result of damage to key economic assets. For example, almost all the container berths and wharves in Japan's largest container port were unusable (Kobe was the sixth-largest cargo port in the world). By 2002, seven years after the disaster, Kobe's port facilities have been completely rebuilt, but trade was reduced by more than 20 per cent because of the disruption of the disaster and the time taken to reconstruct the port facilities. As already noted, the manufacture of plastic shoes has ceased, with 80 per cent of the factories damaged, mainly in Nagata Ward where the fires were extensive (Nagle 1998: 20).

The failure of disaster preparedness measures

The earthquake exposed major gaps in the overall system of disaster mitigation, preparedness and emergency management in both central and local government. These included the omission or inadequacy of the following vital protection measures: building codes, public awareness of seismic risk, preparedness systems, detailed disaster plans, emergency operating procedures, information management systems, damage and needs assessment, physical provision for emergency management (such as fire-fighting), communications and staff competencies (Tierney and Goltz 1997; Heath 1995). It is important to summarise other telling observations by Tierney and Goltz (1997).

- Kobe municipality was incapable of doing a rapid impact assessment: only 20 per cent of municipal employees were available to work after the earthquake, the city's emergency radio system was down, the phones were down, ground transportation was impossible.
- Lacking accurate field information the Prime Minister's office was slow to make an assessment.
- There was no plan for a disaster of this scale, especially no plan for coping with so many homeless people or, very soon, so many volunteers.
- There was little planning for links between municipalities; each had planned in isolation.
- There was a false sense of safety because of the engineering used in buildings and infrastructure.
- By comparison, the Northridge earthquake in Los Angeles in 1994 was a small event, and the USA should not feel complacent. A large earthquake in the Oakland area (across the bay from San Francisco) or Memphis or Charleston (South Carolina) could produce similar levels of damage; furthermore, the USA has not seen this kind of damage since 1906 in San Francisco – even hurricane Andrew in 1991 did not cause the equivalent level of damage to Miami.

Even though a disaster of this scale would have stretched the capacity of the best-prepared city, these failures came as a profound shock to Japanese citizens who believed that they had a secure system in place that would protect them in the event of an earthquake. Tatou Takayama, commentator for the Japanese newspaper, *Yomiuri Shimbun*, wrote:

The response to the earthquake of what many call the faceless and gigantic bureaucracy that controls Japan was dictated by territorialism, passivism and the inclination to follow precedent in times of emergency.

(Quoted in Frampton et al. 1998: 56–57)

The processes by which vulnerability was generated in Kobe are summarised in Box 8.1, which follows similar headings to those used in the PAR model.

Box 8.1: Progression of vulnerability – the Kobe earthquake

ROOT CAUSES
- Prejudice towards ethnic minorities and social groups.
- Unequal distribution of economic power.
- Belief in modernisation and science.

DYNAMIC PRESSURES
Lack of:
- awareness of the severity of earthquake risk due to absence of a major earthquake in Japan since 1923;
- awareness that the Kobe region was as vulnerable to earthquakes as other regions of Japan;
- interest by Japanese earthquake engineers and architects in developing safety for non-engineered wooden structures (due to lack of financial returns);
- earthquake resistance in poor quality housing built cheaply and quickly during reconstruction following damage sustained during the Second World War;
- equitable policies and practices towards the Korean minority and the *Buraku* communities resulting in their poverty and sub-standard living conditions.

Macro-forces:
- Japanese economic strength acting as a powerful magnet for work opportunities resulting in urbanisation due to extensive work opportunities in Kobe industries and port facilities;
- an ageing population

UNSAFE CONDITIONS
Physical environment:
- unsafe, non-engineered dwellings due to design faults and lack of maintenance;
- unsafe lifeline resources (bridges, hospitals, schools, basic services);
- severely overcrowded high-density urban areas in certain city wards;
- unsafe infrastructure due to lack of knowledge concerning certain engineering practices in relation to seismic loading.

Box 8.1 continued

Local economy:
- unsafe conditions in leather and plastic shoe factories;
- little use of home owners' insurance.

Social relations:
- lack of mobility of elderly persons hampering escape from buildings during earthquake ground-shaking
- poverty and sub-standard living conditions for the Korean minority and the *Buraku* communities.

Public actions and institutions:
- lack of effective disaster planning;
- lack of co-ordinated action;
- lack of disaster preparedness at all levels.

HAZARDS

Primary impact:
- earthquake ground-shaking.

Secondary impact:
- fires.

Community resilience

In contrast to manifest governmental weaknesses exposed by the disaster, the earthquake revealed the strength of social mechanisms and institutional capacities. The authorities registered 342,000 evacuees in 1,153 evacuation centres on the day of peak occupancy (23 January). The vast majority of these centres were schools. Interviewing five of the headmasters who received this influx, all regarded it as their responsibility as 'community leaders' to open their doors to the crowds of frightened, disoriented and predominately elderly people seeking shelter (Davis 2001). One likened his role to that of a benevolent priest in a Latin American disaster setting, stating that in Japanese society the teacher had similar social prestige to that of priests in countries where there were strong religious affiliations, and that with this position came far-reaching community responsibilities which all Japanese teachers readily accepted.

One important feature of this use of the schools was the role of the temporary residents, who continued to occupy the schools when (within two weeks) the children returned. Many were able to assist in managing the centres as well as supporting the teaching of children (and adults). This period when the buildings served the dual functions of school and shelter persisted for about three months, until temporary housing had been constructed (Tokuyama 1995; Sugiman 2003; Atsumi and Suzuki 2003).

Further evidence of solidarity was seen in the spontaneous response of more than 630,000 volunteers who descended on Kobe from all over Japan, along with many Japanese students who took leave of absence from their studies in the USA and elsewhere to come and provide support. Throughout Japan, schools sent teachers to assist in the evacuation centres, sharing both welfare and teaching responsibilities, thus requiring colleagues in their normal schools to cover their teaching responsibilities (Tierney and Goltz 1997).

Many of the people made homeless left Kobe voluntarily to find shelter with friends or family in Osaka and elsewhere. The number is unknown, but is likely to be in the region of 100,000 (or about a third of all those made homeless). This vital 'coping mechanism', which took place without any support from the authorities, spread the responsibility of caring for displaced families throughout the country and thus provided considerable relief for an already over-stretched government (Davis 2001).

Gujarat earthquake, 08:45 hours, 26 January 2001

This earthquake has been described as the most damaging in India since that of Calcutta in 1737 (in which 300,000 were killed). It measured 6.9 Richter scale (about 30 times more powerful than the Latur earthquake of 1993 that registered 6.4) and affected a very large area and about 30 per cent of the state's population (of 50 million). Gujarat is located in a Zone V, the most serious seismic classification for India; however, this risk had largely been ignored since the last significant earthquake was in 1956 and thus a rather distant memory. The only earthquake of comparable power to that of 2001 occurred in 1819, and is believed to have been about 7.7 Richter. It caused 2,000 deaths in the Rann of Kutch (north Gujarat) which was a largely uninhabited salt marsh at that time (UNDP 2001a).

Gujarat has recently had a troubled disaster history, with a severe cyclone in 1998 causing 3,000 deaths and widespread property damage on its coast (see Chapter 7). It has also suffered a four-year drought, starting in 1997, during which the Government of Gujarat spent $128 million in drought relief works in Kutch. In 2001 the rains returned to give much needed relief to the rural communities, but this was compounded later by the earthquake devastation (UNDP 2001a; Wisner 2001g).

Social vulnerability: high-risk groups

The disaster resulted in the deaths of 20,000 people (DEC 2001a, 2001b) and hundreds of thousands of injured. Unlike many previous Indian disasters that have caused most suffering to the landless and marginalised poor, the victims in this earthquake were drawn from all strata of society. Like a volcanic eruption which indiscriminately affects all within its range, there were extensive casualties around the epicentre (near the town of Bhuj), without regard to whether people were rich, middle class or poor, high or low caste. However, with regard to the *recovery* programme the normal post-disaster pattern prevailed, as the rich knew how to operate the system to their advantage. This has resulted in social polarisation with regard to the options presented to villages either for relocation or financial compensation.

Rohit Jigyasu, a planning student undertaking field study in Gujarat, noted in 2001 that in response to these options:

> [t]he socially and economically powerful castes got together and purchased their own land and, in this way, decided to get relocated; while the weaker groups were left with no option but to stay back. This is happening in nearly all the villages. In many cases, a single village has got split into as many as 4 parts, at safe distance to each other. This is very serious as physical segregation will further deepen the polarisation.
>
> (Jigyasu 2001)

He then observed that in many cases due to 'good political connections the powerful castes have even managed to attract infrastructure and investment, while the poor and the marginalised are now left as "abandoned hamlets" devoid of even basic facilities'.

Unsafe buildings

Over 1.2 million homes were affected in the disaster. This total included 150,000 houses that were subsequently demolished and 750,000 that needed repairs. The historic town of Bhuj required total reconstruction, and three villages required major reconstruction efforts. Critical 'lifelines', or facilities and structures which were particularly important in dealing with the emergency, fared badly. Three hospitals and 5,000 health units were damaged or destroyed. In addition more than 50,000 school rooms were either damaged or destroyed, including 992 primary schools that collapsed completely (one killing 37 children; another collapsed later as a result of damage and killed 400 children) (GDMSA 2002). In Ahmedabad (the state capital and a city of 4.5 million) it was the urban middle class who suffered, especially those living in the 86 apartment buildings that collapsed (Rutten and Engelshoven 2001).

In urban areas there had been growth of cheap, three- to four-storey apartment buildings where construction quality was poor, with insufficient and incorrectly placed steel reinforcements, and adulterated cement. However, the primary building failures in Gujarat occurred when non-engineered dwellings collapsed in large numbers, frequently killing or injuring their occupants. A local engineer involved in a disaster recovery NGO in Ahmedabad argues that housing must bear the brunt for destruction and damage, especially the non-engineered housing that people construct themselves, generally with the help of the local building artisans and with very little technological support from outside (Desai 2002).

Economic vulnerability

Over 10,000 small and medium-sized industrial units went out of production with the consequent loss of 50,000 jobs. The direct losses have been assessed to be $3,222 million, plus indirect losses of $641 million (due to the closure of businesses) (GDMSA 2002).

Despite the fact that Gujarat is one of the four wealthiest states in India, it contains underclasses of very poor industrial and agricultural workers who are unlikely to earn more than 11,000 rupees ($250) per annum. For their families, formal building codes and their enforcement, or land-use planning controls, are of minimal significance (IFRC 2001c; Humanitarian Initiatives et al. 2001; Wisner 2001g).

The overall patterns of vulnerability are summarised in Box 8.2, which again is designed with headings that parallel those in the PAR model.[6]

Box 8.2: Progression of vulnerability – the Gujarat earthquake

ROOT CAUSES
- Caste system.
- History of conflict between Hindu and Muslim.
- Lack of ethical standards and accountability in public authorities and in the building industry.
- Aspiration to 'modernise' housing (use of heavier concrete block).

DYNAMIC PRESSURES
Lack of:
- building code enforcement;
- effective land-use planning;
- earthquake awareness in local populations;
- understanding or interest in non-engineered building within the engineering profession;
- awareness of earthquake risk in general public and government due to long return period of earthquake incidence in Gujarat;
- governmental commitment to effective disaster planning;
- resilience in many rural communities due to the four-year drought.
Macro-forces:
- migrant workers come to Gujarat for work;
- agricultural modernisation produces landless people who seek work in cities;
- rapid urbanisation;
- rapid industrialisation.

UNSAFE CONDITIONS
Physical environment:
- building code violations;
- inadequate land-use planning;

302

Box 8.2 continued

- low quality of building construction;
- lack of seismic resistance in non-engineered construction;
- lack of supervision of building construction;
- poor building maintenance;
- ageing building stock in towns and cities;
- unsafe 'lifeline facilities': schools, medical buildings, etc.

Local economy:
- lack of insurance protection;
- unsafe industrial buildings and facilities.

Social relations:
- low-caste groups particularly vulnerable;
- low-caste and Muslim groups often excluded from relief and recovery resources.

Public actions and institutions:
- lack of effective disaster plans;
- lack of Emergency Operations Centres (EOCs);
- lack of co-ordinated action;
- lack of search and rescue capacity.

HAZARDS
Primary impact:
- ground shaking.

Secondary impact:
- fires.

Community resilience

The government has been severely criticised for its failure to deliver immediate support to the surviving population. India has all the professional skills and skilled human resources needed for disaster management, and many of these resources were rapidly deployed by many people on a voluntary basis. There was a remarkable response from Indian society, from the survivors themselves, from other parts of India and from many in the Gujarati communities of Europe and North America.

Volcanoes and related hazards

Volcanoes are vents in the crust of the earth through which molten rock is extruded as lava or ejected as ash or coarser debris, sometimes accompanied by steam and hot (often poisonous) gases (Davis and Gupta 1991: 29; Winchester 2003). Other hazards associated with eruptions are earthquakes, tsunamis and mud- and rockslides (which can often occur years afterwards). Volcanic eruptions endanger any person living within the

303

high-risk zone whether rich or poor, landowner or landless farm labourer, man or woman, old or young, member of ethnic minority or majority. As Tomblin has commented:

> Eruptions differ from most other major causes of disaster such as earthquakes, hurricanes and floods, in that they cause virtually total destruction of life and property within relatively small areas which can be easily delineated.
>
> (Tomblin 1987: 17)

Poisonous gas emissions do not differentiate between social groups.[7] But even where these are not the main threat, income levels, the quality of house construction and the type of occupation all seem to have little bearing on people's differential capacity to resist the volcanic arsenal of hot gas emissions, blast impact, lava flows, projectiles, volcanic mudslides (*lahars*) and the deposit of ash.

It can be argued that wealthy people have more access to knowledge, which can include an awareness of volcanic risk, and that they are able to respond to warnings of (and have better transport for) evacuation in the event of an eruption. But there is growing evidence that poor people living near active volcanoes are also aware of the risks. Once they observe signs of volcanic activity they are just as likely to follow evacuation orders as their rich neighbours (Kuester and Forsyth 1985; Tayag n.d.; Zarco 1985).

Also, although volcanoes are a hazard, in the longer term some of their products can be beneficial. Extremely fertile soils result from the weathering of volcanic ashes and pyroclastic materials. Farmers often obtain bumper harvests as a result of a mild sprinkling of volcanic ash on their fields (Wood 1986: 130). In 1992, Cerro Negro erupted near Leon in Nicaragua. A thick layer of volcanic ash was deposited, with gloomy forecasts that the agriculture would be interrupted for years. However, within ten months farmers were already enjoying good crops from the fertile soils mixed with volcanic ash (Baxter 1993).[8]

Such blessings undoubtedly constitute an extremely powerful economic attraction to settle on slopes of volcanoes. It is often suggested that people inhabiting high-risk zones are gamblers by nature, who take big risks to achieve uncertain benefits. But it does not appear to take families very long to decide to face the risk of an eruption, with a return period of perhaps 45 years, in return for economic opportunities that are realised every day. Paradoxically, effective disaster preparedness, with its expectations of good warning and evacuation by the authorities, only adds to the appeal.

The Nevado del Ruiz eruption, 15:15 hours, Colombia, 13 November 1985

This case study returns to the issue of volcanic warnings and evacuations with which we introduced this chapter. It concerns the second most deadly eruption of the twentieth century – the eruption of Nevado del Ruiz in Colombia in 1985, with the loss of 23,000 lives (Wood 1986; United Nations 1985). The town of Armero was built on debris from earlier eruptions in 1595 (when 600 were killed) and 1845 (when 1,000 died) (Bishop 1998: 60). The 1985 eruption occurred at 15:15 hours, and two hours later the residents of the town of Armero (population 29,000) noticed that fine dust was falling. At 17:30, the National Geology and Mining Institute is reported to have advised that the area at risk should be evacuated. By 19:30 the Red Cross attempted to carry out such an evacuation. But perhaps on account of a very heavy thunderstorm, or the lack of evacuation drills, few agreed to leave their homes. At 21:05 a strong earth tremor occurred, followed by a rain of hot pumice and ash. As a result, part of the ice cap of the 5,400 m high volcano melted and caused the river Guali to overflow. This in turn caused a natural dam to burst, releasing a torrent that travelled at speeds estimated at 70 km/h (42.5 m/h), and a massive mudflow enveloped the town of Armero (Tomblin 1985; Siegel and Witham 1991; Parker 1989: 160–167; Bruce 2001).

Within the town one of the few survivors, Rosa Maria Henoa, described how at about 23:35 hours:

> first there were earth tremors, the air suddenly smelt of sulphur, then there was a horrible rumbling that seemed to come from deep inside the earth. Then the avalanche rolled into town with a moaning sound like some kind of monster ... Houses below us started cracking under the advance of the river of mud.
>
> (Quoted in UNDRO 1985: 5)

What transpired in those minutes is a horrific story of people attempting to drive away from the wall of roasting mud, stones and water. Some died in the chaos as terrified people tried to climb aboard moving vehicles (Sigurdson and Carey 1986; Davis 1988; Parker 1989; Siegel and Witham 1991).

In the early months of 1988, a group of lawyers inserted a notice in the local press of the city of Manizales, and in the small towns of Guayabal and Lerida. These are close to Armero, and housed some of the 3,000 survivors. There was little point in pinning any notices in Armero since the town had become a vast deserted 'cemetery' where as many as 23,000 people are buried in the mud that killed them. The lawyers invited anyone who had suffered injury, or the loss of relatives or property in these volcanic mudslides to contact them if they wanted to sue the government for gross negligence in not warning or evacuating them in time.

In response 750 claims were filed, amounting to a total of approximately $80 million. The claim hinged on the government's alleged negligence in failing to develop effective preparedness planning (including evacuation procedures) to enable the population to escape falling debris and mudslides. These are known hazards that have occurred after previous volcanic eruptions in the region.

It was anticipated that government lawyers would argue that the residents were aware of the risks in choosing to occupy what is a hazardous yet highly fertile area. By the time the case reached the court in Tolima, the number of claimants had risen to about a thousand. They sued the Ministry of Mines and Energy, since the geological service (which had responsibility for the issuing of volcanic warnings) was attached to this ministry. Government lawyers argued that the 'ordering of evacuations' was not one of the designated functions of the ministry. But as proof of governmental concern they produced evidence that Civil Defence had conducted a 'door-to-door' campaign to warn people to evacuate during the early stages of the eruption. Three volcanologists were asked as expert witnesses whether the scale, location and timing of the mudflow could have been accurately forecast, and since they said no, the government was cleared of responsibility (Wilches-Chaux 1992b).

The Nevada del Ruiz disaster was a catalyst that had a dramatic impact on the development of disaster protection in Colombia. An effective government preparedness system at central and provincial level was created, which includes detailed warning and evacuation systems. However, whilst preparedness planning exists on paper, in practice economic priorities can easily override safety. For example, the very active Galeras volcano erupted in January 1993, and the nearby town of Pasto was at risk.[9] In 1992 and 1993 the government Disaster Preparedness Agency wanted to issue warnings to the public, but the local authorities refused to act on them. Their refusal stemmed from the economic consequences of a warning given several years previously, which provoked an immediate financial crisis in the locality when credits and loans were closed and investors withdrew (Wilches-Chaux 1993).

Montserrat volcanic eruptions 1995–1998[10]

Having described in outline the two most severe volcanic eruptions of the last century in terms of lives lost, we now turn to two more recent volcanic eruptions. The Montserrat (Caribbean) eruption was selected on account of lessons to be learnt from its long-term impact that stretches from 1995 until 2003. The Goma (eastern Congo) eruption of 2002 was chosen as a tragic example of a disaster that was piled on top of others: genocide and an ongoing bitter civil war.

Geography and vulnerability

Montserrat is a small island country in the Lesser Antilles (Caribbean), measuring 16 km. north–south and 11 km. east–west, 102 sq. km. in total (roughly half the size of Washington DC). It is part of an arc of volcanic islands that includes St Kitts and Nevis, Guadeloupe, Dominica, St Lucia, St Vincent and Martinique (see above on the eruption of Mount Pelée in 1902). Montserrat is a British Dependent Territory, the responsibility of the Foreign and Commonwealth Office (FCO), with a resident Governor who represents the Queen. Britain therefore has responsibility for the well-being of Montserrat's citizens, including their protection and assistance in times of emergency. The island is dominated by the 915 m volcanic peak of Soufriere Hills. However, before 1995 the population of about 10,500 did not regard the volcanic threat as serious. There were no records of an eruption, although volcanologists believe that there had been one in the mid-sixteenth century. Thus, the population had to adapt to life with an active volcano, which was active for three years from 1995. Although the eruptions ended in 1998, at the time of writing (in 2002) normality had yet to return and some communities who evacuated their homes had had to live in sub-standard temporary housing for a full seven years

Social vulnerability: high-risk groups

Volcanoes can be seen as great levellers, potentially deadly to all life and all people, rich or poor, who are within reach of their destructive power. But in the case of Montserrat there were exceptions. Firstly, some people who had insufficient resources or employment re-entered the official danger zone in order to continue farming, and 19 of these people were killed in a sudden eruption. Secondly, not everyone had equal access to alternative livelihoods during the period of eruptions. Many with better education and foreign connections were able to emigrate. Those who could not stayed in temporary housing on the northern part of the island, including 40 elderly persons. Florence Dailey was in charge of two shelters provided for this group, and has described how some of them had been abandoned by their children long before the eruption. Others lived on their own within the exclusion zone without any support, and there was a third group who refused to leave the island when their children left as they felt too old or frail. Later, a few chose to join their evacuated families when a flight to the UK was arranged by the government. This vulnerable community of the elderly were initially distressed, having to live in converted schoolrooms without privacy. They later came to accept the situation, and have now been accommodated in permanent sheltered accommodation (Dailey 2002).

Unsafe conditions

The size of the island complicates risk-reduction measures. Clay et al. (1999: vol. 1: 8) state that 'the smallness of Montserrat raises a special problem of diseconomies of scale'. Wadge et al. have also commented on the special problems associated with the small size of the island:

> [T]he size of Montserrat has been a key factor in the crisis. Had the eruption taken place on a smaller island, total evacuation would have been inevitable. On a much larger island the issue of safety zone boundaries would have been easier to address.
>
> (Wadge et al. 1998: 3.28)

Montserrat has similarities with many South Pacific island countries where there is a high exposure to a range of hazards, but with tiny populations, low tax revenues, and limited skills and human resources (Pelling and Uitto 2002; Lewis 1999). This makes it difficult, if not impossible, to address the disaster risks they face. A characteristic of small island states is that for many hazards, a large percentage of the population is likely to be affected, with few spared to assist in the recovery. In addition, there are diseconomies of scale in the provision of infrastructure and services, and often reliance on a single hospital or a single airport. However, as noted above, Montserrat is different from small, independent, sovereign states since it is a British Dependent Territory, a status that unlocks resources for disaster risk reduction as well as for relief and recovery actions (Pattullo 2000).

Economic vulnerability

The economy of Montserrat before the eruptions was based on subsistence agriculture and fisheries, tourism and light industry. But with the eruptions, all have been lost or severely depleted. The fisheries have suffered because the volcano seriously damaged the coral reefs and reduced fish density. Farming was mainly in the southern part of the island, and had to be abandoned as this area was within the exclusion zone (Panton and Archer 1996; Brosnam 2000).

Failure of protection measures

In 1987, the Pan Caribbean Disaster Preparedness and Prevention Project (PCDPPP) carried out a volcanic hazard risk assessment in Montserrat (Wadge and Isaacs 1987). The subsequent report was not an academic study that remained confined to a university library, since it was addressed to the Government of Montserrat and proposed a series of practical actions. The resulting hazard map projected a number of volcanic scenarios including the potential threat to the capital city, Plymouth. In 1987 the report was handed to the Governor of Montserrat and the Chief of Police who had responsibility for Disaster Preparedness.

Ten years later in 1997, when a UK House of Commons Committee conducted an enquiry into the official response to the Montserrat eruption, senior officials said they had no knowledge of the report's existence. There seems to have been a complete lack of official awareness, interest or practical action to address the volcanic threat that was expressed in a detailed and scientifically substantiated hazard map and volcanic forecast. However, the fact that only 19 people lost their lives in three years of volcanic eruptions reflects well on the system of volcanic risk monitoring, the warning system, emergency management and risk-reduction policies. This achievement is offset by high levels of dissatisfaction with compensation payments, evacuation procedures and emergency shelters among the displaced communities (Clay et al. 1999: vol.1: 5 item 20).

Plymouth was destroyed by the fall of toxic ash, and this was unavoidable. But there is evidence of human failure as well, and certain losses of critical facilities are very definitely not to be regarded as 'Acts of God'. As noted above, Plymouth was identified as being severely 'at risk' in a scientific assessment in 1987, eight years before the start of volcanic activity.

Box 8.3: Chronology of events during the Montserrat eruptions

18 July 1995 — This marked the beginning of volcanic activity, with explosions and ash fall in the Plymouth area as predicted in the assessment. At this time the population was 10,500.

21 August 1995 — The first eruption was followed by the initial evacuation of 5,000 persons from the capital city, Plymouth. During this phase of the emergency, there was a debate over whether to order a total evacuation from the island. A decision was made to evacuate the most dangerous southern parts of the island, adjacent to the volcano.

2 December 1995 — Following further pyroclastic flows, Plymouth was evacuated for the second time and 6,000 people were relocated to temporary camps in the north of the island.

3 April 1996 — Plymouth was evacuated for the third and final time. The Governor declared a state of emergency. From this date the city of Plymouth was abandoned; according to expert medical advice, the ash contained carcinogenic silica including cristoballite (Baxter 1993).

23 April 1996 — A voluntary evacuation scheme was introduced, and islanders were offered the option to move to the UK for two years; only 40 took up the offer.

Box 8.3 continued

July 1996	By this time 2,000 residents had left Montserrat, mainly for Antigua or other Caribbean islands, bringing the population down to 8,096. From September 1996 to July 1997 there was a period of intensive volcanic activity.
17 September 1996	A magmatic explosive event took place causing destruction of houses in Long Ground and other eastern villages.
12 December 1996	Collapse of 'Galways Wall' in the volcanic crater resulting in concern that a tsunami could now cause a threat to neighbouring islands, especially Guadeloupe.
25 June 1997	Pyroclastic flows occurred close to the airport, causing its closure. Three villages were destroyed and 19 were killed whilst farming on the slopes of the volcano within the exclusion zone, against advice and warning.
4 July 1997	Following explosive ash eruptions, a revised volcanic risk map was produced that placed over half the island in the exclusion zone. By this time the population had dropped by a further 3,096 to 5,000 (half the original population).
August 1997	Three towns were evacuated and 1,598 'high-risk' individuals were living in shelters. However the design, siting, timing of delivery and administration of the shelters became the subject of sustained criticism (International Development Committee 1997: para 50).
19 August 1997	At this stage, when the volcanic activity was at its height, the evacuees had two options, both of which they found highly undesirable. Firstly, they could leave the island and travel to Antigua or other parts of the Caribbean, or move to the UK. The second option was to remain in Montserrat, but to move into overcrowded temporary housing in the relatively less dangerous northern part of the island. A total of 700 took the first option leaving 3–4,000 in the 'safe zones'. The British Government offered £2,500 per adult ($3,365) to provide support for resettlement, but this sum was regarded as derisory by the residents, who conveyed their feelings to the British Government through their political representatives.
26 December 1997	The largest explosion of the entire eruption occurred, ejecting 50 million cubic metres of material. By this date the population had dropped by a further 2,000 to just 3,000.

Box 8.3 continued

10 March 1998	After this date there was a significant reduction in volcanic activity. People slowly started to return to the island so that by December 1998 the population had reached 4,400.
February 2002	Some families who were unable to return to their homes in the exclusion zone were still living in temporary housing. A decision on a new airport to replace the original that was destroyed in 1997 had still not been taken. The economic base of the island economy had been seriously disrupted. The future was very much in doubt for those who remained.
October 2002	The exclusion zone was increased due to the growth of the volcanic dome.
May 2003	The volcanic dome was growing again with constant ash falls but no major volcanic activity in the preceding weeks. The population was estimated to have risen to 4,500 (under half the pre-eruption population). A new airport in the north of Montserrat was being built to replace the airport destroyed in the eruption and is due to open by October 2004.[11]

During that period no disaster planning or mitigation steps were taken. Why? Furthermore, the authorities (including the British Government) continued to invest in Plymouth, and refurbished the Glendon Hospital at a cost of £5 million (after it was badly damaged by hurricane Hugo in 1989). The decision to approve this, rather than relocate the entire hospital to a safer part of the island, was approved by a Senior Health Official in the Overseas Development Administration (ODA) about a year after the hazard map was published.[12]

Box 8.4: Progression of vulnerability – the Montserrat eruptions

ROOT CAUSES
- Colonial history and resulting geographically and culturally distant 'overseas' administration.

DYNAMIC PRESSURES
Lack of:
- political interest in the volcanic threat prior to the eruptions;
- harmonious working relationship between the Government of Montserrat and the British Government during the crisis;
- a 'common language' between scientists and decision makers.

Box 8.4 continued

Macro-forces
- patterns of out-migration of skilled people from Montserrat (particularly during the 1960s);
- diseconomies of scale.

UNSAFE CONDITIONS
Physical environment:
- the capital city of Plymouth was sited in a high-risk zone very close to the volcano;
- key 'lifelines' such as the airport and the main hospital were both sited in high-risk zones;
- extended and acute uncertainties: 'waiting on the volcano'.

Local economy:
- economy dependent on agriculture, fisheries and tourism, all of which were devastated by the long series of eruptions and accumulation of toxic ash.

Social relations:
- lack of public awareness programmes concerning the volcanic threat prior to the eruptions;
- differential access to overseas remittance incomes;
- differential access to immigration opportunities other than the temporary one offered by the British government.

Public actions and institutions:
- before the eruptions there was very limited risk assessment or disaster planning in Montserrat by the government;
- lack of adequate emergency shelter accommodation.

HAZARDS
Primary impact:
- pyroclastic flows, magmatic explosions, ash falls.

Secondary impact:
- coral and fish die off;
- respiratory illnesses in humans from ash.

Livelihoods, warnings, governance and volcanoes

In the case of both Nevada del Ruiz and Montserrat there is a complex inter-relationship between economic reasons why people are located near volcanoes, the reasons why authorities do or do not issue timely warnings and whether they are acted upon, and the history and form of governance.[13] Lack of access to alternative land means that many low-income farming families are located on the slopes of active volcanoes elsewhere in the Caribbean and in such places as the Philippines, Indonesia, Mexico and countries in Central America and the Andes. An extreme event that demonstrates these economic and political relationships occurred in eastern Congo at the beginning of 2002.

Goma, Democratic Republic of Congo (DRC), the eruption of Mount Nyiragongo, 17 January 2002

During the morning of 17 January 2002, Mount Nyiragongo erupted with lava flows spewing in several directions, splitting the city of Goma into segments. The volcano had been a threat to Goma before, most recently in 1977 when 2,000 were killed in a violent eruption. It is a mere 11 km from Goma, a city with a population of 500,000. Accurate informal warnings of an impending eruption had been made by two perceptive local volcanologists, but these were not communicated to the public since there was no responsible government to issue warnings, order evacuations or take any emergency management actions. In place of city government, there was an occupying force of armed rebels fighting the central government of the DRC.

Damage was extensive: 14,000 homes out of a total of 100,000 were destroyed; 147 people were reported dead, including at least 50 who died when a petrol station exploded as people looted petrol. About 75 per cent of the population was displaced by the lava flows, mostly to nearby Rwanda. However, as soon as the volcanic eruption subsided there was an immediate reverse flow back into Goma, still smoking from fires ignited by the eruption. This was despite strenuous efforts by the United Nations and local NGOs to prevent a return to the city on safety grounds. The evacuees were very dissatisfied with conditions in the Rwandan relief camps, while reports of the looting of abandoned property made people keen to return and protect their assets. During 1994–1996 many of the city's inhabitants had witnessed the cholera and violence in the nearby camps for Rwandans displaced by the earlier genocide (1994) in their country. They probably associated all UN-run camps with these conditions, and this added to their aversion to the shelter that was offered (Wisner 2002b). About 30,000 remained in the Rwandan relief camps, but by the end of March 2002 most had returned to Goma.

Social vulnerability: high-risk groups

At the time of the eruption, Goma was controlled by a rebel faction, and their presence had led to a progressive decline in living standards, with the economy and all basic services, such as health and education, being decimated. There had been reports of the abuse of children and women by rebel soldiers, who had also made repeated attempts to recruit child soldiers from the local community.

Many citizens of Goma were affected deeply by the traumatic events of the Rwanda genocide of 1994, as they played host to a staggering 800,000 Rwandan refugees seeking sanctuary in and near their city. These memories were embedded in attitudes and the behaviour of the society, and their response to this eruption was undoubtedly affected by this experience. HIV-AIDS was an additional source of danger in Goma, and this threat was intensified by the presence of rebel soldiers and the associated prostitution that arose as part of the local economy.

Unsafe conditions

There are few measures that can be taken to protect buildings and infrastructure from the ravages of a volcanic eruption other than to provide sufficiently strong, incombustible roofs to support hot ash falls. Since Goma is also in an active seismic zone, the materials used, the siting, form and construction of all structures need to be to aseismic standards. This was not the case for normal buildings, and there is also the legacy of the 1994 influx in Goma when housing for the Rwandan refugees was rapidly and badly constructed. These buildings still remained throughout the city. Partly because of the various traumas of the past decade, little attention had been given to hazard-resistant planning or construction in the city. The volcano also destroyed vital 'lifeline' infrastructure that included 47 schools, two hospitals and several key roads that link parts of the city and the airport.

Economic vulnerability

Goma was one of the few centres in eastern Congo with anything resembling a functioning economy. The damage to business spelt financial and economic hardship for the community. The rebel forces were probably as poor as the residents of the city and so did not have any resources to bring to assist in the economic recovery of the community.

Failure of protective measures

All aspects of vulnerability in Goma have been dominated by the fierce conflict between rebel forces and the government of the DRC, and among different rebel groups based in the eastern part of the country. The local government was not democratically elected and failed to serve the interests of its people. The civil strife extended to seven other countries and at least three opposing rebel forces were engaged in this prolonged power struggle that, by 2002, had cost 2.5 million lives.[14] More than 50 million people had had their lives disrupted and jeopardised by the civil war. One of the strongest underlying reasons for the power struggle was the competition for control of the rich mineral resources of the region, especially copper, diamonds, gold and columbite-tantalite (known as coltan, an essential ingredient in electronics, especially mobile phones) (Oxfam, UK 2002).

The rebel forces in Goma were not at all equipped to establish any viable local administration. In this power and administrative vacuum the UN and key international NGOs (such as the Red Cross and Caritas) and local NGOs attempted to provide some protection and basic services. However, there was no basis in this restricted role to allow either warning of the eruption or an orderly evacuation.

Community resilience

The people in Goma had had to become highly resourceful to cope with a series of crises that have threatened their lives and livelihoods. These traumas included volcanic eruption (1977), refugee influx (1994–1996), civil war (most recently since 1998) and then another major eruption (2002). Evidence of their coping strengths and strategies was shown by the spontaneous return to the city of its people, despite the obvious risks from the continuing eruption. Their preferred strategy was to try and protect their livelihoods and assets back home, despite official pressure to remain in safe but uncomfortable refugee camps in Rwanda. In subsequent days people demonstrated further levels of resourcefulness as they cooked food in the 'ovens' of boiling lava that lay just below the new ground surface. Other enterprising residents secured bulldozers to create temporary access routes across the lava flows that had divided the city. The relatively low loss of life itself may be testimony to the people's own capabilities, and may also have been an important factor in maintaining high levels of community morale in the face of this adversity.

Policy response and mitigation

Five approaches to risk reduction in the face of earthquakes and volcanoes will now be suggested (these are elaborated in Chapter 9). Firstly, earthquake and volcanic disasters can be used as opportunities to challenge the root causes of vulnerability. Popular development organisations can capitalise on a disaster event to challenge and possibly change unjust political, social and economic relations that lead to vulnerability. Holloway has suggested that:

> [d]isasters will often set up a dynamic in which social structures can be overturned, and relief and rehabilitation judiciously applied can help change the status quo; while projects will be the models in microcosm that can be used to demonstrate to government the possibilities of a variety of ways of working.
>
> (Holloway 1989: 220)

Thus, in the aftermath of the Mexico City and Gujarat earthquakes, neighbourhood and women's organisations were strengthened, and increased their demands for government services (Robinson et al. 1986; Annis 1988; UNDP 2001a; Bhatt 2003). There is no *direct* relationship between the strength of local organisations and reduction of vulnerability to disaster, but certainly the converse is true: in the absence of grassroots and neighbourhood organisation, vulnerability increases.

Secondly, and following from the first point, local institutions can be strengthened and the capability of families to reduce their own vulnerability can be improved. This is Anderson and Woodrow's (1998) notion of 'rising from the ashes'. However, to achieve this, energy and resources need to be focused on strengthening the self-reliance of the most vulnerable households and their institutions, especially local organisations. We return to the difficult question of identifying and aiding 'the most vulnerable' in Chapter 9.

After an earthquake in Ecuador (March 1987) local artisan builders rebuilt homes in a safe manner in collaboration with an outsider, who reflected on the experience:

> We have learnt that with outside support, but not external control, and with limited technical objectives the people can achieve great things ... [T]rue development, disaster or no disaster, will only take place through the strengthening of indigenous infrastructures directly accountable to local people.
>
> (Dudley 1988: 120)

Accountability builds trust, and trust allows access to the inner workings of local coping mechanisms. When these are translated into architectural form, there is the possibility of designing low-cost, safer shelter together with local people as partners (Maskrey 1989; Aysan and Davis 1992).

Thirdly, reconstruction following disaster offers a chance to tap local knowledge and strengthen livelihoods (Oliver-Smith 1988, 1990). An example of the former comes from Gujarat, where in designing new, earthquake-resistant dwellings, local architects and engineers drew inspiration from the traditional *bhoonga* houses that exist throughout the Kutch region. These circular dwellings (built with a conical roof, mud, sticks and wooden supports) withstood the earthquake on account of their shape and method of construction whilst many modern houses collapsed (UNDP 2001a: 9–10). Livelihood support is evident also in cases such as the reconstruction following the Mexico City earthquake, when thousands of labouring jobs were created by the recovery process (Kreimer and Echeverria 1991).

Fourthly, the disaster provides an opportunity to develop effective risk assessment with good cost–benefit arguments for protective measures. An example is encouragement offered to local authorities by a World Bank team that has been working in La Paz, the capital city of Bolivia, which faces numerous hazards. In a report on the lessons learnt from the project, the team concluded that risks could be evaluated, quantified, programmed and addressed with measures that the city could afford, even with all its pressing demands on the budget.

[W]e calculated that disaster prevention and preparedness would cost US $500,000 in 1987 and total about US $2.5 million, or US $2.50 per capita ... [T]his amount is far exceeded by annual losses from natural disasters estimated at $8 per capita. With this minimal level of funding, disaster mitigation could be affordable, cost-effective and within the realm of La Paz's needs.

(Plessis-Fraissard 1989: 135)

Finally, disasters provide an opportunity to educate political leaders and decision makers about the deeper causes of vulnerability and disaster. Authorities, civic and business leaders may be ignorant, or they may deliberately avoid recognising their own role in increasing vulnerability. They may respond to messages such as that of the financial calculation just discussed. In addition, such leaders may be moved by moral imperatives backed up by political demands. A good example of this is the growing outrage expressed by ordinary people and professionals alike at the collapse of school buildings that have been improperly constructed, often because of corruption. The deaths of schoolchildren in El Salvador and Gujarat (2001), Italy (2002) and Turkey (2003) has led to an increasing outcry both locally and internationally. Thus we may be witnessing the beginning of a pattern that begins with outrage, develops political organisation and eventually leads to legislation and proper enforcement (Wisner 2003b; see Chapter 9 for more on the collapse of school buildings and this general pattern of mounting political pressure for change).

Notes

1 The earthquake disaster in Peru (31 May 1970) killed 70,000 people, left half a million homeless and seriously damaged 152 provincial cities and 1,500 villages. Oliver-Smith (1994) argues that the causes must be traced back 500 years to the Spanish conquest and the destruction of the Inca 'prevention mindset' that had created a remarkable ability to mitigate such hazards.

2 The authors, especially Davis, wish to acknowledge the insights and knowledge derived from members of the joint UK–Mexican Government Research Project into Urban Seismic Risk Reduction, 1988–1990, especially Yasemin Aysan, Andrew Coburn, Robin Spence, Alexandro Rivas-Vidal, Susanna Rubin and Hugo Garcia-Perez.

3 However, in Japan, and particularly in Kobe, over the age of 55 women outnumber men significantly (Ukai 1996), suggesting that the excess mortality of women belonging to older age groups reflects their greater share of the total.

4 Discrimination against the *Burakumin* is not as extreme as is the case for India's *Dalits*, but there are similarities. There are also parallels in attempts by the governments of the two countries to discourage such discrimination. There are laws in India aimed at special assistance for the untouchables; while the Japanese government established special funds in 1969 for the improvement of some 4,500 specific neighbourhoods (called, collectively, the *Dowa*) where *Burakumin* lived. However, there are probably another 2,000 neighbourhoods inhabited by *Buraku* that were not included in the original plans (Laidlaw 2001: ch. 3).

5 Steinberg (2000: 28) points out that San Francisco had been expanding into reclaimed parts of the Bay for more than 100 years. In 1906 the business district suffered as a result of liquefaction of the soil due to its location on landfill that had begun in the 1850s.

6 This application of the PAR model was first drafted as a student project in 2002 by postgraduate students undertaking the Masters in Defence Administration course (MDA) at Cranfield University, under the supervision of Ian Davis. The students were: Henry Cummins, Henry Hailstone, Alex Hall, Jamie Hartley, Amanda Hassel, Mike Healey, Duncan Gregory, Tony Leadbeater, Alison Munro, Suzanne Murray, George Ramshaw, Tim Reynolds, Ron Rowley, Caron Tassel, Liren Xie and Yong Zhang. The authors are grateful for the first draft provided by these students, to which modifications have been made.

7 This point is also evident in another very rare volcanic hazard, mentioned in Chapter 1, which struck the villages scattered to the north of Lake Nyos in Cameroon on 21 August 1986. A cloud of carbon dioxide gas was released from the lake, for reasons that remain unclear. The gas could have been the product of volcanic activity, or it could be an equally rare event in which gas trapped by a cold layer of very deep water was released due to a minor earthquake. In any case, the gas affected everyone living in the area, and asphyxiated 1,700 people (95,000 survived) (Baxter and Kapila 1989; Sigurdson 1988).

8 Eruptions also produce valuable mineral products such as pumice, perlite, scoria, borax and sulphur. Residual heat in volcanic regions can also be tapped to provide cheap geothermal energy. The medicinal and recreational use of hot springs has been recognised throughout the world for thousands of years.

9 It caused the death of six of the world's leading volcanologists, who were conducting scientific studies in the crater at the time.

10 Assistance in writing this case study has been generously provided by Franklyn Michael (Permanent Secretary to the Government of Montserrat during the emergency), Roger Bellers (DFID Advisor on Disaster Management to the Overseas Territories), Professor Geoff Wadge (Reading University) and Dr Peter Baxter (University of Cambridge).

11 The ODA was Britain's foreign aid ministry, now called the Department for International Development (DFID).

12 The May 2003 update was provided by personal communication between Ian Davis and the Montserrat Desk, Overseas Territories Department, Foreign and Commonwealth Office, UK Government, 9 May 2003.

13 Since publication of the first edition of this book there has been much work done on early warning systems. In Chapter 7 we saw that despite much progress in understanding the behaviour of cyclones, many social and economic factors can still intervene to block effective warnings. The same is true of volcanoes. There has been considerable progress in producing accurate warnings (Tilling 2002), and some notable success stories such as the warning that saved many lives before Mount Pinatubo erupted in the Philippines in 1991 (Punongbayan and Newhall 1995). Nevertheless, there continue to be situations, such as the one about to be described in eastern Congo, where no effective municipal government existed to organise warnings, or in Montserrat, where economic pressure forced some to disregard warnings.

14 A report in 2003 estimated that as many as 4.7 million people could have perished due to violence, disease and starvation during four and a half years of conflict in Congo (Astill and Chevallot 2003).

Part III

TOWARDS A SAFER ENVIRONMENT

9

TOWARDS A SAFER
ENVIRONMENT

'Towards a safer environment': are statements of intent merely hot air?

Before we review international efforts to reduce people's vulnerability and make some of our own suggestions, let us engage with some pessimistic views about their efficacy, some of them radical and others post-modern. The story runs something like this. Conferences and the airing of statements of concern, declarations, objectives and principles therein are simply a waste of time. They may stabilise the expectations of international bureaucrats in times of uncertainty, threatened guilt and blame but merely represent manoeuvrings in corridor politics far from the site and sight of death, destruction and destitution (Bellamy Foster 2003; Sachs et al. 2002). Disasters are discursively 'produced' by the well-paid and well-fed in rich countries who indulge in essentialising cultural discourses which denigrate large parts of the world as disease-ridden, poverty stricken and disaster prone (see an example of the deconstruction of famine by Bankoff 2001, mentioned in Chapter 1, and a critique similar to ours in Adams 2001; Broch-Due 2000; Leach and Mearns 1996; Hoben 1995).

The radical structuralist version of pessimism takes the view that conferences such as the World Summit on Sustainable Development in Johannesburg in 2002 (reviewed in the next section) are nothing more than media events for which corporate sponsorship has already purchased the script, and global corporations have already determined what are to be framed as 'problems' or orchestrated as 'silences'. In short, 'he who pays the piper calls the tune' (Bellamy Foster 2003; Sachs et al. 2002). Disasters are, in any case, outcomes of global socio-economic forces ('root causes' and 'dynamic pressures' in the PAR model), and quite outside any transformation which any amount of hot air might offer.

Disasters are essentially historically and spatially specific outcomes of the process of contemporary capitalism. Unsafe practices are pursued by privatised public utilities unconcerned by what anyone might say at the conference table. Structural adjustment policies have been adopted without any consideration at all for their impact on increased vulnerability, and have led to cuts in health provision, education and public services in general. The liberalisation of trade is grotesquely one-sided – in that poorer countries are forced to liberalise their economies and open up to imports from the G7 countries, while the USA and the European Union through its Common Agricultural Policy continue to protect their farmers with subsidies. This hypocrisy has an immediate impact upon the ability of farmers and pastoralists to export their produce to rich countries. Furthermore, such political decisions as the embargo imposed on Iraq over the sale of oil and the maintenance of the No Fly Zone have directly caused a huge increase in the vulnerability of the population. In all these cases, corporate capital, and the nation(s) which promote such interests, will do what it will – without the slightest compunction. What impact can a few conferences on disasters ever make? In short, they are no more than media events, bankrolled by the very villains who create vulnerability, and serve merely to soften the sharp and jagged edges of global inequality. Or so the radical argument runs.

These sentiments may make good press in the eyes of some in the academy but they probably have even less impact worldwide than a much less radical and less overtly political intervention in the conference hall. While public statements may not be a very effective way of shaming decision makers – they will do most of what they feel like doing anyway – it does draw attention to some of the direct implications of their actions. This has happened, although it is difficult to assess the effect. The impact on global vulnerability of the USA's decision to walk away from the conference table regarding the international treaty to ban land mines has had direct effect. The decision by the Ethiopian army to plant mines on the agricultural land of Eritrea as an act of war likewise has long-lasting effects – as intended. The Sudanese government constantly attempts to starve the SPLA and bombs humanitarian attempts to alleviate suffering. The actions of North Korean and Zimbabwean governments in very different ways have directly caused the starvation and under-nourishment of their citizens. These impacts upon disasters and vulnerability are publicised by conferences and international meetings and, via them, by press, radio and television broadcasts.

A radical critique goes further, claiming that corrupt elites in some nations have a self-interest in continuing the cycle of disaster and reconstruction of the status quo because they benefit financially and politically from the flows of foreign relief (Susman et al. 1983). In a similar way, some have suggested that warlords can create a semi-permanent, quasi-stable economy based on the cycle of violence, displacement of populations and receipt of foreign disaster assistance (Duffield 2001, 2002) and that traders may profit hugely from the relief process (Keen 1994).

These pessimistic views have to be confronted, and the limitations of exhortations in international conference halls must clearly be identified along with the considerable opportunities for mobilisation, exchange of information, the creation of new initiatives and new procedures, and commitments of resources. Of course disasters still occur, and in increasing numbers and extent, as this book has shown. But also, there has been a great deal of progress in preventing and alleviating disasters via concerted international efforts. Improvements in the practice of much disaster prevention and mitigation, as well as relief, is technical in nature (as well as political). Some measures have been quite successful. While they may be criticised as a 'tyranny of techniques' and as apolitical, they have often proved very successful, and the symbolic shaming of particular perpetrators of vulnerability and disaster may be significant (see, for example, de Waal's [1997] book *Famine Crimes*).

It is people, rather than a capricious nature, who are to blame for increasing vulnerability. Progress in the prevention of famine and improvements in national food security and famine forecasting have been remarkable, and mostly achieved through technical improvements. Famine is largely a legacy of the past and now only occurs in nations that isolate themselves from such international contact (for example, North Korea for ideological reasons) and countries in the grip of chronic civil war. Therefore, we acknowledge the limitations of what follows, but also the importance, circumscribed though it may be, of international dialogue and action to prevent disasters.

From Yokohama to Johannesburg via Geneva

The first edition of this book was published just before the mid-point of the International Decade for Natural Disaster Reduction (IDNDR), at a pivotal point in the development of disaster risk-reduction policies. This was shortly before the World Conference on Natural Disaster Reduction held in Yokohama in May 1994, the first UN World Conference to address specifically the issue of disaster risk reduction (United Nations 1994). Now, the second edition of this book follows another major UN conference, this time the World Summit on Sustainable Development (WSSD), held in Johannesburg in August/September 2002 (United Nations 2002a, 2002b).

Between these events the millennium occurred, with much fanfare and a number of significant steps taken by the United Nations that may come to be recognised as important milestones in addressing ways to reduce the vulnerability of people and the fragility of their livelihoods. These steps seem to have met with general approval. Whilst there has been a steady stream of critics who have questioned the value and impact of international aid and development assistance, especially in the aftermath of complex emergencies (Vaux 2001; Macrae 2001), there has not been any serious critique of pro-active actions taken to reduce disaster risks.

Before considering a series of *risk-reduction objectives*, we believe that it is important to highlight key commitments and decisions coming from these UN initiatives that relate to the theme of this book and its two models of vulnerability and its reduction.

The Yokohama Conference, 1994

The 1994 Yokohama conference provided an opportunity for countries to focus on disaster risk reduction, and this was the first international conference where the social aspects of vulnerability were given serious consideration. Previously, a strong emphasis of the IDNDR had been on science and technology. Clear evidence of a change in attitude came in the first affirmation of the Yokohama message.

> Those usually most affected by natural and other disasters are the poor and socially disadvantaged groups in developing countries as they are least equipped to cope with them.
>
> (IDNDR 1994)

The sixth affirmation highlighted the idea that community involvement should be actively encouraged in order to gain greater insights into individual and community perceptions of both development and risk. The opinion was expressed that there is a need to understand the social dynamics of hazard-prone communities, and a need for

> a clear understanding of the cultural and organisational characteristics of each society as well as of its behaviour and interactions with the physical and natural environment. This knowledge is of the utmost importance to determine those things which favour and hinder prevention and mitigation or encourage or limit the preservation of the environment for the development of future generations, and in order to find effective and efficient means to reduce the impact of disasters.
>
> (Ibid.)

The IDNDR Programme Forum, 1999

Five years later the UN International Decade ended with the IDNDR Programme Forum in Geneva in July 1999. By then the social agenda of vulnerability reduction had significantly expanded to the point where no fewer than three of the four 'Goals for the International Strategy for Disaster Reduction' were directly concerned with the human dimensions of risk reduction. It is important to note the beginning of a concern for livelihood protection in the conference rhetoric.

Goal 1   Increase public awareness of risks ... posed to modern societies.
Goal 2   Obtain commitments by public authorities to reduce risks to people, their livelihoods, social and economic infrastructure and environmental resources.
Goal 3   Engage public participation at all levels of implementation to create disaster-resistant communities through increased partnership.

One of the Forum's objectives went far beyond the Yokohama Message, expressing the intention to gather accurate data for analysis of the effect of disasters on people:

Objective 11   Establish internationally and professionally agreed standards/methodologies for the analysis and expression of the socio-economic impacts of disasters on societies.

The Forum also established an important link between poverty reduction and mitigation. The Programme Summary reminded delegates:

that the people most vulnerable to disasters are the poor, who have very limited resources to avoid losses. Environmental degradation resulting from poverty exacerbates disaster impacts.... Innovative approaches are needed; emphasis should be given to programmes to promote community level approaches.

(IDNDR 1999)

The Millennium Declaration, 2000

A year later the UN system marked the new millennium with a memorable gesture towards the elimination of poverty. Eight Millennium Development Goals (MDGs) were agreed by world leaders in the Millennium Declaration of September 2000. These goals were further broken down into 18 targets (measured by 48 indicators) to be achieved by 2015 (World Bank 2000).

While critics may regard these goals as empty political rhetoric, their significance lies in the fact that these are now the internationally agreed yardsticks for national development, with numerical targets and quantifiable indicators to assess progress. All the signatory countries now claim to be working toward these goals, and donors are providing sharply focused aid packages to support their endeavours.

Within the Millennium Declaration there are several points where disaster risk reduction is relevant. Under the goal 'Eradicate extreme poverty and hunger' there are a pair of targets: to halve between 1990 and 2015 the proportion of people whose income is less than $1 a day, and also to halve in that same period the proportion of people who suffer from hunger. In this book we have written a good deal (especially in Chapter 3) about the interconnections between household livelihoods and vulnerability. Thus, the relevance of the income goal to vulnerability reduction (and vice versa) is easy to comprehend. Likewise, our discussion of famine in Chapter 4 and chronic hunger and ill-health in Chapter 5 should make their relevance to this second millennium target obvious (and vice versa).

The fourth millennium goal is to 'Reduce child mortality'. We have argued in many places in this book that children, especially those under five years of age, are particularly vulnerable to the impacts of many hazards, as well as the stress of displacement and household livelihood disruption.[1]

Millennium goal number six is to 'Combat HIV-AIDS, malaria, and other diseases'. Here, too, we have shown numerous cross-connections among biological and other natural hazards. Recall, for example, that African populations affected with HIV-AIDS are less able to cope with the stress of drought and other hazards that threaten food production. Also, in Chapter 5 we reviewed evidence that flooding can bring on epidemic increases of malaria and other water-related diseases.

The seventh millennium goal is to 'Ensure environmental sustainability'. There is a clear link between this goal and the impact of hazards, since major disasters – or the regular occurrence of persistent smaller events – can cumulatively wipe out any hope of sustainable urban or rural development. In particular, one of the targets associated with this goal is to achieve a significant improvement in the lives of at least 100 million slum dwellers by the year 2020 (Target 11). We have described the multiple hazards faced by the poor in informal urban settlements. Since most of the global population increase during the next 20 years will be in the cities of LDCs it will be impossible to achieve 'significant improvement' in the lives of the many slum dwellers without dealing with their vulnerability to floods, storms, earthquakes, landslides and epidemic disease.

Among the targets defined as part of the eighth and final millennium goal ('Develop a global partnership for development') are the special challenges faced by small island states (Target 14). Our book has provided numerous examples of the catastrophic damage to island economies and livelihoods brought about by volcanic eruptions, drought, tsunamis and tropical cyclones.

In Section IV of the Millennium Declaration, entitled 'Protecting our common future', there is a further commitment (not expressed or quantified as one of the central eight millennium goals) to intensify collective efforts to reduce the number and effects of natural and man-made disasters. In the so-called 'Road Map' towards implementation of the millennium goals, the watchword is protection of the vulnerable and promotion of human security (United Nations 2001). Whilst the direct concern in the document is with women and children in complex humanitarian emergencies, the enforcement of international undertakings to protect human rights, conflict management and peace-building, these sections of the Road Map can equally be seen to apply, indirectly, to a reduction of vulnerability to natural hazards (Vatsa 2002a). The Road Map for implementation also announces a fundamental change from reaction and response to prevention and mitigation. It is this last-mentioned shift that will be the focus of much of the final chapter of our book.

The Millennium Declaration implies that future disaster risk-reduction structures, plans and policies can no longer be isolated as distinct entities and in future will need to be synchronised with structures, plans and policies concerned with poverty reduction.[2]

The Johannesburg Summit, 2002: World Summit on Sustainable Development

The extensive preparatory work for the Summit included a valuable background document, *Disaster Risk and Sustainable Development: Understanding the Links between Development, Environment and Natural Hazards Leading to Disasters* (ISDR 2002a). This paper was a salutary and rare example of integrated teamwork by various UN agencies (UN ISDR, UNDP, UNEP, UN/Habitat, UN DESA and UN OCHA). It went far beyond any previous official UN document in its review of the scale and complexity of vulnerability. The following quotation is reproduced in full since it demonstrates that the model of vulnerability described in this book is consistent with the emerging international consensus.

> [Vulnerability] ... describes the degree to which a socio-economic system or physical assets are either susceptible or resilient to the impact of natural hazards. It is determined by a combination of several factors, including awareness of hazards, the condition of human settlements and infrastructure, public policy and administration, the wealth of a given society and organised abilities in all fields of disaster and risk management. The specific dimensions of social, economic and political vulnerability are also related to inequalities, to gender relations, economic patterns, and ethnic or racial divisions.
>
> (ISDR 2002a, Item 9: 6)

Despite such enlightened texts produced during the preparatory phase, the final document to emerge from the Johannesburg Summit contained minimal reference to the cause, effect and reduction of vulnerability to disaster. Where references were made to disaster reduction, they tended to revert to technical aspects, in the manner of pre-Yokohama IDNDR deliberations. In item 19 of the Johannesburg Declaration, under the theme 'Our commitment to sustainable development', disaster problems had to take their place alongside no less than 17 other threats to sustainable development:

We reaffirm our pledge to place particular focus on, and give priority attention to, the fight against the worldwide conditions that pose severe threats to the sustainable development of our people. Among these conditions are: chronic hunger; malnutrition; foreign occupation; armed conflicts; illicit drug problems; organised crime; corruption; *natural disasters*; illicit arms trafficking; trafficking in persons; terrorism; intolerance and incitement to racial, ethnic, religious and other hatreds; xenophobia; and endemic, communicable and chronic diseases, in particular HIV-AIDS, malaria and tuberculosis.

(United Nations 2002a: Item 19, our italics)

Of course, we have argued in this book that many of the afflictions and problems listed in this passage are indeed connected to disaster vulnerability as the analysis works its way from the specific to the general; that is, from unsafe conditions, through dynamic pressures, to root causes. Also, moving from the political declaration of the Johannesburg Summit to the detailed programme of action, the relevance of disaster risk reduction becomes clearer.

For instance, two of the commitments detailed in the Johannesburg programme of action are improving access to safe water supplies and more and cleaner energy sources for many millions of people. If implemented, nearly a billion people would gain better access to clean water by 2015. The direct benefits of water and energy in villages and slums would be to cut the appalling present death rate of thousands of children each day from diarrhoea and reduce unnecessary death and illness from chronic exposure to smoke from the wood and dung used for cooking. However, indirectly, water and energy projects could also reduce vulnerability to natural hazards. How? The answer depends on the way in which such measures are implemented (Wisner 1987b, 1992a).

Firstly, the implementation of new water supplies and renewable energy projects in slums and villages would have to be done in a participatory fashion. Such ambitious infrastructure construction would be unfeasible, unsustainable and possibly dangerous if conducted with only a top-down approach, utilising only contractors and private firms. It would be unfeasible because it would simply cost too much. It would be unsustainable because, without the benefit of local knowledge, such schemes would be hard to design and subsequently costly to maintain without local involvement. The danger comes from the creeping privatisation of public utilities. International corporations are now running public water supplies in many countries. However, if one bravely assumed that citizens would be involved in a variety of ways in such new water and energy schemes, the spin-offs for increased local networking and self-organisation could be great. Increased self-organisation in squatter settlements and remote villages is a main prerequisite of disaster risk reduction. People who are better organised can more easily become better prepared, better able to respond to hazard warnings, better able to demand government attention to hazards.

Secondly, there are also some very specific technical links between a variety of common natural hazards and projects to improve access to water and to clean energy. In constructing most local, small-scale water projects, an opportunity arises for residents to become more aware of the local topography and the pattern of flow in the watershed where they live. Early warning of flash flooding could easily be built into this phase of a water project. For example, in Honduras and Guatemala local communities (including women in the highlands) have been trained to monitor stream flow and report increases that could herald flooding (USAID n.d.; Fordham 2000; FEMID 2002; Wisner 2003e). Measures to mitigate drought could also be added on to the construction of a village drinking water supply. When rural electricity is provided by solar, wind or micro-hydro technologies, the basis is there for more widespread and reliable communication systems that could be used for sending warning messages. Use of renewable energy for domestic purposes such as cooking would prevent the cutting of trees that anchor slopes, thus preventing landslides and reducing the risk of flooding.

The two cases of water supply improvements and the substitution of alternatives for wood fuel in the kitchen can be expected to bring great health benefits – a reduction in diarrhoea from water-borne diseases and in respiratory disease from wood smoke. In turn, savings for the household from not having to travel to the health post and buy medicines will contribute to further improvements in nutrition and well-being (see our Access model, Figure 3.1). A healthier labour force can work harder and perhaps more productively, and indeed water and electricity could provide the basis for new rural and home-based income opportunities. Natural hazards research has repeatedly shown that it is not only better-organised localities that have the capacity to resist extreme events and the resilience to recover quickly, but localities composed of well-nourished and healthy individuals and households with diverse and productive livelihoods.

In concluding this discussion concerning the relevance of international agreements, it may be useful to consider the threat posed to overall human development by disaster risk through a familiar metaphor: a large brick building. Constructing and occupying any building is best regarded as a 'process', while it is also possible to observe the completed structure as a 'product'. Similarly, both words can apply to the way human development is created and sustained over time: the walls of our structure comprise bricks, each with a different label. Typical examples could include 'reducing chronic hunger' or 'attacking malnutrition'. Such individual bricks are vital elements in the structure: they provide overall support as well as strengthening each other. However, they can easily be removed, and while their absence may initially be hardly noticed, the cumulative effect of removing bricks will weaken the structure and perhaps contribute to sudden collapse. This will obviously cause the destruction of the building in question as well as causing damage to the wider environment.

The point needs to be made forcefully that if a concern to reduce risk from natural hazards is separated from social development or poverty reduction in a specific context, it is no exaggeration to state that it can put the entire local development process at risk. Thus vulnerability reduction strategies have to be maintained at all costs and clear objectives are needed in order for this to occur.

Risk-reduction objectives

We now turn to the core of this final chapter. Here we crystallise seven *risk-reduction objectives* by reflecting back over Parts I and II of this book. Each of these seven objectives flows from the logic of our Pressure and Release (PAR) and Access models, and each is supported by the body of case material we presented in Part II. The number seven is arbitrary or significant, depending on one's cultural background and point of view,[3] and from a practical point of view, a list with seven items is short enough, we hope, to be memorable. In order to make these seven objectives more memorable, we have assigned each a key word whose first letters spell 'CARDIAC' – the 'heart' or core of our message, as follows:

1. C = **Communicate** understanding of vulnerability
2. A = **Analyse** vulnerability
3. R = focus on **Reverse** of PAR model
4. D = emphasize sustainable **Development**
5. I = **Improve livelihoods**
6. A = **Add recovery**
7. C = extend to **Culture**

First risk-reduction objective

Understand and communicate the nature of hazards, vulnerabilities and capacities. **[Communicate understanding of vulnerability]**

Training and education programmes are concerned with a pair of linked processes. The first is the acquisition of knowledge concerning the nature of hazards, vulnerability and capacities. The second is to build capacities that allow patterns in daily life change in ways that increase personal and social protection. Public awareness followed by informed action is the bedrock requirement to reduce vulnerability and to develop resilient households, localities and societies (Handmer and Penning-Rowsell 1990).

The oft-repeated concern to 'raise the level of public awareness of disaster risks' can be found in most national disaster plans and tends to occupy a prominent place in most training courses in disaster management. However, the expression is often no more than a platitude. The subject needs to be re-examined and redefined so that public awareness reaches far deeper levels, not only increasing in knowledge and information but applying it in modified patterns of behaviour. Ambraseys (1995) has reflected on this requirement.

> The term 'Public Awareness' is often little more than a description of a rag-bag of this or that item of information that is directed to the public; however what is needed is genuine learning to the point where individuals will be prepared to take actions to promote safety which may *not* always be in their own interests.

We take this statement to mean that, like all other aspects of social change and 'development', vulnerability reduction involves conflicts. Short-term hardship may be required in order to achieve more safety in the long run. Slum dwellers who protest a landlord's neglect of building maintenance (which could lead, say, to collapse in an earthquake or wind storm) may be threatened with eviction or be beaten by hired thugs.

What we mean by 'understand' disaster and vulnerability is that ordinary people in unsafe conditions themselves must work through how they got there. Training and consciousness raising should involve their actually working through the 'pressure' and 'release' analysis themselves.[4] It is gratifying, as authors, to be able to report in the second edition of this book that in many parts of the world citizen-based organisations (CBOs) have developed materials and methods based on our PAR model that do precisely that (see 'Second risk-reduction objective' below).

The mention of CBOs here underscores another important point about 'understanding' and what is often referred to as 'risk communication'. Individuals are unlikely to engage in active learning on their own, especially those who are marginalised. Organisation at various levels is a prerequisite for risk communication that will actually result in changed patterns of vulnerability and capacity. This, in turn, requires governance at the national level that encourages (or at least does not discourage) self-organisation and action by civil society. For example, in late 2002 whole regions of Zimbabwe where the population had organised itself and supported an opposition candidate for president were prevented from receiving foreign famine relief by the national government of the incumbent president. Under these conditions it is hard to imagine much effective 'risk communication' taking place on the subject of mitigating the impact of the drought affecting Zimbabwe and several other southern African countries (see Chapter 4).

Conventionally, public awareness programmes have been targeted at individuals, whether in schools, religious buildings, teahouses, workplaces or homes. However, this approach may not be the most effective. Quarantelli (2002) has observed the need for public awareness strategies that focus on institutions rather than individuals:

> It is a good idea to try to raise public awareness. But usually this is thought of in terms of individuals. It is actually much easier to try to get the support of key groups, such as churches, certain social clubs, neighbourhood associations, etc. Convincing key informal or formal leaders in such organisations is much easier to do (relatively speaking) than trying to educate the population at large when that is thought of as individuals. One should take advantage of the fact that communities and societies are made up of groups. ... Getting them to help is very important in any educational effort or attempt to raise awareness or change behaviour.

When considering how people perceive risks and understand vulnerability, it is also very important to remember that ordinary people already have knowledge and experience. Whatever form this takes, it must be the starting point for risk communication in the form of a dialogue, not a monologue. We agree entirely with Twigg (2002: 20), who observes that '[p]eople at risk do make rational choices about protecting themselves from disasters. They do have a view of that risk. We cannot understand responses to EW [early warning] without first having a good understanding of this'. Throughout this book we have mentioned local coping and the potential for building on this as a basis for vulnerability reduction. We do not want to romanticise 'traditional' or 'local' coping, and do not consider it to be necessarily sufficient, the best or the only kind of response to risk. However, the mobilisation of knowledge and efforts at the neighbourhood and village level is critical. The starting point should be the achievements of ordinary people in 'living with' floods, droughts, pest and plant disease attacks, epidemics, storms, earthquakes, steep unstable slopes and volcanoes (Peri Peri 1999). Previous chapters have digested a wealth of evidence to show that people know a good deal about these hazards and are not passive victims.

In fact, our caveat concerning the bland assumptions often made about 'public education' and 'risk communication' must extend further. Often in Parts I and II we have shown that social processes involve conflict. There can be no assumed cosy coincidence of interest between 'government' and 'experts' and 'the people'. The key ingredient that is often missing is trust. In Chapter 1 we acknowledged, but set to one side, a large literature concerned with environmental risks, especially chemical and radiation hazards,[5] a literature very nicely reviewed by Barnes (2002).[6] Unfortunately there have been very few efforts to link such work on risk, accidents, credibility and outrages (Sandman 1989) in industrial society and work which is primarily focused on natural hazards in developing countries.[7] Clearly, the issue of trust and the credibility of governments and official sources of information are also important where slum dwellers have experienced generations of malign neglect or abuse by city authorities (administrations, police and enforcement agencies), or where small farmers have been in long-standing disputes over land and human rights with the very state officials that are supposed to protect them from hurricanes or volcanic eruptions.

These two subjects – local knowledge and trust – are seldom dealt with in discussions of risk communication. However, if our two caveats are taken into account, public awareness of risk can be an effective and efficient tool. Public awareness is a broad process embracing an ongoing exercise in mutual education between authorities and the public. The active involvement in hazard protection of localities at risk is essential, and marks recognition that patterns of vulnerability and most mitigation activities are local in character. The primary level of disaster risk reduction is the neighbourhood or village, and therefore local leaders need to take a primary responsibility.

Second risk-reduction objective

Conduct risk assessment by analysing hazards, vulnerabilities and capacities [Analyse vulnerability]

Integrated hazard and capacity/vulnerability analysis (CVA) rarely happens. In the risk assessment field there is a noticeable bias towards the natural sciences. The majority of disaster managers tend to believe that disaster risk assessment is synonymous with scientifically generated 'hazard mapping', and that this is the sum total of the diagnostic process. This view stems from a technocratic and fundamentally false assumption that once hazards are mapped in terms of their location, duration, frequency, severity and impact characteristics, then the risk assessment process is complete. It is also likely that the perception of officials is that hazard mapping is a politically neutral process in sharp contrast to all the sensitivities of vulnerability assessment. Officials may fear being identified or vilified by a hostile media or political opponents if vulnerability analysis reveals past policies that have consciously or unconsciously placed specific groups of people at risk.

Despite such obstacles, there is evidence of progress in certain areas (Enarson et al. 2003a, 2003b; Cannon et al. 2003; Buckle et al. 2000; Trujillo et al. 2000; Stephen and Downing 2001; King 2001; Brocklesby and Fisher 2003; Beerlandt and Huysman 1999). For example, CVA has been widely used by the International Federation of Red Cross and Red Crescent Societies (IFRC), who call it vulnerability/capacity assessment, or VCA. A useful overview of this progress is set out in the *World Disasters Report* for 2000, sub-titled *Assessing Vulnerabilities and Capacities during Peace and War* (IFRC 2001a). This report highlights the fact that many of the 178 national Red Cross and Red Crescent Societies throughout the world are now routinely undertaking VCAs as an essential part of their disaster assistance and development planning. They have adopted a dual approach: assess the capacities and vulnerabilities of their member organisations as well as the communities in which they work (IFRC 1999b).

Some form of CVA is also practised at a local level in many LDCs, following the path-breaking work of Anderson and Woodrow (1998), who, in turn, built on the efforts of Cuny (1983), Cutler (1984), York (1985) and Maskrey (1989). For example, the Citizens' Disaster Response Network in the Philippines (CDRN) uses a Hazard, Vulnerability and Capacity Assessment Matrix to structure discussion at a local level (Figures 9.1 and 9.2).

In a rural, southern African context, the NGO network Peri Peri, explicitly uses the equation $R = H \times V$, as a way of focusing village discussions (Figure 9.3).

One of the developments in the decade since *At Risk* was first published has been the proliferation of regional networks of disaster reduction NGOs. As time goes on Asian, Latin American and African networks are interacting, and these kinds of citizen-based, participatory methods for CVA are being shared and adapted for local conditions. Many of these methods originated with various members of the Network of Social Science for Disaster Reduction in Latin America (known as *La RED*, The Network, for short).[8]

There is also a growing recognition of the need to address vulnerability in some governmental circles. For example, in 2002 the UK House of Commons International Development Committee reported on 'Global climate change and sustainable development'. The report included a proposal that:

1. Hazard type	2. Warning signs	3. Forewarning	4. Speed of onset	5. Frequency	6. When	7. Duration	8. Extent	9. Elements at risk and reasons why elements are at risk	10. People at risk	11. Locations of people at risk	12. Resources left and capacity for disaster response
← Hazard assessment →								← Vulnerability assessment →			← Capacity assessment →

Figure 9.1 Hazard, Vunerability and Capacity Assessment Matrix

Source: Heijmans and Victoria (2001: 28)

Dominant approach	**Citizen-based and development-oriented disaster response**

Figure 9.2 Cartoon figures used in local discussions by CRDN to contrast vulnerability and conventional approaches

The goal of disaster responses is to alleviate immediate suffering and bring things back to normal as before the disaster event. Disaster responses consist of a sequential series of actions to gain control over disasters, before, during and after the emergency period (disaster cycle model).

1. Disasters are unforeseen events that disrupt normal life and require help from outside.
2. People affected by disasters are helpless victims and passive recipients of external aid.
3. Stress placed on emergency response, relief, and technological

The goal is to reduce people's vulnerability by increasing their capacities to prepare for, to cope with and to mitigate the adverse effects of disasters. Aware and organised communities can pressure government to implement policies and programs recognising people's needs and interests and promoting a safer environment.

1. Disaster is a question of vulnerability. Disasters happen when hazards hit vulnerable communities whose inherent capacity is not enough to protect itself and easily recover from its damaging effects. Disasters are

Figure 9.2 continued

and scientific solutions to address physical vulnerability.
4. Donors decide what victims need.

Assessment of disaster situations is rapidly carried out by external experts just after the disaster event.

5. Responses focus on individual families and on restoring infrastructure that serves national economic interests.
6. Key players are government, aid agencies, scientists, experts and disaster managers.

the product of the social, economic and political environment.
2. People affected by disasters are active actors in rebuilding their life and livelihood. People's existing capacities are recognised and further strengthened.
3. Addresses roots of people's vulnerabilities and contributes to transforming or removing structures generating inequity and underdevelopment.
4. People's participation is essential in all phases of disaster management and contributes to building their capacities.

Assessment of disaster threat situations is a continuous process before, during and after disaster events and involves community members considering class, gender, age, culture, location, etc.

5. Premium on building organisational capacity of most-vulnerable communities through formation of grassroots disaster response organisations.
6. The less vulnerable sectors are mobilised into a partnership with the vulnerable sectors in disaster management and development work.

The Department of International Development (DFID) should sponsor vulnerability assessments in developing countries and use the information to help target work on adaptation [to climate change] where vulnerability is greatest, rather than focusing work on adaptation only on the poorest.

(UK House of Commons 2002: Recommendation No. 16)

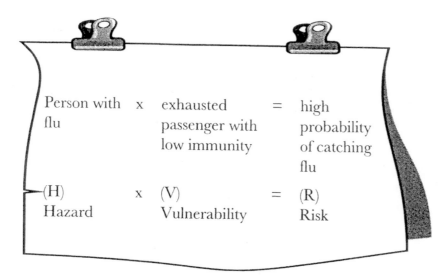

Figure 9.3 Flip-chart example of community discussions
Source: von Kotze and Holloway (1999: 47)

Box 9.1: Emergency Management Australia – extract from a study on the
assessment of personal and community resilience and vulnerability
(adapted from ISDR 2002b: 76).

GUIDELINES FOR MAKING RISK ASSESSMENTS

Contextual aspects:
- analysis of current and predicted demographics;
- recent hazard events;
- economic conditions;
- political structures and issues;
- geophysical location;
- environmental condition;
- access/distribution of information and traditional knowledge;
- community involvement;
- organisations and management capacity;
- linkages with other regional/national bodies;
- critical infrastructures and systems.

Highly vulnerable:
- infants/children;
- frail elderly;
- economically disadvantaged;

Box 9.1 continued

- intellectually, psychologically and physically disabled;
- single-parent families;
- new immigrants and visitors;
- socially/physically isolated;
- seriously ill;
- poorly sheltered social groups.

Identifying basic social needs/values:
- sustaining life;
- physical and mental well-being;
- safety and security;
- home/shelter;
- food and water;
- sanitary facilities;
- social links;
- information;
- sustain livelihoods;
- maintain social values/ethics.

Increasing capacities/reducing vulnerability:
- positive economic and social trends;
- access to productive livelihoods;
- sound family and social structures;
- good governance;
- established networks regionally/nationally;
- participatory community structures and management;
- suitable physical and service infrastructures;
- local plans and arrangements;
- reserve financial and material resources;
- shared community values/goals;
- environmental resilience.

Practical assessment methods:
Constructive frameworks; data sources include:
- local experts;
- focus groups;
- census data;
- survey and questionnaires;
- outreach programmes;
- historical records;
- maps;
- environmental profiles.

Emergency Management Australia, the national-level disaster management agency, and several Australian states routinely implement risk assessment that includes not only a focus on 'highly vulnerable social groups'[9] but also the broader political and socio-economic context we have been arguing for in our book (Box 9.1).

Some progress has also been made in developing an agreed methodology to assess risk factors using standardised checklists.[10] Many authors have suggested methods (Wisner 2003a; Davis 1994, 2003; Cannon 2000a; Morrow 1999; Lavell 1994). However, an agreed inter-agency or inter-governmental methodology for social vulnerability assessment remains elusive, other than the growing general acceptance of the need to firmly link vulnerability to capacity. One reason for the lack of an agreed methodology is the paucity of information concerning the different assessment approaches used as well as concerning their relative effectiveness as assessment tools. This gap will only be closed by applied, interdisciplinary research that compares assessment approaches across different hazard categories within different country and cultural contexts in order to identify the key variables that are needed relative to different hazards.

An example of the difficulty of developing a comprehensive method for assessing vulnerability can be seen in the work of an important international programme called the Global Earthquake Safety Initiative (GESI) (Global Earthquake Safety Initiative 2001). GESI's pilot programme, completed in 2001, examined the seismic safety of 21 cities. In order to define the seismic vulnerability of each city the project team defined a five-part methodology, with data being collected on the following topics:

- *Building Fatality Potential* (soils/building stock/building construction and materials/building occupancy rates).
- *Landslide Fatality Potential* (landslides triggered by the earthquake).
- *Search and Rescue Life-Saving Potential* (numbers of people available to participate/levels of training, etc.).
- *Fire Fatality Potential* (fires induced by earthquakes).
- *Medical Care Life-Saving Potential* (casualty management).

It is notable that this list inexplicably omits a range of critical factors that *also* have a decisive impact on vulnerability and capacity:

- *The level and effectiveness of Public Awareness Programmes* (particularly those that are focused on school children).
- *The level and effectiveness of Disaster Plans* (disaster plans at all levels from national to local).
- *High-Risk Social Groups* (social vulnerability and capacity assessment).
- *Economic Assessment* (urban seismic vulnerability is intimately related to those industrial/commercial and individual livelihoods at risk).

An explanation for such omissions may lie in the difficulty of creating inter-disciplinary projects and approaches. GESI is primarily the work of civil engineers, who are inevitably preoccupied with physical, tangible elements, and such concerns as public awareness, disaster planning, social and economic vulnerability may be unfamiliar and outside their sphere of professional competence and influence. However, as the GESI project continues to take on international urban studies, it is vital, and urgent, for the project organisers to expand their assessment criteria and build interdisciplinary teams to under-take the work in each city being investigated. Such teams will need to include social workers, architects, physical planners, government emergency managers, economists and educationists. Without this wider interdisciplinary frame of reference the project will have fundamental flaws in its design.

The Second Risk-reduction objective offers a framework for comple-menting the engineering approach to risk assessment we have just criticised. Routine assessment activities already exist that focus on construction safety, harvest forecasts for staple crops (using both field observations, market reports and remote sensing), monitoring of the nutritional status of chil-dren, epidemiological surveillance, as well as numerous flood and storm warning systems. However, Part II of the book suggested that vulnerable people often suffer a series of interrelated disasters and that their vulnera-bility often increases through failure of recovery. This has been aptly termed the 'ratchet effect' by Robert Chambers (1983) (see Chapter 3). Existing monitoring systems do not address these cascading, compound problems.

Therefore, our recommendation is that these existing routines incorporate an explicit vulnerability orientation and that their results (interpreted differ-ently due to the vulnerability framework) become the focus of a co-ordinating governmental body ('disaster commission'). The information required is summarised in Table 9.1. It concentrates on five groups that are most likely to have the least protection against hazards and the least reserves for recovery. We have chosen these groups for the purposes of demonstration only. Identi-fication of vulnerable groups will vary from society to society, and situation to situation, where very specific differences based on class, caste (if applicable), gender, health status and disability, age, race, nationality and immigration status, and location may all have a role. Our choice in this example is based on the common ground that the cases in Part II seem to suggest. Thus the five vulnerable groups would be:

- a defined group of the poorest households (taking account of relevant definitions of poverty in that context, and data availability);
- women, particularly women-headed households and those from poor households;
- children and youth;
- the elderly and disabled;
- the poorest and most marginalised of minority groups.

Table 9.1 Types of information required for risk assessment, including vulnerability, capacity and exposure to hazards

Potentially vulnerable group	Natural resources			Physiological and social resources			Financial resources			Hazardousness of home and workplace		
Poorest 33%	+	0	-	+	0	-	+	0	-	+	0	-
Middle 33%	+	0	-	+	0	-	+	0	-	+	0	-
Richest 33%	+	0	-	+	0	-	+	0	-	+	0	-
Women	+	0	-	+	0	-	+	0	-	+	0	-
Children	+	0	-	+	0	-	+	0	-	+	0	-
Elderly and disabled	+	0	-	+	0	-	+	0	-	+	0	-
Minority group A	+	0	-	+	0	-	+	0	-	+	0	-
Minority group B	+	0	-	+	0	-	+	0	-	+	0	-
Minority group C	+	0	-	+	0	-	+	0	-	+	0	-

Note:

The symbols are used to signify whether a particular group is likely to experience increased (+), reduced (-) or no change (0) in its situation in regard to the different factors across the columns.

For each group we need to know several aspects of their 'access to resources' (as discussed in Chapter 3) as well as the hazardousness of the locations they frequent. In all cases we want to determine, at least in a qualitative way, trends over time. Are they gaining or losing access to resources that can help to protect them or help them to cope and recover? Are the spaces they inhabit or where they work becoming more or less hazardous?

We take the reciprocal relation between vulnerability and capacity into account in our use of the term 'resource' in a variety of ways. In each case, an increase in access to a resource (be it natural, physiological, social or financial) suggests an increase in the capacity to cope with and to recover from a hazard event. A decrease in access suggests a decrease in capacity (or an increase in vulnerability).

In Table 9.1 'Natural resources' refers to land, water, forest products and other natural assets (see also Figures 1.1 and 3.1). In many of the preceding chapters we have seen that the most vulnerable have been losing access to such resources over many years, a trend that the Johannesburg Summit on sustainable development also highlighted. There are many reasons for this trend, including privatisation and concentration of land, increased inequality in income distribution and land degradation.

'Physiological and social resources' refer to nutritional status and health, education, access to technology and information. These change in positive or negative directions with time and across generations. For instance, in South Africa the social resource of kinship is reported to be increasingly less important to coping (Sharp and Spiegel 1984). In the USA some reports suggest literacy levels are declining in certain social groups and localities (e.g. Hispanic populations in some of the ghettos of Los Angeles). IMF and World Bank Structural Adjustment Programmes (SAPs) caused many governments to cut back on education and health budgets and staple food subsidies. This policy acted as a dynamic pressure that increased vulnerability because of the erosive effect it had upon knowledge and access to informational resources as well as upon physiological reserves and resources.

'Financial resources' refers to income, market access, banking and other credit facilities. Recalling the Access model outlined in Chapter 3, our discussion of coping and the debates surrounding famine in Chapter 4, it should be clear that the concept of 'financial resources' can be extended to include liquid assets that a household can sell to buy emergency food, health care or to rebuild a house. They would include, for instance, livestock and jewellery. Overgrazing, privatisation or nationalisation (e.g. of game reserves) can deprive herders of pasture, and so interfere with the assets to which people are accustomed.

The final column contains information about the location of home and livelihood activities in relation to various hazard triggers. These could include biological and physical variables affecting the home and workplace such as slope, soil type, micro-climate, and they can be mapped. These variables interact with known (or knowable) hazards, such as torrential rain or the introduction of cholera into the locality. Also relevant is the presence or absence of toxic hazards or the potential for catastrophic industrial accidents (although here we come up against the limit of our self-imposed scope). These 'hazard maps' for vulnerable groups change over time as human groups are forced to move about the landscape and as the spatial distribution of livelihood opportunities changes (Figure 3.3).

In this systematic manner it is possible to gather, display and interpret information of very diverse kinds that may influence the vulnerability to disaster of a specific sub-group in society.

Trends could be established by inquiring about each of these items, by row and column, in regard to the situation in the past, for example two generations ago, 10 years ago and at present. The basic question in each case is whether the specific resource or locational factor is significant for increasing the vulnerability of a group to a hazard, and how this factor affects vulnerability over time.

Third risk-reduction objective

Reduce risks by addressing root causes, dynamic pressures and unsafe conditions[11] **[focus on Reverse in PAR model]**

Our book argues that the environment cannot be made safer by technical means alone. The vulnerability perspective suggests both that it is possible to make the human environment safer and that there are limits set by economic and social inequities, cultural biases and political injustices in all societies. We are optimistic about the possibility of improvement. The 'Pressure' model can be reversed to provide security instead of risk. Vulnerable people's access to resources can be improved, and changes in power relations can be made. Vulnerability can be decreased, and if aid is properly conceived and implemented, even the most vulnerable survivors can recover in such a way that future vulnerability is reduced.

Here we illustrate this by presenting a transformed version of the 'Pressure' model used throughout this book, recalling that the 'R' in the PAR model means 'release'. The new outcome is 'safe' as opposed to 'unsafe' conditions, 'sustainable' versus 'unstable' or 'fragile' livelihoods, and 'resilience' (i.e. 'capable' versus 'vulnerable' people). Figure 9.4 summarises the 'release' process as a reversal of disaster 'pressure'.

The social, economic and political mechanisms (dynamic pressures) that translate root causes into unsafe conditions for specific people can some-times be blocked, changed or even reversed. To illustrate the way the 'pressure' process leading to disaster can be reversed, we should recall some of the good news from chapters in Part II. Village-based and regional grain storage has successfully provided food reserves that rid some African farmers of the burden of indebtedness and the cycle of the 'hungry season' that can escalate into famine. Primary health care at neighbourhood and village levels (including child immunisation) has dramatically increased protection from some diseases, reducing vulnerability to other hazards, and has improved community water supply and sanitation.

Flood mitigation has also been achieved through local efforts, and the 'living with' approach is actively competing with technocratic views. Death and damage from tropical storms continues to be high; however the grass-roots, decentralised efforts of Red Crescent workers in Bangladesh and the success of storm shelters encourages some optimism. At a more macro-level, but with vital links to localities, the development of early warning systems to track the paths of tropical cyclones, using data gained from remote sensing as well as from weather radar systems, has had a major impact in reducing casualties and property losses. Great progress has been made in low-cost housing design and construction that is much safer in earthquakes, and during the past decade major international retrofit programmes have been undertaken to address low-cost dwellings.

All these relate to the middle of the 'Release' model. They are instances of capacity-building measures based on pro-poor economic and political changes. Such actions reverse the mechanisms that translate root causes such as long histories of racial discrimination, unequal access to land and other resources, etc. into specific unsafe conditions. There has been less progress in overcoming the effects of other dynamic pressures. Foreign debt, war, global environmental change and rapid urbanisation still provide formidable challenges. However, such initiatives as debt forgiveness and rescheduling, various debt-for-develop-ment swaps, and the drafting of Poverty Reduction Strategy Papers (PRSPs) by various Highly Indebted Poor Countries (HIPCs) (see Chapter 2) may reduce the influence of some of these dynamic pressures.

THE PROGRESSION OF SAFETY

1   2   3

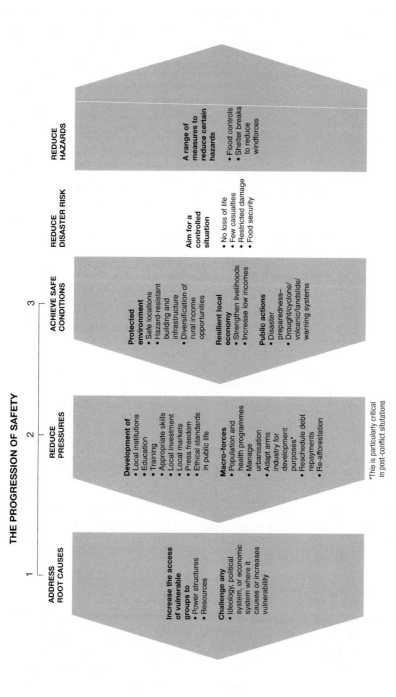

ADDRESS ROOT CAUSES

Increase the access of vulnerable groups to
• Power structures
• Resources

Challenge any
• Ideology, political system, or economic system where it causes or increases vulnerability

REDUCE PRESSURES

Development of
• Local institutions
• Education
• Training
• Appropriate skills
• Local investment
• Local markets
• Press freedom
• Ethical standards in public life

Macro-forces
• Population and health programmes
• Manage urbanisation
• Adapt arms industry for development purposes*
• Reschedule debt repayments
• Re-afforestation

ACHIEVE SAFE CONDITIONS

Protected environment
• Safe locations
• Hazard-resistant building and infrastructure
• Diversification of rural income opportunities

Resilient local economy
• Strengthen livelihoods
• Increase low incomes

Public actions
• Disaster preparedness—
• Drought/cyclone/volcanic/landslide/warning systems

REDUCE DISASTER RISK

Aim for a controlled situation
• No loss of life
• Few casualties
• Restricted damage
• Food security

REDUCE HAZARDS

A range of measures to reduce certain hazards
• Flood controls
• Shelter breaks to reduce windforces

*This is particularly critical in post-conflict statutions

Figure 9.4 The release of 'pressures' to reduce disasters: progression of safety

Root causes (and global dynamic pressures) can be changed, and should not be regarded as immutable and inevitable. In its 1992 report on the state of the world's children, UNICEF recommended that most of Africa's external debts be discounted (UNICEF 1992). Ten years later debt reduction for HIPC countries has become conventional wisdom (although implementation lags). In some cases debts have been cancelled and PRSPs have been written (admittedly in some cases coercively by the World Bank with little local participation). How much these will reduce poverty and therefore indirectly reduce vulnerability remains to be seen.

Peacekeeping efforts by the UN have successfully prioritised humanitarian relief over national sovereignty, and suggest that progress can be made against the influence of war over disaster vulnerability. There is also a considerable focus on the prevention of conflicts (Annan 2002) even though this careful multilateral approach seems, at the time of writing, to be at least temporarily confounded by the unilaterialist doctrine of 'pre-emptive war' developed by the Bush administration in the USA.

In terms of physical hazards, even these can be modified in many cases, although our concern is to ensure that mitigation and modification of the hazards is done in such a way that the investments and science involved do not create other forms of vulnerability, are effective and do not replace proper vulnerability analysis (see 'Second Risk-reduction objective' above).

The central role of governance in reversing the PAR model

In the 10 years since the first edition of this book was published, there has been considerable discussion of 'good governance' in development studies. Improved governance is probably the most important factor in reversing many of the dynamic pressures and even the root causes of vulnerability. We understand good governance not merely as a technical matter of free and fair elections, decentralisation and audits, but as a term that covers the ideologies, power relations, formal and informal networks, and resource flows that determine the relationship between the state (at various levels: national, sub-national, local/municipal) and civil society. 'Good governance' has aspects that are cultural, political, social and economic. Vatsa (2002b) expresses this very well:

> Governance goes much beyond the entity of government. Governance refers to norms, traditions, and processes that impinge on the exercise of formal power and authority. It encompasses government, the private sector, and civil society, and a complex interaction among these segments. The government's policies, resources, and capacities are shaped by the participation and stakeholding of civil society and the private sector in the political process. The distribution of power and authority is actually decided through reconciliation of competing priorities of different segments.

The close interdependence of norms, networks, knowledge and institutions as well as what were once known as 'the means and relations of production' is acknowledged in the literature on 'livelihoods' in development studies (Chambers and Conway 1992; Sanderson 2000) and the literature on disaster and resource management where 'livelihood' and 'access' are centrally important (Blaikie and Brookfield 1987). Our Access model in Chapter 3 depends on such a notion of governance ('structure of domination').

The reader should not mistake our use of the term 'governance' with the one-size fits-all blueprint for Western-style democracy that emerged during the 1990s, as part of development discourse, as 'political conditionality' (Simon 1995; van Spengen 1995). As in the passage by Vatsa quoted above, we are referring to society-wide participation and negotiation that goes well beyond cosmetic institutional changes such as regular national elections. Good governance is exemplified in Cuba's management of hurricane Michelle in 2001, a very large and dangerous storm during which only five Cubans died. Wisner identified a 'golden dozen' characteristics of social relations and governance that seem to explain Cuba's success (Wisner 2001c, quoted in IFRC 2002a: 28–29).

- Social cohesion and solidarity (self-help and citizen-based social protection at the neighbourhood level).
- Trust between the authorities (national or sub-national, e.g. municipal) and civil society (the population at large).
- Investments in economic development that explicitly take into account the potential consequences for disaster risk reduction or increase.
- Investment in human development (basic needs).
- Investment in social capital (e.g. training of neighbourhood-based activists).
- Investment in institutional capital (e.g. capable, accountable and transparently operating government institutions for the prevention and mitigation of disaster risk, not only for response and preparedness; also, investment in scientific capacity for such institutions as Havana's Weather Institute, Cuba's public health service that provided emergency chlorination water, etc.).
- Good co-ordination, information sharing and co-operation among the institutions involved in disaster risk reduction.
- Attention given to the most vulnerable groups of people.
- Attention given to lifeline infrastructure.
- Laws, regulations and directives adequate to support all of the above.
- An effective risk communication system and institutionalised historical memory of disasters.
- A political commitment to disaster risk reduction.

This 'golden dozen' list has many variations and versions in the publications of disaster management worldwide (ISDR 2002b; IFRC 2002a: 24–29; Boyce 2000; Mileti 1999). For example, in one of the most concise and comprehensive statements to appear before the Johannesburg Summit, a joint Four-Point Plan for a Safer World was put forward by three institutions: the Benfield Greig Hazards Research Centre (University College London), the Centre for the Epidemiology of Disaster (Catholic University of Louvain) and the development NGO, ActionAid (Yates et al. 2002) (see Box 9.2). All four points include direct and indirect references to good governance in the broad sense in which we use the term, especially points three and four.

Box 9.2: Four-point plan for a safer world
1 Strengthen society's resilience

- Accelerate global and national poverty eradication initiatives, informed by people's vulnerability to disasters.
- Put the concerns of vulnerable people at the centre of development policy and practice at all levels.
- Review existing long-term development policies and programmes from a vulnerability point of view.
- Incorporate vulnerability indicators in national and international programme reviews and evaluations.
- Make globalisation work for the poor and for poor nations.

2 Integrate and support disaster reduction measures

- Invest in preparedness and mitigation in disaster-prone countries, ideally as a central component of development policy and programming.
- Build the capacity of state institutions, civil society organisations and vulnerable populations.
- Develop and promote regional and sub-regional approaches and assist vulnerable countries in international negotiations.

3 Tackle sources of risk

- Deepen political commitment to environmental protection and renewal – in particular, to full implementation of the Kyoto Protocol and recommendations of the Inter-Governmental Panel for Climate Change.
- Ensure that development efforts at various levels do not increase vulnerability to natural hazards.
- Promote responsive government at all levels to protect citizens' rights to security of life and livelihood.

4 Improve accountability

- Make disaster response and recovery more accountable to the needs of disaster-affected people.
- Promote a culture of respecting the rights of disaster-affected populations.
- Set specific commitments, targets and indicators for disaster reduction.
- Donors should allocate sufficient funding for international initiatives that reduce the impact of emergencies.

A recent analysis and critique of the World Bank's strategic statement, *Attacking Poverty*, called attention to the common problem of 'cultures of secrecy in government, a vacuum of government performance information' that 'helps to make politics an intrigue rather than a competition to improve services and living standards' (*Journal of International Development* 2001). Nowhere is this problem more apparent than in the enforcement of safety codes and standards.

Building codes have proliferated as part of the globalisation of knowledge (Coburn and Spence 2002: 213–217). However, they have been so egregiously unenforced in Turkey, El Salvador, Taiwan, India, Italy and Japan that engineers and others who study disasters have begun to speak about corruption. This was especially apparent after the unnecessary loss of life in earthquakes in the Marmara region of Turkey in 1999, in El Salvador and Gujarat in 2001 and in the collapse of school buildings in earthquakes in Italy (2002)[12] and the Kurdish, south-eastern region of Turkey in 2003. In the last case, in the aftermath of the collapse of a school dormitory, built in 1998, which resulted in the death of 84 sleeping schoolboys, there have been riots and allegations of corruption in the construction process (Millar 2003a; Wisner 2003b). One parent, waiting to see if his child would be found alive in the rubble said, 'Earthquakes don't kill people in Turkey, it's the builders who kill them' (Millar 2003b).

Political scientists such as May (1992), legal experts (Platt et al. 1999) and sociologists (Stallings 1995), as well as some urban and regional planners (Burby 1998), have studied the development and enforcement of building codes and land-use regulations. The danger of continued 'business as usual' – corruption, indifference and lack of adequate personnel – is so great that there have been wider calls for a reform of enforcement mechanisms for such regulations (Wisner 2001e). If such reforms were forthcoming, they would stand as an example of a reversal of the PAR model.

Fourth risk-reduction objective

Build risk reduction into sustainable development[13] **[emphasize sustainable Development]**

Disaster vulnerability is also a function of the way in which humans interact with nature. Part II of this book presented many examples. The perspective of the PAR model also suggests that *local* decisions about land-use are not solely responsible for disastrous results. To assert this would be 'blaming the victim'. Such problems as land degradation and resource exhaustion – processes that may contribute to disaster vulnerability – are closely linked to society-wide 'styles' or modes of production and consumption spatially far removed from the specific site of degradation or depletion. Indeed, before the Johannesburg Summit on Sustainable Development, there was wide

consensus that *global* patterns of production and consumption are implicated, and, more broadly, 'styles' of development. This 'style' includes
energy-intensive, industrial production, and urbanisation based on worldwide trade. We would prefer to use the term 'globalised capitalist mode of
production' for the less-precise word 'style' (Wisner 2001a; Callinicos 2002).
Globalised capitalist production has produced larger and larger scale modifications of ecosystems and the landscape. For instance, in West Virginia
(USA), entire mountain tops are exploded and crushed in a new method of
coal mining: should anyone be surprised that floods and landslides in that
area have increased?

Therefore, when talking about environmental issues and disasters, one has
to discuss them in the context of development policy. The coal mining in
West Virginia produces income. This shows up in financial accounts as
economic growth. Likewise, in the past 50 years, Central America lost half
its forest cover to the steadily increasing production of export crops such as
bananas, cotton, coffee and beef. This shows up as positive economic
growth. However, in neither case does pro-growth economic policy take into
account the economic and human cost of the floods and landslides that
come hand in hand with 'progress'.

In the run up to the Johannesburg Summit, many authors revisited the
connections between land-use and disaster (Voss and Hidajat 2002; McGuire
et al. 2002; ISDR 2002a; Abramowitz 2001; Comfort et al. 1999). They
recalled the lessons of hurricanes Mitch (1998) and Andrew (1992), the
Mississippi floods (1993), and floods in many parts of Europe during the
1990s, as well as enormous floods in China in 1998. In Part II we commented
on these events. Deforestation and other kinds of land-use problems were
implicated in all of these disasters. Also fresh in the Johannesburg delegates'
minds were wildfires in Indonesia, the USA, Australia, Mexico and Brazil;
mudslides in Venezuela in 1999, Algeria and Brazil in 2001; and a deadly
landslide triggered by an earthquake in El Salvador, also in 2001. In all these
cases better land-use planning and enforcement could have prevented the
extreme natural event from becoming a disaster.

So, at a global level, can 'sustainable development' reduce the risks faced
by vulnerable people exposed to natural hazards? Global environmental
change, especially an increase in climate instability, could easily wipe out any
gains in risk reduction derived from improved access to water supply, sanitation and clean energy sources.[14] However, on the positive side, despite
continued opposition by the USA, the Kyoto Protocol on climate change
has come into force with the addition of ratifications by Canada and Russia.
Also, there is a good deal in the agreements reached at the Johannesburg
Summit that could be interpreted and implemented as linking disaster risk
reduction and sustainable development. These specific agreements include
the following (see United Nations 2002b):

- *Water, sanitation and energy* initiatives hold the potential to significantly decrease the vulnerability of millions of urban and rural households, including the 35 per cent of the population of Africa who could benefit from improved clean energy. The global commitment is to supply sufficient energy to achieve the Millennium Goal of halving by 2015 the proportion of people living on $1 per day. This energy is also supposed to be 'reliable, affordable, economically viable, socially acceptable, and environmentally sound'. The indirect effects of improved access to water and energy could be the enhancement of livelihood options as well as savings to households in terms of reduced health care costs (because there is less illness from unclean and insufficient water and the use of smoky fuels) (see the Fifth Risk-reduction objective, below).
- The *Cities Without Slums* initiative contained in the Millennium Goals, with its commitment to improve the lives of 100 million slum dwellers by 2020, also has significant potential to mitigate the risks of earthquake, flood, landslide, storms and epidemic disease faced currently by people living in cities in the LDCs.
- *Biodiversity, forestry and fishery* initiatives are less ambitious, and could portend business as usual. However, even here there is the potential for enhancing household livelihood options and for controlling the large-scale industrial exploitation of resources that continues to displace people and to drive species to extinction.
- *Civil society* was highly visible at the Johannesburg Summit. Eight thousand civil society participants attended the summit. This in itself helps to win credibility for these organisations and offers some protection from hostile opponents in their home countries. As we shall argue in our seventh, and final, risk-reduction objective, strengthening civil society is key to developing a safety culture.

Casting a shadow over these positive aspects of the Johannesburg agreements are the corporate partnerships announced at the Summit. These should cause great concern. For example, nine private energy utilities operating in the European Union have signed agreements with the UN for technical co-operation on energy projects in LDCs. The newly-privatised South African energy company, Eskom, has agreed to 'extend modern energy services to neighboring countries' even as it cuts off services to thousands of low-income households in South Africa because of their inability to pay increased electricity rates. Water supplies in many parts of the world are also being privatised (Petrella 2001; Barlow and Clarke 2002). Is it wise to assume that the market alone can allocate services that are essential to human health and security? Can we expect private owners or managers of water supplies and other lifeline infrastructure to invest in strengthening them against damage from natural hazards when this is likely to reduce their short-term profits?

Fifth risk-reduction objective

Reduce risks by improving livelihood opportunities
[improve Livelihoods]

Livelihoods provide subsistence, cash income, materials and consumables, stores of food, shelter and clothing. A livelihood is a dynamic portfolio which is assembled, constantly reviewed and updated – often in rapid and radical ways in the face of an impending disaster (see Chapter 3). Livelihoods also provide opportunities to convert one type of capital into another. For example, cash income can be 'invested' to educate female children; the education of women is probably one of the most effective strategies in reducing vulnerability. Of course, this way of thinking about livelihoods is economistic and structural (this is why it is complemented by the PAR model, which addresses the political aspects of access to resources and livelihoods). Yet the Access model is fundamental to understanding potential vulnerability in the light of durable and longer-term characteristics of the political economy.

Two aspects of the Access model of livelihood do limit its usefulness. The first is the time dimension. When rapid-onset hazards strike, or longer-term disasters as with famine, the HIV-AIDS pandemic and complex emergencies, the livelihood profiles of different households still shape the damage they suffer and the ways they respond. However, the longer-term iterations of livelihood decisions, determined by the access profile of the household, become less predictable within the Access/Livelihood model outlined in Chapter 3. The model allows iterations through time, which are graphically represented in Box 7 of Figure 3.1 as repeated and changing positions of access profiles and income opportunities as the disaster unfolds. These are shaped by rapidly changing access qualifications and by the preferences of households. For example, stealing and looting (e.g. breaking into grain stores) may be tolerated in ways that they would not be in normal times. Income opportunities may now become necessary which are normally precluded for certain individuals or households on the grounds of gender, class and caste, dignity, respect and custom. These changes are seldom predictable as are the other structural determinants of access and vulnerability. The livelihood approach may become less useful in drawn-out and profoundly damaging disasters where the initial pre-disaster livelihood access profiles of households (and their access qualifications and payoffs) become less and less available and/or relevant.

Secondly, the livelihood approach principally focuses on small units, such as households and groups of individuals with shared characteristics across different households (such as women who are poor, the elderly, children or those excluded on the basis of ethnicity, religion or class). The livelihoods approach is not particularly illuminating with regard to the potential for collective action and community self-protection, nor local politics (which is why it is the complement to our PAR model and third risk-reduction objective, discussed above).

Livelihood patterns that increase vulnerability have the following characteristics. Firstly, they may *lack resilience to shocks*, either because they are not diversified (diversification in itself may be a risk-reduction strategy), or are inherently subject to variability (for example, crop yields on non-irrigated land in the semi-arid tropics), or are characterised by low levels of productivity. Secondly, the main activities in a livelihood portfolio may *provide little cash or consumptive goods* from a narrow range of income opportunities which can only be sold or purchased in a volatile market (for example, pastoralists selling livestock or purchasing grain in a drought). Thirdly, as Chapter 3 and discussion of Table 9.1 have indicated, some livelihood activities *lead people into hazardous time/spaces*. For example, agricultural labourers may have to work in flood-prone lowlands, fishers may have to use small boats during the cyclone season, while urban recyclers are forced to work in and around solid waste sites. Finally, livelihood activities may be *embedded in exploitative or competitive social relations* where one person's disaster is another person's gain.[15]

There is a signal failure in most policy approaches to link livelihood characteristics and vulnerability. The development of certain aspects of livelihoods can reduce risk substantially. For instance, land reform can dramatically increase the access of poorer households to new livelihood options. But even in the absence of such dramatic reform, access by the landless to irrigation, water harvesting or trees can provide streams of income that build household resilience (Chambers et al. 1990). These are the kinds of measures that promote sustainable livelihoods and can lead to a reduction in vulnerability. They flow logically from the Access model in Chapter 3 and its applications as outlined in the body of this book. The following is a list of such measures.

- *Diversifying income sources* for the vulnerable within and outside the agricultural sector, with a view to capital formation and the building up of assets of their own.
- *Diversifying agricultural production* and the crops grown.
- *Increasing food security* by enhancing local subsistence production (returns to labour and land, and reduction of risk through climatic variation).
- *Facilitating local networks* of support and risk awareness.
- *Strengthening local coping mechanisms* through the decentralisation of decision making.
- *Developing 'buffers'* (including food, cash savings and accessible forms of insurance) to cushion the trauma of disasters.

- *Developing crops and seeds storage* (e.g. community grain banks).
- *Securing increased, equitable access to key resources* for those 'at risk' (including natural and financial resources – see below – as well as logistical and informational resources such as timely information about extreme events, transportation where evacuation is required, shelter, emergency health services, communication with relatives, etc.).
- *Challenging the structures of domination* that impede the equitable distribution of livelihood resources (including urban and rural land reform, dissemination of knowledge about land law, and vigorous public oversight and regulation of privatised services such as water and electricity).
- *Developing micro-credit* and small-scale, decentralised banking systems.
- *Public provision of universal education and health care in the longer term* and subsidies allowing universal coverage within privatised or mixed delivery systems in the short term.
- *Recognising the importance of the local state, the municipality and mediating institutions (NGOs) as facilitators* of access to key resources for livelihood sustainability.
- *Giving the necessary encouragement, funding and facilities for women's empowerment*, for example, adult literacy classes, training in interpersonal skills, accounting, managing public meetings, etc., savings clubs, women-only micro-credit and training to combat the threats present after disasters (for example, polluted drinking water, epidemic disease in children).

Sixth risk-reduction objective

Build risk reduction into disaster recovery [Add recovery]

Recovery has received considerable attention in the decade since the first edition of this book. This is due to three developments. Firstly, as we have noted several times, recovery following civil war and other conflicts, often involving the resettlement of large numbers of displaced people, has become a major focus of foreign aid. Some of the concepts that have emerged, such as the 'relief–development continuum' (UNDP 1994b) and 'linking relief with development' (UNDP 1998; Speth 1998) have been extended to recovery following large-scale natural hazard events such as hurricane Mitch and the floods in Mozambique.

Secondly, corporations have begun to invest in measures to ensure 'business continuity' in the face of a wide range of risks, including extreme natural events (Pinkerton 2002). So intense is this corporate interest in business recovery that an internet search for the phrase 'disaster recovery' yields many more entries for consulting firms and workshops focused on the needs of companies than on the recovery needs and opportunities of households, villages and towns.

Thirdly, the World Bank, regional development banks and bilateral donors have invested large amounts in reconstruction following recent earthquakes (e.g. in Colombia, Turkey and Gujarat), tropical cyclones (e.g. in Central America, India and Mozambique) and floods (e.g. in China and south-east Asia). As the amounts of money have increased, more attention has turned to building 'sustainable disaster reduction' into the recovery process so that these lenders and donors will not have to bear the burden of repetitions (Consultative Group 1999). National agencies such as those responsible for emergency management in Australia, New Zealand, Canada and the USA have also begun to seek ways of utilising the 'window of opportunity' created by disasters to implement 'sustainable development' and reduce repetitive losses.

During the 1990s, 'holistic' approaches to recovery (Monday et al. 2001) and 'comprehensive' recovery became fashionable (Wisner and Adams 2003: ch. 5). *In principle*, such recovery would address economic, political and social needs – not only rebuilding infrastructure and housing, but opening the way for more resilient livelihoods. *In practice*, implementation of recovery along these lines requires reversing (or at least substantially palliating) the dynamic pressures and root causes that have contributed to the disaster in the first place. Has this actually happened anywhere? Is it possible?

One high-profile attempt to use a 'developmental' and 'comprehensive' approach following hurricane Mitch in Central America suggests that implementation is very difficult indeed (see Box 9.3). Furthermore, since this approach to recovery based itself on the concept of 'sustainable development', it shares the many theoretical and practical problems associated with this controversial and ambiguous concept. Therefore our thinking about this sixth risk-reduction objective requires careful examination of some fundamental issues. What is 'recovery', and how long does it take? Can recovery strengthen livelihoods? Can recovery contribute to sustainable development? Can recovery address the dynamic pressures and root causes of disaster vulnerability?

These questions that link disasters to social change inevitably require careful definitions of recovery, sustainability, risk reduction and other key terms. These will reflect the context and material needs of those affected in the disasters, and will also encompass the complex set of linkages between vulnerability, the disaster and recovery. The context for such linkages is a politics where the voices of the vulnerable are seldom heard, and possible solutions are structurally excluded from debates. This may well undermine recoverability and sustainability.

Box 9.3: Central America – implementing comprehensive recovery?

After hurricane Mitch in 1998 (see Chapter 7) the countries of Central America met with aid donors in Stockholm. A series of principles for recovery were set out which seemed to involve a fresh approach that would actually build resilience in households, localities and nations. This would reduce losses from the frequent multiple hazards (hurricane, flood, landslide, volcanic eruption, earthquake) that affect the region. The Stockholm Principles were that recovery assistance should (Wisner 2001f):

- reduce the social and ecological vulnerability of the region;

- reconstruct and transform Central America on the basis of transparency and good governance;

- consolidate democracy and good governance, reinforcing the decentralisation of governmental functions and powers, with the active participation of civil society;

- promote respect for human rights, including equality between women and men and the rights of children, ethnic groups and other minorities;

- co-ordinate donor efforts, guided by priorities set by the recipient countries;

- intensify efforts to reduce the external debt carried by countries of the region.

There is little sign of any of these noble goals having been implemented. Instead the emphasis has been on large-scale infrastructure reconstruction projects and the introduction of technical mapping and planning methods, not on building local capacities or strengthening livelihoods

(IFRC 1999b: 52–53).

Of all the countries affected (the others were Honduras, Nicaragua and Guatemala), El Salvador should have been in the best position to take advantage of financial and technical assistance to increase its resilience. It suffered least, and had only 260 of the 21,000 deaths in the region, only 1 per cent of the injured, and its share of those who needed shelter was only 12 per cent. In addition, El Salvador lost only approximately 10,000 homes of the total of 124,068 (ibid.). Yet El Salvador was fully integrated into the post-Mitch Stockholm aid process. It was in a much better position to implement new systems and approaches for prevention and mitigation because the damage it had sustained was relatively light. However, very little changed in El Salvador. A coalition of NGOs proposed a new law to modernise the National Emergency

Box 9.3 continued

Commission (COEN), making it responsible for prevention as well as recovery. The proposed law was rejected twice by the national assembly, and finally passed in 2002; however, at the time of writing, nearly five years after hurricane Mitch, there are few signs of its implementation. Demands by citizen-based groups that they be part of a planning process for future disasters were ignored by the national government.

As a result, when two earthquakes hit El Salvador in January and February 2001, the 'lessons of hurricane Mitch' had not led to any significant institutional development, nor to a better response. More than 1,000 people were killed in these earthquakes, 8,122 injured and more than 150,000 homes destroyed. An additional 185,000 homes were damaged. Infrastructure was also heavily affected, including 23 hospitals, another 121 health care units and 1,566 schools. This represents 40 per cent of hospital capacity and 30 per cent of the nation's schools. The total economic loss was estimated at $1.255 billion (Government of El Salvador 2001).[16]

At a very minimum, the Stockholm Principles should have brought forth vigorous government efforts to protect health facilities and schools. These public facilities are subject to damage in a wide range of extreme natural events. A programme to protect them against hurricane force winds or flooding could easily have incorporated strengthening them against earthquakes. However, none of this had been done following hurricane Mitch.[17] The recovery process after hurricane Mitch, in particular in El Salvador, must be seen as an object lesson in how *not* to apply development rather than only welfare criteria, and how *not* to involve the population in the process.

What is 'recovery' and how long does it take?

In 1977 Haas, Kates and Bowden published work on recovery in post-earthquake Managua, Nicaragua, together with an historical account of recovery in San Francisco after the 1906 earthquake (Haas et al. 1977). Their temporal model seemed to show that 'recovery' proceeds in predictable 'waves' of activity that can be described by bell curve-type graphs. Thus, in the first few days 'response' activities rise and fall; immediate short-term 'relief' activities increase as 'response' tails off. 'Relief' itself reaches a peak after a few weeks, and then medium-term 'reconstruction' takes centre stage, to be replaced after several months by a set of activities that can be described as longer-term economic and social 'recovery' – a part of the process that may take several years.

This model, with its intelligible, even aesthetically pleasing, bell curves has been tested in a number of post-disaster situations (Rovai 1994). Humans crave order and meaning and seek to stabilise their expectations in situations of uncertainty. Perhaps we cannot understand why 'bad things happen to good people', but at least there may be some order to be found in our ability to start over again and to assist in recovery. There is even a hint of deeper order in the fact that Haas and his colleagues proposed a logarithmic relationship, in which each of the stages from 'response' through to 'recovery' seems to take ten times as long as the previous stage.

As appealing and reassuring as this model is, one must ask, is it verifiable? Rovai reviewed experience with the model and concludes as follows:

> The Haas et al. model of disaster recovery has been criticised (Bolin 1994). Berke et al. (1993) argue that the model is a 'value-added' approach, which views a community as going through a series of fixed stages, each stage a necessary development adding value to the final product, which is a recovered community. The problem, as they define it, is that a value-added approach is a linear and orderly representation of irregular and uncertain decision-making processes: community decision making is not a technical exercise where each stage occurs in 'proper sequence' and successful outcomes are guaranteed.
>
> Another study of 14 disaster-stricken communities (Rubin et al. 1985) did not support the model's linearity of phase occurrence. In these communities, the four stages were not necessarily sequential, sometimes occurring simultaneously or in different sequences. However, Stuphen's 1983 study of a single community recovering from a flood does support the basic observation of stages of recovery activities, although her study did not reinforce the clear-cut sequence of recovery periods.
>
> There was, however, much overlap between stages in the recovery timelines in both communities ... [and] ... the length of the phases varied from the logarithmic tenfold increase proposed by Haas et al., although this may be premature ... the temporal duration of the restoration and reconstruction periods was directly related to the temporal duration of the emergency period.
>
> (Ibid: 54–55)

The terminology associated with disaster recovery is biased towards optimism. The key words – 'recovery', 're-establish', 'reconstruction', 'restoration' and 'rehabilitation' – are prefixed with 're', indicating a return to the pre-existing situation. A more realistic view challenges the assumption that such recovery will actually be achieved. Instead, the more pessimistic argument suggests there will be uncertainty, unforeseen events and even the reproduction of vulnerability. A rather depressing implication that is allowed for in both the PAR and Access models is that in some cases the most vulnerable households and individuals do not recover. Many studies demonstrate this failure. Tuareg pastoralists in Sahelian west Africa never managed to restock their herds following the 1967–1973 drought, and were forced to live on the margins of towns (Franke and Chasin 1980). The cyclones in Andhra Pradesh (Chapter 7) also showed that some of the poorest households fell into debt, lost their livelihood options and became landless labourers. Following hurricane Andrew in 1992, many of the low-income Haitian farm workers in southern Dade county simply disappeared. Many thousands who left Montserrat because of the volcanic eruption may never return to the island (Chapter 8).

Even if one does not question the orderliness of 'recovery', a good deal of empirical evidence suggests that people who were marginal, excluded and who enjoyed poor access to power and resources before a disaster may lag behind the recovery model's tidy curves and face greater difficulties in accessing assistance (Fothergill et al. 1999). Such evidence comes from studies following hurricanes Gilbert and Hugo in the Caribbean (Berke and Beatley 1997) and Andrew in south Florida (Peacock et al. 1997) as well as those of the Northridge earthquake (Bolin and Stanford 1998a, 1998b) and the Great Hanshin earthquake that affected Kobe (Hayakawa 2000; see Chapter 8). The Whittier earthquake of 1983 in southern California provides yet more evidence. Miller and Nigg (1993) distinguish between household characteristics which contribute to damage during an 'event' and those which contribute to household disruption as a 'consequence' of an event. They found that race and ethnicity were significantly associated with household disruption. Studying the aftermath of the Northridge earthquake in southern California a decade later, Bolin and Stanford (1998a, 1998b) came to the same conclusion: Hispanics were less likely to obtain official assistance. Similarly, predominantly African-American areas of greater Miami received lower insurance payments and had less private investment after hurricane Andrew (Dash et al. 1997; Peacock and Girard 1997).

A visit to Managua in 2002 revealed a city centre that had never been rebuilt after the 1972 earthquake. In addition, several hundred thousand of those displaced by the 1972 earthquake still live in very poor, under-served conditions in the area outside Managua to which they were moved as a 'temporary' measure. This 'temporary' refuge has, in fact, become an official part of Managua's urban structure, now called Ciudad Sandino. Indeed, Cuidad Sandino – once a 'temporary' zone of shelter for earthquake victims – now has its own newer 'suburb' where those who fled hurricane Mitch are concentrated. Its name, ironically, is Nueva Vida (New Life) (Susman 2003). While the situation described is not sufficient evidence to refute the ordered expectations of Haas et al. (1977), it has to be asked what does 'recovery' mean in a country where the urban and rural poor have to live from disaster to disaster? Hurricane Mitch (1998) and the near-famine in rural areas during the drought of 2001–2002 suggest that there was very little resilience, little buffer or reserve in the 'normal' life of most people in Nicaragua. In a similar way, in the Philippines, 10 years after the eruption of Mount Pinatubo, the people still suffer from annual flows of lahar in the rainy season. Can there be 'recovery' without any capacity to withstand the next crisis or shock?

Peacock, who has worked for years on recovery issues (first in post-earthquake Guatemala and more recently in post-hurricane Miami) is sceptical about the notion of 'recovery':

> Economists and other social scientists continually report 'recovery' levels being reached a few years after an event when talking about a community or region (Wright et al. 1979; Friesma et al. 1979). And yet, one would be hard pressed to suggest that these community level recovery levels were always translated to the household level. Hoover and Bates (1985) clearly showed that communities that experienced earthquake damage in both the highlands and the East [of Guatemala] experienced growth in the number of businesses and in overall community complexity. And yet, Peacock, Killian and Bates (1987), equally clearly showed that a majority of households, particularly those receiving aid in the form of temporary housing, failed to reach recovery levels some four years after the event. Indeed, households that were in communities with higher numbers of *Ladinos* [non-indigenous Guatemalan] and in department capitals, were much better off than more rural and Indian households.
>
> (Peacock 2002)

We would suggest that in order to have 'recovered', a household should have not only re-established its livelihood, physical assets and patterns of access, but should be *more* resilient to the next extreme event. Thus, without the profound changes in social relations and structures of domination suggested by our 'Release' model, it could be argued that recovery never takes place and never can take place. Changes in social relations that define access to land, credit, employment and information are required to make households more resilient to the next hazard event. In theory, these changes could be the result of a series of public, private and citizen-based activities that had utilised the 'window' created by disaster to strengthen capabilities at the local level (Anderson and Woodrow 1998; Berke et al. 1993; IFRC 2001c: 13–23). Also at the level of the locality, recent thinking is that recovery should include mitigation of future extreme events. Examples include financial assistance to relocate houses from the flood plain, conversion of hazardous zones into open space for recreation, enforcement of improved building codes in reconstruction (Mileti 1999: 237; Beatley and Berke 1998). However, viewed in this manner, recovery must be conceived over a much longer time-scale than it has been so far (Mitchell 1996).

Can recovery strengthen livelihoods?

The case studies presented in Part II contained many examples of a *failure* to revive and strengthen livelihoods. Due to unequal access to resources in rebuilding, the gap between the rich and the poor farmers in the Krishna delta of India was greater eight years after the cyclone than before. In Bangladesh, rich landowners are more likely to gain access to new farming land created by floods. Kenya, Nigeria, Sudan and other African countries provide cases where drought is an opportunity for the rich to acquire more land and livestock at the expense of the desperate smallholder. After the 1976 earthquake in Guatemala, official recovery widened the gap between rich and poor, and the resulting resentment, coming after decades of tension over the distribution of land, ushered in a period of bloody repression by the army and death squads (1978–1982).

Other case material reveals how official relief failed to address the marginality of poor rural African Americans in North Carolina after a hurricane, or Hispanic farm and cannery workers in California following an earthquake (Miller and Similie 1992; Laird 1992; Johnston and Schulte 1992). After the 1980 earthquake in Campania (southern Italy) most of the $1.9 billion in relief from the European Community and 18 other countries was paid to Italian contractors and industries from the north, as was the $3 billion the government provided for re-housing. The Italian government's programme for industrialisation in the affected mountain valleys used up scarce agricultural land and led to the pollution of the rivers with factory effluent. Unique mountain village cultures and ecologies were destroyed, further weakening the potential for local livelihoods in the face of future disasters. On balance, the pre-existing inequalities between the north and the south of Italy were strengthened (Chairetakis 1992).

Failures to revive and to reinforce household livelihoods such as these do not invalidate attempts to integrate new kinds of support for livelihoods into recovery programmes. The UNDP's notion of 'curative development' (Speth 1998) focuses precisely on livelihoods.[18] For example, in the aftermath of the 2001 floods in Orissa (India), UNDP (with money from DFID) supported rapid agricultural rehabilitation in the affected coastal districts that enabled the restoration of the farming livelihoods of 15,000 households (UNDP 2001b). An element of increased livelihood resilience was built into this recovery programme: revised rice-planting dates were suggested to the farmers on the basis of a sophisticated agroclimatic model.[19]

While this UNDP project in Orissa may be an example of development-oriented recovery and seems to be moving in the right direction, it does not take account of all the measures listed earlier in the fifth risk-reduction objective. Farming is essential to rural livelihoods. But crop diversification, irrigation and the development of non-farm income sources, the use of micro-credit for small businesses, training and skills enhancement, could also make livelihoods more resilient. Moreover, as our earlier discussion of the fifth risk-reduction objective makes clear, particular aspects of inequality (especially of land ownership, financial and social capital) structure 'normal' access to resources both before and after a disaster. In the case of Orissa, the average size of land holdings was a tiny 0.73 acre/household. The prevailing highly unequal land tenure pattern was not something that was challenged by the UNDP livelihood recovery project. Nevertheless, this and other cases indicate that a livelihood-focused form of comprehensive recovery is possible in principle (see Boxes 9.4 and 9.5).

One of the most effective ways of building more economically resilient communities is through micro-credit schemes being applied to disaster recovery. One must bring banking down to the level of the poorest groups which are normally denied access to small loans other than through the exorbitant interest rates of community money lenders. Loans as small as $50 can be provided to stimulate the reconstruction of small enterprises. The micro-credit movement estimates that 23.6 million people globally were participating in micro-credit schemes in 1999. Significantly, more than half of these were the 'poorest of the poor', the majority being women; 78 per cent of MFI (Micro Finance Institution) clients were in Asia (Gardner 2001; Micro Credit Summit 2000).

The best-known institutionalised micro-finance institution is the Grameen Bank, founded in Bangladesh in the 1970s by the economist Muhamed Yunus. The small-scale lending scheme he devised requires no collateral from participants. Those wishing to secure loans are organised into small units of between two and five persons. New loans are only made when old ones have been repaid. Thus the 'social capital' of neighbourhood linkages, as well as peer pressure and encouragement, provide a form of 'moral collateral'. The system has been one of the success stories of development. The Grameen Bank has worked with 2.3 million members, in 40,000 Bangladesh villages, who maintain a remarkably high 95 per cent loan repayment rate (Grameen Bank 2002).[20]

One value of micro-credit schemes is their potential link to community preparedness programmes, and their importance in assisting disaster recovery. For example, in 1996 Grameen supported rural reconstruction after flooding in Bangladesh by providing small grants for the repair or reconstruction of dwellings. The energy and commitment of groups of women in Bangladesh at one of these gatherings is clear to any visitor, including one of the authors. These programmes are a massive improvement to standing in line passively waiting for relief payments being provided by paternalistic governments and NGOs.

Box 9.4: Flood recovery in Anhui province, China, 1993

After flooding in Anhui province the UNDP allocated $8.3 million as a recovery grant. They assembled an international team of advisers, comprising Chinese and foreign expertise, under the leadership of one of the present authors (Davis et al. 1992). The team was tasked to advise on the allocation of micro-finance schemes to establish and strengthen micro-enterprises in the worst flood-affected areas. UNDP insisted on this 'developmental approach' to recovery planning, and no funds whatsoever were used as relief 'hand-outs'. Thus the finance was to devolve through micro-credit repayments to enable new investment to recur in the region as permanent development support.

For most of the assessment team this task was completely new territory. They were more familiar with traditional 'relief distribution' programmes following disasters, involving cash or 'food for work' or direct grants of goods or cash to those in greatest need. An early task for the team was to determine a priority ranking for the type of micro-credit tasks to be supported. This was necessary to deal with the continual lobbying the team experienced as various groups put in bids for a portion of the recovery grant. Therefore a variety of criteria were discussed for the selection of projects for investment. The options considered included projects that would:

- bring benefits to the most vulnerable communities;
- bring benefits to those worst affected by the flooding;
- use local materials so that the financial inputs would have a 'ripple impact' (or local economic multiplier), bringing economic benefit down the line to providers of raw materials;
- place emphasis on labour rather than capital-intensive approaches to maximise employment in the region;
- build from existing micro-credit schemes already in place in the province;
- provide support to flood recovery (such as brick making for reconstruction).

Following extensive discussions with the potential stakeholders these criteria were put in ranked order and then used for project selection. A parallel task for the assessment team was to identify suitable institutions to manage such micro-credit programmes.

As the assessment team travelled throughout the flood-affected areas they became aware of the rapid recovery of some of the worst-affected families. Many had already rebuilt their homes within three months of the flood water receding. This was largely because they had insurance cover and were able to rebuild rapidly as soon as their compensation payments were made. Many of these families had low annual incomes of under $400 per annum, but their insurance cover was part of a government-organised agricultural insurance scheme throughout Anhui Province. Therefore their insurance premiums were very low and took the form of a provincial tax.

The provincial authorities and the UNDP adopted a policy in which families were able to fund their own recovery through small loans (micro-credit) and/or via the provision of community insurance.

Box 9.5: Flood recovery in Mozambique, 2000

Following the floods of 2000, major donors in Mozambique followed similar policies to those adopted in Anhui (see Box 9.4). USAID had traditionally dispensed tangible relief goods after disasters; in this case, instead, they provided cash grants of $100 to 85,000 families. The Director of USAID in Mozambique, Cynthia Rozzell, explained the rationale for this approach:

People need money to get back to normal life. The traditional donor response is to pass out seeds or clothes. The people sell those and buy what they really need. So give people money and let them determine their own needs.

(Quoted in Christie and Hanlon 2001: 142)

In addition USAID, as well as NGOs such as CARE and World Relief, adopted extensive micro-credit schemes after the flood. USAID set up a $7 million low-interest credit line for wholesalers, and if they took up the offer they were in turn required to extend 60 days credit to their retailers in lieu of the normal 14 days for repayment (ibid.).

Can recovery contribute to sustainable development?

George Nez is an experienced physical planner who played a key role in the reconstruction of two cities that had been destroyed by earthquakes (Skopje in Yugoslavia from 1963 and Managua, Nicaragua from 1972). He has reflected on pre-disaster constraints that impeded the progress of recovery:

When you direct a reconstruction programme everyone tends to blame the disaster for this or that problem. However, gradually you come to realise that ninety percent of the problems you encounter were present before the disaster event, waiting to be tackled. All that has happened is that the disaster has acted like a sharp surgeon's scalpel that has been used to expose all manner of

363

weakness and failure, such as poor government, un-enforced building codes, lack of planning, corruption in all directions, etc. The issue poses a dilemma concerning how far it is possible to go, with the limited resources at your disposal, in rebuilding the *society* as well as its towns and cities.

(Nez 1974, emphasis in original)

Some agencies such as the US Department of Energy (1999) have introduced programmes that promote post-disaster reconstruction with energy-efficient designs; others have found ways for municipalities to convert flood-prone areas into nature reserves and open space (Monday et al. 2001: ch. 7; Burby 1998). Such initiatives are welcome, but they do not constitute 'sustainable development'. They don't begin to recognise, much less address, the scope of 'recovery' identified by Nez. In our earlier discussion of the Johannesburg Summit and our fourth risk-reduction objective, we emphasised the necessity of combining poverty reduction and the other Millennium Goals with the imperatives of ecological integrity.

The words of Nez above highlight the challenge of building 'sustainability' in this broader sense into programmes of recovery. He notes that most of what a planner must confront after a disaster – the obstacles to implementation as well as many of the environmental and social deficiencies encountered – were there *before* the disaster. We would go further, in light of our PAR model, and argue that many of those environmental and social problems actually *contributed to* the disaster because of the way they shaped specific forms of vulnerability (unsafe conditions). In short, recovery can only contribute to sustainable development if it challenges ongoing dynamic pressures and root causes of disaster vulnerability.

Can recovery address dynamic pressures and root causes of disaster vulnerability?

We have said a great deal about how root causes and dynamic pressures involve factors and processes that create the vulnerability of unsafe conditions. But it is vital to recognise that those roots and pressures are also operating during the unfolding of a disaster itself. As the cases presented in Part II show, the structures of domination and social relations remove control over the conditions of life and livelihood away from households and localities, and concentrate it in the hands of others (landlords, political functionaries, corporations, banks, development 'experts' and refugee-camp organisers). Our discussion of the third risk-reduction objective earlier in this chapter asserted that fundamental changes in governance are required to reverse the PAR model. Recall that governance 'encompasses government, the private sector, and civil society, and the complex interaction among these segments' (Vatsa 2002b). Therefore, any recovery programme that shifts power and control back into the hands of households and localities by rearranging the power relations among government, civil society and the private sector will accordingly contribute towards addressing root causes and dynamic pressures. It is not only the sweeping acts of a revolutionary state (such as Cuba's land reform in 1959) that address deeply entrenched institutional arrangements and values (root causes). Slower and smaller reforms can also be effective (Olson and Gawronski 2003).

It is possible that some of what was happening in Gujarat state (India) in 2002 provides an example of recovery which is contributing to the return of control to ordinary people. The United Nations Centre for Regional Development (UNCRD) is working with an Indian NGO called SEEDS (Sustainable Environment and Ecological Development Society) in supporting a series of projects where one of the key objectives is to mobilise and sustain local capacities to reduce vulnerability. In their words:

> Rehabilitation should be empowering. The project team would not, and should not, remain with the community forever. In such a case, the community who are the first responders should be sufficiently equipped to cater for their immediate needs. A well planned rehabilitation exercise can significantly increase the capacity of the community in a more effective response. ... The success of the rehabilitation exercise is judged by the degree to which actions are replicated by the community without intervention from the project team. Inputs on capacity building are therefore important. Additionally, the project team needs to ensure that conditions would continue to exist for easy replication.
>
> (UNCRD and SEEDS 2002: 7–8)

Challenging patriarchy is another way of changing the power structures which increase vulnerability and reproduce it after a disaster. For example, it is important to ensure that women participate fully in decision making concerning the siting as well as the form and safety of their dwellings. The rationale is that in many societies, such as rural Turkey, women and small children are likely to spend far more time than men in their dwellings during daylight hours and thus are far more vulnerable to seismic impact. The men are more likely to be better protected at work in fields or factories. But it has to be recognised that providing such enhanced roles for women would of course challenge deeply ingrained cultural and religious tradition in many societies. In the aftermath of the 2001 earthquake in Gujarat, women's organisations such as the Self-Employed Women's Association (SEWA) became active in recovery work, work that, long after homes have been rebuilt, is likely to result in a more active role for more women in their towns and villages (Bhatt 2003; Enarson 2001; Fordham 2001).

The integration of indigenous coping into relief and recovery is another way to challenge long-established beliefs and a distribution of power that hinders recovery and reinforces patterns of vulnerability. This is never easy, especially in hierarchical systems where there is a long history of antagonism between town and country, peasant and landlord, 'high' culture and 'low' culture, but a disaster may offer opportunities for such challenges to be made. The integration of people's knowledge is not a mechanical activity, but begins with respect for the people concerned, and it requires their trust. NGOs in less developed countries have led the way in demonstrating respect and building trust, and we provided a sample of the methods they use during our presentation of the second risk-reduction objective earlier in this chapter.

The fruit of these efforts has been the emergence in the 1980s of a development framework based on people's knowledge and local organisation (Chambers 1994, 1995b, 1997). This approach requires listening to local people (Pradervand 1989; Narayan and Petesch 2002) and having an awareness of how power relations can block participation of the most vulnerable (Johnson and Mayoux 1998). Indeed, as Chambers (1983) writes, it is essential to 'put the last first'. Doing so opens up a channel of communication between the people and disaster recovery workers that goes beyond 'consultation'. People are able to express their needs and work together with outsiders to overcome obstacles (Lisk 1985; Wisner 1988a).

In a more specialised domain this participatory method of working has been called the 'farmer first' approach (Chambers et al. 1989). In an elaborated form and a more generally applicable approach, it is called a 'deliberative inclusionary process' (Holmes and Scoones 2000). Water projects, sanitation work, reforestation, housing, grain storage design and many other efforts have benefited from participatory or 'action research' methods, in which outsiders and local people are equal learners and teachers (Cuenya et al. 1990; Marchand and Udo de Haes 1990; Wisner et al. 1991). Successes have been registered in Asia and Latin America (Conroy and Litvinoff 1988; Holloway 1989). In Africa this approach to 'development-with' the people has been called 'the silent revolution' (Cheru 1989; cf. Rau 1991; Kiriro and Juma 1989; Salih 2001).

One example highlights the potential of popular control of the recovery process. Following a landslide, residents of the Caracas (Venezuela) neighbourhood of Nazareno refused to move into remote barracks provided by the government. Instead, they broke into and occupied a local school. They demanded land and assistance in building more-secure homes. The government agreed, and a team including architects and social psychologists from the *Escuela Popular de Arquitectura* (People's School of Architecture) worked with the community to develop a permanent recovery plan (Wisner et al. 1991: 282; Sanchez et al. 1988).

Seventh risk-reduction objective

Build a safety culture [extend to Culture]

The framework in Part I of this book, and the evidence produced from the cases studied in Part II, have revealed some generalisable truths about disasters. To summarise, disasters

- are rooted in everyday life;
- are manifestations of development failures;
- have distant and remote precursors;
- are linked to livelihood resilience and household capabilities;
- result in the need to release pressures through changes in institutions, structures of domination and improved access to resources.

The question we address with this final risk-reduction objective concerns the very last of these points. What is to be done to change institutions so that disasters can be prevented, their effects lessened (mitigated), vulnerability reduced, and capabilities and resilience increased? Such institutional change would result in a 'safer environment'. But how do we get there? All of the preceding risk-reduction objectives depend on the process of institutional change.

In the course of Part II we noted a marked tendency in conventional disaster work to treat symptoms rather than causes. The reason for this bias is that vulnerability is deeply rooted, and any fundamental solutions involve political change, radical reform of the international economic system, and the development of public policy to protect rather than exploit people and natural resources.

Some believe, with good reason, that creating a safer environment depends on moral arguments directed at those with power, a minority who have opportunities to make the world safer for the vast majority. The majority are vulnerable because they are subordinated and unable to make choices that would provide safety even when they are aware of safer options. Looking back over the past few centuries, such moral appeals have sometimes borne fruit. But, more often, safety has been the subject of demands for human rights and dignity made by the vulnerable (and their organisations) themselves. Maskrey observed more than a decade ago that the only 'choice' available to the residents of Lima's squatter settlements was 'between different kinds of disaster ... [P]eople seek to minimise vulnerability to one hazard even at the cost of increasing their vulnerability to another' (Maskrey 1989: 12). That situation has not improved; it may even have deteriorated.

At the beginning of the International Decade for Natural Disaster, the editor of the *New Scientist* perceptively observed:

> All the aims of the IDNDR will cost money, and, in particular, money for things that appear to have no immediate benefit. Few politicians will support something that could bring no visible benefit for 10 or 20 years, let alone a century. Add to this the fact that many of the measures that could cut the death toll from disasters will disrupt people's lives, and you have a very good excuse for doing nothing.
>
> (*New Scientist* 1989: 3)

In preparing for the IDNDR, an international committee of experts anticipated this difficulty for governments in 'selling' protection policies to their constituencies. So it was proposed that:

> efforts to reduce the impacts of natural disasters will gain far wider acceptance if they are perceived as a means to protect economic development and improve living standards rather than to mitigate some hypothetical, localised and infrequent event.
>
> (Austin 1989: 65)

That economic argument has now become a matter of consensus. The World Bank, Inter-American Development Bank and the ProVention Consortium agree that the billions of dollars in losses (from events such as hurricanes Andrew and Mitch, the Kobe and Northridge earthquakes, the Mississippi and River Elbe floods) are extremely serious and substantially reduce economic performance. However, despite this, little has been done to reverse the root causes and dynamic pressures causing widespread vulnerability, even in G7 countries.

There are parallels here with the development of 'green politics', where some governments are persuaded to change policy because of carefully-framed economic arguments, often against powerful vested interests. These are also long-term measures that cost large sums of money and cause disturbance. However, while there is now a global political lobby that is pressing for environmental protection (even for 'environmental justice'), no comparable lobby yet exists for protection from disasters. Also, environmental demands have widespread popular support in many countries where millions of ordinary people have returned 'green' candidates to city councils, state and national parliaments. Disaster vulnerability has not yet become so widely accepted as a popular issue, partly because of the persistence in the media and education system of the myth of the 'naturalness' of disasters.

At the political level, then, a reduction in disaster vulnerability must become a 'green' issue as well as an organic part of a poverty reduction agenda. This may require more work in drawing out the similarities in vulnerability to natural hazards and to technological and environmental hazards. The latter are already a significant part of green consciousness, arising from disasters such as those in Bhopal, Chernobyl, Sellafield, Three Mile Island, Love Canal and the sinking of the *Exxon Valdez* in Prince William Sound. To date there has been little contact between academics and activists working in these two areas. Links must also be made between the various bureaucracies that deal separately with disaster risk reduction and environmental protection.

Achieving radical changes to address the fundamental causes of vulnerability is extremely difficult. The mitigation strategy outlined in this chapter focuses on policies and procedures that will reduce some risks even if they leave many vulnerabilities resolutely intact. It is perhaps best described as the 'art of the possible'. The more difficult measures that will be needed to reduce vulnerability significantly involve changes in power relations and economic systems.

Popular grassroots involvement in assessing risks and in designing and implementing mitigation measures can have the further, long-term effect of giving people the self-confidence and organisation to demand more. The words of Boyden and Davis still ring true.

> [D]isaster mitigation has implications which are quite different – and much further-reaching – than those of disaster relief. First, relief by its very nature creates dependency between donor and recipient. Mitigation on the other hand aims to increase the self-reliance of people in hazard-prone environments to demonstrate that they have the resources and organisation to withstand the worst effects of the hazards to which they are vulnerable. In other words, disaster mitigation – in contrast to dependency creating relief – is empowering.
>
> (Boyden and Davis 1984: 2)

Their view was echoed in 1999 by Ariyabandu, who saw the arguments for investment in mitigation as the following.

- Mitigation reduces loss of life and human suffering.
- Not all disasters are 'emergencies'. Many disasters including drought and floods are predictable. Investment in mitigation can minimise their impact.
- Mitigation reduces the risk element in development investments, thus making development initiatives more stable.
- Treating disasters separately from development action can contribute towards increased levels of hazards and vulnerability, and can jeopardise future development initiatives.

- Mitigation reduces dependency on relief and aid, and strengthens local capacities for preparedness.
- Mitigation reduces the resources required for emergency responses; therefore in the longer run more resources become available for development (Ariyabandu 1999: 10).

Critics of status quo development efforts have for some time called attention to the potential of what they call the 'third system'. This refers to members of civil society who organise to demand their rights and to protect their communities.[21]

Ideally, the IDNDR would have captured the imagination of civil society and brought forth an upsurge of such citizen action; but it did not. However, the 1990s did see significant popular protest focused on perceived injustices and the risks of the concentration of economic power brought about by globalisation. There was a very effective campaign for debt amnesty in favour of the poorest, highly indebted countries (the Jubilee campaign). Also, in the past the worldwide movement of consumers' unions has successfully confronted the power of the pesticide and baby food industries. In the USA, activist Lois Gibbs (whose family became sick because of industrial pollution from Love Canal in her town near Buffalo, New York) now co-ordinates some 7,000 citizen groups who make up a national toxic monitoring network. Survivors of Chernobyl have become part of the rapidly growing citizen's environmental movement in the territories of the former Soviet Union. Survivors of the poisoning from the Bhopal chemical factory explosion have also become activists. There are 4,000 member organisations of the Environmental Liaison Centre (based in Nairobi), the majority of which are in LDCs, and more than 2,000 cities worldwide are involved with Local Agenda 21.

In many parts of the world, school leavers and university graduates are returning to their home localities to use their knowledge. They struggle alongside their neighbours to reclaim sustainable agricultural technologies half-forgotten or buried by the influence of the Green Revolution (Murwira et al. 2000; Marglin 1998). They continue to fight against 'mega-projects' such as large-scale dams in India, Brazil, Mexico, the Philippines, Canada, China and elsewhere. In other urban and rural communities, churches, mosques and other bodies form the focus for a citizen response to economic dislocation and crisis. Food banks, community kitchens and pantries have sprung up all over the USA and in many Latin American countries such as Argentina, to assist and involve poor and hungry people. People's health centres and public health movements have also emerged in the slums of many of the world's mega-cities from Brooklyn and the Bronx to Rio de Janeiro, Mexico City and Manila.

Such formal and informal organisations are woefully under-utilised by authorities responsible for disaster mitigation. A four-year comparative study of relations between civil society and more than 220 municipalities in six mega-cities found a consistent pattern of disconnection or antagonism between NGOs and government (Wisner 2002a).[22] NGOs have been quicker to recognise the potential of such groups. The people themselves can campaign to secure their livelihoods and life spaces (Anderson and Woodrow 1998), to recognise the 'untapped power of people's science' (Wisner et al. 1977) and the effectiveness of 'community-based' mitigation (Maskrey 1989). People organise themselves spontaneously in many areas of life. Organisations such as tenant unions, squatters' councils, purchasing co-operatives and urban gardening groups often arise as a way of protecting less powerful people from the power of landlords, urban officials or profiteering retailers, organisations that already exist for reasons of self-protection and perceived threats in the local environment.

In these cases it is quite typical for the organisations to turn their attention to the physical hazards that affect them. In Rio de Janeiro, for instance, one of the most powerful centres of community organisation in the *favelas* is the samba school. Each year members of a 'school' plan for months, organising dancers and making elaborate costumes in advance of a great carnival parade and competition. These events are a focus of poor people's hope, pride and energy. Could the samba schools become a major force for reducing vulnerability to urban floods and mudslides in Rio? They could be ideally placed to press municipal authorities to clear the rubbish that often blocks drainage channels and leads to floods (see Chapter 6). In fact, in at least one of Rio's *favelas*, a grassroots organisation that began campaigning for public health improvements, has taken up the issue of floods (SINAL 1992; E. Williams 1992).

Figure 9.5 suggests the sequence through which the kinds of local organisation just discussed can enter the national legislative process and eventually bring about forms of hazard mitigation. It has happened before in regard to automobile safety, fire safety in public places, chemicals in the environment and a host of other safety issues so common now that they are barely noticed.

Preconditions for citizen participation

Popular organisations – as diverse as trade unions, consumers' associations, churches, mosques and samba schools – can contribute to the entire range of possible preventive and mitigating action. However, such citizen protest and lobbying only finds full expression under conditions of economic and political democracy and respect for human rights. A minimum level of needs satisfaction is also required before the poorest are able to engage in organised activities to increase security against natural hazards. Where they have jurisdiction, governments must be responsible for the constitutional protection of rights, and for a safety net for the

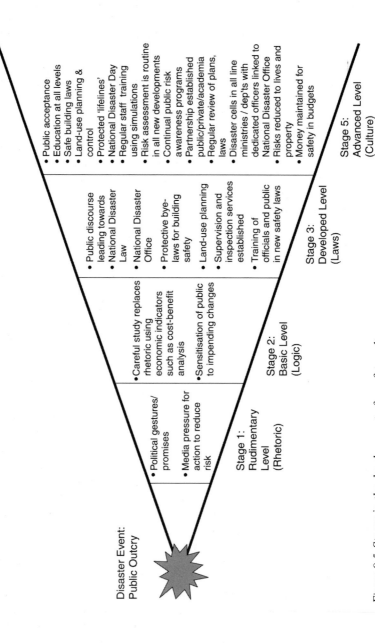

Figure 9.5 Stages in the development of a safety culture

satisfaction of basic needs sufficient to allow the most vulnerable to partic-ipate (Doyal and Gough 1991; Wisner 1988a). If the nation-state has any moral ground for legitimacy, it must be the provision of such basic protec-tion and empowerment.

There must also be a certain minimum level of social peace before the most vulnerable can risk becoming publicly involved. When we wrote the first edition of this book Sudan, Mozambique, El Salvador and Lebanon were beset by war. Of these, only Sudan is still at war, but Colombia, Congo, Sierra Leone, Liberia, Ivory Coast, Chechnya, Afghanistan, Iraq and many others have been added to the list. Even as the second edition is being completed, the USA has named an 'Axis of Evil', and announced and acted upon a doctrine of 'pre-emptive war' that could further destabilise large areas of Asia, the Middle East and Africa.

We are haunted by Hewitt's sombre vision:

> The gravest developments in insecurity, untimely death, uprooting, and impoverishment of peoples in our century have derived from totalitarian violence or monopolistic deployments of science and technology and a war paradigm. They have continued to dominate the landscape during the first half of the IDNDR. There is no indi-cation that such grave dangers arise from too much civil freedom, high quality and equitably distributed social services, devolution of powers, or traditional arrangements for social security and disaster response.
>
> (Hewitt 1998: 91)

Under such circumstances, it is very difficult, if not impossible, for ordi-nary people to exert local, co-operative effort to deal with natural hazards. Lives are stretched to the limit by the struggle to make ends meet, and to deal with the continuing disaster of war. War continues to work powerfully as a dynamic pressure, pushing people into deeper vulnerability. Where war makes national government jurisdiction ambiguous, difficult or impossible, new forms of international intervention are needed, not only to bring an end to the hostilities, but also to ensure that the peace is constructed with a built-in concern for vulnerability to disasters.

But, we have argued, war is not the only global pressure that belongs to the chain of disaster causation. There are other root causes and dynamic pressures that are less intractable, for which there is less excuse for inaction or inadequate responses. If our analysis in this book is persuasive, it must lead to the understanding that disasters are reduced only by releasing people from the unsafe conditions that derive from social, economic and political pressures. These are pressures that will not disappear simply because the new and old economic and technocratic elites concede that they should be abol-ished. Organised, ordinary people must make them disappear.

We began this chapter with some reflections on the way vulnerability to disasters has been addressed in a series of UN conferences and initiatives that span the past decade. Considerable space was devoted to this review since we regard these deliberations as being significant pointers to the stance taken by the UN and member governments regarding the reduction of risks from natural hazards. However, we also questioned the seriousness of this rhetoric and whether it will lead to a reduction of vulnerability. The world that has given rise to the statements about reducing risk is also the world in which political and economic power is increasingly centralised and concentrated.

Against this background, the skills, capacities and growing political consciousness of ordinary people are important in two ways. Throughout our book we have documented the significance of local knowledge and action. This is the first and most straightforward way in which a 'bottom-up' approach is important. Yet such local efforts are often stunted or even blocked by policies that come 'from above'. Secondly, therefore, local initiatives are important as building blocks toward a 'global civil society'[23] that is able to insist on concrete implementation of the goals proclaimed in international forums. Local technical knowledge plays a role in disaster vulnerability reduction; however, there is also an important role to be played by local political understanding and consciousness. The latter can serve to bring to bear organised popular pressure on legislators, officials and businesses for the reforms required to reverse the direction of our PAR model. Exhortation by UN bodies (or textbook authors) is insufficient without significant pressure for change coming from those most 'at risk'.

Notes

1 School-aged children are also very much at risk, as the tragic deaths of pupils in a school in Italy in 2002 and a school dormitory in Turkey in 2003, reminded us all.

2 While such linkages are imperative, we are concerned that once disaster-risk reduction is absorbed into the wider picture, there is a possibility that it will be forever lost from view. Our fear comes from listening, rather too often, to development specialists describe disaster mitigation and preparedness dismissively as 'an integral part of development'. Their analysis is, of course, hard to fault, but the practical outcome of this perspective is the failure to include such concerns in their work. This gives rise to the destruction of development initiatives in disasters as well as the inadvertent increase in vulnerability that can be caused by 'development'.

3 One is reminded of Gandhi's list of 'Seven Deadly Social Sins' and similar catalogues in other traditions. Perhaps we are proposing the opposite: seven social virtues.

4 We mean 'consciousness raising' in the sense that the great Brazilian adult educator, Paulo Freire describes in *Pedagogy of the Oppressed* (1972), that is, as a 'study of the (social and physical) environment'.

5 Douglas and Wildavsky (1982); Brown (1987); Wynne (1987); Beck (1992); Slovic (1993).

6 Both Barnes (2002) and Gilbert (1998: 16) describe the social disorder following disaster as due, in part, to an upsetting of the system of meaning. We see the issue of 'trust' in the context of the system of meaning. It is therefore not only an issue of perceived probity, even-handedness in administration and transparency of bureaucratic process, but goes deeper into the language with which the dominant techno-social order seeks to 'understand' cause and effect.

7 Kasperson and Pijawka (1985) may be a notable exception.

8 See the work of Soto (1998); Turcios et al. (2000); Trujillo et al. (2000); Chiappe and Fernandez (2001); Wilches-Chaux and Wilches-Chaux (2001).

9 The Emergency Management Australia framework is described in ISDR (2002b: 76); see also Buckle et al. (2000).

10 We are aware that there are views of vulnerability assessment that favour replacing checklists with a 'situational' analysis (Wisner 2003a) or more participatory approaches that privilege the 'voices of the poor' (Hewitt 1998; Enarson and Morrow 1997; Narayan and Petesch 2002). While we are largely in agreement from a theoretical and methodological standpoint, we can still see practical situations – for example, refugee camps, centres for internally displaced people, shelters – where checklists are invaluable. These can, in fact, be developed and validated in quieter times on the basis of more situationally specific and participatory methods. In addition, we should note that there are ethical concerns which suggest that research and assessment results should, at a very minimum, be shared with the vulnerable people concerned.

11 We will not repeat the many catalogues of specific measures that have been developed over the years by engineers to construct flood- and wind-proof structures, to strengthen buildings against earthquakes, to map, analyse, and in some cases, to provide early warning of hazards (Zschau and Kueppers 2002; Alexander 1993, 2002; Smith 2001; Coburn and Spence 2002; Smith and Ward 1998; Pielke and Pielke 2000; Parker 2000; Parker and Handmer 1992; Chen 2002). We could only scratch the surface of this rich knowledge base, and, in the end, the problem as we see it is not lack of such specific technical understanding, but of the root causes and dynamic pressures that combine to block its effective application. See the section of RADIX devoted to the theme 'Knowing versus doing' http://online.northumbria.ac.uk/geography_research/radix/knowingvsdoing.htm

12 In total, 26 schoolchildren plus their teacher died in San Giuliano di Puglia, in southern Italy, when the reinforced concrete second storey of their schoolhouse fell in on them during a moderate earthquake. The first storey was built in the 1950s, prior to strict building codes, while the second was a recent addition – most probably an addition which the original structure could not support safely (Soares 2002).

13 Some of this section is based on Aragon and Wisner (2002). We will not here be making a critique of the term 'sustainable development'. Many others have pointed out the ambiguity and difficulties of this phrase (Adams 2001; Dovers et al. 2001; Stott and Sullivan 2000; Redclift and Benton 1995; Bramwell 1994; Worster 1993; Dovers and Handmer 1993).

14 Walter and Simms (2002) even write, apocalyptically, of the 'end of development' and 'the great reversal of human progress' as the result of inattention to global warming.

15 Such a 'zero sum' situation is an extreme case, but as we pointed out at the beginning of this chapter, there is ample evidence that traders, war lords and others very clearly benefit from displacement and famine. It is tempting to see such exploitative relations as extreme extensions of 'normal' social relations in capitalist society and not as aberrations due to individual misbehaviour. US corporations are lined up to make billions of dollars from the reconstruction of a conquered Iraq (Burkeman

2003; Penman 2003; Dow 2003) – more than $8 billion (Dow 2003), moreover, to come from selling the oil of the vanquished, which makes perfect sense in terms of the logic of capitalism in the twenty-first century.

16 Data on losses vary considerably. For example, the Economic Commission for Latin America's estimate of total economic loss is closer to $1.5 billion, and the count of homes destroyed has been as high as 300,000, depending on the source. We use Government of El Salvador (GOES) data. This estimate is likely to be on the low side.

17 See Wisner (2001f) for more detail.

18 The term 'curative development' is used by UNDP in the context of its post-conflict programme. However, this agency's programmes that deal with recovery from (and prevention of) disasters triggered by natural hazard events and those that arise because of violent conflict are housed under the same division (Emergency Response Division) and their approach is very similar. This involves 'rebuilding infrastructure and production systems' as well as 'resettlement and reintegration of displaced people, restoration of health and education services, local planning and participatory systems, and environmental rehabilitation' (UNDP n.d.: 2). Indeed, the operational co-ordination role mandated to the UNDP Resident Representative in each country means that recovery from all forms of emergency and disaster potentially fall under the brief of UNDP, just as, in principle, all immediate UN disaster response is co-ordinated by the UN Office for Co-ordination of Humanitarian Activity (OCHA). Its 'compendium of the UNDP record in crisis countries' (ibid.), describes an 'area-focused' programme to resettle people displaced by drought in northern Sudan (ibid.: 6) and also the use of the same approach aimed 'at alleviating war-induced deprivation' (ibid.).

19 Of course, we must hope that this revised crop calendar was the product of a dialogue between the outsiders and the farmers, and not simply imposed from above.

20 Some scepticism has been expressed concerning the very high loan repayment rate reported – more than 90 per cent; however, there is no doubt that the Grameen Bank has provided a valuable service to millions of people in Bangladesh (see Lawson-Tancred 2002).

21 See Nerfin (1990); Wignaraja (1993); Stiefel and Wolfe (1994); Hall (1995); Burbach et al. (1997); Smith et al. (1997); Alvarez et al. (1998); Mohanty and Mukerji (1998); Fox and Brown (1998); Kothari (1999); de Senillosa (1999); UNRISD (2000).

22 Sponsored by the United Nations University, this study examined perceptions of urban social vulnerability held by municipal officials, citizen-based groups and NGOs in sample districts of six mega-cities (conurbations with more than 10 million inhabitants). The cities surveyed were Tokyo, Manila, Mumbai, Johannesburg, Mexico City and Los Angeles (see Wisner 2002a).

23 On the question of the development of 'global civil society' in relation to a wide range of risks, see the Irmgard Coninx Foundation (Berlin) initiative (n.d.).

BIBLIOGRAPHY

Abbott, A. 2001. *Time Matters: On Theory and Method*. Chicago: University of Chicago Press.

Abramovitz, J. 2001. Averting Unnatural Disasters. In: L. Brown et al., *State of the World*, pp.123–142. New York: Norton.

Achebe, A., Hyden, G., Magadza, C. and Pala Okeyo, A. (eds) 1990. *Beyond Hunger in Africa*. Nairobi and London: Heinemann Kenya and James Currey.

Action Against Hunger 1999. *Geopolitics of Hunger 1998–1999*. Paris: Presse Universitaires de France.

Action Against Hunger 2001. *The Geopolitics of Hunger 2000–2001*. Boulder, CO: Lynne Rienner.

Adams, J. 1995. *Risk*. London: UCL Press.

Adams, R. and Spence, R. 1999. Earthquake. In: J. Ingleton (ed), *Natural Disaster Management*, pp.52–54. Leicester: Tudor Rose.

Adams, W. 2001. *Green Development: Environment and Sustainability in the Third World*. 2nd edn. London: Routledge.

Adamson, P. 1982. The Rains. In: J. Scott (ed), *The State of the World's Children*, 1982–1983, pp. 59–128. New York: UNICEF.

Adedeji, A. 1991. Will Africa Ever Get Out of its Economic Doldrums? In: A. Adedeji, O. Teriba and P. Bugembe (eds), *The Challenge of African Economic Recovery and Development*, pp.763–782. London: Frank Cass.

Adger, N. 1999. Social Vulnerability to Climate Change and Extremes in Coastal Vietnam. *World Development* 27: 249–269.

Adnan, S. 1993. Social and Environmental Aspects of the Flood Action Plan in Bangladesh: A Critical Review. Paper presented at Conference on the Flood Action Plan in Bangladesh, European Parliament, Strasbourg, May.

Africa World Press 1997. Africa's Debt Burden. *Worldviews*
http://worldviews.igc.org/awpguide/debt.html .

Agarwal, A., Kimondo, J., Moreno, G. and Tinker, J. 1989. *Water, Sanitation, Health – For All?* London: International Institute for Environment and Development/Earthscan.

Agarwal, B. 1986. *Cold Hearths and Barren Slopes*. New Delhi and London: Allied Publishers and Zed Press.

Agarwal, B. 1990. Social Security and the Family: Coping with Seasonality and Calamity in Rural India. *Journal of Peasant Studies* 17, 3: 341–412.

377

Agence France-Press (AFP) 2000. Mozambique Flood Damage Worth A Billion Dollars. World Bank, 28 March.
http://wwww.reliefweb.int/w/rwb.nsf/0/7283d3233c244468c12568b0005897d8? OpenDocument

Agence France-Presse (AFP) 2002a. Devastating Floods Start Receding as Indonesians Ask Who's to Blame. ReliefWeb, 4 February.
http://www.reliefweb.int/w/rwb.nsf/s/0E36B7FFD232F7A249256B560026DB13.

Agence France-Presse (AFP) 2002b. At Least 75 killed by Floods in Indonesia's East Java. ReliefWeb, 7 February.
http://www.reliefweb.int/w/rwb.nsf/s/F1031FDF2BB7F7E249256B59002D0A7C.

Agence France-Presse (AFP) 2002c. Infrastructure Damage Bill Soars from Indonesian Floods. ReliefWeb, 17 February.
http://www.reliefweb.int/w/rwb.nsf/s/D42004F28D20118C49256B640018C15D.

Agence France-Presse (AFP) 2002d. Ingushetia Stops Food Deliveries to Chechen Refugees. ReliefWeb, 1 March.
www.reliefweb.int/w/rwb.nsf/9ca65951ee22658ec125663300408599/9fbdfccdb16c1 68c85256b6f007c0d2a?OpenDocument.

Agence France-Presse (AFP) 2002e. Tanzania Donates 2,000 Tonnes of Maize to Famine-Hit Malawi. ReliefWeb, 28 March.
http://wwww.reliefweb.int/w/rwb.nsf/480fa8736b88bbc3c12564f6004c8ad5/c4fc71e f3f2c49e8c1256b8a00524797?OpenDocument.

Agence France-Presse (AFP) 2002f. Famine in Southern Africa Worsening: Experts. ReliefWeb, 12 May.
http://www.reliefweb.int/w/rwb.nsf/480fa8736b88bbc3c12564f6004c8ad5/531b8e 10f0e2f8adc1256bb8003551b7?OpenDocument

Aglionby, J. 2001. 46,000 to be Moved from Avalanche Path. *Guardian*, 4 September.

AIDS ACTION 2001. Mother to Child HIV Transmission in Africa. Policy Facts (January). http://www.aidsaction.org/legislation/pdf/mom2child.pdf

Akong'a, J. 1988. Drought and Famine Management in Kitui District, Kenya. In: D. Brokensha and P. Little (eds), Anthropology of Development and Change in East Africa, pp.99–120. Boulder, CO: Westview.

Alamgir, M. 1980. *Famine in South Asia*. Cambridge, MA: Oelgeschlager, Gunn & Hain.

Alamgir, M. 1981. An Approach Towards a Theory of Famine. In: J. Robson (ed), *Famine: Its Causes, Effects and Management*, pp.19–44. New York: Gordon & Breach.

Alexander, D. 1985. Death and Injury in Earthquakes. *Disasters* 9,1: 57–60.

Alexander, D. 1989. Urban Land Slides. *Progress in Physical Geography* 13, 2: 157–191.

Alexander, D. 1993. *Natural Disasters*. London: UCL Press.

Alexander, D. 2000. *Confronting Catastrophe: New Perspectives on Natural Disasters*. New York: Oxford University Press.

Alexander, D. 2002. *Principles of Emergency Planning and Management*. Harpenden, Herefordshire, UK: Terra Publications.

Ali, A. 1987. Personal communication with Ben Wisner in Dhaka.

Ali, M. 1987. Women in Famine. In: B. Currey and G. Hugo (eds), *Famine as a Geographical Phenomenon*, pp.113–134. Dortrecht: D. Reidel.

Allan, W. 1965. *The African Husbandman*. London: Oliver & Boyd.

378

Allen, E. 1994. Political Responses to Flood Disaster: The Case of Rio De Janeiro, 1988. In: A. Varley (ed), *Disasters, Development and Environment*, pp. 99–108. Chichester: Wiley.

Altieri, G. 1987. *Agroecology*. Boulder CO: Westview.

Alvarez, R. 1999. Tropical Cyclone. In: J. Ingleton (ed), *Natural Disaster Management*, pp.34–36. Leicester: Tudor Rose.

Alvarez, S., Dagnino, E. and Escobar, A. (eds) 1998. *Cultures of Politics, Politics of Cultures: Re-Visioning Latin American Social Movements*. Boulder, CO: Westview.

Ambraseys, N. 1988. Unpublished Notes of a Presentation on the Mexican Earthquake of 1985 to a Workshop on Disaster Management 4–5 July, p.2. Oxford: Disaster Management Centre.

Ambraseys, N. 1995. Observation Made during a Seminar on the Reduction of Disaster Vulnerability. Organized by the UK IDNDR, Committee at the Royal Society in May 1995. Quotation taken from notes by Ian Davis, Chairman of the seminar discussion (unpublished).

Anderson, J. 1987. Lands at Risk, People at Risk: Perspectives on Tropical Forest Transformations in the Philippines. In: P. Little and M. Horowitz (eds), *Lands At Risk*, pp.249–268. Boulder, CO: Westview.

Anderson, M. 1990. Which Costs More: Prevention or Recovery? In: A. Kreimer and M. Munasinghe (eds), *Managing Natural Disasters and the Environment*, pp.17–27. Washington, DC: World Bank.

Anderson, M. 1999. *Do No Harm: How Aid Can Support Peace – or War*. Boulder, CO: Lynne Rienner.

Anderson, M. and Woodrow, P. 1989. *Rising from the Ashes: Development Strategies in Times of Disaster*. Boulder, CO: Westview.

Anderson, M. and Woodrow, P. 1998. *Rising from the Ashes: Development Strategies in Times of Disaster*. Boulder, CO and London: IT Publications and Lynne Rienner.

Anderson, R. and May, R. (eds) 1982. *Population Biology of Infectious Diseases*. Berlin: Springer Verlag.

Andrae, G. and Beckman, B. 1985. *The Wheat Trap: Bread and Underdevelopment in Nigeria*. London: Zed Press.

Andrews, L. and Nelkin, D. 2001. *Body Bazaar: The Market for Human Tissue*. New York: Crown Publishers.

Annan, K. 2002. *Prevention of Armed Conflict: Report of the Secretary General*. New York: United Nations.

Annis, S. 1988. What Is Not The Same About The Urban Poor: The Case of Mexico City. In: J. P. Leavis (ed), *Strengthening the Poor: What Have We Learned?* Washington, DC: Overseas Development Council.

Anton, P., Arnold, K., Truong, G. and Wong, W. 1981. Bacterial Enteric Pathogens in Vietnamese Refugees in Hong Kong. *Southeast Asian Journal of Tropical Medicine and Public Health* 12: 151–156.

Aragon, F. and Wisner, B. 2002. Mitigating Disasters and Conflicts. Chapter for CD based Sustainable Development Module produced for the World Summit on Sustainable Development by Leadership for Environment and Development (LEAD). www.lead.org.

Ariyabandu, M. 1999. *Defeating Disasters: Ideas for Action*. Colombo: Duryog Nivaran and Intermediate Technology Development Group.

379

Arnold, D. 1988. *Famine: Social Crisis and Historical Change*. Oxford: Blackwell.

Arnold, D. 1999. Hunger in the Garden of Plenty: The Bengal Famine of 1770. In: A. Johns (ed), *Dreadful Visitations: Confronting Natural Catastrophe in the Age of Enlightenment*, pp.81–111. New York: Routledge.

Article 19 1990. *Starving in Silence: A Report on Famine and Censorship*. London, International Centre on Censorship.

Article 19 1993. *Silent Violence*. London: Article 19.

Astill, J. and Chevallot, I. 2003. Conflict in Congo Has Killed 4.7 M., Charity Says. *Guardian Unlimited*, 8 April.
http://www.guardian.co.uk/international/story/ 0,3604,931997,00.html.

Atsumi, T. and Suzuki, I. 2003. A Nationwide Network of Disaster NPOs to Cope with Regional Vulnerability in Japan. Paper presented at the DPRI-IIASA Third Symposium on Integrated Disaster Risk Management (IDRM-03). Kyoto: Kyoto University, 4 July.

Austin, J. and Bruch, C. 2000. *The Environmental Consequences of War*. Cambridge: Cambridge University Press.

Austin, T. 1989. Decade for Natural Disaster Reduction. *Civil Engineering*, December: 64–65.

Aykroyd, W. 1974. *The Conquest of Famine*. London: Chatto & Windus.

Aysan, Y. and Davis, I. (eds) 1992. *Disasters and the Small Dwelling: Perspectives for the UN IDNDR*. London: James & James Science Press.

Aysan, Y. F., Coburn A. W., Davis I. R. and Spence R. J. S. 1989. *Mitigation of Urban Seismic Risk: Actions to Reduce the Impact of earthquakes on Highly Vulnerable Areas of Mexico City*. Report of Bilateral Technical Co-operation Agreement between the Governments of Mexico and the United Kingdom. Oxford and Cambridge: Disaster Management Centre, Oxford Polytechnic and University of Cambridge.

BBC 1999. China's Floods: Is Deforestation to Blame? August.
http://news.BBC.co.uk/hi/english/world/asia-pacific/newsid_413717.stm.

BBC NEWS 1999. Your Experiences of the Venezuelan Flood. BBC Talking Point, 29 December. http://news.bbc.co.uk/hi/english/talking_point/newsid.

Baird, A., O'Keefe, P., Westgate, A. and Wisner, B. 1975. *Toward an Explanation of Disaster Proneness*. Occasional Paper No. 11. Disaster Research Unit, University of Bradford.

Ballard, P. 1984. The Miskito Indian Controversy. *Antipode* 16, 2: 54–64.

Balter, M. 1998. On World AIDS Day: A Shadow Looms over Southern Africa. *Science* 282, Dec. 4: 1790–1791.

Bankoff, G. 2001. Rendering the World Safe: 'Vulnerability' as Western Discourse. *Disasters* 25, 10: 19–35.

Barker, D. and Miller, D. 1990. Hurricane Gilbert: Anthropomorphizing a Natural Disaster. *Area* 22, 2: 107–116.

Barlow, M. and Clarke, T. 2002. *Blue Gold: The Fight to Stop the Corporate Theft of the World's Water*. New York: The New Press.

Barnaby, F. (ed) 1988. *The Gaia Peace Atlas*. New York: Doubleday.

Barnes, P. 2002. Approaches to Community Safety: Risk Perception and Social Meaning. *Australian Journal of Emergency Management* 17, 1: 15–23.

Barnett, A. S. and Blaikie, P. M. 1989. Aids and Food Production in East and Central Africa: A Research Outline. *Food Policy* 14, 1: 2–7.

Barnett, A. and Blaikie, P, 1992. *Aids in Africa: Its Present and Future Impact*. London: Belhaven.

Barnett, A. and Blaikie, P. 1994. AIDS as a Long-Wave Disaster. In: A. Varley (ed), *Disasters, Development and Environment*, pp. 139–162. Chichester: Wiley.

Barnett, A. and Whiteside, A. 2001. AIDS in the Twenty-first Century: Disease and Globalization. Basingstoke, UK: Palgrave/Macmillan.

Barth-Eide, W. 1978. Rethinking Food and Nutrition Education under Changing Socio-economic Conditions. *Food and Nutrition Bulletin* 2, 2: 23–28.

Bates, R. 1981. *Markets and States in Tropical Africa*. Berkeley: University of California Press.

Bates, R. 1986. Postharvest Considerations in the Food Chain. In: A. Hansen and D. McMillan (eds), *Food in Sub-Saharan Africa*, pp.239–253. Boulder, CO: Lynne Rienner.

Baxter, P. 1993. Personal communication with Ian Davis on the evacuation before Pinatubo volcanic eruption and on recovery after Cerro Negro eruption.

Baxter, P. 1997. Volcanoes. In: E. Noji (ed), *The Public Health Consequences of Disasters*, pp.179–206. New York: Oxford University Press.

Baxter, P. and Kapila, M. 1989. Acute Health Impact of the Gas Release at Lake Nyos, Cameroon, 1986. *Journal of Volcanology and Geothermal Research* 39: 266–275.

Bay, C. 1988. Human Needs as Human Rights. In: R. Coate and J. Rosati (eds), *The Power of Human Needs in World Society*, pp.59–100. Boulder, CO: Lynne Rienner.

Bebbington, A. and Perrault, T. 1999. Social Capital and Political Ecological Change in Highland Ecuador: Resource Access and Livelihoods. *Economic Geography* 75, 4: 395–419.

Bebbington, A. and Thiele, G. 1993. *Non-Governmental Organizations and the State in Latin America: Rethinking Roles in Sustainable Agricultural Development*. London: ODI.

Beck, U. 1992. *Risk Society: Towards a New Modernity*. London: Sage.

Beck, U. 1995. *Ecological Politics in an Age of Risk*. Cambridge: Polity Press.

Beck, U. 1998. The Politics of Risk Society. In: J. Franklin (ed), *The Politics of Risk Society*, pp.9–22. Cambridge: Polity Press.

Beck, U., Giddens, A. and Lash, S. 1994. *Reflexive Modernisation: Politics, Tradition and Aesthetics in the Modern Social Order*. Cambridge: Polity Press.

Becker, J. 1996. *Hungry Ghosts: China's Secret Famine*. London: John Murray.

Beerlandt H. and Huysman, S. 1999. *Manual for Bottom-up-Approach in Food Security Interventions*. Rome: IFAD/ Belgian Survival Fund. http://www.ifad.org/gender/tools/hfs/bsfpub/manual_toc.htm.

Bell, B., Kara, G. and Batterson, C. 1978. Service Utilization and Adjustment Pattern of Elderly Tornado Victims in an American Disaster. *Mass Emergencies* 3: 71–81.

Bellamy Foster, J. 2003. A Planetary Defeat: The Failure of Global Environmental Reform. *Monthly Review* 54, 8 (January). http://www.monthlyreview.org/0103jbf.htm

Bellington, A. 1999. Capitals and Capabilities: A Framework for Analysing Peasant Vitality, Rural Livelihoods and Poverty. *World Development* 27, 12: 2021–2044.

Benedick, R. 1991. *Ozone Diplomacy: New Directions in Safeguarding the Planet*. Cambridge, MA: Harvard University Press.

Bennett, J. 2000. Inter-agency Coordination in Emergencies. In: D. Robinson, T. Hewitt and J. Harris (eds), *Managing Development: Understanding Inter-organizational Relationships*, pp.167–192. London: Sage.

Benson, C. 2002. Personal communication between Ben Wisner and Dr. C. Benson, Overseas Development Institute, London, 4 June, 2002.

Benson, C. 2003. Macroeconomic Concepts of Vulnerability: Dynamics, Complexity and Public Policy. In: G. Bankoff, G. Frerks and T. Hilhorst (eds), *Vulnerability: Disasters, Development and People*, ch. 11. London: Earthscan.

Benson, C. and Clay, E. 1998. *The Impact of Drought on Sub-Saharan African Economies*. World Bank Technical Paper No. 401. Washington, DC: World Bank.

Berke, P. and Beatley, T. 1997. *After the Hurricane: Linking Recovery to Sustainable Development in the Caribbean*. Baltimore, MD: Johns Hopkins University Press.

Berke, P., Kartez, J. and Wenger, D. 1993. Recovery after Disaster: Achieving Sustainable Development, Mitigation and Equity. *Disasters* 17, 2: 93–109.

Bernstein, H. 1977. Notes on Capital and Peasantry. *Review of African Political Economy* 10: 60–73.

Bernstein, H. 1990. Taking the Part of Peasants? In: H. Bernstein, B. Crow, M. MacKintosh and C. Martin (eds), *The Food Question: Profit versus People?*, pp.69–79. London: Earthscan.

Berry, S. 1984. The Food Crisis and Agrarian Change in Africa: A Review Essay. *African Studies Review* 27, 2: 59–112.

Bhatt, E. 2003. SEWA's Approach to Poverty Removal. Ahmedabad, India: Self-Employed Women's Association (SEWA). http://www.sewa.org/sewa-approach.htm.

Biehl, J. 1991. *Rethinking Ecofeminist Politics*. Boston: South End Press.

Bishop, V. 1998. *Hazards and Responses*. London: Collins Educational.

Black, R. 1998. *Refugees, Environment and Development*. Harlow, Essex: Longman.

Blackstock, C. 2002. Minister Hints at Flood Tax on Households at Risk. *Guardian*, 14 February.
http://www.guardian.co.uk/guardianpolitics/story/0,3605,649940,00.html.

Blaikie, P. 1985a. Natural Resources and the World Economy, In: R. Johnston and P. Taylor (eds), *A World in Crisis: Geographical Perspectives on Global Problems*, pp.107–126. London: Blackwell.

Blaikie, P. 1985b. *The Political Economy of Soil Erosion in Developing Countries*. London: Longman.

Blaikie, P. 1989. Explanation and Policy in Land Degradation and Rehabilitation for Developing Countries. Land Degradation and Rehabilitation 1, 1: 23–38.

Blaikie, P. and Brookfield, H. 1987. *Land Degradation and Society*. London. Methuen.

Blaikie, P. Cameron, J. and Seddon, J. 1977. *Centre, Periphery and Access in West-Central Nepal: Social and Spatial Relations of Inequality*. Monographs in Development Studies No. 5, University of East Anglia, UK.

Blaikie, P. and Coppard, D. 1998. Environmental Change and Livelihood Diversification in Nepal: Where is the Problem? *Himalayan Research Bulletin* 18, 2.

Blaikie, P., Harriss, J. and Pain, A. 1985. *Public Policy and the Utilization of Common Property Resources in Tamil Nadu, India*. Report to Overseas Development Administration, Research Scheme R3988.

Blaikie, P. and Sadeque, Z. 2000. *Policy in High Places: Environment and Development in the Himalayas.* Kathmandu, Nepal: International Centre for Integrated Mountain Development.

Blaut, J. 1993. *The Colonizer's Model of the World.* New York: Guilford.

Bolin, R. 1994. *Household and Community Recovery After Earthquakes.* Monograph No. 56. Boulder: University of Colorado.

Bolin, R. and Bolton, P. 1986. *Race, Religion and Ethnicity in Disaster Recovery.* Boulder: University of Colorado.

Bolin, R. and Klenow, D. 1983. Response of the Elderly to Disaster: An Age Stratified Analysis. *International Journal of Ageing and Human Development* 16, 4: 283–296.

Bolin, R. and Stanford, L. 1998a. The Northridge Earthquake, Community-based Approaches to Unmet Recovery Needs. *Disasters* 22, 1: 21–38.

Bolin, R. and Stanford, L. 1998b. The Northridge Earthquake, Vulnerability and Disaster. London and New York: Routledge.

Bolton, P., Liebo, E. and Olson, J. 1993. Community Context and Uncertainty Following a Damaging Earthquake: Low-Income Latinos in Los Angeles, California. *The Environmental Professional* 15: 240–247.

Bommer, J. 1985. The Politics of Disaster – Nicaragua. *Disasters* 9, 4: 270–278.

Bondestam, L. 1974. People and Capitalism in the North Eastern Lowlands of Ethiopia. *Journal of Modern African Studies* 12, 3: 423–439.

Booth, D., Lugangira, F., Masanja, P., Mvungi, A. Mwaipopo, R., Mwami, J. and Redmayne, A. 1993. *Social, Cultural and Economic Change in Contemporary Tanzania: A People-oriented Focus.* Stockholm: SIDA.

Borton, J. 1984. *Disaster Preparedness and Response in Botswana.* Report to the Ford Foundation. London: Relief and Development Institute.

Borton, J. 1988. *Evaluation of ODA Emergency Provision to Africa 1983–86.* Environmental Report EV425, August, London: ODA.

Borton, J., Brusset, E. and Hallam, A. 1996. The International Response to Conflict and Genocide: Lessons from the Rwanda Experience. *Journal of Humanitarian Assistance*, May. http://www.reliefweb.int/library/nordic/book3/pb022.html.

Boseley, S. 1999. WHO Calls for Action on Killer Diseases. *Guardian*, 18 June. http://www.guardian.co.uk/print/0,3858,3875987,00.html.

Bowbrick, P. 1986. The Causes of Famine: A Refutation of Professor Sen's Theory. *Food Policy* 11, 2: 105–132.

Boyce, J. 1987. *Agrarian Impasse in Bengal.* Oxford: Oxford University Press.

Boyce, J. 1990. Birth of a Megaproject: The Political Economy of Flood Control in Bangladesh. *Environmental Management* 14, 4: 419–428.

Boyce, J. 1992. *Land and Crisis in the Philippines.* London: Macmillan.

Boyce, J. 2000. Let Them Eat Risk. *Disasters* 24, 3 (September): 254–261. http://online.northumbria.ac.uk/geography_research/radix/resources/boyce-disasters.doc (see also PERI Working Paper No. 4, University of Massachusetts. http://www.umass.edu/peri/pdfs/WP4.pdf).

Boyden, J. and Davis, I. 1984. *Editorial: Getting Mitigation on the Agenda.* Bulletin No.18, October. University of Reading Agricultural Extension and Rural Development Centre.

Bradley, D. 1977. The Health Implications of Irrigation Schemes and Man-made Lakes in Tropical Environments. In: R. Feachem, M. McGarry and D. Mara (eds), *Water, Wastes and Health in Hot Climates*, pp.18–29. Chichester: John Wiley & Sons.

Brammer, H. 1989. Report on the International Conference on the Greenhouse Effect and Coastal Areas of Bangladesh. *Disasters* 13, 1: 95.

Brammer, H. 1990a. Floods in Bangladesh: I. Geographical Background to the 1987 and 1988 Floods. *The Geographical Journal* 156, 1: 12–22.

Brammer, H. 1990b. Floods in Bangladesh: II. Flood Mitigation and Environmental Aspects. *The Geographical Journal* 156, 2: 158–165.

Brammer, H. 1992. Floods in Bangladesh: Vulnerability and Mitigation Related to Human Settlement. In: Y. Aysan and I. Davis (eds), *Disasters and the Small Dwelling*, pp.110–118. London: James & James Science Press.

Brammer, H. 1993. Protecting Bangladesh. *Tiempo: Global Warming and the Third World*. No. 8, April, pp.7–10.

Brammer, H. 2000. Controversies Surrounding the Bangladesh Flood Action Plan. In: D. Parker (ed) *Floods*, pp.302–315, Vol. 1. London: Routledge.

Bramwell, A. 1994. *The Fading of the Greens: The Decline of Environmental Politics in the West*. New Haven, CT: Yale University Press.

Branford, S. and Kucinski, B. 1988. *The Debt Squads: The U.S., the Banks and Latin America*. London: Zed Press.

Brandt, W. 1986. *World Armament and World Hunger*. London: Gollancz.

Brauer, M. and Hisham-Hashim, J. 1998. Fires in Indonesia: Crisis and Reaction. *Environmental Science and Technology* 32, 17: 404A–407A.

Bread for the World (ed) 1991. *Food as a Weapon*. Washington, DC: Bread for the World.

Brecher, J. 1999. *Panic Rules: Everything You Want to Know about the Global Economy*. London: Southend Press.

Briggs, P. 1973. *Rampage: The Story of Disastrous Floods, Broken Dams and Human Fallibility*. New York: David McKay.

British Overseas Development 1990. Holding back the Flood: Action Planned to Help Save Bangladesh. *British Overseas Development* 10 (February): 1 & 4.

Broch-Due, V. 2000. Producing Nature and Poverty in Africa: An Introduction. In: V. Broch-Due and R. Schroeder (eds), *Producing Nature and Poverty in Africa*, pp. 9–52. Stockholm: Nordiska Afrikainstitutet.

Brocklesby, M. and Fisher, E. 2003. Participatory Vulnerability Analysis. A Step-by-Step Field Guide. Swansea: Centre for Development Studies. Mimeo.

Brokensha, D., Warren, D. and Werner, O. (eds) 1980. *Indigenous Knowledge Systems and Development*. Lanham, MD: University Press of America.

Brosnam, D. 2000. *The Montserrat Volcano: Sustainable Development in Montserrat*. Portland, OR: Sustainable Ecosystems Institute. http://www.sei.org/sustainable_development.html.

Brown, E. 1991. Sex and Starvation: Famine and Three Chadian Societies. In: R. Downs, D. Kerner and S. Reyna (eds), *The Political Economy of African Famine*, pp.293–321. Philadelphia: Gordon & Breach Science Publishers.

Brown, L 1995. *Who Will Feed China? Wake-up Call for a Small Planet*. New York: W. W. Norton.

Brown, L 1996. *Tough Choices; Facing the Challenge of Food Security*. New York: W. W. Norton.

Brown, P. 1987. Popular Epidemiology: Community Response to Toxic Waste – Induced Disease in Woburn, Massachusetts. *Science, Technology and Human Values* 12, 3&4: 78–85.

Brownlea, A. 1981. From Public Health to Political Epidemiology. *Social Science and Medicine* 15D: 57–67.

Bruce, V. 2001. *No Apparent Danger*. New York: HarperCollins.

Brush, L. Wolman, M. and Bing-Wei, H. 1989. *Taming the Yellow River: Silt and Floods*. Dordrecht: Kluwer.

Bryant, E. 1991. *Natural Hazards*. Cambridge: Cambridge University Press.

Bryceson, D. 1989. Nutrition and the Commoditization of Food in Sub-Saharan Africa. *Social Science and Medicine* 28, 5: 425–440.

Bryceson, D. 1999. *Sub-Saharan Africa Betwixt and Between: Rural Livelihood Practices and Policies*. Working Paper No.43. De-agrarianisation and Rural Employment Network, Leiden: Afrika-Studiecentrum.

Buchanan-Smith, M. and Davies, S. 1995. *Famine Early Warning and Response: The Missing Link*. London: Intermediate Technology Publications.

Buckle, P. 1998/99. Re-defining community and vulnerability in the context of emergency management. *Australian Journal of Emergency Management* 13, 4, pp. 21-26.

Buckle, P., Marsh, G. and Smale, S. 2000. New Approaches to Assessing Vulnerability and Resilience. *Australian Journal of Emergency Management* 15, 2: 8–14.

Buckle, P., Marsh, G. and Smale, S. 2001. Assessment of Personal and Community Resilience & Vulnerability. Report (EMA Project 15/2000).

Bujra, J. 2000. Risk and Trust: Unsafe Sex, Gender and AIDS in Tanzania. In: P. Caplan (ed), *Risk Revisited*, pp.9–84. London: Pluto Press.

Bullard, R. 1990. *Dumping in Dixie: Race, Class and Environmental Quality*. Boulder, CO: Westview.

Burbach, R., Nunez, O. and Kagarlitsky, B. 1997. *Globalisation and its Discontents: The Rise of Postmodern Socialisms*. London: Pluto Press.

Burby, R. (ed) 1998. *Cooperating with Nature: Confronting Natural Hazards with Land Use Planning for Sustainable Communities*. Washington, DC: Joseph Henry Press.

Burgo, E. and Stewart, H. 2002. IMF Policies 'Led to Malawi Famine', *Guardian*, 29 October. http://www.guardian.co.uk/business/story/0,3604,821285,00.html.

Burkeman, O. 2003. Cheney Oil Firm Widens Iraq Role. *Guardian*, 8 May. http://www.guardian.co.uk/international/story/0,3604,951313,00.html.

Burton, I., Kates, R. and White, G. 1978. *The Environment as Hazard*. New York: Guilford.

Burton, I., Kates, R. and White, G. 1993. *The Environment as Hazard*. 2nd edn. New York: Guilford.

Bush, R. 1985. Drought and Famines. *Review of African Political Economy* 33: 59–63.

Button, G. 1992. When Marsians Take Over: The Politics of Symbolic Resistance in Mars Cove, Alaska. Paper presented at the 51st Annual Meeting of the Society for Applied Anthropology, Memphis, TN.

Buvinic, M., Vega, G., Bertrand, M., Urban, A., Grynspan, R. and Truitt, G. 1999. *Hurricane Mitch: Women's Needs and Contributions*. Washington, DC: World Bank, Women in Development Program Unit.

Byrne, L. 1988. Tree Felling Blamed for Rio Disaster. *Observer* (London), 28 February: 19.

Cairncross, S., Hardoy, J. and Satterthwaite, D. (eds) 1990a. *The Poor Die Young: Housing and Health in Third World Cities*. London: Earthscan.

Cairncross, S., Hardoy, J. and Satterthwaite, D. 1990b. New Partnerships for Healthy Cities. In: S. Cairncross et al. (eds), *The Poor Die Young: Housing and Health in Third World Cities*, pp.245–68. London: Earthscan.

Caldwell, J. Reddy, P. and Caldwell, P. 1986. Period High Risk as a Cause of Fertility Decline in a Changing Rural Environment: Survival Strategies in the 1980–1983 South Indian Drought. *Economic Development and Cultural Change* No. 34, 4: 677–701.

Callinicos, A. 2002. Marxism and Global Governance. In: D. Held and A. McGraw (eds), *Governing Globalization: Power, Authority and Global Governance*, pp.249–66. Cambridge: Polity Press.

Cammack, P. 2002. Attacking the Poor. *New Left Review* 13 (January/February): 125–134.

Campbell, D. 1987. Participation of a Community in Social Science Research: A Case Study from Kenyan Maasailand. *Human Organization* 46, 2: 160–167.

Campbell, J. 1984. *Dealing with Disaster, Hurricane Responses in Fiji*. Suva: Pacific Islands Development Programme, East–West Center and the Government of Fiji.

Cannon, T. 1991. Hunger and Famine: Using a Food Systems Model to Analyse Vulnerability. In: H. Bohle, T. Cannon, G. Hugo and F. Ibrahim (eds), *Famine and Food Scarcity in Africa and Asia: Indigenous Responses and External Intervention to Avoid Hunger*, pp.291–312 (see also *Bayreuther Geowissenschaftliche Arbeiten*, Vol. 15. Bayreuth: Naturwissenschaftliche Gesellschaft Bayreuth).

Cannon, T. 1994. Vulnerability Analysis and the Explanation of 'Natural' Disasters. In: A. Varley (ed), *Disasters, Development and Environment*, pp.13–30. Chichester: Wiley.

Cannon, T. 2000a. Vulnerability Analysis and Disasters. In: D. Parker (ed), *Floods*, pp.43–55. London: Routledge.

Cannon, T. (ed) 2000b. *China's Economic Growth: The Impact on Regions, Migration and The Environment*, Basingstoke: Macmillan.

Cannon, T. 2002. Gender and Climate Hazards in Bangladesh. In: R Masika (ed), *Gender, Development and Climate Change*, pp.45–50. Oxford: Oxfam.

Cannon, T. 2003. Food Security, Food Systems and Livelihoods: Competing Explanations of Hunger. *Die Erde* 133, 4: 345–362.

Cannon, T., Twigg, J., and Rowell, J. 2003. *Social Vulnerability, Sustainable Livelihoods and Disasters*. Report to DFID, Conflict and Humanitarian Assistance Department (CHAD) and Sustainable Livelihoods Support Office. Chattam, Kent: Livelihoods and Institutions Group, Natural Resources Institute, University of Greenwich.

Caplan, P. (ed) 2000. *Risk Revisited*. London: Pluto Press.

Carney, D. (ed) 1998. *Sustainable Rural Livelihoods: What Contribution Can We Make*. London. Department for International Development.

Carney, J. 1988. Struggles Over Crop Rights and Labour within Contract Farming Households in a Gambian Irrigated Rice Project. *Journal of Peasant Studies* 15, 3: 334–349.

Carrasco, F. and Garibay, B. 2000. *Guia communitária para la prevención de desastres*. Mexico City: Government of the City of Mexico and Institute of Social Research, National Autonomous University of Mexico (UNAM).

Carroll, R. 2003. 40 Million Starving 'As the World Watches Iraq'. *Guardian* (London), 9 April. http://www.guardian.co.uk/famine/story/0,12128,932778,00.html

Carson, R. 1962. *Silent Spring*. Harmondsworth: Penguin.

Carter, R. 1987. Man's Response to Sea-Level Change. In: R. J. N. Devoy (ed), *Sea Surface Studies*, pp.464–498. London: Croom Helm.

Cassen, R. 1994. *Population and Development: Old Debates, New Conclusion*. Oxford: Transactions Publications.

Castel, R. 1991. From Dangerousness to Risk. In: G. Burchell, C. Gordon and P. Miller (eds), *The Foucault Effect: Studies in Governmentality*, pp.281–298. Chicago, IL: University of Chicago Press.

Castells, M. 1996. *The Information Age: Economy, Society and Culture Volume 1. The Rise of the Network Society*. Oxford: Blackwell.

Cater, N. 1986. *Sudan The Roots of Famine*. Oxford: Oxfam.

Cawthorne, A. 2001. Hurricane Michelle Pounds Cuba, Heads for Bahamas. Reuters, 5 November. ReliefWeb, http://www.reliefweb/int

Cedeno, J. 1986. Rainfall and Flooding in the Guayas River Basin and its Effects on the Incidence of Malaria 1982–1985. *Disasters* 10, 2: 107–111.

Chairetakis, A. 1992. Past as Present: History and Reconstruction after the 1980 Earthquake in Campania, Southern Italy. Paper presented at the 51st Annual Meeting of the Society for Applied Anthropology, Memphis, TN.

Chambers, R. 1983. *Rural Development: Putting the Last First*. London: Longman.

Chambers, R. 1989. Editorial Introduction: Vulnerability, Coping and Policy. *IDS Bulletin* 20, 2: 1–7.

Chambers, R. 1994. The Origins and Practice of Participatory Rural Appraisal. *World Development* 22, 7: 953–969.

Chambers, R. 1995a. Paradigm Shifts and the Practice of Participatory Research and Development. In: N. Nelson and S. Wright (eds), *Power and Participatory Development Theory and Practice*, pp.30–42. London: Intermediate Technology Publications.

Chambers, R. 1995b. *Poverty and Livelihoods: Whose Reality Counts?* IDS Discussion Paper No. 347. Falmer, Sussex, UK: Institute for Development Studies, University of Sussex.

Chambers, R. 1997. *Whose Reality Counts? Putting the First Last*. London: Intermediate Technology Publications.

Chambers, R. and Conway, G. 1992. *Sustainable Rural Livelihoods: Practical Concepts for the 21st Century*. IDS Discussion Paper No. 296. Brighton, UK: IDS (Institute for Development Studies), University of Sussex.

Chambers, R., Pacey, A. and Thrupp, L. (eds) 1989. *Farmer First*. London: Intermediate Technology Publications.

Chambers, R., Saxena, N. and Shah, T. 1990. *To the Hands of the Poor: Water and Trees*. Boulder, CO: Westview.

Chambers, R. et al. (eds) 1981. *Seasonal Dimensions to Rural Poverty*. London: Francis Pinter.

Chapin, G. and Wasserstrom, R. 1981. Agricultural Production and Malaria Resurgence in Central America and India. *Nature* 293, 5829 (17 September): 181–185.

Cavanagh, J. et al. 2002. *Alternatives to Globalization*. New York: Berrett-Koehler.

Chen, L. (ed) 1973. *Disaster in Bangladesh*. Oxford: Oxford University Press.

Chen, R. 2002. Assessment of High-Risk Disaster Hotspots. World Bank and Center for Hazards and Risk Management Workshop on Assessment of High-Risk Disaster Hotspots http://www.proventionconsortium.org/files/highrisk/chen.pdf.

Cheru, F. 1989. *The Silent Revolution in Africa: Debt, Development and Democracy*. Harare and London: Anvil Press and Zed Press.

Cheru, F. 2002. *African Renaissance: Roadmaps to the Challenge of Globalization*. London: Zed Books.

Chiappe, I. and Fernandez, M. 2001. *Manual del capacitador: 7 Módulos para capacitadores en gestión de riesgo*. La Paz, Bolivia: Programa Nacional de Prevención de Riesgos y Atención de Desastres.

Childers, C. 1999. Elderly Female-headed Households in the Disaster Loan Process. *International Journal of Mass Emergencies and Disasters* 17, 1: 99–110.

Chowdhury, J. 1991. Flood Action Plan: One Sided Approach? *Bangladesh Environmental Newsletter* 2, 2: 1 & 3 (Dhaka: Bangladesh Centre for Advanced Studies).

Christensen, N. 2002. Causes of Failure: Getting to the Bottom of Murphy's Law. *Dealing with Disasters*. Duke Environmental Leadership Forum (November) http://www.env.duke.edu/forum02/christensen.pdf.

Christian Aid UK 2002. Afghanistan: The Key Figures. http://www.ccmep.org/hotnews/afghanistan103101.html.

Christie, F. and Hanlon, J. 2001. *Mozambique and the Great Flood of 2000*. London, Oxford and Bloomington, IN: The International African Institute, James Currey and Indiana University Press.

Christoplos, I. (ed) 2001. Theme Issue on the Role of NGOs in Disaster Mitigation. *Disasters* 25, 3.

Chung, J. 1987. Fiji, Land of Tropical Cyclones and Hurricanes: A Case Study of Agricultural Rehabilitation. *Disasters* 11, 1: 40–48.

CIESIN 2003. *The Relationship of Skin Cancer Prevalence and the Increase in Ultraviolet-B Exposure due to Ozone Depletion*. CIESIN Thematic Guides. New York: Columbia University, Earth Institute, Center for Earth Science Information Network (CIESIN). http://www.ciesin.org/TG/HH/ozskin1.html

CIIR (Catholic Institute of International Relations) 1975. *Honduras: Anatomy of a Disaster*. London: CIIR.

CIMADE, INODEP, MINK 1986. *Africa's Refugee Crisis*. London: Zed Books.

Clark, C. 1982. *Flood*. Alexandria, VA: Time-Life Books.

Clark, T. 2001. Bangladeshis to Sue over Arsenic Poisoning. *Nature*, 11 October. http://www.nature.com/nsu/011011/011011-14.html.

Clark, W. 1989. Managing Planet Earth. *Scientific American* 262, 3: 46–57.

Clarke, J. I. (ed) 1989. *Population and Disaster*. Oxford: Blackwell.

Clay, E. 1985. The 1974–1984 Floods in Bangladesh: From Famine to Food Crisis Management. *Food Policy* 10, 3: 202–206.

Clay, E. et al. 1999. *An Evaluation of HMG's Response to the Montserrat Volcanic Emergency*. Evaluation Report EV635, Vol. 1, London: DFID.

Clay, J. and Holcomb, B. 1985. *Politics and The Ethiopian Famine 1984–1985.* Cambridge, MA: Cultural Survival.

Clay, J., Steingraber, S. and Niggli, P. 1988. *The Spoils of Famine.* Cambridge, MA: Survival International.

Cliff, J. 1991. The War on Women in Mozambique: Health Consequences of South African Destabilization, Economic Crisis and Structural Adjustment. In: M. Turshen (ed), *Women and Health in Africa*, pp.15–34. Trenton: Africa World Press.

Cliffe, L. and Moorsom, R. 1979. Rural Class Formation and Ecological Collapse in Botswana. *Review of African Political Economy* 15/16: 35–52.

CND 2002. 1.8 Million People Relocated to Make Way for Floods. CND (China News Digest), 20 February. http://www.cnd.org/Global/02/02/20/020220-91.html.

Coburn A. and Spence R. 1993. *Earthquake Protection.* Chichester: John Wiley.

Coburn, A. and Spence, R. 2002. Earthquake Protection. 2nd edn. Chichester: John Wiley & Sons.

Coburn, A., Spence R. and Pomonis, A. 1991. *Vulnerability and Risk Assessment. Trainers and Trainees Guide.* Disaster Management Training Programme (DMTP) Geneva: UNDRO/UNDP.

Coburn, A., Hughes, R., Illi, D., Nash, D. and Spence, R. 1984. The Construction and Vulnerability to Earthquakes of Some Building Types in Northern Areas of Pakistan. In: K. Miller (ed), *The International Karakoram Project*, Vol. 2, pp.228–237. Cambridge: Cambridge University Press.

Cochrane, H. 1975. *Natural Hazards and their Distributive Effects.* National Science Foundation Program on Technology, Environment and Man Monograph NSF-RA-E-75–003. Boulder, CO: University of Colorado, Institute of Behavioral Science.

Cohen, J. 2000a. AIDS Research in Africa. *Science* 288 (23 June): 2150–2153.

Cohen, J. 2000b. South Africa's New Enemy. *Science* 288 (23 June): 2168–2170.

Cohen, J. M. and Lewis, D. 1987. Role of Government in Combatting Food Shortages: Lessons from Kenya 1984–85. In: M. Glantz (ed), *Drought and Hunger in Africa*, pp.269–296. Cambridge: Cambridge University Press.

Cohen, M. 1977. *The Food Crisis in Prehistory: Overpopulation and the Origins of Famine.* New Haven, CT: Yale University Press.

Cohen, S. and Raghavulu, C. 1979. *The Andhra Cyclone of 1977.* New Delhi: Vikas.

Collins, J. 1989. *Fire on the Rim.* San Francisco, CA: Food First.

Collins, J. and Rau, B. 2000. *AIDS in the Context of Development.* Geneva: UNRISD.

Comfort, L., Wisner, B., Cutter, S., Pulwarty, R., Hewitt, K., Oliver-Smith, A., Weiner J., Fordham, M., Peacock, W. and Krimgold, F. 1999. Reframing Disaster Policy: The Global Evolution of Vulnerable Communities. *Environmental Hazards* 1, 1: 39–44.

Conroy, C. and Litvinoff, M. 1988. *The Greening of Aid.* London: Earthscan.

Consultative Group for the Reconstruction and Transformation of Central America 1999. *Reducing Vulnerability to Natural Hazards: Lessons Learned from Hurricane Mitch.* Washington, DC: Inter-American Development Bank. http://www.iadb.org/regions/re2/consultative_group/groups/ecology_workshop_1.htm.

Cooper, D., Velve, R. and Hobbelink, H. 1992. *Growing Diversity: Genetic Resources and Food Security.* London: IT Publications.

Copans, J. (ed) 1975. *Secheresses et famines du Sahel*, 2 Vols. Paris: Maspero.

Copans, J. 1983. The Sahelian Drought: Social Science and the Political Economy of Underdevelopment. In K. Hewitt (ed), *Interpretations of Calamity*, pp.83–97. Boston, MA: Allen & Unwin.

Corbett, J. 1988. Famine and Household Coping Strategies. *World Development* 16: 1099–2012.

Cornia, G., Jolly, R. and Stewart, F. (eds) 1987. *Adjustment with a Human Face*, 2 vols. New York: Oxford University Press.

Coulson, A. 1982. *The Political Economy of Tanzania*. Oxford: Oxford University Press.

CRED (Centre for Research in the Epidemiology of Disasters) 2002. *EM-DAT 2002*. Brussels: Université Catholique de Louvain. www.credbe/emdat

Crichton, D. 2001. A Scottish Lead in Managing Flood Risk. *Town and Country Planning Journal* 70, June: 188–189.

Crosby, A. 1986. *Ecological Imperialism: The Biological Expansion of Europe, 900–1900*. Cambridge: Cambridge University Press.

Crosby, A. 1991. The Biological Consequences of 1492. *Report on the Americas* 25, 2: 6–13.

Cross, S. 2002. *Customary Land Tenure, Taxes and Service Delivery in Rural Malawi: A Review of Institutional Features of Rural Livelihoods*. LADDER Working Paper No. 21. University of East Anglia, Norwich. Mimeo.

Crossette, B. 1992. Sudan is Said to Force 400,000 People into Desert. *New York Times*, 22 February: 5.

Crow, B. 1984. Warnings of Famine in Bangladesh. *Economic and Political Weekly* 19, 40: 1754–1758.

Crow, B. 1990. Moving the Lever: A New Food Aid Imperialism? In: H. Bernstein, B. Crow, M. Mackintosh and C. Martin (eds), *The Food Question: Profits versus People*, pp.32–42. London: Earthscan.

Crush, J., (ed) 1995. *Power of Development*. London: Routledge.

Cuenya, B., Almada, H., Armus, H., Castells, J., di Loreto, M. and Penalva, S. 1990. Community Action to Address Housing and Health Problems: The Case of San Martin in Buenos Aires, Argentina. In: S. Cairncross et al. (eds), *The Poor Die Young: Housing and Health in Third World Cities*, pp.25–55. London: Earthscan.

Cuny, F. 1983. *Disasters and Development*. New York: Oxfam and Oxford University Press.

Cuny, F. 1987. Sheltering the Urban Poor, Lessons and Strategies of the Mexico City and San Salvador Earthquakes. *Open House International* 12, 3: 16–20.

Cuny, F. and Hill, R. 1999. *Famine, Conflict and Response: A Basic Guide*. West Hartford, CN: Kumarian Press.

Currey, B. 1978. The Famine Syndrome: its Definition for Preparedness and Prevention in Bangladesh. *Ecology of Food and Nutrition* 7, 1: 87–98.

Currey, B. 1981. The Famine Syndrome: Its Definition for Relief and Rehabilitation in Bangladesh. In: J. Robson (ed), *Famine: Its Causes, Effects and Management*, pp. 123–133. New York: Gordon & Breach.

Currey, B. 1984. Coping with Complexity in Food Crisis Management. In: B. Currey and G. Hugo (eds), *Famine as Geographical Phenomenon*, pp.183–202. Dordrecht: D. Reidal.

Currey, B. 2002. Strategic Directions for Vulnerability Analysis and Mapping in the 21st Century: Examples from Bangladesh. Medford, MA: Tufts University, Food Policy and Applied Nutrition Program, Discussion Paper 16.
http://nutrition.tufts.edu/pdf/publications/fpan/wp16-strategic_directions.pdf.

Curtis, D., Hubbard, M. and Sheppard, A. (eds) 1988. *Preventing Famine: Policies and Prospects for Africa.* London: Routledge.

Cutler, P. 1984. Famine Forecasting: Prices and Peasant Behaviour in Northern Ethiopia. *Disasters* 8,1: 48–55.

Cutler, P. 1985. Detecting Food Emergencies: Lessons from the Bangladesh Crisis. *Food Policy* 10: 17–36.

Cutler, P. 1986. The Response to Drought of Beja Famine Refugees in Sudan. *Disasters*, 10, 3: 181–187.

Cutter, S. 1995. The Forgotten Casualties: Women, Children and Environmental Change. *Global Environmental Change* 5, 3: 181–194.

da Cruz, J. 1993. *Disaster and Society: The 1985 Mexican Earthquakes.* Lund: Lund University Press.

Dahl, G. and Hjort, G. 1976. Having Herds: Pastoral Herd Growth and Household Economy. *Stockholm Studies in Social Anthropology* 2. Stockholm. Department of Social Anthropology, University of Stockholm.

Dai Qing (ed) 1998. *The River Dragon has Come!* New York: M. E. Sharpe.

Dailey, F. 2002. Conversation with Ian Davis, May 2002.

Dalal-Clayton, B. 1990. *Environmental Aspects of the Bangladesh Flood Action Plan.* Gatekeeper Series No. 1. London: International Institute for Environment and Development (IIED).

Dando, W. 1980. *The Geography of Famine.* London: Arnold.

Dankelman, I. and Davidson, J. 1988. *Women and Environment in the Third World.* London: Earthscan.

Dartmouth College 1999. *1998 Global Register of Large River Flood Events.* Dartmouth Flood Observatory Archives. Dartmouth College.
http://www.dartmouth.edu/artsci/geog/floods/1999sum.html

Dash, N., Peacock, W. and Morrow, B. 1997. And the Poor Get Poorer: A Neglected Black Community. In: W. Peacock, B. Morrow and H. Gladwin (eds), *Hurricane Andrew: Ethnicity Gender and the Sociology of Disasters*, pp.171–190. London: Routledge.

Davenport, W. 1960. Jamaican Fishing: A Game Theory Analysis. *Yale University Publications in Anthropology* 59. New Haven, CT: Yale University Press.

Davidson, R. 2002. Personal communication between Ben Wisner and Dr Rachel Davidson, Department of Civil Engineering, Cornell University, Ithaca, New York, 5 June.

Davidson, R., Gupta, A., Kakhandiki, A. and Shah, H. 1997. Urban Earthquake Disaster Risk Assessment and Management. *Journal of Seismology and Earthquake Engineering* 1, 1: 59–70.

Davidson, R., Villacis, C., Cardona, O. and Tucker, B. 2000. A Project to Study Urban Earthquake Risk Worldwide. *Proceedings of the 12th World Conference on Earthquake Engineering.* Paper No. 791. Auckland, New Zealand, 30 January–4 February.

Davis, I. 1977a. Emergency Shelter. *Disasters* 1, 1: 23–40.

Davis, I. 1977b. The Intervenors. *New Internationalist* 53: 21–23.

Davis, I. 1978. *Shelter After Disaster*. Oxford: Oxford Polytechnic Press.

Davis, I. (ed) 1981. *Disasters and the Small Dwelling*. Oxford: Pergamon Press.

Davis, I. 1984a. The Squatters Who Live Next Door to Disaster. *Guardian*, 7 December: 7.

Davis, I. 1984b. A Critical Review of the Work Method and Findings of the Housing and Natural Hazards Group. In: K. Miller (ed), *The International Karakoram Project*, Vol. 2, pp.200–227. Cambridge: Cambridge University Press.

Davis, I. 1986. The Planning and Maintenance of Urban Settlements to Resist Extreme Climatic Forces. In: T. Oke (ed), *Urban Climatology and its Applications with Special Regard to Tropical Areas*, pp.277–312. World Climate Programme – Proceedings of the Technical Conference, Mexico. Geneva: World Meteorological Organisation.

Davis, I. 1987. Safe Shelter within Unsafe Cities: Disaster Vulnerability and Rapid Urbanisation. *Open House International* 12, 3: 5–15.

Davis, I. 1988. Acts of God Increasingly Amount to Acts of Criminal Negligence. *Guardian*, 30 December: 7.

Davis, I. 1994. Assessing Community Vulnerability. In: UK National Coordination Committee for the International Decade for Natural Disaster Reduction (IDNDR) (ed), *Medicine in the International Decade for Natural Disaster Reduction (IDNDR)*, pp. 11–13. Proceedings of a Workshop at the Royal Society, London, 19 April 1993.

Davis, I. 2001. Location and Operation of Evacuation Centres and Temporary Housing Policies. *Global Assessment of Earthquake Countermeasures*. Kobe: Hyogo Prefecture.

Davis, I. 2003. Assessing Progress in the Analysis of Social Vulnerability and Capacity of the Impact of Natural Hazards. In: G. Bankoff, G. Frerks and T. Hilhorst, (eds), *Vulnerability: Disasters, Development and People*, ch. 8. London: Earthscan.

Davis, I. and Bickmore, D. 1994. Data Management for Disaster Planning. In: The Royal Society, Proceedings of Conference: Natural Disasters – Protecting Vulnerable Communities 13–15 October 1993. London: Royal Society.

Davis, I. and Gupta, S. 1991. Technical Background Paper. In: Asian Development Bank (ed), *Disaster Mitigation in Asia and the Pacific*, pp.23–69. Manila: Asian Development Bank.

Davis, I., Kishigami, H., Takei, S., Yaoxian, Y. and Johansson, M. 1992. *Rehabilitation Assistance to Anhui Province Following the Flood Disaster, May–July 1991*. Report of UNDP Appraisal Mission, December 4–16, 1991. CPR/91/712. Beijing: UNDP.

Davis, M. 2000. *Late Victorian Holocausts*. London: Verso.

Davis, S. and Hodson, J. 1982. *Witnesses to Political Violence in Guatemala. The Suppression of a Rural Development Movement*. Boston: OXFAM America.

de Beer, C. 1986. *The South African Disease*. Trenton NJ: Africa World Press.

de Blij, H. 1994. *Nature on the Rampage*. Washington, DC: Smithsonian Books.

de Boer, J. and Dubouloz, M. (eds) 2000. *Handbook of Disaster Medicine*. Utrecht, The Netherlands: International Society of Disaster Medicine and Van der Wees Uitgeverij.

De Castro, J. 1952. *Géopolitique de la faim*. Paris: Editions Ouvieres.

De Castro, J. 1966. *Death in the Northeast*. New York: Vintage.

De Castro, J. 1977. *Geopolitics of Hunger*. Translation of de Castro 1952. New York: Monthly Review.

de Senillosa, I. 1999. A New Age of Social Movements. In: M. Kothari (ed), *Development and Social Action*, pp.87–103. Oxford: Oxfam Publications.

de Ville, C. and Lechat, M. 1976. Health Aspects in Natural Disasters. *Tropical Doctor*, October: 168–170.

de Waal, A. 1987. The Perception of Poverty and Famines. *International Journal of Moral and Social Studies* 2, 3: 57–68.

de Waal, A. 1989a. *Famine That Kills. Darfur, Sudan, 1984–1985*. Oxford: Clarendon Press.

de Waal, A. 1989b. Famine Mortality: A Case Study of Darfur, Sudan 1984–5. *Population Studies* 43, 1: 5–24.

de Waal, A. 1991. Famine and Human Rights. *Development in Practice: An Oxfam Journal* 1, 2: 77–83.

de Waal, A. 1997. *Famine Crimes: Politics and the Disaster Relief Industry in Africa*. Oxford: James Currey.

de Waal, A. 1999. Democratic Political Process and the Fight Against Famine. In: A. de Waal (ed), *Who Fights? Who Cares?*, pp. 129–158. Trenton, NJ: Africa World Press.

de Waal, A. 2000. *Democratic Political Process and the Fight against Famine*. IDS Working Paper No. 107, Brighton, UK: IDS (Institute for Development Studies), University of Sussex.

de Waal, A. 2001. *'AIDS-Related National Crises': An Agenda for Governance, Early-Warning and Development Partnership*. AIDS and Governance Issue Paper No. 1, September. London: Africa Justice.

de Waal, A. 2002. What AIDS Means in a Famine. *New York Times*, 19 November.

de Waal, A. and Amin, M. M. 1986. *Report on Save the Children Fund Activities in Darfur*. Nyala: Save the Children Fund.

Dean, C. 1999. *Against the Tide: The Battle for America's Beaches*. New York: Columbia University Press.

DEC (Disasters Emergency Committee) 2001a. The Earthquake in Gujarat, India: Report of a Monitoring Visit for the DEC, March. www.dec.org.uk/dec_standard/upload/Report%20280401.doc.

DEC (Disasters Emergency Committee) 2001b. *Independent Evaluation of Expenditure of DEC India Earthquake Appeal Funds January 2001–October 2001*. London: DEC.

Debach, P. 1974. *Biological Control by Natural Enemies*. Cambridge: Cambridge University Press.

Degg, M. 1989. Earthquake Hazard Assessment after Mexico 1985. *Disasters* 13, 3: 237–254.

del Ninno, C., Dorosh, P., Smith, L. and Roy, D. 2001. *The 1998 Floods in Bangladesh: Disaster Impacts, Household Coping Strategies and Response*. Washington, DC: International Food Policy Research Institute, Research Report 122.

Delica, Z. 2002. Community Mobilisation for Early Warning in the Philippines. In: J. Zschau and A. Kueppers (eds), *Early Warning Systems for Natural Disaster Reduction*, pp.37–47. Berlin: Springer Verlag.

Deng, L. 1999. *The 1998 Famine in the Sudan: Causes Preparedness and Response.* IDS Discussion Paper 369. Brighton, UK: IDS (Institute for Development Studies), University of Sussex.

Dercon, S. and Krishnon, P. 1996. Income Portfolios in Rural Ethiopia and Tanzania: Choices and Constraints. *Journal of Development Studies* 32, 6: 850–875.

Desai, R. 2002. *Field Shake Table Program For Confidence Building in Quake Resistant Building Technology*, Ahmedabad: National Centre for People's Action in Disaster Preparedness (NCPDP).

Deutsche Presse Agentur 2002. Michelle storm abates, Castro says Cuba "will survive". http://www.reliefweb.int/w/rwb.nsf/6686f45896f15dbc852567ae00530132/94987f742181f60b85256afc006ad1af?OpenDocument

Devereux, S. 1987. FAO and FED = Famine: Not a Refutation of Professor Sen's Theory. Paper at Workshop 'The Causes of Famine', Queen Elizabeth House, Oxford, 9 May.

Devereux, S. 1993. *Theories of Famine.* Hemel Hempstead, UK: Harvester Wheatsheaf.

Devereux, S. 2000. *Famine in the Twentieth Century.* IDS Working Paper No. 105. Brighton, UK: IDS (Institute for Development Studies), University of Sussex.

Devereux, S. 2001. Famine in Africa. In: S. Devereux and S. Maxwell (eds), *Food Security in Sub-Saharan Africa*, ch. 5. London: ITDG Publishing.

Devereux, S. and Hay, R. 1986. *Origins of Famine: A Review of the Literature.* Oxford: University of Oxford (Food Studies Group).

Devereux, S. and Maxwell, S. (eds) 2001. *Food Security in Sub-Saharan Africa.* London: ITDG Publishing.

Dey, J. 1981. Gambian Women: Unequal Partners in Rice Development Projects? *Journal of Development Studies* 17, 3: 109–122.

Dinham, B. and Sarangi, S. 2002. The Bhopal Gas Tragedy 1984 to ?, *Environment and Urbanization* 14, 1: 89–100.

Diriba, K. 1991. *Famines and Food Security in Kembatana Hadiya, Ethiopia. A Study in Household Survival Strategies.* Ph.D. Thesis, University of East Anglia.

Disaster Relief 1998a. China Floods Exacerbated by Man's Impact, 19 August. http://www.disasterrelief.org/Disasters/980819China12/.

Disaster Relief 1998b. Thousands Missing after Major Levee Collapses, 8 August. http://www.disasterrelief.org/Disasters/980808China7/.

Disaster Relief 1998c. Chinese Blow Levees in Last-ditch Attempt to Prevent New Floods, 10 August. http://www.disasterrelief.org/Disasters/980810China8/.

Disaster Relief 1998d. Death Toll from Indian Cyclone Rises. American Red Cross/CNN Interactive/IBM. http://www.disasterrelief.org/Disasters/980611indiaupdate/.

Doerner, W. 1985. Last Rites for a Barrio: A Crushing Mud Slide Kills Hundreds in Puerto Rico. *Time*, 21 October: 32.

Douglas, M. 1985. *Risk Acceptability According to the Social Sciences.* New York: Routledge and Russell Sage Foundations.

Douglas, M. and Wildavsky, A. 1982. *Risk and Culture: An Essay on the Selection of Technical and Environmental Dangers.* Berkeley: University of California Press.

Dovers, S. and Handmer, J. 1993. Contradictions in Sustainability. *Environmental Conservation* 20: 217–222.

Dovers, S., Norton, T. and Handmer, J. 2001. Ignorance, Uncertainty and Ecology: Key Themes. In: J. Handmer, T. Norton and S. Dovers (eds), *Ecology, Uncertainty and Policy: Managing Ecosystems for Sustainability*, pp.1–25. Harlow, UK: Prentice Hall-Pearson Education.

Dow, S. 2003. List of Iraq Reconstruction Contracts. *La.Indymedia.Org*, 19 April. http://www.la.indymedia.org/news/2003/04/51989.php.

Downing, T. 1991 *Assessing Socioeconomic Vulnerability to Famine: Frameworks, Concepts and Applications*. Alan Shawn Feinstein World Hunger Programme. Providence, RI: Brown University.

Downing, T. 1992. *Climate Change and Vulnerable Places: Global Food Security and Country Studies in Zimbabwe, Kenya, Senegal and Chile*. Environmental Change Unit (ECU) Research Report No. 1. Oxford: University of Oxford.

Downing, T.E., Butterfield, R., Cohen, S., Huq, S., Moss, R., Rahman, A., Sokaona, Y. and Stephen, L. 2001. *Climate Change Vulnerability*. Oxford: Oxford Environmental Change Institute.

Downing, T., Gitu, K. and Kamau, C. (eds) 1989. *Coping with Drought in Kenya: National and Local Strategies*. Boulder, CO: Lynne Rienner.

Downs, R., Kerner, D. and Reyna, S. (eds) 1991. *The Political Economy of African Famine*. Philadelphia, PA: Gordon & Breach Science Publishers.

Doyal, L. and Gough, I. 1991. *A Theory of Human Need*. London and New York: Macmillan and Guilford.

Drabek, T. 1986. *Human Systems Response to Disaster*. London: Pergamon.

Drèze, J. 1988. *Famine Prevention in India*. Development Economics Research Programme, No.3. London. London School of Economics.

Drèze, J. and Sen, A. 1989. *Hunger and Public Policy*. Oxford: Clarendon Press.

Drinkwater, M. and McEwan, M. 1994. Household Food Security and Environmental Sustainability in Farming Systems Research; Developing Sustainable Livelihoods. *Journal of Farming Systems Research* 4, 2: 111–126.

D'Souza, F. 1984. The Socio-economic Cost of Planning for Hazards. an Analysis of Barculti Village, Yasin, Northern Pakistan. In: K. Miller (ed), *The International Karakoram Project*, Vol. 2: 289–322. Cambridge: Cambridge University Press.

D'Souza, F. 1988. Famine: Social Security and an Analysis of Vulnerability. In: G. Harrison (ed), *Famine*, pp.1–56. Oxford: Oxford University Press.

Dudley, E. 1988. Disaster Housing: Strong Houses or Strong Institutions? *Disasters* 12, 2: 111–121.

Duffield, M. 2001. *Global Governance and the New Wars: The Merger of Development and Security*. London: Zed Books.

Duffield, M. 2002. Reprising Durable Disorder: Network War and the Securitisation of Aid. In: B. Hettne and B. Oden (eds), *Global Governance in the 21st Century: Alternative Perspectives on World Order*, pp.75–105. Stockholm: Expert Group on Development Initiatives (EGDI), Swedish Ministry for Foreign Affairs.

Dupree, H., and Roder, W. 1974. Coping with Drought in a Preindustrial, Preliterate Farming Society. In: G. White (ed), *Natural Hazards*, pp.115–119. Oxford University Press: New York.

During, A. 1989. Mobilizing at the Grassroots. In: L. Brown et al. (eds), *State of the World*, pp.154–173. New York: W. W. Norton.

Dynes, R., DeMarchi, B. and Pelanda, C. (eds) 1987. *Sociology of Disaster*. Milan: Franco Agneli Libri.

Dyson, T. 1996. *Population and Food: Global Trends and Future Prospects*. ESRC Global Environmental Change Programme, London: Routledge.

Eade, D. 1998. *Capacity Building*. Oxford: Oxfam.

Eade, D. and Williams, S. 1995. *The Oxfam Handbook of Development and Relief.* Vol 1. Oxford: Oxfam.

Earickson, R. and Meade, M. 2000. *Medical Geography*. 2nd edn. New York: Guilford.

Easterling, D., Meehl, G., Parmesan, C., Changnon, S., Karl, T. and Mearns, L. 2000. Climate Extremes, Observations, Modeling and Impacts. *Science* 289: 2068–2074.

Ebert, C. 1993. *Disasters: Violence of Nature, Threats by Man*. Dubuque, IO: Kendall/Hunt.

Echeverria, E. 1991. Decentralising Mexico's Health Care Facilities. In: A. Kreimer and M. Munasinghe (eds), *Managing Natural Disasters and the Environment*, pp. 60–61. Washington, DC: World Bank.

Eckholm, E. 1976. *Losing Ground*. Oxford: Pergamon.

Eckholm, E. 2003. Tide of China's Migrants. *New York Times*, 29 July, pp. A1 & A8.

Economist 1989. Score One for the Trees. 14 January: 53.

Economist 2002a. With the Wolf at the Door: Southern Africa's Food Shortage. 1 June: 45–46.

Economist 2002b. Stop Denying the Killer Bug. 23 February: 49–51.

Edmonds, R. 2000. Recent Developments and Prospects for the Sanxia (Three Gorges) Dam. In: T. Cannon (ed), *China's Economic Growth: The Impact on Regions, Migration and the Environment*, pp.161–183. Basingstoke, UK: Macmillan.

Eide, A., Eide, W., Goonatilake, S., Gussow, J. and Omawale (eds) 1984. *Food as a Human Right*. Tokyo: United Nations University Press.

Ekejuiba, F. 1984. Contemporary Households and Major Socio-economic Transitions in Eastern Nigeria: Toward a Reconceptualisation of the Household. In: J. Guyer and P. Peters (eds), *Conceptualising in the Household*, pp.9–13. Harvard, MA: Harvard University Press.

Elahi, K. 1989. Population Displacement due to Riverbank Erosion of the Jamuna in Bangladesh. In: J. Clarke (ed), *Population and Disaster*, pp.81–97. Oxford: Blackwell.

Ellis, F. 1988. *Peasant Economics: Farm Households and Agrarian Development*. Cambridge: Cambridge University Press.

Ellis, F. 2000. *Rural Livelihoods and Diversity in Developing Countries*. Oxford: Oxford University Press.

Ellis, F. 2001. *Rural Livelihoods, Diversity and Poverty Reduction Policies: Uganda, Tanzania, Malawi and Kenya*. LADDER Working Paper No.1. University of East Anglia, Norwich.

Elsberry, R. 2002. Track Forecast Guidance Improvements for Early Warnings of Tropical Cyclones. In: J. Zschau and A. Kueppers (eds), *Early Warning Systems for Natural Disaster Reduction*, pp.167–183. Berlin: Springer Verlag.

Emel, J. and Peet, R. 1989. Resource Management and Natural Hazards. In: R. Peet and N. Thrift (eds), *New Models in Geography*, Vol. 1, pp.49–76. London: Unwin Hyman.

Emergency Management Australia 2001. Framework for Risk Assessment. In: ISDR 2002, *Living with Risk: A Global Review of Disaster Reduction Initiatives*, p.76. Preliminary edn. Geneva: ISDR.

Enarson, E. 2001. We Want Work: Rural Women in the Gujarat Drought and Earthquake.
http://online.northumbria.ac.uk/geography_research/radix/resources/surendranag ar.doc.

Enarson, E., Childers, C., Morrow, B. and Wisner, B. 2003a. *Vulnerability Approach to Emergency Management*. Instructor's Guide, FEMA, Higher Education Project. Emmitsburg, MD: FEMA.
http://166.112.200.141/emi/edu/aem_courses.htm

Enarson, E. with Meyreles, L., Gonzalez, M, Morrow, B., Mullings, A., Soares, J. 2003b. *Working with Women at Risk: Practical Guidelines for Assessing Local Disaster Risk*. Miami: International Hurricane Center, Florida International University (June) http://www.ihc.fiu.edu/lsbr.

Enarson, E. and Morrow, B. (eds) 1997. *Gendered Terrains of Disaster: Through Women's Eyes*. New York: Praeger.

Enarson, E. and Morrow, B. (eds) 2001. *The Gendered Terrain of Disasters: Through Women's Eyes*. Miami: International Hurricane Center.

Ennew, J. and Milne, B. 1989. *The Next Generation: Lives of Third World Children*. London: Zed Books.

EPOCA (The Environmental Project on Central America) 1990. *Guatemala: A Political Ecology*. Green Paper No. 5. San Francisco, CA: Earth Island Institute.

EQE 1995a. *The January 17, 1995 Kobe Earthquake. An EQE Summary Report*. Houston, TX: EQE (ABS Consulting).

EQE 1995b. Fire Following Earthquake. In: *EQE, The January 17, 1995 Kobe Earthquake. An EQE Summary Report*. Houston, TX: EQE (ABS Consulting). http://www.eqe.com/publications/kobe/firefoll.htm.

Escobar, A. 1995. *Encountering Development: The Making and Unmaking of the Third World*. Princeton: Princeton University Press.

Ewald, P. 1993. *The Evolution of Infectious Disease*. New York: Oxford University Press.

Faber, D. 1993. *Environment under Fire*. New York: Monthly Review.

Faber, D. (ed) 1998. *The Struggle for Ecological Democracy: Environmental Justice Movements in the United States*. New York: Guilford.

FAO 1996. http://www.fao.org/docrep/003/w3613e/w3613e00.htm

FAO 2001. *Food: a fundamental human right*.
http://www.fao.org/FOCUS/E/rightfood/right1.htm.

FAO 2002a. International Treaty on Plant Genetic Resources for Food and Agriculture. http://www.ukabc.org/ITPGRe.pdf .

FAO 2002b. UN Agencies Warn of Massive Southern Africa Food Crisis: 10 million People Threatened by Famine. ReliefWeb, 29 May.
http://www.reliefweb.int/w/rwb.nsf/UNID/A8402C0BB9938839C1256BC8002DD4 AA?OpenDocument.

FAO n.d. AIDS – A Threat to Rural Africa. http://www.fao.org/Focus/E/aids/aids1-e.htm

FAO/IAWG 1998. *Food Insecurity and Vulnerability Information and Mapping Systems.* http://www.fao.org/News/1998/980502-e.htm and http://www.fivims.org/static.jspx?lang=en&page=overview .

Farooque, M. 1993. A Legal Perspective on the FAP. *Tiempo: Global Warming and the Third World.* No. 8, April: 17–19.

Farrington, J. and Lewis, D. 1993. *Non-Governmental Organizations and the State in Asia: Rethinking Roles in Sustainable Agricultural Development.* London: ODI.

Fazul, W. n.d. Bangladesh: Power of Humanity. http://www.ifrc.org/publicat/commsguide/html/Eight/WFazlulBox.PDF.

Fire Globe 2003. Global Fire Monitoring Center (GFMC) http://www.fire.uni-freiburg.de/.

Feachem, R., McGarry, M. and Mara, D. (eds) 1978. *Water, Wastes and Health in Hot Climates.* Chichester: John Wiley & Sons.

Feierman, S. 1985. Struggles for Control: The Social Roots of Health and Healing in Modern Africa. *African Studies Review* 28, 2/3: 73–147.

FEMA. (United States Federal Emergency Management Agency) 1997. *Multihazard Identification and Risk Assessment.* Washington, DC: FEMA.

FEMID 2002. Proyecto para fortalecimiento de estructuras en la mitigación de desastres (Project for Strengthening Structures for Mitigation of Disaster). http://www.disaster-info.net/cepredenac/femid/web/paginas/femid.html.

Ferguson, J. 2002. The Tragedy of St. Pierre. *Geographical Magazine* 74, 5: 14–19.

Fernandes, E. and Varley, A. (eds) 1998. *Illegal Cities: Law and Urban Change in Developing Countries.* London: Zed Books.

Fernandes, W. and Menon, G. 1987. Tribal Women and Forest Economy: Deforestation, Exploitation and Status Change. New Delhi: Indian Social Institute.

Fernandez, M. (ed) 1999. *Cities at Risk: Environmental Degradation, Urban Risk and Disaster.* Lima: LA RED.

Fernando, P. and Fernando, V. 1997. *South Asian Women: Facing Disasters, Securing Life.* Colombo: Duryog Nivaran.

Ferraz, B. and Munslow, B. (eds) 1999. *Sustainable Development in Mozambique.* London: James Currey.

Field, J. O. 1993. *The Challenge of Famine.* Connecticut: Kumarian Press.

First, R. 1983. *Black Gold: The Mozambican Miner, Proletarian and Peasant.* Brighton, UK: The Harvester Press.

Firth, R. 1959. *Social Change in Tikopia.* London, Allen & Unwin.

Fiselier, J. 1990. *Living Off the Floods: Strategies for the Integration of Conservation and Sustainable Resource Utilization in Floodplains.* Leiden: Environmental Database on Wetland Interventions.

Flash Flood Lab n.d.: www.cira.colostate.edu/fflab/recentff.htm .

Florida Resources and Environmental Analysis Center (FREAC) 2001. *Atlas of Florida.* Gainsville: FREAC, Florida State University. http://www.freac.fsu.edu/FloridaAtlas/Population/pop5.html

Foote, K. 1997. *Shadowed Ground: America's Landscapes of Violence and Tragedy.* Austin: University of Texas.

Ford, K. 1989. Personal communication between Ian Davis and Keith Ford, Programme Officer in the Office of Disaster Preparedness (ODP), Government of Jamaica.

Forde, D. 1972. *Trypanosomiasis in Africa*. Oxford: Oxford University Press.

Fordham, M. 1998. Making Women Visible in Disasters: Problematising the Private Domain. *Australian Journal of Emergency Management* Summer: 27–33.

Fordham, M. 1999. The Intersection of Gender and Social Class in Disasters: Balancing Resiliency and Vulnerability. *International Journal of Mass Emergencies and Disasters* 17, 1: 15–36.

Fordham, M. 2000. Approaches to Participatory Planning for Flood Management. In D. Parker (ed), *Flood Hazards and Disasters*, pp.66–79. London: Routledge.

Fordham, M. 2001. RADIX Field Report: Gujarat. http://online.northumbria.ac.uk/ geography_research/radix/gujarat-fieldreport/gujarat-fieldreport.htm .

Fordham, M. 2003. Gender, Development and Disaster: The Necessity for Integration. In: M. Pelling (ed), *Natural Disasters and Development in a Globalizing World*, pp.57–74. London: Routledge.

Fothergill, A. 1996. Gender, Risk and Disaster. *International Journal of Mass Emergencies and Disasters* 14, 1: 33–56.

Fothergill, A. 1999. An Exploratory Study of Woman Battering in the Grand Forks Flood Community: Responses and Policies. *International Journal of Mass Emergencies and Disasters* 17, 1: 79–98.

Fothergill, A., Maestras, E. and Darlington, J. 1999. Race, Ethnicity and Disasters in the United States: A Review of the Literature. *Disasters* 23, 2: 156–173.

Fowler, C. and Mooney, P. 1990. *Shattering: Food, Politics and the Loss of Genetic Diversity*. Tucson: University of Arizona Press.

Fox, J. and Brown, L. (eds) 1998. *The Struggle for Accountability: The World Bank, NGOs and Grassroots Movements*. Cambridge, MA: MIT Press.

Frampton, S., Chaffey, J., McNaught, A. and Hardwick, J. 1998. *Natural Hazards Causes, Consequences and Management. 'The Kobe Earthquake'*, pp.56–7. London: Hodder & Stoughton.

Francis, E. 2000. *Making a Living: Changing Livelihoods in Rural Africa*. London: Routledge.

Franke, R. 1984. Tuareg of West Africa. In: D. Stea and B. Wisner (eds), *The Fourth World: The Geography of Indigenous Struggles*. Thematic issue of *Antipode* 16, 2: 45–53.

Franke, R. and Chasin, B. 1980. *Seeds of Famine: Ecological Destruction and the Development Dilemma in the West African Sahel*. Montclair, NJ: Allenheld, Osmun.

Franke, R. and Chasin, B. 1989. *Kerala: Radical Reform as Development in an Indian State*. San Francisco, CA: Institute for Food and Development Policy.

Frazier, K. 1979. *The Violent Face of Nature*. New York: William Morrow.

FREAC (Florida Resources and Environmental Analysis Center) 2001a. http://www.freac.fsu.edu/FloridaAtlas/Population/pop32.html.

FREAC (Florida Resources and Environmental Analysis Center) 2001b. http://www.freac.fsu.edu/FloridaAtlas/Population/pop31.html.

Freeman, P., Martin, L., Mechler, R. and Warner, K. 2002. *Catastrophes and Development: Integrating Natural Catastrophes into Development Planning*. Disaster Risk Management Working Paper 4. Disaster Management Facility. Washington, DC: World Bank. http://www.worldbank.org/dmf/files/catastrophes_complete.pdf.

399

Freire, P. 1972. *Pedagogy of the Oppressed.* London: Penguin.

French, R. 1989. Houses Built on Sand. *Geographical Magazine* March: 32–34.

Friedman, T. 2000. *The Lexus and the Olive Tree: Understanding Globalization.* Newly updated and expanded edn. New York: Anchor Books.

Friesema, H., Caparaso, J., Goldstein, G., Lineberry, R. and McClearly, R. 1979. *Aftermath: Communities After Natural Disasters.* Beverly Hills, CA: Sage.

Gadgil, M. and Guha, R. 1995. *Ecology and Equity: The Use and Abuse of Nature in Contemporary India.* London: Routledge.

Gardner, G. 2001. Microcredit Expanding Rapidly. In: L. Brown (ed), *Vital Signs. 2001–2002*, pp.110–111. London: Earthscan and World Watch Institute.

Gardner, G. 2002. The Challenge for Johannesburg: Creating a More Secure World. In: L. Starke (ed), *State of the World 2002*, pp.3–23. New York: W. W. Norton.

Garnsey, P. 1988. *Famine and Food Supply in the Graeco-Roman world: Responses to Risks and Crises.* Cambridge: Cambridge University Press.

Gass, V. 2002. *Democratizing Development: Lessons from Hurricane Mitch Reconstruction.* Washington, DC: Washington Office on Latin America (WOLA).

GDMSA (Government of Gujarat Disaster Management Authority) 2002. *Disaster Report.* Ahmedabad: GDMSA.

Geilfus, F., 1997. *Ochenta herramientas para el desarrollo participativo: diagnóstico, planificación, monitoreo, evaluación.* Bogotá, Colombia: Servicio Jesuita a Refugiados (SJR).

GeoHazards International and UNCRD 2001. *Global Earthquake Safety Initiative.* Pilot Project Final Report. Kobe: GeoHazards International and the United Nations Centre for Human Development (UNCRD).

George, S. 1988. *A Fate Worse Than Debt: The Third World Financial Crisis and the Poor.* London: Penguin.

Gerrity, E. and Flynn, B. 1997. Mental Health Consequences of Disasters. In: E. Noji (ed), *The Public Health Consequences of Disasters*, pp.101–121. New York: Oxford University Press.

Gess, D. and Lutz, W. 2002. *Firestorm at Peshtigo.* New York: Henry Holt.

Gheorghe, A. 2003. Complexity Induced Vulnerability. Paper presented at the DPRI-IIASA Third Symposium on Integrated Disaster Risk Management (IDRM-03). Kyoto: Kyoto University, 3 July.

Giddens, A. 1990. *The Consequences of Modernity.* Cambridge: Polity Press.

Giddens, A. 1992. *The Transformation of Intimacy.* Cambridge: Polity Press.

Gilbert, C. 1998. Studying Disaster: Changes in the Main Conceptual Tools. In: E. Quarantelli (ed), *What is a Disaster?*, pp.11–18. London: Routledge.

Gilbert, R. and Kreimer, A. 1999. *Learning from the World Bank's Experience of Natural Disaster Related Assistance.* Disaster Management Facility, Working Paper Series 2. Washington, DC: World Bank.
http://www.worldbank.org/dmf/ files/learningfromwb.pdf.

Gini, C. and De Castro, J. (eds) 1928. *Materiaux pour l'étude des calamités.* Geneva: League of Nations.

Gittings, J. 2002. The Forgotten 800 Million: How Rural Life is Dying in the New China. *Guardian*, 19 October.
http://www.guardian.co.uk/international/story/0,3604,815051,00.html .

Glantz, M. (ed) 1987. *Drought and Hunger in Africa* Cambridge: Cambridge University Press.

Glantz, M. 2001. *Once Burned, Twice Shy? Lessons Learned from the 1997–98 El Niño.*Tokyo: United Nations University Press.

Glantz, M. (ed) 2002. *La Niña and Its Impacts: Facts and Speculations.* Tokyo: United Nations University Press.

Global Earthquake Safety Initiative 2001. *GESI Global Earthquake Safety Initiative,* Palo Alto, CA and Kobe: GeoHazards International and United Nations Centre for Regional Development (UNCRD).

Goheen, M. 1991. Ideology, Gender and Change: Social Relations of Production and Reproduction in Nso, Cameroon. In: R. Downs, D. Kerner and S. Reyna (eds), *The Political Economy of African Famine,* pp.273–292. Philadelphia, PA: Gordon & Breach Scientific Publishers.

Goldblatt, D. 1999. Risk Society and the Environment. In: M. Smith (ed), *Thinking Through the Environment,* pp.373–382. London: Routledge.

Goldenberg, S. 2002. Billions in Aid Bypass Afghanistan's 'Hunger Belt'. *Guardian Weekly,* 7–13 February: 3 (see also http://www.guardian.co.uk/Archive/Article/0,4273,4349166,00.html).

Goldstein, A. 2003. Bioterrorism Effort Aiding Response to SARS. *Washington Post,* 25 April: B6. http://www.washingtonpost.com/wp-dyn/articles/A35467-2003Apr24.html.

Gomez, M. 1991. Reducing Urban and Natural Risks in Mexico City. In: A. Kreimer and M. Munasinghe (eds), *Managing Natural Disasters and the Environment,* pp.56–57. Washington, DC: World Bank.

Goodfield, J. 1991. *The Planned Miracle.* London: Cardinal Books.

Gould, P. 1969. Man Against His Environment: A Game Theoretic Framework. In: A. Vayda (ed), *Environment and Cultural Behavior,* pp.234–251. Garden City, NY: The Natural History Press.

Government of El Salvador (GOES) 2001. *Recovery Plan from the Damage Caused by the Earthquakes of Jan 13th and Feb 13th, 2001.* GOES, Consultative Group Meeting, Coordinated by the Inter-American Development Bank, Madrid, Spain, 7 March 2001.

Government of India 1999. *Super Cyclone Orissa.* Ministry of Agriculture and Cooperation, Natural Disaster Management Division. http://www.ndmindia.nic.in/cycloneorissa/

Government of Jamaica 2002. *Jamaica: External Medium and Long-term Public Debt and Indicators, 1990–2001.* Debt Management Unit, Ministry of Finance and Planning. http://www.mof.gov.jm/dmu/download/2002/extdebt/edi0112cy.pdf.

Government of Jamaica 2003. *Looking back at Hurricane Gilbert.* Office of Disaster Preparedness and Emergency Management.
http://207.21.234.161/articles/gilbert.html

Goyder, H. and Goyder, C. 1988. Case Studies of Famine: Ethiopia. In: D. Curtis, M. Hubbard and A. Shepherd (eds), *Preventing Famine: Policies and Prospects for Africa,* pp.73–110. London: Routledge.

Graff, T. and Wiseman, R. 1978. Changing Concentrations of Older Americans. *Geographical Review* 68: 379–393.

Grainger, A. 1990. *The Threatening Desert: Controlling Desertification.* London: Earthscan.

Grameen Bank 2002. www.grameen-info.org/bank/cds.html.

Gray, W. 2000. General Characteristics of Tropical Cyclones. In: R. Pielke and R. Pielke (eds), *Storms,* Vol. 1, pp.145–163. London: Routledge.

Green, R. 1994. The Course of the Four Horsemen: The Costs of War and its After-math in Sub-Saharan Africa. In: J. Macrae and A. Zwi (eds), *War and Hunger*, pp.37–49. London: Zed Books.

Green, R. I. 1996. *Land Utilisation Study: Customary Land Sector*. Lilongwe: Government Printer.

Greenough, P. 1982. *Prosperity and Misery in Modern Bengal: The Famine of 1943–44*. Oxford: Oxford University Press.

Griggs, G. and Gilchrist, J. 1983. *Geological Hazards, Resources and Environmental Planning*. 2nd edn. Belmont, CA: Wadsworth.

Gueri, M., Gonzalez, C. and Morin, V. 1986. The Effect of the Floods Caused by El Niño on Health. *Disasters* 10, 2: 118–124.

Guillette, E. 1991. The Impact of Recurrent Disaster on the Aged of Botswana. Paper presented at the 50th Annual Meeting of the Society for Applied Anthro-pology, Charleston, SC.

Gunson, P. 2000. Flood Erodes Resistance to Moving. *Christian Science Monitor*, 4 January. http://www.csmonitor.com/durable/2000/01/04/p1s3.htm.

Gupta, A., Kakhandiki, A. and Davidson, R. 1996. Multidisciplinary Approach to Urban Earthquake Disaster Risk Assessment and Management. *The John A. Blume Earthquake Engineering Center Newsletter* 8 (Summer Quarter): 2–3. http://blume.stanford.edu/pdffiles/Summer%201996.pdf.

Gurdilek, R. 1988. Sniffer Dogs Search for Landslide Victims as Hopes Fade. *The Times* (London), 25 June: 7.

Guyer, J. 1981. The Household in African Studies. *African Studies Review* 24, 2–3: 87–137.

Guyer, J. and Peters, P. (eds) 1984. *Conceptualising the Household*. Cambridge, MA: Harvard University Press.

Haas, E., Kates, R. and Bowden, M. 1977. *Reconstruction Following Disaster*. Cambridge, MA: MIT Press.

Haeuber, R. and Michener, W. 1998. Natural Flood Control. *Issues in Science and Technology*, 15, 1: 74–81.

Haghebaert, B. 2001. Roundtable Comments. Workshop on Vulnerability Theory and Practice, Wageningen Agricultural University, Netherlands, May.

Haghebaert, B. 2002. *Perspectieven op pro-actief beheer van natuurrampen in het Zuiden. Een onderzoek naar technocratische, behavioristische, structurele en neopopulistische benaderingen* (Perspectives on pro-active management of natural disasters in the South. An enquiry into technocratic, behaviouristic, structural and neo-populist approaches). Unpublished Ph.D dissertation, Political and Social Sciences, Centre for Third World Studies. Ghent: University of Ghent.

Hagman, G. 1984. *Prevention Better than Cure: A Swedish Red Cross Report on Human and Environmental Disasters in the Third World*. Stockholm: Swedish Red Cross.

Hall, J. (ed) 1995. *Civil Society: Theory, History, Comparison*. Cambridge: Polity Press.

Hall, N., Hart, R. and Mitlin, D. (eds) 1996. *The Urban Opportunity: The Work of NGOs in Cities of the South*. London: IT Publications.

Handmer, J. and Penning-Rowsell, E. (eds) 1990. *Hazards and the Communication of Risk*. Aldershot: Gower Publishing.

Handmer, J., Norton, T. and Dovers, S. 2001. *Ecology, Uncertainty and Policy: Managing Ecosystems for Sustainability*. Harlow, UK: Prentice Hall/Pearson Education.

Hangzhou Declaration 1999. International Workshop on Coastal Mega-cities: Challenges of Growing Urbanisation of the World's Coastal Areas, Hangzhou, People's Republic of China, 27–30 September.
http://icm.noaa.gov/globalinfo/ hangzhou.html.

Hanlon, J. 1991. *Mozambique: Who Calls the Shots?* Bloomington: Indiana University Press.

Hanlon, J. 1996. *Peace without Profit: How the IMF Blocks Rebuilding in Mozambique*. London: James Currey.

Hansen, A. and Oliver-Smith, A. (eds) 1982. *Involuntary Migration and Resettlement: The Problems and Responses of Dislocated People*. Boulder, CO: Westview.

Hansen, E. (ed) 1987. *Africa: Perspectives on Peace and Development*. London: Zed Books.

Hansen, S. 1988. Structural Adjustment Programs and Sustainable Development. Paper commissioned by UNEP for the annual session of the Committee of International Development Institutions on the Environment (CIDIE), Washington, 13–17 June.

Hanson, W. J. 1967. *East Pakistan in the Wake of the Cyclone*. London: Longmans.

Haq, K. and Kirdar, U. (eds) 1987. *Human Development, Adjustment and Growth*. Islamabad: North South Roundtable.

Haque, C. 1997. *Hazards in a Fickle Environment: Bangladesh*, Dordrecht, Netherlands: Kluwer.

Haque, C. E. and Blair, D. 1992. Vulnerability to Tropical Cyclones: Evidence from the April 1991 Cyclone to Coastal Bangladesh. *Disasters* 16, 3: 217–229.

Hardin, G. 1968. The Tragedy of the Commons. *Science* 162: 1243–1248.

Hardin, G. 1974. Living on a Lifeboat. *BioScience* 24: 561–568.

Harding, L. 2001. Indian Quake Widens Rifts between the Castes. *Guardian Weekly*, 22–28 February: 5.

Harding, L. 2002. Frantic Search for British Sons Lost in Gujarat Riots.*Guardian* 24 April. http://www.guardian.co.uk/international/story/0,3604,689499,00.html.

Hardoy, J. and Satterthwaite, D. 1989. *Squatter Citizen: Life in the Urban Third World*. London: Earthscan.

Hardt, M. and Negri, A. 2000. *Empire*. Cambridge, MA: Harvard University Press.

Harley, R. 1990. *Breakthroughs on Hunger*. Washington, DC: Smithsonian Institute Books.

Harrell-Bond, B. 1986. *Imposing Aid: Emergency Assistance to Refugees*. Oxford: Oxford University Press.

Harremoes, P., Gee, D., MacGarvin, M., Stirling, A., Keys, J., Wynne, B., and Guedes Vaz, S. 2002. *The Precautionary Principle in the 20th Century*. London: Earthscan and the European Environment Agency.

Harris, B. 1998. After Mitch, the Street Children, Forgotten by Everyone, Casa Alianza. http://www.boes.org/actions/america/cenral/casa16.html.

Harrison, P. 1987. *The Greening of Africa*. London: Penguin.

Harrison, P. and Palmer, R. 1986. *News out of Africa: Biafra to Band Aid*. London: Hilary Shipman.

Hart, J. 1971. The Inverse Care Law. *The Lancet* i: 405–412.

Hartmann, B. 1995. *Reproductive Rights and Wrongs: The Global Politics of Population Control.* Revised Edn. Boston, MA: South End Press.

Hartmann, B. and Boyce, J. 1983. *A Quiet Violence: View from a Bangladesh Village.* London: Zed Books.

Hartmann, B. and Standing, H. 1989. *The Poverty of Population Control: Family Planning and Health Policy in Bangladesh.* London: Bangladesh International Action Group.

Harvey, D. 1996. *Justice, Nature and the Geography of Difference.* Oxford: Blackwell.

Havlick, S. 1986. Third World Cities at Risk: Building for Calamity. *Environment* 28, 9 (November): 6.

Hayakawa, Z. 2000. Why Do We Not Learn from the Victims? http://www.ywca.or.jp/kobe/sinsai/ENGLISH/hayakawa.html.

Hays, S. 1987. *Beauty, Health, and Permanence: Environmental Politics in the United States, 1955–1985.* Cambridge: Cambridge University Press.

Hays, W. 1990. Perspectives on the International Decade for Natural Disaster Reduction. *Earthquake Spectra* 6: 125–143.

Heath, R. 1995. The Kobe Earthquake: Some Realities of Strategic Management of Crises and Disasters, *Disaster Prevention and Management* 4, 5: 11–24.

Heijmans, A. and Victoria, L. 2001. *Citizen-based and Development-oriented Disaster Response.* Quezon City, Philippines: Center for Disaster Preparedness (contacts:cdp@info.com.ph).

Heilig, G. 1999. Can China Feed Itself? A System for Evaluation of Policy Options. Laxenburg, Austria: International Institute for Applied Systems Analysis (IIASA). http://www.iiasa.ac.at/Research/LUC/ChinaFood/index_m.htm.

Heiman, M. 1996. Race, Waste, Class: New Perspectives on Environmental Justice. *Antipode* 28, 2: 111–121.

Hellden, U. 1984. Drought Impact Monitoring: A Remote Sensing Study of Desertification in Kordofan, Sudan. Lund Universitets Naturgeografiska Institution, Lund, Sweden. pp.61. Mimeo.

HelpAge International 2000. *Older People in Disasters and Humanitarian Crises: Guidelines for Best Practices.* ReliefWeb. http://www.reliefweb.int/library/documents/HelpAge_olderpeople.pdf.

Hervio, G. 1987. *Appraisal of Early Warning Systems in the Sahel* (Main Report). Paris: OECD/CILSS. (see also: http://www.oecd.org/dataoecd/34/4/1907520.pdf).

Hewitt, K. 1981/1982. Settlement and Change in Basal Zone Ecotones: An Interpretation of the Geography of Earthquake Risk. In: B. Jones and M. Tomazevic (eds), *Social and Economic Aspects of Earthquakes*, pp.15–42. Proceedings of the Third International Conference: The Social and Economic Aspects of Earthquakes and Planning to Mitigate their Impacts. Bled, Yugoslavia and Ithaca: Institute for Testing in Materials and Structures Ljubljana and Cornell University.

Hewitt, K. (ed) 1983a. *Interpretations of Calamity.* Boston, MA: Allen & Unwin.

Hewitt, K. 1983b. The Idea of Calamity in a Technocratic Age. In: K. Hewitt (ed), *Interpretations of Calamity*, pp.3–32. Boston, MA: Allen & Unwin.

Hewitt, K. 1994. When the Great Planes Came and Made Ashes of our City...: Towards an Oral Geography of the Disasters of War. *Antipode* 26, 1: 1–34.

Hewitt, K. 1995. Sustainable Disasters? Perspectives and Powers in the Discourse of Calamity. In: J. Crush (ed), *Power of Development*, pp.115–128. London: Routledge.

Hewitt, K. 1997. *Regions of Risk: A Geographical Introduction to Disasters*. Harlow: Longman.

Hewitt, K. 1998. Excluded Perspectives in the Social Conception of Disaster. In: E. Quarantelli (ed), *What is a Disaster*, pp.75–91. London: Routledge.

Hill, A. and Cutter, S. 2001. Methods for Determining Disaster Proneness. In: S. Cutter, (ed), *American Hazardscapes: The Regionalization of Hazards and Disasters*. Washington, DC: Joseph Henry Press.

Hines, C. 2000. *Localization: A Global Manifesto*. London: Earthscan.

H. John Heinz Center 2000. *The Hidden Costs of Coastal Hazards: Implications for Risk Assessment and Mitigation*. Washington, DC: Island Press.

Hoag, H. 2002. Escaping the Rising Waters? Environmental 'Crises' and Resettlement in the Rufiji District of Tanzania, 1968–1974. Paper presented at the Annual Meeting of the African Stuidies Association, Washington, DC, 8 December.

Hoben, A. 1995. Paradigms and Politics: The Cultural Construction of Environmental Policy in Ethiopia. *World Development* 23, 6: 1007–1022.

Hofrichter, R. (ed) 1993. *Toxic Struggles: The Theory and Practice of Environmental Justice*. Philadelphia, PA: New Society Publishers.

Holland, G. (ed) 2000. *Global Guide to Tropical Cyclone Forecasting*. Melbourne, Australia: Bureau of Meteorological Research Centre, Bureau of Meteorology, Commonwealth of Australia.
http://www.bom.gov.au/bmrc/pubs/tcguide/globa_guide_intro.htm.

Holloway, R. 1989. *Doing Development – Governments, NGOs and the Rural Poor in Asia*. London: Earthscan.

Holmes, T. and Scoones, I. 2000. *Participatory Environmental Policy Processes: Experiences from North and South*. IDS Working Paper No. 113. Brighton, UK: IDS (Institute for Development Studies), University of Sussex.

Homer-Dixon, T. 2001. *The Ingenuity Gap: How Can We Solve the Problems of the Future?* London: Vintage.

Hoover, G. and Bates, F. 1985. The Impact of Natural Disaster on the Division of Labor in Twelve Guatemalan Communities: A Study of Social Change in a Developing Country. *International Journal of Mass Emergencies and Disasters* 3, 3: 7–26.

Hopkins, R. 1987. The Evolution of Food Aid: Toward a Development-First Regime. In: J. Gittinger, J. Leslie and C. Hoisington (eds), *Food Policy: Integrating Supply, Distribution and Consumption*, pp.246–259. Baltimore, MD: Johns Hopkins University Press.

Horn, J. 1965. *Away with All Pests*. New York: Monthly Review.

Horowitz, M. 1989. Victims of Development. *Development Anthropology Network* 7, 2: 1–8.

Horowitz, M. and Salem-Murdock, M. 1990. Management of an African Floodplain: A Contribution to the Anthropology of Public Policy. In: M. March and and H. Udo de Haes (eds), *The People's Role in Wetland Management*, pp.229–236. Leiden: Centre for Environmental Studies.

Hossain, H., Dodge, C. and Abed, F. (eds) 1992. *From Crisis to Development: Coping with Disasters in Bangladesh*. Dhaka: The University Press.

Houghton, J. et al. (eds) 2001. *Climate Change 2001: The Scientific Basis*. Cambridge: Cambridge University Press.

Hughes, C. and Hunter, J. 1970. Disease and 'Development' in Africa. *Social Science and Medicine* 3: 443–493.

Human Rights Watch 1995. The Three Gorges Dam in China: Forced Resettlement, *Suppression of Dissent and Labor Rights Concerns* 7, 2 (February). http://www.hrw.org/summaries/s.china952.html.

Humanitarian Initiatives, UK, Disaster Mitigation Institute, India (DMI) and Mango, UK 2001. *Independent Evaluation of Expenditure of DEC India Earthquake Appeal Funds January 2001–October 2001*. London: Disasters Emergency Committee (DEC).

Hussein, A. 1976. The Political Economy of Famine in Ethiopia. In: A. Hussein (ed), *Rehab: Drought and Famine in Ethiopia*, pp.9–43. London: International African Institute.

ICIHI (Independent Commission on International Humanitarian Issues) 1988. *Winning the Human Race*. London: Zed Books.

IDNDR (International Decade for Natural Disaster Reduction) 1994. Yokohama Strategy and Plan of Action for a Safer World: Guidelines for Natural Disaster Prevention, Preparedness and Mitigation. World Conference on Natural Disaster Reduction, Yokohama, Japan, 23–27 May. http://www.unisdr.org/unisdr/resyokohama.htm .

IDNDR (International Decade for Natural Disaster Reduction) 1999. *Proceedings: Programme Forum*. Geneva: IDNDR.

IDNDR and QUIPUNET 1996. *Solutions for Cities at Risk*. Internet Conference, 26 August 1996 - 25 October.

IFRC (International Federation of Red Cross and Red Crescent Societies) 1994. India: Earthquake Myths and Realities. In: *World Disasters Report 1994*, pp.119–24, Geneva: IFRC.

IFRC (International Federation of Red Cross and Red Crescent Societies) 1996. Earthquake Perceptions and Survival. In: *World Disasters Report 1996*, pp.65–75, Geneva: IFRC.

IFRC (International Federation of Red Cross and Red Crescent Societies) 1998. Must Millions More Die from Traffic Accidents? In: *World Disasters Report*, ch. 2, pp.20–31. Geneva: IFRC.

IFRC (International Federation of Red Cross and Red Crescent Societies) 1999a. *World Disasters Report 1999*. Geneva: IFRC.

IFRC (International Federation of Red Cross and Red Crescent Societies) 1999b. *Vulnerability and Capacity Assessment: An International Federation Guide*. Geneva: IFRC.

IFRC (International Federation of Red Cross and Red Crescent Societies) 2000a. La Niña Storms Savage Venezuela. In: *World Disasters Report 2000*, p.10. Geneva: IFRC.

IFRC (International Federation of Red Cross and Red Crescent Societies) 2000b. Turkish Earthquakes Leave Long-term Legacy. In: *World Disasters Report 2000. Focus on Public Health*, Box 1.5, pp.26–27. Geneva: IFRC.

IFRC (International Federation of Red Cross and Red Crescent Societies) 2000c. *World Disasters Report 2000. Focus on Public Health*. Geneva: IFRC

IFRC (International Federation of Red Cross and Red Crescent Societies) 2001a. *World Disasters Report 2000: Assessing Vulnerabilities and Capacities in Peace and War*. Geneva: IFRC.

IFRC (International Federation of Red Cross and Red Crescent Societies) 2001b. Trapped in the Gap: Post-Landslide Venezuela. In: *World Disasters Report 2001*, ch. 4. Geneva: IFRC.

IFRC (International Federation of Red Cross and Red Crescent Societies) 2001c. Relief, Recovery and Root Causes. In: *World Disaster Report 2001: Focus on Recovery*, pp.9–33. Geneva: IFRC.

IFRC (International Federation of Red Cross and Red Crescent Societies) 2002a. *World Disasters Report 2002: Focus on Reducing Risk*. Geneva: IFRC.

IFRC (International Federation of Red Cross and Red Crescent Societies) 2002b. Assessing Vulnerabilities and Capacities – During Peace and War. In: *World Disasters Report 2000: Focus on Reducing Risk*, pp.129–147. Geneva: IFRC.

IFRC (International Federation of Red Cross and Red Crescent Societies) 2002c. Senegal: Floods Information Bulletin No. 01. ReliefWeb, 2 February. http://www.reliefweb.int/w/rwb.nsf/s/274388013ED6CDF249256B570009DBFC.

IFRC (International Federation of Red Cross and Red Crescent Societies) 2002d. Federation to Target Aids-affected as Food Crisis Worsens in Southern Africa. Geneva/Harare. ReliefWeb, 2 May. http://www.reliefweb.int/w/rwb.nsf/480fa8736b88bbc3c12564f6004c8ad5/122dedb 13e10e07c85256bad005969d3?OpenDocment.

IFRC (International Federation of Red Cross and Red Crescent Societies) and Centre for Research in the Epidemiology of Disasters (CRED) 1993. *World Disasters Report*. Geneva: (IFRCS).

IIED 2002. *Breaking New Ground: Mining, Minerals, and Sustainable Development*. Report of the MMSD Project. London: IIED. http://www.iied.org/mmsd/finalreport/ .

IMF (International Monetary Fund) 2002. World Economic Outlook 2002 http://www.imf.org/external/pubs/ft/weo/2002/01/pdf/chapter2.pdf.

Indian Express 2000. Arunachal Floods – Dam Breach in Tibet, China 'Hushed Up'. 11 July. www.expressindia.com/ie/daily/20000711/iin11063.html.

Ingleton, J. (ed) 1999. *Natural Disaster Management*. London: Tudor Rose.

International Centre, Cities on Water 1989. *Impact of Sea Level Rise on Cities and Regions*. International Centre, S. Marco, Venice, Italy.

International Development Committee 1997. *Montserrat*. First Report House of Commons Session 1997–1998, London: Stationery Office, 28 July.

Irmgard Coninx Foundation (Berlin). n.d. Inititative. Transnational Risks: The Responsibility of the Media and the Social Sciences. http://www.irmgard-coninx-stiftung.de/en/roundtables.htm.

Isaza, P., de Quinteros, Z., Pineda, E., Parchment, C., Aguilar, E. and McQuestion, M. 1980. A Diarrheal Disease Control Programme among Nicaraguan Refugee Children in Campo Luna, Honduras. *Bulletin of the Pan American Health Organization* 14: 337–342.

ISDR (International Strategy for Disaster Reduction) 1999. International Strategy for Disaster Reduction. A Safer World for the 21st Century: Disaster and Risk Reduction. *ISDR Newsletter for Latin America and the Caribbean* No. 15: 3–5.

ISDR (International Strategy for Disaster Reduction) 2001. *The Socio-economic Consequences of the Earthquake at Izmit, Turkey*, New York and Geneva: UN ISDR.

ISDR (International Strategy for Disaster Reduction) 2002a. *Disaster Risk and Sustainable Development: Understanding the Links between Development, Environment and Natural Hazards Leading to Disasters.* Background document for the World Summit on Sustainable Development (WSSD). Geneva: ISDR.

ISDR (International Strategy for Disaster Reduction) 2002b. *Living with Risk: A Global Review of Disaster Reduction Initiatives.* Preliminary edn. Geneva: ISDR.

ISDR (International Strategy for Disaster Reduction)/RADIUS 2001. *United Nations Initiative towards Earthquake Safe Cities.* Geneva: ISDR.

Islam, M. 1974. Tropical Cyclones: Coastal Bangladesh. In: G. White (ed), *Natural Hazards*, pp.19–25. New York: Oxford University Press.

Ives, J. and Messerli, B. 1989. *The Himalayan Dilemma: Reconciling Development and Conservation.* London: Routledge.

Jabry, A. 2003. *Children in Disasters.* London: PLAN International. http://www.plan-uk.org/action/childrenindisasters/

Jackson, T. 1982. *Against the Grain: The Dilemma of Project Food Aid.* London: Oxfam.

Jacobs, D. 1987. *The Brutality of Nations.* New York: Paragon.

Jacobs, M. (ed) 1998. *The Politics of Risk.* London: Polity Press.

Jacobson, J. 1988. *Environmental Refugees: A Yardstick of Habitability.* Worldwatch Paper 86. Washington, DC: Worldwatch Institute.

Jamaica at a Glance 1999. Key Economic Indicators and Long-Term Trends. http://wbln0018.worldbank.org/LAC/lacinfoclient.nsf/5996dfbf9847f67d85256736 005dc67c/6ed022bb169c089e852569b30055ccad/$FILE/Annex%20A%20Country %20at%20a%20Glance%20(page1.pdf.

Jan-Hwa, C. (ed) 1998. *Tropical Cyclone Forecasters' Reference Guide.* Monterey, CA: Naval Research Laboratory, Marine Meteorology Division. http://www.nrlmry.navy.mil/~chu/tropcycl.htm.

Jasanoff, S. (ed) 1994. *Learning from Disasters: Risk Management After Bhopal.* Philadelphia: University of Pennsylvania Press.

Jeffrey, S. 1980. Universalistic Statements About Human Social Behaviour. *Disasters* 4, 1: 111–112.

Jeffrey, S. 1982. The Creation of Vulnerability to Natural Disaster: Case Studies from the Dominican Republic. *Disasters* 6, 1: 38–43.

Jeggle, T. and Stephenson, R. 1994. Concepts of Hazards and Vulnerability Analysis. In V. Sharma (ed) *Disaster Management*, pp. 251-257. New Delhi: Indian Institute of Public Administration.

Jiggins, J. 1986. Women and Seasonality: Coping with Crisis and Calamity. *IDS Bulletin* 17, 3: 9–18.

Jigyasu, R. 2001. *Field Assessment: Post-earthquake Rehabilitation in Gujarat – 9 Months After.* Tronheim: Department of Town and Regional Planning, Norwegian University of Science and Technology.

Jodha, N. 1991. *Rural Common Property Resources: A Growing Crisis.* Gatekeeper Series No. 24. London: International Institute for Environment and Development.

Johns, A. (ed) 1999. *Dreadful Visitations: Confronting Natural Catastrophe in the Age of Enlightenment.* New York: Routledge.

Johnson, H. and Mayoux, L. 1998. Investigation as Empowering: Using Participatory Methods. In: A. Thomas, J. Chataway and M. Wuyts (eds), *Finding Out Fast:*

Investigative Skills for Policy and Development, pp.147–171. London: Sage and Open University.

Johnston, B. (ed) 1994. *Who Pays the Price? The Socio-cultural Context of Environmental Crisis*. Washington, DC: Island Press.

Johnston, B. (ed) 1997. *Life and Death Matters: Human Rights and the Environment at the End of the Millennium*. Walnut Creek, CA: Alta Mira Press.

Johnston, B. and Schulte, J. 1992. Natural Power and Power Plays in Watsonville, California and the U.S. Virgin Islands. Paper presented to the Society for Applied Anthropology, March 26, Memphis, TN.

Johnston, P. and Simmonds, M. 1991. Green Light for Precautionary Science. *New Scientist* 3 August: 4.

Jones, N. 2001. Bubbling Up. *New Scientist* 24 February.
http://www.newscientist.com/ news/news.jsp?id=ns9999393.

Jones, N. 2003. Lake's Silent Killer to be Disarmed. *New Scientist*, January.
http://www.newscientist.com/news/news.jsp?id=ns9999393.

Journal of International Development 2001. Theme issue, *Attacking Poverty: A Strategic Dilemma for the World Bank*. Based on seminars at the Overseas Development Institute and at the International Development Department, University of Birmingham, in Autumn, 2000
http://www.odi.org.uk/speeches/ and http://www.bham.ac.uk/IDD/.

Jowett, J. 1990. People: Demographic Patterns and Policies. In: T. Cannon and A. Jenkins (eds), *The Contemporary Geography of China: The Impact of Deng Xiaoping's Decade,* pp.102–132. London: Routledge.

Juma, C. 1989. *The Gene Hunters*. London and Princeton, NJ: Zed Books and Princeton University Press.

Kailes, J. 1996. *Living and Lasting on Shaky Ground: An Earthquake Preparedness Guide for People with Disabilities*. www.jik.com/resource.html.

Kalsi, S. and Gupta, M. 2002. Success and Failure of Early Warning Systems: A Case Study of the Gujarat Cyclone of June, 1998. In: J. Zschau and A. Kueppers (eds), *Early Warning Systems for Natural Disaster Reduction*, pp.199–202. Berlin: Springer Verlag.

Kane, P. 1988. *Famine in China: Demographic and Social Implications*. London: Macmillan.

Kasperson, R. and Kasperson, J. (eds) 2000. *Global Environmental Risk*. London: Earthscan.

Kasperson, R. and Pijawka, D. 1985. Societal Response to Hazards and Major Hazard Events: Comparing Natural and Technological Hazards. Special issue, *Public Administration Review* 45: 7–18.

Kebbede, G. 1992. *The Ethopian Predicaments: State-dictated Development, Ecological Crisis, Famine and Mass Displacement*. Atlantic Heights, NJ: Humanities Press.

Keen, D. 1994. *The Benefits of Famine: A Political Economy of Famine and Relief in Southwestern Sudan 1983–1989*. Princeton, NJ: Princeton University Press.

Keen, M., Ross, H. and Handmer, J. 1988. The cultural dimension of hazard management: flooding in Alice Springs. *The International Panel for Risk Reduction in Hazard Prone Areas Newsletter* (renamed Hazard and Disaster Management) No. 3 (November): 23-27.

Keller-Herzog, A. 1996. Globalisation and Gender: Development Perspectives and Interventions. Discussion Paper prepared for Women in Development and Gender

Equity Division, Policy Branch, Canadian International Development Agency. http://www.acdicida.gc.ca/cida_ind.nsf/8949395286e4d3a58525641300568be1/a14f0 eb58e1b97138525694f005aa5d5?OpenDocument.

Kemp, P. 1991. For Generations to Come: The Environmental Catastrophe. In: P. Bennis and M. Moushabeck (eds), *Beyond the Storm: A Gulf Crisis Reader*, pp.325–334. New York: Olive Branch Press.

Kent, G. 1987. *Fish, Food and Hunger*. Boulder, CO: Westview.

Kent, G. 1988. Nutrition Education as an Instrument of Empowerment. *Journal of Nutrition Education* 20, 4: 193–195.

Kent, G. 2003. The Human Right to Adequate Food. Forthcoming. http://www2.hawaii.edu/%7Ekent/00HRAF2003ENTRYWAY.doc.

Kent, R. 1987. *Anatomy of Disaster Relief: The International Network in Action*. London and New York: Pinter Publishers.

Kerner, D. and Cook, K. 1991. Gender Hunger and Crisis in Tanzania. In: R. Downs, D. Kerner and S. Reyna (eds), *The Political Economy of African Famine*, pp.257–272. Philadelphia, PA: Gordon & Breach Scientific Publishers.

Kerr, J. 2002. *From 'WID' to 'GAD' to Women's Rights: The First Twenty Years of AWID*. Occasional Paper 9. Association for Women in Development. http://www.awid.org/publications/OccasionalPapers/occasional9.html

Khan, M. and Shahidullah, M. 1982. The Role of Water and Sanitation in the Incidence of Cholera in Refugee Camps. *Transactions of the Royal Society of Tropical Medicine and Hygiene* 76: 373–377.

Khan, M. I. 1991. The Impact of Local Elites on Disaster Preparedness Planning: The Location of Flood Shelters in Northern Bangladesh. *Disasters* 15, 4: 340–354.

Kibreab, G. 1985. *African Refugees*. Trenton, NJ: Africa World Press.

Kiessling, K. and Landberg, H. (eds) 1994. *Population, Economic Development and the Environment: The Making of Our Common Future*. Oxford, Oxford University Press.

Kiljunen, K. (ed) 1984. *Kampuchea: Decade of Genocide. Report of a Finnish Inquiry Commission*. London: Zed Books.

Kimambo, I., Maddox, G. and Gibblin, J. 1995. *Custodians of the Land: Environment and Hunger in Tanzanian History*. Athens: Ohio University Press.

King, D. 2001. Uses and Limitations of Socioeconomic Indicators of Community Vulnerability to Natural Hazards: Data and Disasters in Northern Australia. *Natural Hazards* 24, pp. 147-156.

Kirby, A. 1990a. On Social Representations of Risk. In: A. Kirby (ed), *Nothing to Fear: Risks and Hazards in American Life*, pp.1–16. Tucson: University of Arizona Press.

Kirby, A. 1990b. Toward a New Risk Analysis. In: A. Kirby (ed), *Nothing to Fear*, pp.281–298. Tucson: University of Arizona Press.

Kirby, A. (ed) 1990c. *Nothing to Fear: Risks and Hazards in American Life*. Tucson: University of Arizona Press.

Kiriro, A. and Juma, C. (eds) 1989. *Gaining Ground: Institutional Innovations in Land-use Management in Kenya*. Nairobi: ACTS Press (African Centre for Technology Studies).

Kjekshus, H. 1977. *Ecological Control and Economic Development in East African History*. Berkeley: University of California Press.

410

Klee, G. (ed) 1980. *World Systems of Traditional Resource Management*. New York: Halstead.

Klinenberg, E. 2002. *Heat Wave: A Social Autopsy of a Disaster*. Chicago, IL: University of Chicago Press.

Kloos, H. 1982. Development, Drought and Famine in the Awash Valley of Ethiopia. *African Studies Review* 25, 4: 21–48.

KMDA (Kolkata Metropolitan Development Authority) 2003. http://www.cmdaonline.com/evolution.html.

Kobe City 1995. Damage. http://www.city.kobe.jp/cityoffice/06/013/report/1–2.html .

Kolkata Metropolitan Development Authority n.d. Sewerage, Drainage, and Sanitation. http://www.cmdaonline.com/sewdrainsan.htm.

Kotch, N. 2002. Cruel Nature Not The Only Cause Of Africa's Hunger. Reuters, 30 May. Viewed on ReliefWeb, 4 June. http://www.reliefweb.int/w/rwb.nsf/480fa8736b88bbc3c12564f6004c8ad5/d8ad703cd1 23932fc1256bc9003f453d?OpenDocument.

Kothari, S. 1999. Inclusive, Just, Plural, Dynamic: Building a 'Civil' Society in the Third World. In: M. Kothari (ed), *Development and Social Action*, pp.34–53. Oxford: Oxfam Publications.

Kreimer, A. and Echeverria, E. 1991. Case Study: Housing Reconstruction in Mexico City. In: A. Kreimer and M. Munasinghe (eds), *Managing National Disasters and the Environment*, pp.53–61. Washington, DC: World Bank.

Kriner, S. 2000. Three Months after Super Cyclone, Orissa Begins to Rebuild. http://www.disasterrelief.org/Disasters/000201Orissaupdate/.

Kriner, S. 2002. At Least 622 Die in Southern India Heat Wave, 17 May. http://www.disasterrelief.org/Disasters/020516IndiaHeatwave/.

Kristof, N. 1991. In Bangladesh Storms, Poverty More than Weather is the Killer. *New York Times*, 11 May: A1 & A5.

Kruks, S. 2002. *Retrieving Experience*. Ithaca, NY: Cornell University Press.

Kruks, S. and Wisner, B. 1989. Ambiguous Transformations: Women, Politics and Production in Mozambique. In: S. Kruks, R. Rapp and M. Young (eds), *Promissory Notes: Women in the Transition to Socialism*, pp.148–171. New York: Monthly Review.

Kuester, I. and Forsyth, S. 1985. Rabaul Eruption Risk: Population Awareness and Preparedness Survey. *Disasters* 9, 3: 179–182.

Kumar, G. 1987. *The Ethiopian Famine and Relief Measures: An Analysis and Evaluation*. Addis Ababa: UNICEF.

La RED 1998. *Revista Desastres y Sociedad* 9. Especial: El Niño. Lima: La RED. http://www.desenred ando.org/public/revistas/dys/rdys09/index.html.

Laidlaw, I. 2001. *The Origin and Future of the Burakumin*. Unpublished MA Thesis, Otago University, New Zealand. http://www.geocities.com/gaijindo4dan/thesis/ index.html

Laidlaw, I. 2002. Personal correspondence with B. Wisner via email, March and April 2002.

Laird, R. 1992. Private Troubles and Public Issues: The Politics of Disaster. Paper presented to the Society for Applied Anthropology, 26 March, Memphis, TN.

Langlands, B. (ed) 1968. *The Medical Atlas of Uganda*. Kampala: Makerere University, Department of Geography.

Latin America 1976. 9 April: 115.

Lavell, A. 1994. Prevention and Mitigation of Disasters in Central America: Vulnerability to Disasters at the Local Level. In: A. Varley (ed), *Disasters, Development and Environment*, pp.49–63. Chichester: Wiley.

Lavell, A. 2001. Roundtable Comments. Workshop on Vulnerability Theory and Practice, Wageningen Agricultural University, Netherlands, May.

Lavell, A. 2002. After Hurricane Mitch. Text prepared for *World Vulnerability Report*. Geneva: UNDP/ERD.

Lawson-Tancred, A. 2002. Microcredit Scheme Comes Under Fire. *Financial Times*, 21 May: 6.

Le Moigne, G., Barghouti, S. and Plusquellec, H. (eds) 1990. *Dam Safety and the Environment*. Technical Paper No. 115. Washington, DC: World Bank.

Leach, G. and Mearns, R. 1989. *Beyond the Woodfuel Crisis*, London: Earthscan.

Leach, M. and Mearns, R. (eds) 1996. *The Lie of the Land: Challenging Received Wisdom on the African Environment*. London: International African Institute and James Currey.

Leach, M., Mearns, R. and Scoones, I. 1997. *Environmental Entitlements: A Framework for Understanding the Institutional Dynamics of Environmental Change*. IDS Working Paper No. 359. Brighton, UK: IDS (Institute for Development Studies), University of Sussex.

Leaf, M. 1997. Local Control Versus Technocracy: The Bangladesh Flood Response Study. *Journal of International Affairs* 51, 1: 179–200.

Leaning, J. 2000. The Public Health Approach to the Impact of War. In: J. Austin and C. Bruch (eds), *The Environmental Consequences of War*, pp.384–401. Cambridge: Cambridge University Press.

Learmonth, A. 1988. *Disease Ecology*. Oxford: Basil Blackwell.

Lee, B. and Davis, I. (eds) 1998. *Forecasts and Warnings*. IDNDR Flagship Programme. UK National Coordination Committee for the IDNDR. London: Thomas Telford.

Lee, T. 1999. *Water Management in the 21st Century: The Allocative Imperative*. Cheltenham, UK: Edward Elgar.

Lefebvre, H. 1991. *The Production of Space*. Oxford: Blackwell.

Leftwich, A and Harvie, D. 1986. *The Political Economy of Famine*. York: University of York, Institute for Research in the Social Services.

Lemma, H. 1985. The Politics of Famine in Ethiopia. *Review of African Political Economy* 33: 44–58.

Levin, R. and Weiner, D. 1997. *'No More Tears…': Struggles for Land in Mpumalanga, South Africa*. Trenton, NJ: Africa World Press.

Lewis, J. 1981. Some Perspectives on Natural Disaster Vulnerability in Tonga. *Pacific Viewpoint* 22, 2: 145–162.

Lewis, J. 1984a. A Multi-hazard History of Antigua. *Disasters* 8, 3: 190–197.

Lewis, J. 1984b. *Disaster Mitigation Planning: Some Lessons from Island Countries*. Occasional Paper, Centre for Development Studies. Bath: University of Bath.

Lewis, J. 1987. Vulnerability and Development – and the Development of Vulnerability: A Case for Management. Development Studies Association, Annual Conference, University of Manchester, 16–18 September.

Lewis, J. 1989. *Sea-level rise: Tonga, Tuvalu (Kiribati)*. Commonwealth Expert Group on Climate Change and Sea Level Rise. London: Commonwealth Secretariat.

Lewis, J. 1990. The Vulnerability of Small Island-states to Sea Level Rise: The Need for Holistic Strategies. *Disasters* 14, 3: 241–248.

Lewis, J. 1999. *Development in Disaster-prone Places*. London: IT Books.

Lewis, N. 1991. String of Crises Overwhelms Relief Agencies and Donors. *New York Times*, 4 May: A5.

Lindell, M. and Perry, R. 1992. *Behavioral Foundations of Community Emergency Planning*. Washington, DC: Hemisphere.

Lindsay, J. 2002. The Determinants of Disaster Vulnerability: Achieving Sustainable Mitigation through Population Health. Unpublished manuscript. Winnipeg, Canada: Manitoba Health (courtesy of the author via personal communication with B. Wisner).

Lipton, M. and Longhurst, R. 1989. *New Seeds and Poor People*, Baltimore, MD: Johns Hopkins University Press.

Lisk, F. (ed) 1985. *Popular Participation in Planning for Basic Needs*. Aldershot: Gower.

Little, P. and Horowitz, M. (eds) 1987. *Lands at Risk in the Third World: Local-level Perspectives*. Boulder, CO: Westview.

Liverman, D. 1989. Vulnerability to Global Environmental Change. Paper presented to International Workshop on Understanding Global Environmental Change, Clark University, Center for Technology, Environment and Development, 11–13 October.

Liverman, D., Moran, E., Rindfuss, R. and Stern, P. (eds) 1998. *People and Pixels: Linking Remote Sensing and Social Science*. Washington, DC: National Academy Press.

Lomborg, B. 2001. *The Skeptical Environmentalist*. Cambridge: Cambridge University Press.

Longhurst, R. 1986. Household Food Strategies in Response to Seasonality and Famine. *IDS Bulletin* 17: 27–35.

Longshore, D. 2000. *Encyclopedia of Hurricanes, Typhoons and Cyclones*. New York: Checkmark Books.

Lopez, A. 2001. *Talking Sense on Colombia*. Philadelphia, PA: American Friends Service Committee, September, 12 pp.
http://www.afsc.org/latinamerica/peace/talkingsense.pdf.

Lopez, M. 1987. The Politics of Lands at Risk in a Philippine Frontier. In: P. Little and M. Horowitz (eds), *Lands At Risk in the Third World*, pp.230–248. Boulder, CO: Westview.

Luna, M. 2001. Examples from the Philippines. http://online.northumbria.ac.uk/geography_research/radix/sustainabledev.htm#Examples%20from%20the%20Philippines.

Lupton, D. 1999. *Risk*. London: Routledge.

Lyngdoh, J. 1988. Disaster Management: A Case Study of Kosi Security System in North-east Bihar. *Journal of Rural Development* (Hyderabad) 7, 5: 519–540.

Macaulay, A., Commanda, L., Freeman, W., Gibson, N., McCabe, M., Robbins, C. and Twohig, P. 1999. Participatory Research Maximises Community and Lay Involvement. *British Medical Journal* 319 (18 September): 774–778.

McCarthy, J. et al. (eds) 2001. *Climate Change 2001: Impacts, Adaptation and Vulnerability*. Cambridge: Cambridge University Press.

413

McCully, P. 1996. *Silenced Rivers: The Ecology and Politics of Large Dams*, (2nd edn 2001). London: Zed Books.

McGlothlen, M., Goldsmith, P. and Fox, C. 1986. Undomesticated Animals and Plants. In: A. Hansen and D. McMillan (eds), *Food in Sub-Saharan Africa*, pp.222–238. Boulder, CO: Lynne Rienner.

McGranahan, G., Jacobi, G., Songsore, J., Surjadi, C. and Kjellen, M. 2001. *The Citizen at Risk: From Urban Sanitation to Sustainable Cities*. London: Earthscan.

MacGregor, K. 2002. Twenty Million at Risk from Famine in Southern Africa. *Independent on Sunday*, 26 May.

McGuire, B., Mason, I. and Kilburn, C. 2002. *Natural Hazards and Environmental Change*. London: Arnold.

McIntire, J. 1987. Would Better Information From an Early Warning System Improve African Food Security? In: D. Wilhite and W. Easterling (eds), *Planning for Drought*, pp.283–293. Boulder, CO: Westview.

McKeown, T. 1988. *The Origins of Human Disease*. Oxford: Basil Blackwell.

McMichaels, A., Haines, A., Slooff, R. and Kovats, S. (eds) 1996. *Climate Change and Human Health*. Geneva: WHO/ WMO/ UNEP.

McNeil, W. 1979. *Plagues and Peoples*. Harmondsworth: Penguin.

Macrae, J. 2001. *Aiding Recovery? The Crisis of Aid in Chronic Political Emergencies*, pp.24–47, ch. 3, Aid beyond the State: The Emergence of a New Aid Orthodoxy. London: Zed Books and ODI.

Macrae, J. and Zwi, A. (eds) 1994. *War and Hunger: Rethinking International Responses to Complex Emergencies*. London: Zed Books.

Maddox, G., James Giblin, J. and Isaria Kimambo, I. (eds) 1996. *Custodians of the Land: Ecology and Culture in the History of Tanzania*. Athens: University of Ohio Press.

Madeley, J. 1999. *Big Business, Poor Peoples: The Impact of Transnational Corporations on the World*. London: Zed Books.

Mafeje, A. 1987. Food for Security and Peace in the SADCC Region. In: E. Hansen (ed), *Africa: Perspectives on Peace and Development*, pp.183–212. London: Zed Books.

Mahmud, A. 1988. Navies Hunt for Victims of Cyclone. *Guardian* (London), 2 December.

Mahmud, A. 2002. Cash and Carry On: Battle Rages over Bangladesh Government's Ban on 'Killer' Plastic Bags *Guardian*, 27 March.
http://society.guardian.co.uk/ societyguardian/story/0,7843,674289,00.html

Malaria Foundation International 1999. DDT, Target of Global Ban, Finds Defenders in Experts on Malaria. http://www.malaria.org/DDT_NYTimes_29_VIII.html

Mallory, W. 1926. *China: Land of Famine*. New York: American Geographical Publishing Society.

Maltby, E. 1985. *Peat Mining in Jamaica*. SIEP 2. Gland, Switzerland: IUCN.

Maltby, E. 1986. *Waterlogged Wealth: Why Waste the World's Wet Places*. London: Earthscan.

Mamdani, M. 1985. Disaster Prevention: Defining the Problem. *Review of African Political Economy* 33: 92–96.

Mamdani, M. 2002. Misrule Britania. *Guardian*, 8 February.
http://www.guardian.co.uk /Archive/Article/0,4273,4352001,00.html.

414

Mander, J. and Goldsmith, E. (eds) 1996. *The Case Against the Global Economy*. San Francisco: Sierra Club Books.

Marchand, M. and Udo de Haes, H. (eds) 1990. *The People's Role in Wetland Management*. Leiden: Centre for Environmental Studies.

Marchant, J. 2001. Flood Alert: Attempts to Drain Pinatubo's Lake Haven't Removed the Danger. *New Scientist*, 15 September: 6.

Marglin, F. 1998. *The Spirit of Regeneration: Andean Culture Confronting Western Notions of Development*. London: Zed Press.

Margolis, M. 1988. The Deadly Rains of Rio. *Newsweek*, 7 March: 23.

Mariam, M. 1986. *Rural Vulnerability to Famine in Ethiopia 1958–1977*. London: Intermediate Technology Publications.

Marks, G. and Beatty, W. 1976. *Epidemics*. New York: Scribners.

Marshall, A. and Woodroffe, D. 2001. *Policies to Roll Back the State and Privatise? Poverty Reduction Strategy Papers Investigated*. London: World Development Movement.

Martin, B., Capra, M., van der Heide, G., Stoneham, M. and Lucas, M. 2001. Are Disaster Management Concepts Relevant in Developing Countries? The Case of the 1999–2000 Mozambican Floods. *Australian Journal of Emergency Management* 16, 4 (Summer): 25–33.

Mascarenhas, A. 1971. Agricultural Vermin in Tanzania. In: S. Ominde (ed), *Studies in the Geography and Development of East Africa*, pp.259–267. Nairobi: Heinemann.

Maskrey, A. 1989. *Disaster Mitigation: A Community Based Approach*. Development Guidelines No.3. Oxford: OXFAM.

Maskrey, A., 1998. *Navegando entre Brumas: La aplicación de los sistemas de información geográfica al análisis de riesgo en América Latina*. Lima and Peru: La RED and IT Peru.

Maskrey, A. 1999. Reducing Global Disasters. In: J. Ingelton (ed), *Natural Disaster Management*, pp.84–86. Leicester: Tudor Rose.

Maskrey, A. and Romero, G. 1983. *Como entender los desastres naturales*. Lima: PREDES.

Maslow, A. 1970. *Motivation and Personality*. 2nd edn. New York: Harper & Row.

Mass, W. 1970. *The Netherlands at War: 1940–1945*. New York: Aberlard-Schumann.

Matthiessen, C. 1992. The Day the Poison Stopped Working. *Mother Jones*, March/April: 48–55.

Maxwell, S. (ed) 1991. *To Cure all Hunger. Food Policy and Food Security in Sudan*. London: Intermediate Technology Publications.

May, P. 1992. Policy Learning and Policy Failure. *Journal of Public Policy* 12: 331–354.

Maybury, R. (ed) 1986. *Violent Forces of Nature*. Mt Airy, MD: Lomond Publications & UNESCO.

Mazumder, D. and Chakrabarty, A. 1973. Epidemic of Smallpox among the Evacuees from Bangladesh in Salt Lake Area near Calcutta'. *Journal of Indian Medical Association* 60: 275–280.

Mbithi, P. and Wisner, B. 1973. Drought and Famine in Kenya: Magnitude and Attempted Solutions. *Journal of East Africa Research and Development* 3, 2: 113–143.

415

Mearns, R. 1991. *Environmental Implications of Structural Adjustment: Reflections on Scientific Method*. IDS Discussion Paper No. 28. Falmer, Sussex, UK: Institute of Development Studies, University of Sussex.

Médecins Sans Frontières 1997. *Refugee Health: An Approach to Emergency Situations*. London: Macmillan.

Médecins Sans Frontières 2001. *The Top Ten Under Reported Humanitarian Crises of 2001*. www.msf.org/content/page.cfm?articleid=7B5D6023-75EA415A80CC71C8E6B90DCF.

Meillassoux, C. (ed) 1973. *Qui se nourrit de la famine en Afrique?* Paris: Maspéro.

Meillassoux, C. 1974. Development or Exploitation: Is the Sahel Famine Good Business? *Review of African Political Economy* 1, 1: 27–33.

Mekendamp, M., van Tongeren, P. and van de Veen, H. (eds) 1999. *Searching for Peace in Africa: An Overview of Conflict Prevention and Management in Africa*. Utrecht: European Platform for Conflict Prevention and Transformation.

Melnick, M. and Logue, J. 1985. The Effect of Disaster on the Health and Well-being of Older Women. *International Journal of Aging and Human Development* 21, 1: 27–38.

Mellor, J. and Gavian, S. 1987. Famine, Causes, Prevention and Relief. *Science* 235: 539–545.

Mengisteab, K. and Ikubolajeh Logan, B. 1995. *Beyond Economic Liberalization in Africa: Structural Adjustments and the Alternatives*. London: Zed Books.

Merchant, C. 1989. *The Death of Nature: Women, Ecology and the Scientific Revolution*. San Francisco, CA: HarperSanFrancisco.

Messer, E. 1991. *Food Wars: Hunger as a Weapon of War in 1990*. Alan Shawn Feinstein World Hunger Program Research Report. Providence, RI: Brown University.

MSF-USA 1999. www.doctorswithoutborders.org/publications/reports/1999/top10.shtml

MSF-USA 2001. www.doctorswithoutborders.org/publications/reports/2001/top10.htm.

Michaels, J. 1988. Rains Pour Torrent of Woe on Rios Poor. *Christian Science Monitor*, 6 March: 7.

Micro Credit Summit 2000. *Empowering Women with Micro Credit: 2000 Micro Credit Summit Campaign Report*. www.microcreditsummit.org/campaigns/ report00.html.

Miculax, R. and Schramm, D. 1989. Earthquake Housiing Reconstruction and Rural Development: Joyabaj, Quiche, Guatemala. In: M. Anderson and P. Woodrow (eds), *Rising from the Ashes: Development Strategies in Times of Disaster*, pp. 223-240. Boulder, CO: Westview [Reprinted in Anderson and Woodrow 1998].

Middleton, N. and O'Keefe, P. 1998. *Disasters and Development: The Politics of Humanitarian Aid*. London: Pluto Press.

Mihill, C. 1996. Killer Diseases Making a Comeback, Says WHO. *Guardian*, 20 May: 3.

Mileti, D. 1999. *Disasters by Design: A Reassessment of Natural Hazards in the United States*. Washington, DC: Joseph Henry Press.

Mileti, D., Drabek, T. and Haas, J. 1975. *Human Systems in Extreme Environments*. Monograph 21. Boulder: University of Colorado Institute of Behavioral Science, Program on Environment and Behavior.

Millar, S. 2003a. Quake Victims Pursue Building 'Mafia'. *Guardian Unlimited*, 4 May. http://www.guardian.co.uk/turkey/story/0,12700,949143,00.html.

Millar, S. 2003b. First the Suffering, then the Hope, then the Anger and Entry of the Army. *Guardian Unlimited*, 3 May.

416

http://www.guardian.co.uk/turkey/story/ 0,12700,948666,00.html

Miller, K. and Nigg, J. 1993. Event and Consequence Vulnerability: Effects on the Disaster Recovery Process. Preliminary Paper No. 217. Disaster Research Center, University of Delaware. http://www.udel.edu/DRC/preliminary/217.pdf

Miller, K. and Simile, C. 1992. They Could See Stars from their Beds: The Plight of the Rural Poor in the Aftermath of Hurricane Hugo. Paper presented to the Society for Applied Anthropology, 27 March, Memphis, TN.

Mills, C. 1959. *The Sociological Imagination*. New York: Oxford University Press.

Milne, A. 1986. *Floodshock: The Drowning of Planet Earth*. Gloucester: Alan Sutton.

Minear, L 1991. *Operation Lifeline Sudan*. Trenton, NJ: Red Sea Press.

Mitchell, J. 1990. Human Dimensions of Environmental Hazards. In: A. Kirby (ed), *Nothing to Fear*, pp.131–175. Tucson: University of Arizona Press.

Mitchell, J. (ed) 1996. *The Long Road to Recovery: Community Responses to Industrial Disaster*. Tokyo: United Nations University Press.

Mitchell, J. (ed) 1999a. *Crucibles of Hazard: Mega-cities and Disasters in Transition*. Tokyo: United Nations University Press.

Mitchell, J. 1999b. Natural Disasters in the Context of Mega-cities. In: J. Mitchell (ed), *Crucibles of Hazard: Mega-cities and Disasters in Transition*, pp.15–55. Tokyo: United Nations University Press.

Mkandawire, T. and Soludo, C. 1999. *Our Continent, Our Future: African Perspectives on Structural Adjustment*. Trenton, NJ: Africa World Press.

Mohanty, M. and Mukerji, P. (eds) 1998. *People's Rights: Social Movements and the State in the Third World*. New Delhi: Sage.

Momsen, J. and Townsend, J. (eds) 1987. *Geography of Gender in the Third World*. London: Hutchinson.

Monan, J. 1989. *Bangladesh: The Strength to Succeed. A Report for Oxfam*. Oxford: Oxfam.

Monbiot, G. 2003. *The Age of Consent: A Manifesto For a New World Order*. London: HarperCollins/Flamingo.

Monday, J., Eadie, C., Emmer, R., Esnard, A.-M., Michaels, S., Philipsborn, C., Phillips, B. and Salvesen, D. 2001. *Holistic Disaster Recovery: Ideas for Building Local Sustainability after a Natural Disaster*. Boulder, CO: Natural Hazards Research and Applications Center. http://www.colorado.edu/hazards/holistic_recovery/.

Morgan, R. 1988. Drought-relief Programmes in Botswana. In: D. Curtis et al. (eds), *Preventing Famine*, pp.112–120. London: Routledge.

Morris, J., West, G., Holck, S., Blake, P., Echeverria, P. and Karaulnik, M. 1982. Cholera Among Refugees in Rangsit, Thailand. *Journal of Infectious Diseases* 1: 131–134.

Morris, R. and Sheets, H. 1974. *Disaster in the Desert*, Special Report. Washington, DC: Carnegie Endowment for International Peace.

Morrow, B. 1999. Identifying and Mapping Community Vulnerability. *Disasters* 23, 1: 1–18.

Morrow, B. and Phillips, B. (eds) 1999. Special Issue on Women and Disasters. *International Journal of Mass Emergencies and Disasters* 17, 1.

Mortimore, M. 1989. *Adapting to Drought: Farmers, Famines and Desertification in West Africa*. Cambridge: Cambridge University Press.

Moser, C. 1998. The Asset Vulnerability Framework: Re-assessing Ultra Poverty Reduction Strategies. *World Development* 26, 1: 1–19.

417

Mount, J. 1998. Levees More Harm than Help. *Engineering News Record* 240, 5: 59.

Muhema, B. 1972. The Impact of Flooding in Rufiji. *Journal of the Geographical Association of Tanzania* 7: 49–64.

Munasinghe, M., Menezes, B. and Preece, M. 1991. Case Study: Rio Flood Reconstruction and Prevention Project. In: A. Kreimer and M. Munasinghe (eds), *Managing Natural Disasters and the Environment*, pp.28–31. Washington, DC: World Bank.

Munich Re 1997. *Annual Review of Natural Catastrophes, 1997*. Munich, Germany: Munich Re Group.

Munich Re 1998. *Annual Review of Natural Catastrophes, 1998*. Munich, Germany: Munich Reinsurance Company.

Munich Re 1999. *Topics 2000 Natural Catastrophe: The Current Position*. Munich, Germany: Munich Re Group.

Munich Re 2002. http://www.munichre.com/.

Murphy, L. and Moriarty, A. 1976. *Vulnerability, Coping and Growth from Infancy to Adolescence*. New Haven, CT: Yale University Press.

Murray, C. 1981. *Families Divided: The Impact of Migrant Labour in Lesotho*. Cambridge: Cambridge University Press.

Murray, M. J., Murray, A., Murray, N. and Murray, M. B. 1978. Diet and Cerebral Malaria: The Effect of Famine and Refeeding. *American Journal of Clinical Nutrition* 31: 57–61.

Murwira, K., Wedgwood, H., Watson, C. and Win, E., with Tawney, C. 2000. *Beating Hunger: The Chivi Experience: A Community-based Approach to Food Security in Zimbabwe*. London: IT Publications.

Mustafa, D. 2001. Post-modernism and Natural Hazards. Paper presented at the Annual Meeting of the Association of American Geographers, New York City.

Nafziger, E. 1988 *Inequality in Africa*. Cambridge: Cambridge University Press.

Nagle, G. 1998. *Hazards*. Walton on Thames: Nelson.

Naik, G., Fuhrmans, V., Karp, J., Millman, J., Fassihi, F. and Slater, J. 2003. Fertility 'Revolution' Lowers Birth Rates. *Wall Street Journal 20th Anniversary The World in 2003*, 24 January. http://online.swj.com/article/0,,SB104334616761507264,00.html

Narayan, D. and Petesch, P. (eds) 2002. *Voices of the Poor*. New York: Oxford University Press and World Bank.

NCDC (National Climate Data Centre) 1998. Flooding in China Summer 1998. 20 November. http://lwf.ncdc.noaa.gov/oa/reports/chinaflooding/chinaflooding.html.

Neefjes, K. 2000. *Environments and Livelihoods: Strategies for Sustainability*. Oxford: Oxfam.

Nerfin, M. 1990. Environment and Development: Listen to the South Citizen. *IFDA Dossier* 77 (May/June): 1–2. Nyon: International Foundation for Development Alternatives.

New Scientist 1989. Editorial. 28 October: 3.

Newell, K. 1988. Selective Primary Health Care: The Counter Revolution. *Social Science and Medicine* 26, 9: 903–906.

Newman, L. (ed) 1990. *Hunger in History: Food Shortage, Poverty and Deprivation*. Oxford: Basil Blackwell.

Newman, S. 1989. Earthweek: Diary of the Planet (Week ending 13 October). San Francisco, CA: Cronicle Features.

Nez, G. 1974. Private communication with Ian Davis following a meeting to discuss reconstruction policies in Denver, Colorado in May 1974.

Nicaraguan Ecumenical Group 1988. Statement on Hurricane Joan. *Lucha/Struggle: A Journal of Christian Reflection on Struggles for Liberation* 12, 6: 38–39.

Nichols, N. 1988. *Food Information Systems in Sub-Saharan Africa: Effective Tools or Illusions of Preparedness?* M.Sc. Thesis. Norwich, University of East Anglia.

Niehof, A. 2001. The Application of Vulnerability Theory to People's Coping with Disaster. Paper at workshop, Vulnerability in Disaster Theory and Practice, Wageningen Agricultural University, 28–30 June.

NOAA (US National Oceanic and Atmospheric Agency) and Organization of American States (OAS) 2002. Vulnerability Assessment Techniques and Applications. http://www.csc.noaa.gov/vata/.

Noble, J. 1981. Social Inequity in the Prevalence of Disability. *Assignment Children* 53/54: 23–32.

Noji, E. (ed) 1997. *The Public Health Consequences of Disasters.* New York: Oxford University Press.

Noland, M., Robinson, S. and Tao Wang. 1999. *Famine in North Korea: Causes and Cures.* Working Paper 99, 2. Washington, DC: Institute for International Affairs.

Nordheimer, J. 2002. Nothing's Easy for New Orleans Flood Control. *New York Times*, 30 April. www.hurricane.lsu.edu/_in_the_news/april30_ny_times.htm.

Norman, A. 2002. Investigating the Needs of Disabled People in Flood Warning and Response. http://online.northumbria.ac.uk/geography_research/radix/

Norman, A. 2003. *Investingating the Needs of Disabled People in Flood Warning and Response.* Masters of Science. Cambridge: Department of Geography, Anglia Polytechnic University (May).

NSO (National Statistical Office, Republic of Philippines) 2003. *Census 2000 Final Counts* http://www.census.gov.ph/.

Nuguid, A. 1990. Environmental Abuse Getting Worse. *Business Star* (Bangkok), 2 March.

O'Brien, C. and O'Brien, M. 1972. *The Story of Ireland.* New York: Viking.

O'Brien, J. and Gruenbaum, E. 1991. A Social History of Food, Famine and Gender in Twentieth-century Sudan. In: R. Downs, D. Kerner and S. Reyna (eds), *The Political Economy of African Famine*, pp.177–203. Philadelphia, PA: Gordon & Breach Science Publishers.

O'Brien, M. 2000. *Making Better Environmental Decisions: An Alternative to Risk Assessment.* Cambridge, MA: MIT Press.

OCHA (United Nations Office for the Coordination of Humanitarian Affairs) 2001. Caribbean – Hurricane Michelle OCHA Situation Report No. 7. ReliefWeb, 6 November. http://www.reliefweb.int/w/rwb.nsf/s/F710706ED85F3BB785256AFC006E4F92.

OCHA (United Nations Office for the Coordination of Humanitarian Affairs) 2002a. Bolivia – Floods OCHA Situation Report No. 4. ReliefWeb, 27 February: http://www.reliefweb.int/w/rwb.nsf/s/CA6E5D3C9689230285256B6D005FBCFE.

OCHA (United Nations Office for the Coordination of Humanitarian Affairs) 2002b. Inundaciones Bolivia – Misión UNDAC Informe Final. ReliefWeb, 2 March: http://www.reliefweb.int/w/rwb.nsf/s/EEA62485F36911E0C1256B76005787BC.

OCHA (United Nations Office for the Coordination of Humanitarian Activity) 2002c. Malawi: Women Worst Hit By Hunger. ReliefWeb, 3 June.

http://www.reliefweb.int/w/rwb.nsf/480fa8736b88bbc3c12564f6004c8ad5/5e11c473
1e9bf07249256bce0009f419?OpenDocument.

OCHA (United Nations Office for the Coordination of Humanitarian Activity)
2002d. Uganda: Minister Highlights Water, Sanitation Problems. UN OCHA
Integrated Regional Information Network. ReliefWeb.
http://www.reliefweb.int/w/rwb.nsf/9ca65951ee22658ec125663300408599/b4854c3
0f0a55a0585256b61007e08b9?OpenDocument.

OCHA (United Nations Office for the Coordination of Humanitarian Activity)
2002e. Malawi: Bubonic Plague Adds to Woes. ReliefWeb, 31 May.
http://www.reliefweb.int/w/rwb.nsf/480fa8736b88bbc3c12564f6004c8ad5/65c64abc
437a44be49256bcd000b30cd?OpenDocument.

Odegi-Awuondo, C. 1990. *Life in the Balance: Ecological Sociology of Turkana
Nomads*. Nairobi, ACTS Press.

Odhiambo, T., Anyang'Nyong'o, P., Hansen, E., Lardner, G. and Wai, D. (eds) 1988.
Hope Born out of Despair: Managing the African Crisis. Nairobi: Heinemann
Kenya.

O'Keefe, P. and Wisner, B. 1975. *African Drought: The State of the Game*. In: P.
Richards (ed), *African Environment: Problems and Perspectives*, pp.31–39.
London. International African Institute.

O'Keefe, P., Westgate, K. and Wisner, B. 1976. Taking the 'Naturalness' out of
'Natural Disaster'. *Nature* 260 (15 April): 566–567.

Oliver-Smith, A. 1986a. Disaster Context and Causation: An Overview of Changing
Perspectives in Disaster Research. In: A. Oliver-Smith (ed), *Natural Disasters and
Cultural Responses*, pp.1–34. Studies in Third World Societies No. 36. Williams-
burg, VA: College of William and Mary.

Oliver-Smith, A. 1986b. *The Martyred City: Death and Rebirth in the Andes*. Albu-
querque: University of New Mexico Press.

Oliver-Smith, A. 1988. Planning Goals and Urban Realities: Post-disaster Reconstruction in a Third World City. *City and Society* 2, 2: 105–126.

Oliver-Smith, A. 1990. Post-disaster Housing Reconstruction and Social Inequality –
a Challenge to Policy and Practice. *Disasters* 14, 1: 7–19.

Oliver-Smith, A. 1994. The Five Hundred Year Earthquake: Natural and Social
Hazards in the Third World (Peru), pp.31–48. In A. Varley (ed), *Disasters, Development and Environment*. Chichester: Wiley.

Oliver-Smith, A. 1996. Anthropological Research on Hazards and Disasters. *Annual
Review of Anthropology* 25: 303–328.

Oliver-Smith, A. 1999a. Peru's Five Hundred-Year Earthquake: Vulnerability in
Historical Context. In: A. Oliver-Smith and S. Hoffman (eds), *The Angry Earth:
Disasters in Anthropological Perspective*, pp.74–88. New York: Routledge.

Oliver-Smith, A. 1999b. What is a Disaster? In: A. Oliver-Smith and S. Hoffman
(eds), *The Angry Earth: Disasters in Anthropological Perspective*, pp.18–34. New
York: Routledge.

Oliver-Smith, A. and Hoffman, S. (eds) 1999. *The Angry Earth: Disasters in Anthropological Perspective*. New York: Routledge.

Olson, R. 2000. The 1985 Mexico Earthquake Disaster and Emergent Organizations.
Natural Hazard Research and Applications Annual Workshop, University of
Colorado, Boulder, 12 July. www.colorado.edu/hazards/ss/ss00/s37.html.

Olson, R. S., Olson, R. A. and Gawronski, V. 1999. *Some Buildings Just Can't Dance: Politics, Life Safety, Disaster*. Stamford, CN: JAI Press.

O'Neill, B. 1990. Cities against the Seas. *New Scientist* 125, 1702: 3.

Onibokun, A. 1999. Synthesis and Recommendations. In: A. Onibukun (ed), *Managing the Monster: Urban Waste and Governance in Africa*. Ottawa: IDRC.

Onimode, B. (ed) 1989. *The IMF, the World Bank and the African Debt*. 2 vols. London: Zed Books and the Institute for African Alternatives.

Organization of American States (OAS) 1991. *Primer on Natural Hazard Management in Integrated Regional Development Planning*. Washington, DC: OAS/Department of Regional Development and Environment.

O'Riordon, T. 1986. Coping with Environmental Hazards. In: R. Kates and I. Burton (eds), *Geography, Resources and Environment*. Vol. 2, pp.272–309. Chicago, IL: University of Chicago Press.

Osmani, S. 1996. Famine, Demography and Endemic Poverty. *Journal of International Development* 8, 5: 597–623.

Owusu, K. and F. Ng'ambi 2002. Structural Damage: The Causes and Consequences of Malawi's Food Crisis. London: World Development Movement.

Oxfam 1988. *Debt Crisis Case Study: Jamaica*. People in Crisis Campaign Leaflet. Oxford: Oxfam.

Oxfam, UK 2002. Poverty in the Midst of Wealth. Policy Briefing Paper, January. Oxford: Oxfam. http://www.oxfam.org.uk/policy/papers/drc/povertywealth.htm.

Pacific Islands Development Program n.d. *Agricultural Development and Disaster Preparedness*. Honolulu: East–West Center.

Packard, R. 1989. *White Plague, Black Labor: Tuberculosis and the Political Economy of Health in South Africa*. Berkeley: University of California Press.

Packard, R. and Epstein, R. 1987. Ecology and Immunology: The Social Science Context of AIDS in Africa. *Science for the People* 19, 1: 10–17.

Packard, R., Wisner, B. and Bossart, T. (eds) 1989. Political Economy of Health and Disease in Africa and Latin America. *Social Science and Medicine* 28, 5: 405–530.

PAHO (Pan American Health Organization) 1982. *Epidemiologic Surveillance after Natural Disaster*. Washington, DC: Pan American Health Organization.

PAHO (Pan American Health Organization) 1994. Empowering Local Communities to Reduce the Effects of Disasters. *Disasters: Preparedness in the Americas* 60 (October): 1 & 7.

PAHO (Pan American Health Organization) 1999. Health Achievements in Central America. http://165.158.1.110/english/DPI/r990225a.htm.

PAHO (Pan American Health Organization) 2000a. Supplement on Venezuela Disaster. *Disasters Preparedness and Mitigation* 79 (January).

PAHO (Pan American Health Organization) 2000b. *Natural Disasters: Protecting the Public's Health*. Washington, DC: PAHO.

Palm, R. 1990. *Natural Hazards: An Integrative Framework for Research and Planning*. Baltimore, MD: Johns Hopkins University Press.

Pankhurst, R. 1974. *The History of Famine and Epidemics in Ethiopia Prior to The Twentieth Century*. Addis Ababa: Relief and Rehabilitation Commission.

Panos Institute 2002. *Reducing Poverty: Is the World Bank's Strategy Working?* London: Panos Institute.

Panton, J. and Archer, R. 1996. *Waiting on the Volcano: A Visit to Montserrat.* London: Christian Aid and Montserrat Action Committee.

Parker, D. (ed) 2000. *Floods*, 2 Vols. London: Routledge.

Parker, D. and Budgen, P. 1998. The Tropical Cyclone Warning Dissemination System in Mauritius. In: B. Lee and I. Davis (eds), *Forecasts and Warnings*, pp.1.1–1.87 London: Thomas Telford.

Parker, D. and Handmer, J. (eds) 1992. *Hazard Management and Emergency Planning.* London: James & James Science Publishers.

Parker, D. 1992. The Flood Action Plan: Social Impacts in Bangladesh. *Natural Hazards Observer* 16, 4: 3–4.

Parker, R. 1989. Proyecto Nueva Vida Armero. In: M. Anderson and P. Woodrow (eds) *Rising from the Ashes: Development Strategies in Times of Disaster*, pp. 157–183. Boulder, CO: Westview [Reprinted in Anderson and Woodrow 1998].

Parr, A. 1987. Disaster and Disabled Persons: An Examination of the Safety Needs of a Neglected Minority. *Disasters* 11, 2: 81–159.

Parr, A. 1997. Disasters and Human Rights of Persons with Disabilities: A Case for an Ethical Disaster Mitigation Policy. *Australian Journal of Emergency Management* 9, 3: 173–179.

Parry, M. and Carter, T. 1987. Climate Impact Assessment: A Review of Some Approaches. In: D. Wilhite and W. Easterling (eds), *Planning for Drought: Toward a Reduction of Societal Vulnerability*, pp.165–187. Boulder, CO: Westview.

Pattullo, P. 2000. *Fire from the Mountain: The Tragedy of Montserrat and the Betrayal of its People*, London: Constable.

Paul, B. 1984. Perception of and Agricultural Adjustments to Floods in Jamuna Floodplain, Bangladesh. *Human Ecology* 12, 1: 3–19.

Pavlovsky, Y. (ed). n.d. *Human Diseases with Natural Foci.* Moscow: Foreign Languages Publishing House.

Peacock, W. 2002. Personal communication with B. Wisner.

Peacock, W. and Girard, C. 1997. Ethnic and Racial Inequalities in Hurricane Damage and Insurance Settlements. In: W. Peacock, B. Morrow and H. Gladwin (eds), *Hurricane Andrew: Ethnicity Gender and the Sociology of Disasters*, pp. 171–190. London: Routledge.

Peacock, W., Killian, C. and Bates, F. 1987. The Effects of Disaster Damage and Housing Aid on Household Recovery Following the 1976 Guatemalan Earthquake. *International Journal of Mass Emergencies and Disasters* 5, 1: 63–88.

Peacock, W., Morrow, B. and Gladwin, H. (eds) 1997. *Hurricane Andrew: Ethnicity Gender and the Sociology of Disasters.* London: Routledge.

Peacock, W., Morrow, B. and Gladwin, H. (eds) 2001. *Hurricane Andrew and the Reshaping of Miami.* Miami: International Hurricane Center.

Pearce, F. 1991. The Rivers That Won't Be Tamed. *New Scientist* 13 (April): 38–41.

Pearce, F. 1992. *The Dammed: Rivers, Dams and the Coming World Water Crisis.* London: Bodley Head.

Pearce, F. 2001. Dams and Floods. Research Paper, World Wide Fund for Nature, 21 June. www.panda.org/livingwaters/pubs.html.

Pearce, F. 2002a. Earthquake Aftermath, On Shaky Ground. *Geographical Magazine* 74, 6: 32–34.

Pearce, F. 2002b. Landslide Alert. *New Scientist*, 2 March: 11.

Pearce, F. 2002c. Meltdown! *New Scientist*, 2 November: 45–48.

422

Pellanda, C. 1981. *Disaster and Socio-systemic Vulnerability*. Third International Conference on the Social and Economic Aspects of Earthquakes and Planning to Mitigate their Impacts. Bled, Yugoslavia.

Pelling, M. 1997. What Determines Vulnerability to Floods? A Case Study of Georgetown, Guyana. *Environment and Urbanisation* 9: 203–226.

Pelling, M. 1999. Participation, Social Capital and Vulnerability to Urban Flooding in Guyana. *International Journal of Development* 10: 469–486.

Pelling, M. 2001. 'Natural Disasters?' In: N. Castree and B. Braun (eds), *Social Nature: Theory, Practice and Politics*, pp.170–188. Oxford: Blackwell.

Pelling, M. 2002. Assessing Urban Vulnerability and Social Adaptation to Risk. *International Development Planning Review* 24, 1: 59–76.

Pelling, M. 2003a. Paradigms of Risk. In: M. Pelling (ed), *Natural Disasters and Development in a Globalizing World*, pp.3–16. London: Routledge.

Pelling, M. 2003b. *The Vulnerability of Cities: Natural Disasters and Social Resilience*. London: Earthscan.

Pelling, M. and Uitto, J. 2002. Small Island Developing States: Natural Disaster Vulnerability and Global Change. *Environmental Hazards* 3: 49–62.

Penman, D. 2003. US Firms Set for Post War Contracts. *Guardian Unlimited*, 11 March. http://www.guardian.co.uk/usa/story/0,12271,911943,00.html.

Peri Peri (ed) 1999. *Risk, Sustainable Development and Disasters: Southern Perspectives*. Cape Town, South Africa: Peri Peri Publications/Disaster Mitigation for Sustainable Livelihoods, Department of Environmental and Geographical Sciences, University of Cape Town.

Perrow, C. 1984. *Normal Accidents: Living with High Risk Technologies*. New York: Basic Books.

Perry, R. and Mushkatel, A. 1986. *Minority Citizens in Disasters*. Athens: University of Georgia Press.

Petak, W. and Atkisson, A. 1982. *Natural Hazard Risk Assessment and Public Policy*. New York: Springer Verlag.

Peterson, M. (ed) 1977. *The Portable Thomas Jefferson*, pp.23–232. London: Penguin (see also T. Jefferson. 1781. *Notes on the the the State of Virginia*).

Petrella, R. 2001. *The Water Manifesto: Arguments for a World Water Contract*. London: Zed Books.

Phillips, B. 1990. Gender as a Variable in Emergency Response. In: R. Bolin (ed), *The Loma Prieta Earthquake: Studies of the Short-Term Impacts*, Boulder: Institute of Behavioral Science, University of Colorado.

Pickles, J. (ed) 1995. *Ground Truth: The Social Implications of Geographic Information Systems*. New York: Guilford.

Pielke, R. and Pielke, R. (eds) 2000. *Coastal Storms*. 2 vols. London: Routledge.

Piller, C. 1991. *The Fail Safe Society*. New York: Basic Books.

Pinkerton 2002. *Online Security and Business Interruption/Top Security Concerns in Asia-Pacific*. http://www.ci-pinkerton.com/news/prAsiaTST11.6.html.

Pinstrup-Anderson, P. 1985. The Nutritional Effects of Export Crop Production: Current Evidence and Policy Implications. In: M. Biswas and P. Pinstrup-Andersen (eds), *Nutrition and Development*, pp. 43–59. Oxford: Oxford University Press.

Pinstrup-Anderson, P. (ed) 1988. *Food Subsidies in Developing Countries: Costs, Benefits and Policy Options*. Baltimore, MD: Johns Hopkins University Press and IFPRI.

Pirotte, C., Husson, B. and Grunewald, F. (eds) 1999. *Responding to Emergencies and Fostering Development: The Dilemmas of Humanitarian Aid*. London: Zed Books.

Plant, R. 1978. *Guatemala: A Permanent Disaster*. London: Latin America Bureau.

Platt, R. et al. 1999. *Disasters and Democracy: The Politics of Extreme Natural Events*. Washington, DC: Island Press.

Platteau, J. 1988. *The Food Crisis in Africa: A Comparative Structural Analysis*. WIDER Working Paper No. 44. Helsinki: World Institute for Development Economic Research (WIDER).

Plessis-Fraissard, M. 1989. Mitigation Efforts at the Municipal Level: The La Paz Municipal Development Project. In: A. Kreimer and M. Zador (eds), *Colloquium on Disasters, Sustainability and Development: A Look to the 1990s*, pp.132–135. Washington, DC: World Bank.

Plummer, J., 2000. *Municipalities and Community Participation: A Sourcebook for Capacity Building*. London: Earthscan.

Poirteir, C. (ed) 1995. *The Great Irish Famine*. Cork: Mercier Press.

Pomfret, J. 2003. Rural China Struggling with SARS: Hospitals Unequipped to Handle Infections. *Washington Post*, 25 April: A17.
http://www.washingtonpost.com/ac2/wp-dyn?pagename=article&node=∓conten tId=A349112003Apr24&notFound=true.

Poniatowksa, E. 1998. *Nadie, Nadie: Las Voces del Temblor*. Mexico City: Editora Era.

Poore, D. 1989. *No Timber Without Trees: Sustainability in the Tropical Forest*. London: Earthscan.

Popkin, R. 1990. The History and Politics of Disaster Management in the United States. In: A. Kirby (ed), *Nothing to Fear*, pp.101–130. Tuscon: University of Arizona Press.

Population Reference Bureau 2002. *World Population Data Sheet*. www.prb.org/Content/NavigationMenu/Other_reports/2000–2002/sheet4.html .

Porter, P. 1979. *Food and Development in the Semi-Arid Zone of East Africa*. Foreign and Comparative Studies/African Studies 32. Syracuse, NY: Maxwell School of Citizenship and Public Service, Syracuse University.

Pradervand, P. 1989. *Listening to Africa: Developing Africa from the Grassroots*. New York: Praeger.

Prah, K. (ed) 1988. *Food Security in Southern Africa*. Southern African Studies Series No. 4. Roma, Lesotho: Institute of Southern African Studies, National University of Lesotho.

Prasad, M. 2002. *Community Preparedness*. Contribution to ISDRe-conference, Earth Summit 2002 Debate, 25 April. http://www.unisdr.org.

Prindle, P. 1979. Peasant Society and Famines: A Nepalese example. *Ethnology* 18: 49–60.

Prothero, M. 1965. *Migrants and Malaria*. London: Longman.

Pryer, J. and Crook, N. 1988. *Cities of Hunger: Urban Malnutrition in Developing Countries*. Oxford: Oxfam.

Pryor, L. 1982. *Ecological Mismanagement in Natural Disasters.* Commission on Ecology Papers 2. Gland, Switzerland: IUCN Commission on Ecology/League of Red Cross Societies.

PSOJ (Private Sector Organization of Jamaica) 2002. Jamaica: Selected Interest Rates (Bank of Jamaica data). http://www.psoj.org/data%5Cinterest.pdf.

Puente, S. 1999. Urban Social Vulnerability in Mexico. Paper presented at the Annual Meeting of the American Association of Geographers, Honolulu, Hawaii.

Pulwarty, R. and Riebsame, W. 1997. The Political Ecology of Vulnerability to Hurricane Related Hazards. In: H. Diaz and R. Pulwarty (eds), *Hurricanes: Climate and Socioeconomic Impacts*, pp.185–214. Berlin: Springer Verlag.

Punongbayan, R. 2002. Pinatubo crater wall collapses, triggers mild lahar flow. *Science and Technology Post* (Department of Science and Technology, Republic of Philippines) 20, 8 (August)
http://www.stii.dost.gov.ph/sntpost/frames/aug2k2/pg1b.htm.

Punongbayan, R. 2003. Mitigating Crater Outbreak Flood on Mt. Pinatubo, Philippines. Paper presented at the DPRI-IIASA Third Symposium on Integrated Disaster Risk Management (IDRM-03). Kyoto: Kyoto University, 5 July.

Punongbayan, R. and Newhall, G. 1995. Warning the Public about Mount Pinatubo: A Story Worth Repeating. *Stop Disasters* 25, 3: 11–14.

Putzel, J. 1992. *A Captive Land: The Politics of Agrarian Reform in the Philippines.* London and New York: Catholic Institute of International Relations (CIIR) and Monthly Review Books.

Putzel, J. 2000. *Land Reform in Asia: Lessons from the Past for the 21st Century.* Development Studies Institute Working Paper 4. London: London School of Economics and Political Science.

Quarantelli, E. (ed) 1978. *Disasters: Theory and Research.* Studies in International Sociology 13. Beverley Hills, CA. Sage.

Quarantelli, E. 1984. Perception and Reactions to Emergency Warnings of Sudden Hazards. *Ekistics* 51, 309: 511–515.

Quarantelli, E. 1995. What is a Disaster? (Editor's Introduction), *International Journal of Mass Emergencies and Disasters* 13, 3: 221–230.

Quarantelli, E. (ed) 1998. *What is a Disaster?* New York: Routledge.

Quarantelli, E. 2002. Research Professor, Disaster Research Center, University of Delaware, personal communication with Ian Davis.

Quarantelli, E. and Dynes R. 1972. *Images of Disaster Behaviour: Myths and Consequences.* Preliminary Paper No 5. Columbus: Disaster Research Center, Ohio State University.

Quarantelli, E. and Dynes, R. 1977. Response to Social Crisis and Disaster. *Annual Review of Sociology* 3: 23–49.

RADIX 2002. Southern African Food Emergencies.
http://online.northumbria.ac.uk/geography_research/radix/malawi.htm

Rahmato, D, 1988. *Peasant Survival Strategies.* Geneva: International Institute for Relief and Development/Food for the Hungry International.

Rahnema, M. and Bawtree, V. (eds) 1997. *The Post-Development Reader.* London: Zed Books.

Raikes, P. 1988. *Modernizing Hunger.* London: James Currey.

Rajalakshmi, T. 2003. The Big Freeze: A Cold Wave Takes a Heavy Toll of the Poor in North and Central India. *Frontline* 20, 3 (1–14 February). http://www.flonnet.com /fl2003/stories/20030214005512500.htm.

Rakodi, C. 1999. A Capital Assets Framework for Analysing Household Livelihood Strategies: Implications for Policy. *Development Policy Review* 17: 315–42.

Ramos-Jimenez, P., Chiong-Javier, M. and Sevilla, J. 1986. *Philippine Urban Situation Analysis*. Manila: UNICEF.

Rangasami, A. 1985. Failure of Exchange Entitlements Theory of Famine: A Response. *Economic and Political Weekly* 20, 41: 1747–1752.

Rangasami, A. 1986. Famine: The Anthropological Account; An Evaluation of the Work of Raymond Firth. *Economic and Political Weekly* 21, 36: 1591–1601.

Rao, N. 1974. Impact of Drought on the Social System of a Telengana Village. *The Eastern Anthropologist* 27, 4: 299–315.

Raphael, B. 1986. *When Disaster Strikes*. London: Hutchinson.

Rashid, H. 1977. *Geography of Bangladesh*. Boulder, CO: Westview.

Rau, B. 1991. *From Feast to Famine: Official Cures and Grassroots Remedies to Africa's Food Crisis*. London: Zed Books.

Ravallion, M. 1985. The Performance of Rice Markets in Bangladesh During the 1974 Famine. *Economic Journal* 95: 15–29.

Ravallion, M. 1987. *Markets and Famines*. Oxford: Clarendon Press.

Reacher, M., Campbell, C., Freeman, J., Doberstyn, E. and Brandling-Bennett, A. 1980. Drug therapy for Plasmodium Falciparum Malaria Resistant to Pyrimethamine-Sulfadoxine (Fansidar): A Study of Alternative Regimens in Eastern Thailand. *The Lancet* 3: 1066–1069.

Read, B. 1970. *Healthy Cities: A Study of Urban Hygiene*. Glasgow: Blackie.

Reardon, T. 1997. Using Evidence of Household Income Diversification to Inform Study of Rural Non-farm Labour Market in Africa. *World Development* 25, 5: 735–747.

Redclift, M. and Benton, T. (eds) 1995. *Social Theory and Global Environment*. New York: Routledge.

Reddy, A. 1991. Unpublished notes of a presentation on the Andhra Pradesh cyclone of 1990 to a Workshop on Disaster Management, June 1991 by the Director of the Centre for Disaster Management, Hyderabad. Oxford: Disaster Management Centre.

Regan, C. 1983. Underdevelopment and Hazards in Historical Perspective: An Irish case Study. In: K. Hewitt (ed), *Interpretations of Calamity*, pp.98–120. Boston, MA and London: Allen & Unwin.

Rekacewicz, P. 2000. Millions of Refugees in Africa. *Le Monde Diplomatique* May. http://www.en.monde-diplomatique.fr/maps/africarefugeesmdv51.

ReliefWeb. 1997a. Southern Sudan – Present Displaced and Returnee Populations (Map). 1 August, Source, USAID. http://www.reliefweb.int/w/map.nsf/wByCLatest/B59EF170BED77DCF85256A1D0075 441B?Opendocument.

ReliefWeb. 1997b. Southern Sudan – Districts Affected by the Dry Spell (Map). 1 July, Source, USAID Famine Early Warning System (FEWS). http://www.reliefweb.int/w/map.nsf/wByCLatest/8ADDDB68FCF979F385256A1 D00759298?Opendocument.

Rich, B. 1994. *Mortgaging the Earth: The World Bank, Environmental Impoverishment and the Crisis of Development*. Boston, MA: Beacon.

Richards, P. 1975. 'Alternative' Strategies for the African Environment: 'Folk Ecology' as a Basis for Community Oriented Agricultural Development. In: P. Richards (ed), *African Environment: Problems and Perspectives*, pp.102–117. London: International African Institute.

Richards, P. 1983. Ecological Change and the Politics of African Land Use. *African Studies Review* 26: 1–72.

Richards, P. 1985. *Indigenous Agricultural Revolution*. London: Hutchinson Education.

Richards, P. J. and Thomson, A. 1984. *Basic Needs and the Urban Poor*. London: Croom Helm.

Riddell, R. and Robinson, M., with de Coninck, J., Muir, A. and White, S. 1995. *Non-Governmental Organizations and Rural Poverty Alleviation*. London: ODI.

Rinpokan Renraku Kyogikai 1995. Record of the Great Hanshin Earthquake (in Japanese: *Hanshin Daishinsai no Kiroku 'Asu e no Ibuki'*). Ashiya City, Japan: Hyogo-ken Rinpokan Renraku Kyogikai (Social Service Facility Liaison Council of Hyogo Prefecture).Rivers, J. 1982. Women and Children Last: An Essay on Sex Discrimination. *Disasters* 6, 4: 256–267.

Rivers, J., Holt, J., Seaman, J. and Bowden, M. 1974. Lessons for Epidemiology from the Ethiopian Famine. *Annales de la Société Belge de Medecine Tropicale* 56: 345–367.

ROAPE (*Review of African Political Economy*) 1990. *What Price Economic Reform?* Theme Issue, No. 47.

Rob, M. 1990. Flood Hazard in Bangladesh: Nature, Causes and Control. *Asian Profile* 18, 4: 365–378.

Robinson, S., Franco, Y., Castrejon, R. and Bernard, H. 1986. It Shook Again: The Mexico City Earthquake of 1985. In: A. Oliver-Smith (ed), *Natural Disasters and Cultural Responses*, pp.81–122. Studies in Third World Societies No. 36. Williamsburg, VA: College of William and Mary.

Rodenbeck, M. 2000. *Cairo: The City Victorious*. New York: Vintage.

Rogers, B. 1980. *The Domestication of Women*. London: Tavistock.

Rogers, P., Lydon, P. and Seckler, D. 1989. *Eastern Waters Study: Strategies to Manage Flood and Drought in the Ganges–Brahmaputra Basin*. Report prepared by Irrigation Support Project for Asia and the Near East. Washington, DC: USAID.

Rogge, J. and Elahi, K. 1989. *The Riverbank Erosion Impact Study Bangladesh*. Final Report to the International Development Research Centre. Ottawa: IDRC.

Rosa, E. 1998. Meta-theoretical Foundations for Post-normal Risk. *Journal of RiskResearch* 1, 1: 15–44.

Ross, L. 1984. Flood Control Policy in China: The Policy Consequences of Natural Disasters. *Journal of Public Policy* 3, 2: 209–232.

Rostow, W. 1991. *The Stages of Economic Growth*. 3rd edn (1st edn, University of Chicago Press, 1960). Cambridge: Cambridge University Press.

Rothfeder, J. 2001. *Every Drop for Sale*. New York: Jeremy P. Tarcher/Putnam.

Roundy, R. 1983. Altitudinal Mobility and Disease Hazards for Ethiopian Populations. *Economic Geography* 52: 103–115.

Rovai, E. 1994. The Social Geography of Disaster Recovery. *Association of Pacific Coast Geographers Yearbook* 56: 49–74.

Royal Society of London et al. 2000. *Transgenic Plants and World Agriculture*. A Report Prepared Under the Auspices of the Royal Society of London, the US National Academy of Sciences, the Brazilian Academy of Sciences, the Chinese Academy of Sciences, the Indian Academy of Sciences, the Mexican Academy of Sciences and the Third World Academy of Sciences. London: The Royal Society of London (July).

Rubin, C. 1995. Physical Reconstruction. In: *Wellington After the Quake: The Challenge of Rebuilding Cities*. Christchurch, NZ: Centre for Advanced Engineering and the New Zealand Earthquake Commission.

Rubin, C. 2000. *Emergency Management in the 21st Century: Coping with Bill Gates, Osama bin-Laden and Hurricane Mitch*. Natural Hazards Research Working Paper 104. Boulder: Natural Hazards Research and Applications Information Center, Institute of Behavioral Science, University of Colorado. www.colorado.edu/hazards/wp/wp104/wp104.html.

Rubin, C. and Palm, R. 1987. National Origin and Earthquake Response: Lessons from the Whittier Narrows Earthquake. *International Journal of Mass Emergencies and Disasters* 5, 3: 347–355.

Rubin, C., Saperstein, M. and Barbee. D. 1985. *Community Recovery From a Major Natural Disaster*. Monograph No. 41. Boulder: University of Colorado.

Ruffié, J. 1987. *The Population Alternative: A New Look at Competition and the Species*. London: Penguin.

Russell, J. 1968. That Earlier Plague. *Demography* 5: 174–184.

Rutten, M. and Engelshoven, M. 2001. Earthquake in India: Weak State, Strong Middle Class? *IIAS Newsletter* (Leiden), No. 25, July.
http://iias.leidenuniv.nl/iiasn/25/index.html .

Ryan, A. et al. 1999. *Genetically Modified Crops: The Ethical and Social Issues*. London: Nuffield Council on Bio-ethics Working Party.

Sachs, W. (ed) 1999. *Planet Dialectics*. London: Zed Books.

Sachs, W. et al. 2002. *The Jo'burg Memo*. Berlin: Heinrich Boell Foundation. http://www.joburgmemo.org/ .

Said, E. 1988. *Orientalism*. New York: Vintage.

Salih, M. (ed) 2001. *Local Environmental Change and Society in Africa*. Second edition. Dordrecht: Kluwer Academic Publishers.

Sanchez, E., Cronick, K. and Wiesenveld, E. 1988. Psychological Variables in Participation: A Case Study. In: D. Canter, M. Krampen and D. Stea (eds), *New Directions in Environmental Participation*, Vol. 3. Aldershot: Gower.

Sanderson, D. 2000. Cities, disasters and livelihoods. *Environment and Urbanization* 12, 2: 93–102.

Sandman, P. 1989. Hazard versus Outrage in the Public Perception of Risk. In: V. Covello, D. McCallum and M. Pavlova (eds), *Effective Risk Communication*, pp.45–49. New York: Plenum.

Sapir, D. and Lechat M. 1986. Reducing the impact of Natural Disasters: Why aren't we better prepared? *Health Policy and Planning* 1, 2: 118–26.

SAPRIN (Structural Adjustment Participatory Review International Network) 2003. http://www.saprin.org/.

Sattaur, O. 1991. Counting the Cost of Catastrophe. *New Scientist* (29 June): 21–3.

Scarth, A. 1997. *Savage Earth*. Pelean Eruptions, pp.138–41. London: Harper-Collins.

Schmink, M. and Wood, C. 1992. *Contested Frontiers In Amazonia*. New York: Columbia University Press.

Schmuck-Widmann, H. 1996. *Living with the Floods: Survival Strategies of Char-dwellers in Bangladesh*. Berlin: FDCL (Forchungs und Dokumentationszentrum Chile-Latinamerika).

Schmuck-Widmann, H. 2001. *Facing the Jumuna River: Indigenous and Engineering Knowledge in Bangladesh*. Dhaka: Bangladesh Resource Centre for Indigenous Knowledge (BARCIK).

Schmuck-Widmann, H. 2002. Empowering women in Bangladesh. International Federation of Red Cross and Red Crescent Societies, 25 Feb 2002, via ReliefWeb: http://www.reliefweb.int/w/rwb.nsf/s/570056EB0AE62524C1256B6B00587224 .

Schoepf, B. 1992. Gender Relations and Development: Political Economy and Culture. In: A. Seidman and F. Anang (eds), *21st Century Africa: Towards a New Vision of Self-Sustainable Development*, pp.203–41. Atlanta, GA and Trenton, NJ: African Studies Association and Africa World Press.

Schoepf, B. and Schoepf, C. 1990. Gender, Land and Hunger in Eastern Zaire. In: R. Huss-Ashmore and S. Katz (eds), *African Food Systems in Crisis: Contending with Change*, pp.75–106. Philadelphia: Gordon & Breach Science Publishers.

Schroeder, R. 1987. *Gender Vulnerability to Drought: A Case Study of the Hausa Social Environment*. Natural Hazards Working Paper No. 58. Institute of Behavioral Science. Boulder: University of Colorado.

Scoones, I. 1998. *Sustainable Rural Livelihoods: A Framework for Analysis*. IDS Working Paper No. 72. Brighton, UK: University of Sussex, Institute for Development Studies.

Scott, J. 1976. *The Moral Economy of the Peasant: Rebellion and Subsistence in Southeast Asia*. New Haven, CT and London: Yale University Press.

Scott, J. 1985. *Weapons of the Weak: Everyday Forms of Peasant Resistance*. New Haven, CT: Yale University Press.

Scott, J. 1990. *Domination and the Arts of Resistance: Hidden Transcripts*. New Haven, CT: Yale University Press.

Scott, J. 1998. *Seeing Like a State*. New Haven, CT: Yale University Press.

Scott, M. 1987. The Role of Non-Governmental Organizations in Famine Relief and Prevention. In: M. Glantz (ed), *Drought and Hunger in Africa*, pp.349–366. Cambridge: Cambridge University Press.

Scott, M. and Mpanya, M. 1991. *We Are the World: An Evaluation of Pop Aid for Africa*. Petaluma, CA: World College West and USA for Africa.

Scudder, T. 1980. River-basin Development and Local Initiative in African Savanna Environments. In: D. Harris (ed), *Human Ecology in Savanna Environments*, pp.383–406. London: Academic Press.

Scudder, T. 1989. Conservation vs. Development: River Basin Projects in Africa. *Environment* 31, 2: 4–9 & 27–32.

Seager, J. 1992. Operation Desert Disaster: Environmental Costs of the War. In: C. Peters (ed), *Collateral Damage: The 'New World Order' at Home and Abroad*, pp.197–216. Boston, MA: South End.

Seaman, J. and Holt, J. 1980. *Markets and Famine in the Third World*. Disasters 4, 3: 233–297.

Seaman, J., Holt, J. and Allen, P. 1993. *A New Approach to Vulnerability Mapping for Areas at Risk of Food Crisis*. London: Save the Children Fund.

429

Seaman, J., Leivesley, S. and Hogg, C. 1984. *Epidemiology of Natural Disasters*. Basel: Karger.

Seidman, A. and Anang, F. (eds) 1992. *21st Century Africa: Towards a New Vision of Self-Sustainable Development*. Atlanta, GA and Trenton, NJ: African Studies Association and Africa World Press.

Sen, A. 1981. *Poverty and Famines: An Essay on Entitlement and Deprivation*. Oxford: Oxford University Press.

Sen, A. 1985. *Food, Economics and Entitlements*. WIDER WP-1 28.8.85.

Sen, A. 1988. Family and Food: Sex Bias in Poverty. In: T. Srinivasan and P. K. Bardhan (eds), *Rural Poverty in South Asia*, pp.453–472. New York: Columbia University Press.

Sen, A. 1990. Gender and Cooperative Conflict. In: I. Tinker (ed), *Persistent Inequalities: Women and World Development*, pp.123–149. Oxford: Oxford University Press.

Sen, A. 2000. *Development as Freedom*. New York: Oxford University Press.

Sen, G. and Grown, C. 1987. *Development, Crisis and Alternative Visions*. New York and London: Monthly Review and Earthscan.

Sen, G., Germain, A. and Chen, L. C. (eds) 1994. *Population Policies Reconsidered: Health, Empowerment and Rights*. Cambridge, MA: Harvard University Press.

Seth, S., Das, D., Gupta, G. 1981. Floods in Arid and Semi-arid Areas – Rajasthan. New Delhi: Ministry of Agriculture. Mimeo.

Shakow, D. and O'Keefe, P. 1981. Yes, We Have No Bananas: The Economic Effects of Minor Hazards in the Windward Islands. *Ambio* 10, 6: 344.

Shakur, T. 1987. *An Analysis of Squatter Settlements in Dhaka*, Bangladesh. Ph.D. Thesis, University of Liverpool.

Shanin, T. (ed) 1971. *Peasants and Peasant Society*. London: Penguin.

Sharma, V. and Mehrotra, K. 1986. Malaria Resurgence in India: A Critical Study. *Social Science and Medicine* 22, 2: 835–845.

Sharp, J. and Spiegel, A. 1984. Vulnerability to Impoverishment in South African Rural Areas: The Erosion of Kinship and Neighborhood as Social Resources. Carnegie Conference Paper 52, Cape Town University.

Shepherd, A. 1988. Case Study of Famine: Sudan. In: D. Curtis, M. Hubbard and A. Shepherd (eds), *Preventing Famine. Policies and Prospects for Africa*, pp.28–72. London: Routledge.

Shindo, E. 1985. Hunger and Weapons: The Entropy of Militarisation. *Review of African Political Economy* 33 (August): 6–22.

Shiva, V. 1989. *Staying Alive: Women, Ecology and Development*, London: Zed Books.

Shiva, V. 1991. *The Violence of the Green Revolution: Third World Agriculture, Ecology and Politics*. London and Penang: Zed Books and Third World Network.

Shiva, V. (ed) 1994. *Close to Home: Women Reconnect Ecology, Health and Development Worldwide*. Philadelphia, PA: New Society.

Shrivastava, P. 1992. *Bhopal: Anatomy of a Crisis*. 2nd edn. London: Paul Chapman Publishing.

Shoaf, K. 2002. Personal communication between Ben Wisner and Dr K. Shoaf, University of California at Los Angeles, School of Public Health, 6 June.

Shore, K. 1999. Food Security Under Siege? The Emerging Alliance between Micro-Credit Lenders and Transnational Corporations. International Development Research Centre, 12 July.
http://idrc.ca/reports/read_article_english.cfm?article_num=465

SID (Society for International Development) 2003a. *Participatory Action for Capacity Building and Food Security.* http://www.sidint.org/programmes/food/ .

SID (Society for International Development) 2003b. Impact of Food Aid Programmes and Moving towards Sustainable Livelihoods: A Kenyan community experience in Coast Province. http://www.sidint.org/programmes/food/Kenya.pdf .

Sidel, R. and Sidel, V. 1982. *The Health of China: Current Conflicts in Medical and Human Services for One Billion People.* Boston, MA: Beacon Press.

Siegel, S. and Witham, P. 1991. Case Study Colombia. In: A. Kreimer and M. Munasinghe (eds), *Managing Natural Disaster and the Environment*, pp.170–171. Washington, DC: World Bank.

Sigurdson, H. 1988. Gas Burst from Cameroon Crater Lakes: A New Natural Hazard. *Disasters* 12, 2: 131–146.

Sigurdson, H. and Carey, S. 1986. Volcanic Disasters in Latin America and the 13th November 1985 Eruption of Nevado del Ruiz Volcano in Colombia. *Disasters* 10, 3: 205–216.

Sikander, A. 1983. Floods and Families in Pakistan – a Survey. *Disasters* 7, 2: 101–106.

Simmonds, S., Vaughan, P. and Gunn, S. 1983. *Refugee Community Health Care.* Oxford: Oxford University Press.

Simon, D. 1995. Debt, Democracy and Development: Sub-Saharan Africa. In: D. Simon, W. van Spengen, C. Dixon and A. Naerman (eds), *Structurally Adjusted Africa: Poverty, Debt and Basic Needs*, pp.17–44. London: Pluto Press.

SINAL (Boletim Informativo do Sinal/Sistema de Informações a Nivel Local) 1992. O Rio e as enchentes. (Rio and floods). *SINAL*, Nov.–Dec. 1991–Jan. 1992: 3.

Singer, H., Wood, J. and Jennings, T. 1987. *Food Aid: The Challenge and the Opportunity.* Oxford: Clarendon Press.

Singh, S. 1975. *The Indian Famine, 1967* New Delhi: People's Publishing House.

SIPRI (Stockholm International Peace Research Institute) 1976. *Ecological Consequences of the Second Indochina War.* Stockholm: Almqvist & Wiksell International.

SIPRI (Stockholm International Peace Research Institute) 1980. *Warfare in a Fragile World: Military Impact on the Human Environment.* London: Taylor & Francis.

Sivard, R. 2001. *World Military and Social Expenditures.* Washington, DC: World Priorities.

Six-S (Se Servir de la Saison Sèche en Savane et au Sahel) n.d.
http://iisd1.iisd.ca/50comm/commdb/desc/d12.htm .

Sjöberg, L. (ed) 1987. *Risk and Society: Studies in Risk Generation and Reactions to Risk.* London: Allen & Unwin.

Sklair, L. 2001. *The Transnational Capitalist Class.* Oxford: Basil Blackwell.

Slim, H. and Mitchell, J. 1990. Towards Community Managed Relief: A Case Study from Southern Sudan. *Disasters* 14, 3: 265–268.

Slovic, P. 1993. Perceived Risk, Trust and Democracy. *Risk Analysis* 13, 6: 675–682.

Smith, D. 2000. Floodplain Management: Problems and Opportunities. In: D. Parker (ed), *Floods.* Vol. 1, pp.254–267. London: Routledge.

Smith, J., Chatfield, C. and Pagnucco, R. (eds) 1997. *Transnational Social Movements and Global Politics: Solidarity Beyond the State.* Syracuse, NY: Syracuse University Press.

Smith, K. 1992. *Environmental Hazards: Assessing Risk and Reducing Disaster.* London: Routledge.

Smith, K. 1996. *Environmental Hazards: Assessing Risk and Reducing Disaster.* 2nd edn. London: Routledge.

Smith, K. 2001. *Environmental Disasters.* 3rd edn. London: Routledge.

Smith, K. and Ward, R. 1998. *Floods.* Chichester: John Wiley.

Smith, M. 2000. Southern and Eastern Asian Watersheds
http://www.geocities.com/bhwater2000/mowater/MS.html .

Smith, S. 1990. *Front Line Africa.* Oxford: Oxfam.

Smyth, C. and Royle, S. 2000. Urban Landslide Hazards: Incidence and Causative Factors in Niteroi, Rio de Janeiro State, Brazil. *Applied Geography* 20: 95–117.

Soares, C. 2002. Earthquake School 'Built on the Cheap'. *Guardian Unlimited,* 2 November. http://www.guardian.co.uk/international/story/0,3604,824515,00.html .

Sopher, K. 1981. *On Human Needs.* Brighton, UK: Harvester Press.

Sorokin, P. 1975. *Hunger as a Factor in Human Affairs.* Gainsville: University of Florida Press.

Soto, L. 1998. *Módolos para la capacitación: guia de LA RED para la gestión local del riesgo.* Quito, Ecuador: LA RED and IT Peru.

Southern, R. 1978. The Global Socio-economic Impact of Tropical Cyclones. *Australian Meteorological Magazine* 27: 175–195.

Spence, R. 2003. Earthquake Risk Mitigation In Europe: Progress Towards Upgrading The Existing Building Stock. Paper presented at the Fifth National Conference on Earthquake Engineering, 26-30 May. Istanbul, Turkey.

Speth, J. 1998. Relief and Development: Twin Foundations for Sustainable Recovery and Peace Building. Statement by James Gustave Speth, Administrator, UNDP, to the UN Economic and Social Council, 15 July. New York: UNDP.

Sphere Project 2000. *Humanitarian Charter and Minimum Standards in Disaster Response.* Geneva: Sphere Project. www.sphereproject.org/ .

Spielman, A. and D'Antonio, M. 2001. *Mosquito: The Story of Man's Deadliest Foe.* London: Faber and Faber.

Spitz, P. 1976. *Famine-risk and Famine Prevention in the Modern World: Studies in Food Systems Under Conditions of Recurrent Scarcity.* Geneva: United Nations Research Institute for Social Development (UNRISD).

Stallings, R. 1995. *Promoting Risk: Constructing the Earthquake Threat.* New York: Aldine de Gruyter.

Stallings, R. 1997. Sociological Theories and Disaster Studies. Preliminary Paper 247. Disaster Research Center, University of Delaware.
http://www.udel.edu/ DRC/preliminary/249.pdf .

Stark, K. and Walker, G. 1979. Engineering for Natural Hazards with Particular Reference to Tropical Cyclones. In: R. Heathcote and B. Thom (eds), *Natural Hazards in Australia,* pp.189–203. Canberra: Australian Academy of Sciences.

Stehlik, D., Lawrence, G. and Gray, I. 2000. Gender and Drought: Experiences of Australian Women in the Drought of the 1990s. *Disasters* 24, 1: 38–53.

Steinberg, T. 2000. *Acts of God: The Unnatural History of Natural Disaster in America.* Oxford: Oxford University Press.

432

Stephen L. and Downing T. 2001. Getting the Scale Right: A Comparison of Analytical Methods for Vulnerability Assessment and Household-level Targeting. *Disasters* 25, 2, pp.113-135.

Stewart, F. 1987. Should Conditionality Change? In: K. Havnevik (ed), *The IMF and the World Bank in Africa: Conditionality, Impact and Alternatives*, pp.29–46. Seminar Proceedings No. 18. Uppsala: Scandinavian Institute of African Studies.

Stewart, F. and Fitzgerald, V. (eds) 2000. *War and Underdevelopment: The Economic and Social Consequences of Conflict*. Oxford: Oxford University Press.

Stiefel, M. and Wolfe, M. 1994. *A Voice for the Excluded: Popular Participation in Development – Utopia or Necessity?* London: Zed Books and UNRISD.

Stiglitz, J. 2002. *Globalization and its Discontents*. New York: W. W. Norton.

Stock, R. 1976. *Cholera in Africa*. London: International African Institute.

Stolberg, S. 2002a. Buckets for Bioterrorism, But Less for a Catalog of Ills. *New York Times*, 5 February. http://www.ph.ucla.edu/epi/bioter/bucketsforbioter.html .

Stolberg, S. 2002b. Annan Asks U.S. for More Money for AIDS Fund. *New York Times*, 14 February. http://www.stoptb.org/material/news/press/NYtimes.020214.htm .

Stott, P. and Sullivan, S. 2000. Introduction. In: P. Stott and S. Sullivan (eds), *Political Ecology: Science, Myth and Power*, pp.1–11. London: Arnold.

Streeten, P. 1984. Basic Needs and Human Rights. *Development: Seeds of Change* 3: 10–12.

Sutphen, S. 1983. Lake Elsinore Disaster – The Slings and Arrows of Outrageous Fortune. *Disasters* 7, 3: 194–201.

Sugiman, T. 2003. Citizen Participation and a Role of NPO in Disaster Prevention and Relief Activities. Paper presented at the DPRI-IIASA Third Symposium on Integrated Disaster Risk Management (IDRM-03). Kyoto: Kyoto University, 4 July.

Susman, P. 2001. Personal communication with Dr Paul Susman, Department of Geography, Bucknell University, Lewisburg, PA, USA, May, 2001, about his ongoing research on dengue in Cuba.

Susman, P. 2003. Personal communication with Ben Wisner concerning Susman's observations in March, 2003, while accompanying students from Bucknell University to various sites in Nicaragua [contact: Department of Geography, Bucknell University susman@bucknell.edu].

Susman, P., O'Keefe, P. and Wisner, B. 1983. Global Disasters, a Radical Interpretation. In: K. Hewitt (ed), *Interpretations of Calamity*, pp.274–276. London: Allen & Unwin.

Swedish Rescue Services Agency 1997. *Quito, Ecuador: Risk Mitigation Project*. Karlsrad, Sweden: Swedish Rescue Services Agency.

Swift, J. 1989. Why are Rural People Vulnerable to Famine? *IDS Bulletin* 20, 2: 8–15.

Swift, J. and Hamilton, K. 2001. Household Food Security. In: S. Devereaux and S. Maxwell (eds), *Food Security in Sub-Saharan Africa*, pp.67–92. London: IT Publications.

Swiss Re 1995. *The Great Hanshin Earthquake: Trial, Error, Success*. Zurich: Swiss Reinsurance Company.

Tanida, N. 1996. What Happened to Elderly People in the Great Hanshin Earthquake? *British Medical Journal* 313, 7065: 1133–1135.

Tayag, J. C. n.d. *How The People Escaped and Coped with Mayon Volcanos Fury – A Case Study of Institutional and Human Response to the 1984 Mayon Volcano Eruption.* Quezon City: Philippine Institute of Volcanology and Seismology.

Temcharoen, P., Viboolyavatana, J., Tongkoom, B., Sumethanurugkul, B., Keittivuti, B. and Wanaratana, I. 1979. A Survey of Intestinal Parasitic Infections in Laotian Refugees at Ubon Province, with Special Reference to Schistosomiasis. *Southeast Asian Journal of Tropical Medicine and Public Health* 10: 552–554.

Thébaud, B. 1988. *Elevage et développement au Niger.* Geneva: ILO.

Thompson, G. 2003. In Grip of AIDS, South Africa Cries for Equity. *New York Times*, 10 May: A4.

Thompson, J. et al. 2002. *Drawers of Water: 30 Years of Change in Domestic Water Use and Environmental Health Overview.* London: IIED. http://www.iied.org/sarl/dow/summary/ .

Thompson, M. and Warburton, M. 1988. Uncertainty on a Himalayan Scale. In: J. Ives and D. Pitt (eds), *Deforestation: Social Dynamics in Watersheds and Mountain Ecosystems*, pp.1–53. London: Routledge.

Thompson, P. and Penning-Rowsell, E. 1994. Socio-Economic Impacts of Floods and Flood Protection: A Bangladesh Case Study, pp.81–98. In: A. Varley (ed), *Disasters, Development and Environment.* Chichester: Wiley.

Ticehurst, S., Webster, R., Carr, V. and Lewin, T. 1996. The Psychosocial Impact of an Earthquake on the Elderly, *International Journal of Geriatric Psychiatry* 11, 11: 943–951.

Tierney, K. 1992. Politics, Economics and Hazards. Paper presented to the Society for Applied Anthropology, 26 March, Memphis, TN.

Tierney, K. and Goltz, J. 1997. *Emergency Response: Lessons from the Kobe Earthquake, Overview of Disaster Impacts and Emergency Response Issues*, Disaster Research Center, University of Delaware.

Tierney, K., Petak, W. and Hahn, H. 1988. *Disabled Persons and Earthquake Hazards.* Boulder, CO: Natural Hazards Research and Applications Center.

Tilling, R. 2002. Volcano Monitoring and Eruption Warnings. In: J. Zschau and A. Kueppers (eds), *Early Warning Systems for Natural Disaster Reduction*, pp.505–510. Berlin: Springer Verlag.

Timberlake, L. 1988. *Africa in Crisis.* 2nd edn. London: Earthscan.

Timmerman, P. 1981. *Vulnerability, Resilience and the Collapse of Society.* Environmental Monograph No. 1, Institute for Environmental Studies. Toronto: University of Toronto.

Tinker, I. (ed) 1990. *Persistent Inequalities: Women and World Development.* Oxford: Oxford University Press.

Tobin, G. and Montz, B. 1997. *Natural Hazards: Explanation and Integration.* New York: Guilford.

Tobriner, S. 1988. The Mexico Earthquake of September 19, 1985: Past Decisions, Present Danger: An Historical Perspective on Ecology and Earthquakes in Mexico City. *Earthquake Spectra* 4, 3: 469–479.

Tokuyama, A. 1995. *The Role of School and Teacher in the Shelter of the South Hyogo Earthquake and the Role of the School in the Total System of Disaster Mitigation.* Hyogo University of Teacher Education.

Tomblin, J. 1985. Armero: The Day Before. *UNDRO News*, Nov./Dec.: 4–6.

434

Tomblin, J. 1987. Management of Volcanic Emergencies. *UNDRO News*, July/August: 17.

Toulmin, C. and Wisner, B. (eds) 2003. *Toward a New Map of Africa*. London: Earthscan.

Trainer, T. 1989. *Developed to Death: Rethinking Third World Development*. London: Green Print.

TransAfrica Forum 2003.
http://transafricaforum.org/monitor/Summer2003V1Issue2.pdf .

Trujillo, M., Ordóñez, A. and Hernández, C. 2000. *Risk-Mapping and Local Capacities*. Oxfam Working Papers. Oxford: Oxfam Publications.

Tuan, Y.-F. 1979. *Landscapes of Fear*. New York: Pantheon.

Turcios, A., Jaraguin, U. and Riviera, D. 2000. *Hacia una gestión ecológica de los riesgos: bases conceptuales y metodolgicas para un sistema nacional de prevención y mitigación de desastres, y de protección civil*. San Salvador: Lutheran World Federation and Unidad Ecologica Salvadoreña.

Turner, B. et al. 2003. A framework for vulnerability analysis in sustainability science. *Proceedings of the National Academy of Sciences* 100, 4 (July), pp.8074–8079 http://cesp.stanford.edu/docs/turner_matson_2003.pdf .

Turner, S. and Ingle, R. (eds) 1985. *New Developments in Nutrition Education*. Nutrition Education Series, Vol. 11. Paris: UNESCO.

Turshen, M. 1989. *The Politics of Public Health*. New Brunswick, NJ: Rutgers University Press.

Twigg, J. 2001. *Sustainable Livelihoods and Vulnerability to Disasters*. London: Benfield Greig Hazard Research Centre Working Paper 2 http://www.bghrc.com .

Twigg, J. 2002. The Human Factor in Early Warnings: Risk Perception and Appropriate Communications. In: J. Zschau and A. Kueppers (eds), *Early Warning Systems for Natural Disaster Reduction*, pp.19–26. Berlin: Springer Verlag.

Twigg, J. and Bhatt, M. 1998. *Understanding Vulnerability: South Asian Perspectives*, London and Colombo: IT Publications.

Twose, N. 1985. *Fighting the Famine*. London and San Francisco, CA. Pluto and Food First.

Uemura, T. n.d. Sustainable Rural Development in Western Africa: The Naam Movement and the Six 'S'. FAO, Sustainable Development Department http://www.fao.org/waicent/faoinfo/sustdev/ROdirect/ROan0006.htm .

UK House of Commons International Development Committee 2002. *Global Climate Change and Sustainable Development*. Third Report of Session 2001–2002 Recommendation No.16, p.74, Vol.1. London: The Stationery Office.

Ukai, T. 1996. Problems of Emergency Medical Care at the Time of the Great Hanshin-Awaji Earthquake. *Annals of Burns and Fire Disasters* 9, 4 (December): 235. http://www.medbc.com/annals/review/vol_9/num_4/text/vol9n4p235.htm .

Ullah, Md. S. 1988. Cyclonic-surge Resistant Housing in Bangladesh: The Case of Urir Char. *Open House International* 13, 2: 44–49.

UN ACC (Administrative Committee on Coordination Network on Rural Development and Food Security) 2000. Food Insecurity and Vulnerability Information and Mapping Systems (FIVIMS): A Progress Report (March)
http://www.rdfs.net/OLDsite/en/themes/FIVpro-e.htm .

UN Economic Commission on Africa (UNECA) 2001. *HIV/AIDS in Sub-Saharan Africa: An Overview*. Addis Ababa: UNECA.

UNAIDS 2000. *Report on the Global HIV/AIDS Epidemic June 2000.* Geneva: UNAIDS.

UNCRD (United Nations Centre for Regional Development) and Sustainable Environment and Ecological Development Society (SEEDS) 2002. *The Sustainable Community Rehabilitation Handbook.* Kobe: UNCRD and SEEDS.

UNDP (United Nations Development Programme) 1994a. *Human Development Report 1994.* New York: UNDP.

UNDP (United Nations Development Programme) 1994b. Role of the United Nations Development Programme in Humanitarian Affairs. Executive Board of the United Nations Development Programme and of the United Nations Population Fund 2nd Regular Session on Programme-Level Activities and Special Programmes of Assistance, 5 April, DP/1994/13. New York: United Nations.

UNDP (United Nations Development Programme) 1998. *Human Development Report.* New York: Oxford University Press.

UNDP (United Nations Development Programme) 2001a. *From Relief to Recovery: The Gujarat Experience.* New York: UNDP, Emergency Response Division.

UNDP (United Nations Development Programme) 2001b. Orissa Flood 2001 Recovery. New Delhi: UNDP India.
http://www.undp.org.in/orissa/flood_recovery.htm

UNDP (United Nations Development Programme) 2003a. *World Vulnerability Report.* New York: UNDP.

UNDP 2003b. *Human Development Report 2004.* New York: Oxford University Press.

UNDP (United Nations Development Programme). n.d. *Building Bridges Between Relief and Development.* New York: UNDP.
http://www.undp.org/erd/archives/ bridges.htm .

UNDRO 1982a. *Disaster Prevention and Mitigation: A Compendium of Current Knowledge, Vol. 8 – Sanitation Aspects.* New York: United Nations.

UNDRO 1982b. *Disaster and the Disabled.* New York: United Nations.

UNICEF 1985. *Within Human Reach: A Future for Africa's Children.* New York: UNICEF.

UNICEF 1989. *Children on the Front Line: The Impact of Apartheid, Destabilisation and Warfare on Children in Southern and South Africa.* New York: UNICEF.

UNICEF 1992. *The State of the World's Children 1992.* New York: Oxford University Press.

UNICEF 1999. *The State of the World's Children 2000.* New York.
http://www.reliefweb.int/library/documents/sowc00.pdf .

Union News. 1988. Second Typhoon in Two Weeks Pounds Central Philippines. *Union News* (Springfield, Massachusetts), 8 November: 3.

United Nations 1985. *Volcanic Emergency Management.* New York: Office of the United Nations Disaster Relief Co-ordinator.

United Nations 1994. *Yokohama Strategy and Plan of Action for a Safer World: Guidelines for Natural Disaster Prevention, Preparedness and Mitigation.* Yokohama: World Conference on Natural Disaster Reduction.

United Nations 1995. *The United Nations and Mozambique 1992–1995.* New York: UN Department of Public Information.

United Nations 1999. *Key Findings. World Urbanization Prospects: The 1999 Revision.* New York: United Nations.

United Nations 2001. *Road Map towards the Implementation of the United Nations Millennium Declaration*. Report of the Secretary General to the Fifty-Sixth Session (Ref A/56/150). New York: United Nations General Assembly. http://www/undp.org/mdg/roadmap.doc .

United Nations 2002a. *The Johannesburg Declaration on Sustainable Development*. New York: United Nations Department of Economic and Social Affairs. www.johannesburgsummit.org/html/documents/summit_docs/1009wssd_pol_decl aration.doc .

United Nations 2002b. *Key Outcomes of the Summit*. Johannesburg Summit 2002: World Summit on Sustainable Development. Johannesburg, South Africa, 26 August–4 September 2002. http://www.johannesburgsummit.org .

United Nations. 2002c. *United Nations Population Information Network*. http://www.un.org/popin/ .

United Nations Economic and Social Department and ISDR 2001. *Environmental Management and the Mitigation of Natural Disasters: A Gender Perspective*. Geneva: ISDR. www.un.org/womenwatch/daw/csw/env_manage/ .

United Nations Environment Programme (UNEP) 2002. *GEO: Global Environmental Outlook 3: Past, Present and Future Perspectives*. Nairobi: United Nations Environment Programme. http://www.unep.org/geo/geo3/index.htm .

UNRISD (United Nations Research Institute for Social Development) 2000. *Visible Hands: Taking Responsibility for Social Development*, Geneva: UNRISD.

USAID (United States Agency for International Development) n.d. Unsung Heroines: Women and Natural Disasters. http://www.genderreach.com/pubs/ib8.htm .

US-BR (Bureau of Reclamation) 2003. The Failure of Teton Dam. http://www.usbr.gov/pn/about/Teton.html .

US Committee for Refugees. 2002. Worldwide Refugee Information. http://www.refugees.org/world/worldmain.htm .

US Department of Energy (DOE) 1999. Model Guidelines for Incorporating Energy Efficiency and Renewable Energy into State Energy Emergency Plan. http://www.eere.energy.gov/buildings/state_energy/pdfs/emerg_plan_guide.pdf .

US Embassy. 1998. Yangtze Floods and the Environment. http://www.panasia.org.sg /icimod/focus/risks_hazards/fldrpt_chi.htm and http://www.usembassy-china.org.cn/ english/s andt/fldrpt.htm .

USGS (US Geological Survey) 2003. Earthquakes with 1,000 or More Deaths from 1900. http://neic.usgs.gov/neis/eqlists/eqsmajr.html .

Vaghjee, R. and Lee Man Yan, M. 2002. Improving the Tropical Cyclone Warning System and its Effective Dissemination in Mauritius. In: J. Zschau and A. Kueppers (eds), *Early Warning Systems for Natural Disaster Reduction*, pp.209–211. Berlin: Springer Verlag.

van der Wusten, H. 1985. The Geography of Conflict since 1945. In: D. Pepper and A. Jenkins (eds), *The Geography of Peace and War*, pp.13–28. Oxford: Basil Blackwell.

Van Spengen, W. 1995. Conclusions: On the Context of Structural Adjustment in a Not So New World Order. In: D. Simon, W. van Spengen, C. Dixon and A. Naerman (eds), *Structurally Adjusted Africa: Poverty, Debt and Basic Needs*, pp.229–35. London: Pluto Press.

Van Wilkligen, M. 2001. Riding Out the Storm: The Experiences of the Physically Disabled during Hurricanes Bonnie, Denis and Floyd. Paper presented at the Natural Hazards Research and Applications Workshop, Boulder, CO, July.

Vatsa, K. 2002a. A Note on Millennium Development Goals. Unpublished research note commissioned by UNDP/ERD as background to the *World Vulnerability Report*. Geneva: UNDP/ERD.

Vatsa, K. 2002b. A Note on Governance. Unpublished research note commissioned by UNDP/ERD as background to the *World Vulnerability Report*. Geneva: UNDP/ERD.

Vaughan, M. 1987. *The Story of an African Famine: Gender and Famine in Twentieth-century Malawi*. Cambridge: Cambridge University Press.

Vaux, T. 2001. *The Selfish Altruist*. London: Earthscan.

Velasquez, G. 1995. Situation of Foreigners Affected by the Disaster. In: UNCRD, *Comprehensive Study of the Great Hanshin Earthquake*, pp.218–222. Nagoya: UNCRD.

Velasquez, J. et al. 1999. A New Approach to Disaster Mitigation and Planning in Mega-cities: The Pivotal Role of Social Vulnerability in Disaster Risk Management. In: T. Inoguchi, E. Newman and G. Paoletto (eds), *Cities and the Environment: New Approaches to Eco-societies*, pp.161–184. Tokyo: United Nations University Press.

Veltrop, J. 1990. Water, Dams and Civilisation. In: G. Le Moigne, S. Barghouti and H. Plusquellec (eds), *Dam Safety and the Environment*, pp.5–27. Washington, DC: World Bank.

von Braun, J., Teklue, T. and Webb, P. 1998. *Famine in Africa: Causes, Responses and Prevention*. Baltimore: Johns Hopkins University Press.

von Kotze, A. and Holloway, A. 1998. *Reducing Risk: Participatory Learning Activities for Disaster Mitigation in Southern Africa*. Durban, South Africa: International Federation of Red Cross and Red Crescent Societies and Department of Adult and Community Education, University of Natal (distributed by Oxfam, UK).

von Kotze, A. and Holloway, A. 1999. *Living with Drought: Drought Mitigation for Sustainable Livelihoods*. Cape Town and London: David Philip Publishers and Intermediate Technology Publications.

Voss, H. and Hidajat, R. (eds) 2002. *From a Culture of Reaction to a Culture of Prevention: Joining Forces for a Sustainable Development*. International Symposium on Disaster Reduction and Global Environmental Change. Bonn: German Committee for Disaster Reduction and German National Committee of Global Change Research.

Wadge, G. and Isaacs, M. 1987. *Volcanic Hazards from the Soufriere Hills Volcano, Montserrat West Indies*. Reading, UK: A Report to the Government of Montserrat and the Pan Caribbean Disaster Preparedness and Prevention Project, Department of Geography, University of Reading.

Wadge, G., Woods, A., Jackson, P., Bower, S., Williams, C. and Hulsemann, F. 1998. A Hazard Evaluation System for Montserrat. In: B. Lee and I. Davis (eds), *Forecasts and Warnings*, pp.3.1–3.32. London: Thomas Telford and UK National Co-ordination Committee for IDNDR.

Wainwright, H. 2003. 'People's UN' Marches to a Different Tune. *Guardian Weekly* 168, 6 (30 January–5 February): 5.

438

Walford, C. 1879. *Famines of the World: Past and Present*. New York: Burt Franklin.

Walgate, R. 1990. *Miracle or Menace? Biotechnology and the Third World*. Budapest: Panos Institute.

Walker, B. (ed) 1994. *Women in Emergencies*. Oxford: Oxfam.

Walker, P. 1989. *Famine Early Warning Systems: Victims and Destitution*. London, Earthscan.

Wallensteen, P. and Sollenberg, M. 2001. Armed Conflict, 1989–2000. *Journal of Peace Research* 38, 5: 629–644.

Wallrich, B. 1998. Disaster Needs of People with Disabilities. Transcript of Round Table Discussion 16 April, Emergency Management Forum (EMForum), Emergency Information Interchange Project (EIIP). http://www.emforum.org/ vlibrary/needs.htm .

Walter, J. and Simms, A. 2002. *The End of Development? Global Warming, Disasters and the Great Reversal of Human Progress*. London and Dhaka: New Economics Foundation and Bangladesh Centre for Advanced Studies. http://www.neweconomics.org/uploadstore/pubs/e_o_d.pdf .

Walton, J. and Seddon, D. 1994. *Free Markets and Food Riots*. Oxford: Blackwell.

Warmbrunn, W. 1972. *The Dutch under German Occupation*. Stanford, CA: Stanford University Press.

Washington Post 1998. Chinese Rivers of Death: Untamed Waterways Kill Thousands Yearly. 17 August.

Watanabe, M. 2002. *Hazardless Environment in the Area of Metro Manilla: JICA-MMDA/ PHILVOCS Initiatives for Safety and Sustainability*. Tokyo: Japanese International Cooperation Agency (JICA). http://www.mmeirs.org/pdf/C-3Watanabe.pdf .

Watson, R., Zinyowera, M. and Moss, R. (eds) 1998. *The Regional Impacts of Climate Change*. Cambridge: Cambridge University Press/Inter-Governmental Panel on Climate Change.

Watson, R. et al. (eds) 1999. *The Regional Impacts of Climate Change: An Assessment of Vulnerability*. Cambridge: Cambridge University Press.

Watts, M. 1983a. *Silent Violence: Food, Famine and Peasantry in Northern Nigeria*. Berkeley: University of California Press.

Watts, M. 1983b. On the Poverty of Theory: Natural Hazards Research in Context. In: K. Hewitt (ed), *Interpretations of Calamity*, pp.231–262. Boston, MA: Allen & Unwin.

Watts, M. (ed) 1986. *State, Oil and Agriculture in Nigeria*. Berkeley, CA: Institute of International Studies.

Watts, M. 1991. Entitlements or Empowerment? Famine and Starvation in Africa. *Review of African Political Economy* 51: 9–26.

Watts, M. and Peluso, N. (eds) 2001. *Violent Environments*. Ithaca, NY: Cornell University Press.

WCED (World Commission on Environment and Development) [Brundtland Commission] 1987. *Our Common Future*. Oxford: Oxford University Press.

WEDO (Women's Environment and Development Organization) 2002. Women's Action for a Healthy and Peaceful Planet 2015. http://www.wedo.org/ sus_dev/waa2015broweb.pdf .

Weir, D. 1987. *The Bhopal Syndrome*. London: Earthscan.

Weiss, G. and Haber, H. 1999. *Perspectives on Embodiment*. New York: Routledge.

439

Wellard, K. and Copestake, J. 1993. *Non-Governmental Organizations and the State in Africa: Rethinking Roles in Sustainable Agricultural Development*. London: ODI.

Wells, S. and Edwards, A. 1989. Gone with the Waves. *New Scientist*, 1 November: 47–51.

Wescoat, J. 1992. The Flood Action Plan: A New Initiative Confronted by Basic Questions. *Natural Hazards Observer* 16, 4: 1–2.

West, R. 1989. Richard West on Floods and Fear in Thailand. *Independent Magazine*, 28 October: 18.

Western, J. and Milne, G. 1979. Some Social Effects of a Natural Hazard: Darwin Residents and Cyclone Tracy. In: R. Heathcote and B. Thom (eds), *Natural Hazards in Australia*, pp.488–502. Canberra: Australian Academy of Science.

Westfall, M. 2001. *On-site Integrated Urban Upgrading for Vulnerable Slum Communities of Payatas*. Project Overview. Manila: Asian Development Bank.

Westing, A. (ed) 1984a. *Herbicides in War: The Long-term Ecological and Human Consequences*. London: Taylor & Francis.

Westing, A. (ed) 1984b. *Environmental Warfare: A Technical, Legal and Policy Appraisal*. London: Taylor & Francis.

Westing, A. (ed) 1985. *Explosive Remnants of War: Mitigating the Environmental Effects*. London: Taylor & Francis.

WFP (World Food Programme) (nd) Vulnerability Analysis and Mapping: a tentative methodology (Annex III).
http://www.proventionconsortium.org/files/wfp_vulnerability.pdf .

Whatmore, S. 1999. Geography's Place in the Life-Science Era? *Transactions of the Institute of British Geographers* 24: 259–260.

Whelan, K. 1996. *The Tree of Liberty: Radicalism, Catholicism and the Construction of Irish Identity, 1760–1830*. Cork: Cork University Press.

Whitaker, J. 1988. *How Can Africa Survive?* New York: Harper & Row.

White, A. 1981. *Community Participation in Water and Sanitation*. Technical Paper 17. Rijswijk, Netherlands: International Reference Centre for Community Water Supply and Sanitation.

White, A. 1974. Global Summary of Human Response to Natural Hazards: Tropical Cyclones. In: G. White (ed), *Natural Hazards*, pp.255–265. New York: Oxford.

White, G. 1942. *Human Adjustment to Floods*. Research Paper 29. Chicago, IL: University of Chicago, Department of Geography.

White, M. 1999. Wars, Massacres and Atrocities of the Twentieth Century Wars.
http://users.erols.com/mwhite28/war-1900.htm .

Whitehead, A. 2000. *Continuities and Discontinuities in Sustaining Rural Livelihoods in North-eastern Ghana*. Multiple Livelihoods and Social Change Working Paper Series No. 18. University of Manchester, Institute for Policy Management and Development.

Whitfield, J. 2002. Link Between Climate and Malaria Broken. *Nature*, 21 February.
http://www.nature.com/nsu/020218/020218–12.html .

Whittow, J. 1980. *Disaster: The Anatomy of Environmental Hazards*. Athens and Harmondsworth: University of Georgia Press and Penguin.

WHO (World Health Organisation) 1990. *Health For All When a Disaster Strikes*. Vol. 2. Geneva: WHO/EPR.

WHO (World Health Organisation) 1997. *Earthquakes and People's Health: Proceedings of a WHO Symposium*, Kobe, 27–30 January. Kobe: WHO Centre for Health Development.

WHO (World Health Organisation) 2001a. Orissa 1999 Cyclones. WHO, South East Asia Regional Office (SEARO), Department of Sustainable Development and Healthy Environment. http://w3.whosea.org/sde/cyclone.htm .

WHO (World Health Organisation) 2001b. Cholera in Nigeria – Update. Communicable Disease Surveillance and Response, 3 December.
http://www.who.int/disease-outbreak-news/n2001/december/3december2001.html .

WHO (World Health Organisation) 2002. Cholera Situation in Malawi. ReliefWeb, 24 April.   http://www.reliefweb.int/w/rwb.nsf/480fa8736b88bbc3c12564f6004c8ad5/7be8106c56177c16c1256bab005834bc?OpenDocument .

WHO (World Health Organisation) 2003. Severe Acute Respiratory Syndrome (SARS) – Multi-Country Outbreak – Update 51. Communicable Disease Surveillance and Response (CDR). Geneva: WHO, CDR. http://www.who.int/csr/don/2003_05_09b/en/ .

Wieczorek, G., Larsen, M., Eaton, L., Morgan, B. and Blair, J. 2001. Debris-flow and Flooding Hazards Associated with the December 1999 Storm in Coastal Venezuela and Strategies for Mitigation. United States Geological Survey, Open file report 01–0144. http://geology.cr.usgs.gov/pub/open-file-reports/ofr-01–0144/ / .

Wiest, R., Mocellin, J. and Motsisi, T. 1994. *The Needs of Women in Disasters and Emergencies*. Final Report to UNDP. Winnipeg, Manitoba: University of Manitoba. Available on-line through the Gender and Disaster Network.
http://online.northumbria.ac.uk/geography_research/gdn/resources/women-in-disaster-emergency.pdf .

Wignaraja, P. (ed) 1993. *New Social Movements in the South: Empowering the People*. London: Zed Books.

Wijkman, A. and Timberlake, L. 1984. *Natural Disasters: Acts of God or Acts of Man?* London: Earthscan.

Wilches-Chaux, G. 1992a. The Global Vulnerability. In: Y. Aysan and I. Davis (eds), *Disasters and the Small Dwelling*, pp.30–35. London: James & James Science Press.

Wilches-Chaux, G. 1992b. Personal communication with I. Davis regarding Nevado del Ruiz court case.

Wilches-Chaux, G. 1993. Personal communication with I. Davis regarding Galeras volcanic eruption of 15 January 1993.

Wilches-Chaux, G. and Wilches-Chaux, S. 2001. *¡Ni de Riesgos!* Bogotá, Colombia: Fondo Para la Reconstucción y Desarrollo de Eje Cafetero.

Wilhite, D. (ed) 2000. *Drought*. 2 vols. London: Routledge.

Wilhite, D. and Easterling, W. (eds) 1987. *Planning for Drought: Toward a Reduction of Societal Vulnerability*. Boulder, CO: Westview.

Wilken, G. 1988. *Good Farmers*. Boulder, CO: Westview.

Wilkie, W. and Neal, A. 1979. Meteorological Features of Cyclone Tracy. In: R. Heathcote and B. Thom (eds), *Natural Hazards in Australia*, pp.473–87. Canberra: Australian Academy of Science.

Williams, E. 1992. Popular Organization in the *Favela*. Seminar presentation. Graduate School of Education, University of Massachusetts.

Wills, J., Wyatt, T. and Lee, B. 1998. Warnings of High Winds in Densely Populated Areas. In: B. Lee and I. Davis (eds), *Forecasts and Warnings*, pp.4.42–4.53. London: Thomas Telford.

Wilmsen, E. 1989. *Land Filled with Flies: A Political Economy of the Kalahari*. Chicago, IL: University of Chicago Press.

Wilson, E. (ed) 1988. *Biodiversity*. Washington, DC: National Academy Press.

Wilson, E. 1989. Threats to Biodiversity. *Scientific American* 261, 3: 108–116.

Wilson, F. and Ramphele, M. 1989. *Uprooting Poverty: The South African Challenge*. New York: W. W. Norton.

Wilson, M. and Rachman, G. 1989. Are Hurricanes Growing in the Greenhouse? *The Sunday Correspondent* (London), 24 September: 11.

Winchester, P. 1986. *Cyclone Vulnerability and Housing Policy in the Krishna Delta, South India, 1977–83*. Ph.D. Thesis, School of Development Studies, University of East Anglia, Norwich.

Winchester, P. 1990. Economic Power and Response to Risk. In: J. Handmer and E. Penning-Rowsell (eds), *Hazards and the Communication of Risk*, pp.95–110. Aldershot: Gower Publishing.

Winchester, P. 1992. *Power, Choice and Vulnerability: A Case Study in Disaster Mismanagement in South India*. London: James & James Science Publishers.

Winchester, S. 2003. *Krakatoa: The Day the World Exploded August 27, 1883*. New York: HarperCollins.

Wisner, B. 1975. An Example of Drought-induced Settlement in Northern Kenya. In: I. Lewis (ed), *Abaar: The Somali Drought*, pp.24–25. London: International African Institute.

Wisner, B. 1976a. Health and the Geography of Wholeness. In: G. Knight and J. Newman (eds), *Contemporary Africa: Geography and Change*, pp.81–100. Englewood Cliffs, NJ: Prentice-Hall.

Wisner, B. 1976b. Man-made Famine in Eastern Kenya: The Interrelationship of Environment and Development. Discussion Paper 96 (July). Brighton: Institute of Development Studies at the University of Sussex.

Wisner, B. 1977. Constriction of a Livelihood System: The Peasants of Tharaka Division, Meru District, Kenya. *Economic Geography* 53, 4: 353–357.

Wisner, B. 1978a. Letter to the Editor. *Disasters* 2, 1: 80–82.

Wisner, B. 1978b. *The Human Ecology of Drought in Eastern Kenya*. Ph.D. Thesis, Graduate School of Geography, Clark University.

Wisner, B. 1979. Flood Prevention and Mitigation in the People's Republic of Mozambique. *Disasters* 3, 3: 293–306.

Wisner, B. 1980. Nutritional Consequences of the Articulation of Capitalist and Non-capitalist Modes of Production in Eastern Kenya. *Rural Africana* 8/9: 99–132.

Wisner, B. 1982. Review of Dando's *The Geography of Famine*. *Progress in Human Geography* 6, 2: 271–277.

Wisner, B. 1984. Ecodevelopment and Ecofarming in Mozambique. In: B. Glaeser (ed), *Ecodevelopment: Concepts, Projects, Strategies*, pp.157–168. Oxford: Pergamon.

Wisner, B. 1985. Natural Disasters. *Puerto Rico Libre*, Fall: 16–17.

Wisner, B. 1987a. Doubts about Social Marketing. *Health Policy and Planning* 2, 2: 178–179.

Wisner, B. 1987b. Rural Energy and Poverty in Kenya and Lesotho: All Roads Lead to Ruin? *IDS Bulletin* 18, 1: 23–29.

Wisner, B. 1988a. *Power and Need in Africa: Basic Human Needs and Development Policies.* London and Trenton, NJ: Earthscan and Africa World Press.

Wisner, B. 1988b. GOBI vs. PHC: Some Dangers of Selective Primary Health Care. *Social Science and Medicine* 26, 9: 963–969.

Wisner, B. 1992a. Health of the Future/The Future of Health. In: A. Seidman and F. Anang (eds), *21st Century Africa: Towards a New Vision of Self-Sustainable Development*, pp.149–181. Atlanta, GA and Trenton, NJ: African Studies Association and Africa World Press.

Wisner, B. 1992b. Too Little to Live On, Too Much to Die From: Lesotho's Agrarian Options in the Year 2000. In: A. Seidman, K. Mwanza, N. Simelane and D. Weiner (eds), *Transforming Southern African Agriculture*, pp.87–104. Trenton, NJ: Africa World Press.

Wisner, B. 1993. Disaster Vulnerability: Scale, Power and Daily Life. *GeoJournal* 30, 2: 127–140.

Wisner, B. 1997. Environmental Health and Safety in Urban South Africa. In: B. Johnston (ed), *Life and Death Matters*, pp.265–286. Walnut Creek, CA: Altamira.

Wisner, B. 1999. There are Worse Things than Earthquakes: Hazard Vulnerability and Mitigation Capacity in Greater Los Angeles. In: K. Mitchell (ed), *Crucibles of Hazard*, pp.375–427. Tokyo: United Nations University Press.

Wisner, B. 2000a. The Political Economy of Hazards: More Limits to Growth? *Environmental Hazards* 20: 59–61.

Wisner, B. 2000b. From 'Acts of God' to 'Water Wars'. In: D. Parker (ed), *Floods*, Vol.1, pp.89–99. London: Routledge.

Wisner, B. 2001a. Capitalism and the Shifting Spatial and Social Distribution of Hazard and Vulnerability. *Australian Journal of Emergency Management* 16, 2 (Winter): 44–50. http://www.ema.gov.au/5virtuallibrary/pdfs/vol16no2/wisner.pdf .

Wisner, B. 2001b. Floods and Mudslides in Algiers: Why No Warning? Why Poor Drainage? Why? http://online.northumbria.ac.uk/geography_research/radix/ .

Wisner, B. 2001c. Storms and Socialism. *Guardian* (London), 14 November. http://www.guardian.co.uk/comment/story/0,3604,592992,00.html.

Wisner, B. 2001d. Same Old Story. *Guardian On-Line*, 1 February. www.guardian.co.uk/Archive/Article/0,4273,4128181,00.html .

Wisner, B. 2001e. Disaster Management: What the United Nations Is Doing, What It Can Do. *United Nations Chronicle* 37, 4: 6–9. http://www.un.org/Pubs/chronicle/2000/issue4/0400p6.htm .

Wisner, B. 2001f. Risk and the Neoliberal State: Why Post-Mitch Lessons Didn't Reduce El Salvador's Earthquake Losses. *Disasters* 25, 3: 251–268.

Wisner, B. 2001g. Some Root Causes of Disaster Vulnerability in Gujarat. Philadelphia, PA: American Friends Service Committee. http://www.afsc.org/emap/help/india131.htm

Wisner, B. 2002a. Disaster Risk Reduction in Megacities: Making the Most of Human and Social Capital. Paper presented at the workshop *The Future of Disaster Risk: Building Safer Cities.* Washington, DC: World Bank and the ProVention Consortium, 4–6 December.
http://www.proventionconsortium.org/files/conference_papers/wisner.pdf .

Wisner, B. 2002b. Goma, Congo: City Air Makes Men Free? RADIX web site. http://online.northumbria.ac.uk/geography_research/radix/nyiragongo.htm .

Wisner, B. 2002c. Radix – Violent Conflict and Disasters. RADIX web site. http://online.northumbria.ac.uk/geography_research/radix/violent-conflict.html .

Wisner, B. 2003a. The Communities Do Science! Proactive and Contextual Assessment of Capability and Vulnerability in the Face of Hazards. In: G. Bankoff, G. Frerks and T. Hilhorst (eds), *Vulnerability: Disasters, Development and People*, ch.13. London: Earthscan.

Wisner, B. 2003b. RADIX: Bingol, Turkey, 1 May.
http://online.northumbria.ac.uk/ geography_research/radix/turkey.htm .

Wisner, B. 2003c. Disability and Disaster: Victimhood and Agency in Earthquake Risk Reduction. In: C. Rodrigue and E. Rovai (eds), *Earthquake*. London: Routledge.

Wisner, B. 2003d. Comment. *Benfield Greig Hazard Research Centre Alert 9*, p. 4.

Wisner, B. 2003e. Disaster Preparedness and Response: Why is the Phone Off the Hook? Invited paper for the European Telecommunications Resilience & Recovery Association Inaugural Conference (ETR2A), Newcastle-upon-Tyne, UK, 11-13 June 2003.

Wisner, B. 2003f. Swords, Plowshares, Earthquakes, Floods, and StormsIn an Unstable, Globalizing World. Invited keynote presentation at the DPRI-IIASA Third Symposium on Integrated Disaster Risk Management (IDRM-03). Kyoto: Kyoto University, 4 July.

Wisner, B. and Adams, J. 2003. *Environment and Health in Disasters and Emergencies*. Geneva: World Health Organisation.
http://www.who.int/water_sanitation_health/hygiene/emergencies/emergencies2002/en/ .

Wisner, B. and Mbithi, P. 1974. Drought in Eastern Kenya: Nutritional Status and Farmer Activity. In: G. White (ed), *Natural Hazards*, pp.87–97. New York: Oxford University Press.

Wisner, B., O'Keefe, P. and Westgate, K. 1977. Global Systems and Local Disasters: The Untapped Potential of People's Science. *Disasters* 1, 1: 47–57.

Wisner, B., Stea, D. and Kruks, S. 1991. Participatory and Action Research Methods. In: E. Zube and G. Moore (eds), *Advances in Environment, Behavior and Design*, Vol. 3, pp.271–295. New York: Plenum.

Wisner, B., Westgate, K. and O'Keefe, P. 1976. Poverty and Disaster. *New Society* 9 (September): 547–548.

Wisner, B., Neigus, D., Mduma, E. and Kruks, S. 1979. Designing Storage Systems with Villagers. *African Environment* 3 (3–4): 85–95.

Wisner, B. et al. 1999. *Toward a Culture of Prevention in Two Coastal Communities: A Comparison of the Hazardousness of San Pedro and Laguna Beach*. Research Report. Long Beach: California State University at Long Beach, Department of Geography.

Wood, R. 1986. *Earthquakes and Volcanos*. London: Mitchell Beazley.

Woodham-Smith, C. 1962. *The Great Hunger: Ireland 1845–9*. London: Hamish Hamilton.

Woodruff, B. et al. 1990. Disease Surveillance and Control after a Flood: Khartoum, Sudan 1988. *Disasters* 14, 2: 151–163.

World Bank 1990. *Flood Control in Bangladesh: A Plan for Action*. Asia Region Technical Department. World Bank Technical Paper No. 119. Washington, DC: World Bank.

World Bank 1994. *Averting the Old Age Crisis: Policies to Protect the Old and Promote Growth*. Washington, DC: World Bank.

World Bank 2000. *The United Nations Millennium Goals*. Washington, DC: World Bank. http://www.developmentgoals.org/ .

World Bank 2001. *World Development Report 2000–2001*. Washington, DC: World Bank.

World Bank 2002. *Poverty and Climate Change: Reducing the Vulnerability of the Poor*. Consultation Draft/Discussion Document prepared by AfDB, ADB, DFID, DGIS, EC, BMZ, OECD, UNDP, UNEP and the World Bank, October. Washington, DC: World Bank. http://www.worldbank.org/povcc .

World Commission on Dams 2000a. China Country Review Paper: Experience with Dams in Water and Energy Resource Development in the People's Republic of China. Vlaeberg, Cape Town. http://www.dams.org/kbase/studies/cn/ .

World Commission on Dams 2000b. *Dams and Development: A New Framework for Decision Making*. 16 November. http://www.dams.org/report/ .

World Commission on Dams 2000c. http://www.dams.org/

World Energy Council 2003. *Energy Information Centre: Peat*. http://www.worldenergy.org/wec-geis/publications/reports/ser/peat/peat.asp .

World Food Programme 2002a. Hunger in Southern Africa: The Unfolding Crisis, ReliefWeb, 4 May.
http://www.reliefweb.int/w/rwb.nsf/480fa8736b88bbc3c12564f6004c8ad5/a0a9e369665122bc1256bb2002b1937?OpenDocument.

World Food Programme 2002b. WFP Emergency Report No. 19 of 2002. ReliefWeb, 10 May.
http://www.reliefweb.int/w/rwb.nsf/480fa8736b88bbc3c12564f6004c8ad5/df89fe80e501dcba85256bb50062c4cf?OpenDocument.

World Food Programme 2002c. Delivering Food Aid Afghanistan: Beating the Clock. http://www.wfp.org/newsroom/in_depth/afghanistan-11_9_01.html .

World Resources Institute 1986. *World Resources 1986*. New York: Basic Books.

World Social Forum 2002. The Treaty Initiative to Share the Genetic Commons, agreed at the World Social Forum in Porto Alegre, Brazil, 1 February 2002. http://www.ukabc.org/genetic_commons_treaty.htm .

World Vision 2002a. Indonesia: Flood Victims Threatened by Disease. 13 February. http://www.reliefweb.int/w/rwb.nsf/s/F157E5F40F34FA3949256B610008EA9C .

World Vision 2002b. Indonesia: Flooding puts 20,000 Babies at Risk of Famine. 27 February.
http://www.reliefweb.int/w/rwb.nsf/s/AC96246DF9443EFF49256B6800280A02 .

World Vision 2002c. More Flooding in Caranavi. 12 March.
http://www.reliefweb.int/w/rwb.nsf/s/42A3780DCF2F3F95C1256B73003FAD43 .

Worldwatch Institute 1998. *Beyond Malthus: Sixteen Dimensions of the Population Problem*, World Watch Paper No.143. Washington, DC: Worldwatch Institute.

Worster, D. 1993. The Shaky Ground of Sustainability. In: W. Sachs (ed), *Global Ecology: A New Arena of Political Conflict*, pp.132–145. London: Zed Books.

Wright, A. 1997. *Toward a Strategic Sanitation Approach: Improving the Sustainability of Urban Sanitation in Developing Countries.* UNDP-World Bank Water and Sanitation Program. Washington, DC: World Bank. http://www.wsp.org/pdfs/global_ssa.pdf .

Wright, G. 2003. WHO Warns of Iraq Cholera Outbreak. *Guardian Unlimited,* 8 May. http://www.guardian.co.uk/Iraq/Story/0,2763,951853,00.html .

Wright, J., Rossi, P., Wright, S. and Weber-Burdin, E. 1979. *After the Clean-up: Long Range Effects of Natural Disasters.* Beverly Hills, CA: Sage.

Wynne, B. 1987. *Risk Management and Hazardous Waste: Implementation and the Dialectics of Credibility.* Berlin: Springer Verlag.

Yamano, T., Jayne, T. and McNeil, M. 2002. *Measuring the Impacts of Adult Death on Rural Households in Kenya,* Working Paper 6. Tegemeo Institute of Agricultural Policy and Development, Egerton University, Njoro and Nairobi.

Yamazaki, F., Nishimura, A. and Ueno, Y. 1996. Estimation of Human Casualties due to Urban Earthquakes. Eleventh World Conference on Earthquake Engineering, Paper 443, Acapulco, Mexico, 23–28 June. Proceedings available on CD via International Institute of Earthquake Engineering and Seismology. http://www.iiees.ac.ir/English/Library/eng_lib_cd.html .

Yang, D. 1996. *Calamity and Reform in China: State, Rural Society and Institutional Change since the Great Leap Famine.* Stanford, CA: Stanford University Press, 1996.

Yarnal, B. et al. 2002. *Human Environment Regional Observatory.* http://hero.geog.psu.edu/HERO_REU/ REUindex.htm

Yates, R., Alam, K., Twigg, J., Guha-Sapir, D. and Hoyois, P. 2002. *Development at Risk: Brief for the World Summit on Sustainable Development.* London: Benfield Greig Hazard Research Centre. http://www.bhrc.com .

Yellow Fever Collection 2003. Phillip S. Hench Walter Reed Yellow Fever Collection. http://yellowfever.lib.virginia.edu/reed/ .

Yi Si 1998. The World's Most Catastrophic Dam Failures: The August 1975 Collapse of the Banqiao and Shimantan Dams. In: Dai Qing (ed), *The River Dragon has Come!* New York: M.E. Sharpe.

York, S. 1985. Report on a Pilot Project to Set Up a Drought Information Network in Conjunction with the Red Crescent Society in Darfur. *Disasters* 3: 73–179.

Zaman, M. 1991. Social Structure and Process in Char Land Settlement in the Bramaputra–Jamuna Floodplain. *Man* 26, 4: 549–566.

Zarco, R. 1985. *Anticipation, Reaction and Consequences: A Case Study of the Mayon Volcano Eruption.* Quezon City: Philippine Institute of Volcanology and Seismology.

Zebroski, E. 2002. *The Last Days of St. Pierre: The Volcanic Disaster that Claimed 30,000 Lives.* New Brunswick, NJ: Rutgers University Press.

Zeigler, D., Johnson, J. and Brunn, S. 1983. *Technological Hazards.* Resource Publications in Geography. Washington, DC: Association of American Geographers.

Ziegler, P. 1970. *The Black Death.* London: Pelican.

Zimmerman, R. 1995. *Science, Non-science, and Nonsense.* Baltimore: Johns Hopkins University Press.

Zschau, J. and Kueppers, A. (eds) 2002. *Early Warning Systems for Natural Disaster Reduction.* Berlin: Springer Verlag.

INDEX

Aberfan, Wales: landslide from coal waste tip (1966) 214
Abidjan, Côte d'Ivoire 183
Aboriginal people, Australia 238
absentee land ownership 31, 147, 250–1, 269; nineteenth-century Ireland 178
Acapulco: effect of hurricane Pauline 266
access: idea of 103; patterns 112, 163
Access model 12, 31, 37, 50, 52, 88–97, *89, 92*, 110; constructive criticisms 97–8; developments since introduction of 95–8; illustrating development of famine 149–59, *152–3, 156–7*, 253; illustrating impact of AIDS 189, *190–2*; of Mexico City earthquake 287–90, *288–9*; in relation to hurricanes 252–3; as research framework 121–3; resources during normal life 98–112, *99*, 174–5; time periods for components of 108; transition to disaster *104–5*, 107, 109–10, 238; and understanding vulnerability 110, 222, 276–7
Action Plan for the Reduction of Absolute Poverty (Mozambique) 260
ActionAid 347
Afghanistan: displaced persons due to war 45n, 46n, 175; mines and loss of limbs 74, 145; risk from Lake Sarez, Tajikistan 213; war and drought 5, 27, 172; war as root cause of vulnerability 53, 373
Africa: adverse agrarian trends in sub-Saharan region 79; biological hazards and vulnerability 183–92; cholera outbreaks 38; cities 73, 182; colonial settlement of coastal areas 247; debt

repayment crises 76, 345; disaster reduction networks 334; drought 42n, 117; epidemics in refugee camps 177; famines 119, 120, 128, 129, 144, 160, 183; FEWS developed for 161; 'flood-retreat' agriculture 235; floods in recent years 201; land clearances 81; local grain storage 343; magnitude of HIV-AIDS disaster 26, 115, 188; participatory development projects 366; timber industry 81; vulnerability analysis of HIV-AIDS 183, 185–92, 326; wars during last decades 5, 183, 188, *see also* southern Africa; West Africa
African Americans: bias against in disaster relief 245, 249, 358, 360
age factor 54, 68–9, 80; in social causation of disasters *6*, 7, 36, 268
Agenda 21 22, 29
agriculture *see* farming
agro-chemicals 181–2, 195
agro-industries 249, 257
Ahmedabad, Gujarat 301
aid *see* humanitarian aid
aid agencies and organisations 31, 131, 279, 354
AIDS *see* HIV-AIDS
Airs, Waters and Places (Hippocrates) 196
Alamgir, M. 143
Alexander, D. 50–5, 214
Algeria: mudslide (2001) 23, 47n, 349
Alice Springs, Australia: flood (1985) 238
Amazonian region: migrants from Brazil 176; rainforest 117
Ambraseys, N.N. 331

Anderson, M. 120, 316, 333
Andes 83, 117, 176, 178, 312
Andhra Pradesh, south-east India:
    coastal livelihoods 250; cyclones
    92–3, 246, 252, 252–6, *254, 255,*
    262–3, 262, 269–70, 358
Angola: displaced people 175, 184;
    mines and loss of limbs 74, 145;
    refugees from civil war 73–4, 133;
    violent conflict 29, 45n, 73–4, 199n;
    war as root cause of vulnerability 53,
    133, 145
animals: habitats provided by flooding
    203–4; health risks 175, 181; loss of
    in disasters 220, 224, 264; with
    potential to damage crops 167
Annan, Kofi 171
anthrax: spores sent in US mail (2001)
    171
Argentina 70
Arguetta, Manual Colon 280
Ariyabandu, M 369–70
Armenia, Colombia 72, 277
Armero, Colombia 305
arsenic poisoning: Bangladesh 170
Asia: colonial settlement of coastal
    areas 247; disaster reduction
    networks 334; effect of climate
    change on monsoon 136; epidemics
    in refugee camps 177; participatory
    development projects 366; timber
    industry 81; vulnerability to disease
    in mega-cities 182, *see also* South
    Asia; south-east Asia
Asian Development Bank 65
Asians: expelled from Uganda 186
Aswan dam 235
At Risk: first edition 20, 23, 26, 30–2, 83,
    95; present edition 32–41
Attacking Poverty (World Bank) 348
Austin, T. 368
Australia: Emergency Management
    337–8, 339, 354; mortality from
    tropical cyclones 245; recent floods
    83, 201; wildfires 23, 83, 196, 349, *see
    also* Aboriginal people
Aztecs 282

baby food industries 370
Bangkok 72
Bangladesh 10, 42n, 60, 83, 117, 236,
    271n; arsenic poisoning 170; attempt

to solve flood problem 225–35, 239;
    averted famines 129; cyclones 73, 218,
    235, 246, 247, 262, 263–4; famine
    120, 127, 128, 137; flooding 31, 56–7,
    73, 201, 202, 203, 205–6, 220–1, 221,
    229, 235, 252; Grameen Bank's
    support for livelihoods 162, 361;
    grassroots efforts for protection
    against storms 343; IFPRI study of
    floods 222–5; inequality in access to
    land 360; problems with food price
    rises 140–1; problems of population
    pressure 69, 70; vulnerability to
    storms 244, 245, 247–8, 249; war as
    root cause of vulnerability 145, *see
    also* Dhaka
Bangladesh Flood Action Plan (BFAP)
    226–33, 235
banking *see* credit schemes
Bankoff, G. 19
Banqiao dam, Henan province (China)
    207
Barnes, P. 332
Barnett, A. 188, 189, 192
Basra, Iraq: cholera 197n
Bates, F. 359
Bay of Bengal 225, 245, 253
Beck, Ulrich 16–18, 19, 39
Belgium: Centre for Research in the
    Epidemiology of Disasters (CRED)
    65
Belize 197n
Benfield Greig Hazards Research Centre
    (UCL) 347
Bengal 141, 146
Bhopal chemical disaster 38–9, 369, 370
Bhuj, Gujarat 41, 72, 300–1, 301
Biafra 42n, 127, 199n
Bihar 225
biodiversity: Johannesburg Summit
    agreements 350; loss of 81, 177, 195
biological disasters 37–8, 175, 188
biological hazards 36, 168–72; caused by
    extreme events 172–3; defences
    against 180–3; and health 169, 172–3,
    176–7; livelihoods, resources and
    disease 174–5, 183; risk reduction in
    vulnerability to 193–6; root causes
    and pressures 183–92; triggering
    disasters 167, 169; vulnerability to
    167–8, 172–3, 175–80, 183
biological terrorism 169, 170, 171

settlers in Siberia 176; tornado (1984) 271n
Space Shuttle Columbia: loss of 17
Spanish colonialism 107, 282
species loss 81, 82, 193
squatter settlements 41, 56–7, 70, 73, 98, 266, 279, 290, 367
Sri Lanka 29, 70, 250, 251, 262, 262
Stallings, R. 348
Stanford, L. 245
starvation 128, 141, 142–3, 146, 172, 176; governments' deliberate policies of 322, 331
the state: in Access model 94–5, 155; functions/dysfunctions as root cause of vulnerability 52–3, 256, 269; responses to cyclones in Mozambique 257–60
Stewart, F. 76
Stockholm Principles 355–6
Stop Disasters 21
storm surges 60, 92, 106, 245, 246, 253, 256, 261, 262–3, 269
storms *63*, 83, 201–2, 214, 219, 270, *see also* coastal storms; hurricanes; tropical cyclones
Streeten, P. 196n
structural adjustment programmes (SAPs) 26, 31, 53–4, 76–7, 79, 80, 132, 160, 185, 321–2, 342
structures of domination: Access model *89*, 94–5, 102, 108, 110, 174, 285–6; and famine 155, 158, 253; and recovery 252–3, 359, 364; and vulnerability to floods 238
Sutphen, S. 357
subjective impacts 14–15
Sudan 194, 360; civil war 165n, 199n, 373; colonial history 149, 159; drought 5, 130; exporting of food despite famine 138; famines 120, 128, 130, 135, 145, 147–8, 148–9, *150–1*, 158; government's deliberate actions to increase suffering 322; refugees and displaced people 46n, 173, 175; spraying of cotton crop in Gezira scheme 182; war as root cause of vulnerability 5, 53, 145, 148–9, *see also* Khartoum
Sudan dioch birds 184
Sudan People's Liberation Army (SPLA) 322

Sundarbans 249
sustainable development 9, 27; recovery contributing to 363–4; and risk reduction 22–3, 270, 348–50, 354; UK International Development Committee Report (2002) 334; UN millennium goal of ensuring 326, *see also* World Summit on Sustainable Development
sustainable livelihoods (SL) approach 95–6, 138, 161–2
Swaziland 187
Sweden: Red Cross initiative 34
Swedish Rescue Services Agency 71
Swift, J. 118, 162
Swiss Re (reinsurance company) 65
Sydney 196n

Taiwan 348
Tajikistan 213
Takayama, Tatou 297–8
Tangshan, China: earthquake (1976) 127, 276
Tanzania 10, 60, 165n, 186, 194, *see also* Dar es Salaam
taxation 139, 155
technocratic approaches 23–4, 270, 323, 333; to flood problems 230–1, 232–3, 239–40
technological hazards 16, 38–41
terrorism 17, 38; sabotaging reconstruction of Guatemala 280, 280–1; US war on 23, 38, 171, *see also* biological terrorism
Thailand 196n, 215, 236, 251, *see also* Bangkok
Thatcher, Margaret 204
Third World: as term 41n
Three Gorges dam, China 207, 211–12
Three Mile Island 369
Tianjin, China 72
Tibet 212, 214
Tikopia, Polynesia: famine 158
timber industry: and destruction of forests 81, 262, *see also* logging
time: dimension in vulnerability to disasters 8–9, 12, 56, 59–60, 106–9, 108, 351
time–space factors: natural hazards 56–9, *89*, 89–90, 103–6, 111, 276, 277, 286; in preventive strategies 115
Tobin Tax 24